Sustainable Civil Infrastructures

Editor-in-Chief
Hany Farouk Shehata, SSIGE, Soil-Interaction Group in Egypt SSIGE, Cairo, Egypt

Advisory Editors
Khalid M. ElZahaby, Housing and Building National Research Center, Giza, Egypt
Dar Hao Chen, Austin, TX, USA

Sustainable Civil Infrastructures (SUCI) is a series of peer-reviewed books and proceedings based on the best studies on emerging research from all fields related to sustainable infrastructures and aiming at improving our well-being and day-to-day lives. The infrastructures we are building today will shape our lives tomorrow. The complex and diverse nature of the impacts due to weather extremes on transportation and civil infrastructures can be seen in our roadways, bridges, and buildings. Extreme summer temperatures, droughts, flash floods, and rising numbers of freeze-thaw cycles pose challenges for civil infrastructure and can endanger public safety. We constantly hear how civil infrastructures need constant attention, preservation, and upgrading. Such improvements and developments would obviously benefit from our desired book series that provide sustainable engineering materials and designs. The economic impact is huge and much research has been conducted worldwide. The future holds many opportunities, not only for researchers in a given country, but also for the worldwide field engineers who apply and implement these technologies. We believe that no approach can succeed if it does not unite the efforts of various engineering disciplines from all over the world under one umbrella to offer a beacon of modern solutions to the global infrastructure. Experts from the various engineering disciplines around the globe will participate in this series, including: Geotechnical, Geological, Geoscience, Petroleum, Structural, Transportation, Bridge, Infrastructure, Energy, Architectural, Chemical and Materials, and other related Engineering disciplines.

SUCI series is now indexed in SCOPUS and EI Compendex.

More information about this series at https://link.springer.com/bookseries/15140

Amin Akhnoukh · Kamil Kaloush ·
Magid Elabyad · Brendan Halleman ·
Nihal Erian · Samuel Enmon II ·
Cherylyn Henry
Editors

Advances in Road Infrastructure and Mobility

Proceedings of the 18th International Road Federation World Meeting & Exhibition, Dubai 2021

Set 2

Editors
Amin Akhnoukh
Construction Management
East Carolina University
Greenville, NC, USA

Magid Elabyad
International Road Federation
Alexandria, VA, USA

Nihal Erian
KemetPro, LLC
Greenville, NC, USA

Cherylyn Henry
Charlotte, NC, USA

Kamil Kaloush
School of Engineering
Arizona State University
Tempe, AZ, USA

Brendan Halleman
International Road Federation
Brussels, Belgium

Samuel Enmon II
Street Transportation Department
City of Phoenix
Phonix, AZ, USA

ISSN 2366-3405 ISSN 2366-3413 (electronic)
Sustainable Civil Infrastructures
ISBN 978-3-030-79800-0 ISBN 978-3-030-79801-7 (eBook)
https://doi.org/10.1007/978-3-030-79801-7

© The Editor(s) (if applicable) and The Author(s), under exclusive license
to Springer Nature Switzerland AG 2022

This work is subject to copyright. All rights are solely and exclusively licensed by the Publisher, whether the whole or part of the material is concerned, specifically the rights of translation, reprinting, reuse of illustrations, recitation, broadcasting, reproduction on microfilms or in any other physical way, and transmission or information storage and retrieval, electronic adaptation, computer software, or by similar or dissimilar methodology now known or hereafter developed.

The use of general descriptive names, registered names, trademarks, service marks, etc. in this publication does not imply, even in the absence of a specific statement, that such names are exempt from the relevant protective laws and regulations and therefore free for general use.

The publisher, the authors and the editors are safe to assume that the advice and information in this book are believed to be true and accurate at the date of publication. Neither the publisher nor the authors or the editors give a warranty, expressed or implied, with respect to the material contained herein or for any errors or omissions that may have been made. The publisher remains neutral with regard to jurisdictional claims in published maps and institutional affiliations.

This Springer imprint is published by the registered company Springer Nature Switzerland AG
The registered company address is: Gewerbestrasse 11, 6330 Cham, Switzerland

Contents

Pavement Testing

A New Low-Activity Nuclear Density Gauge for Compaction Control of Asphalt, Concrete, Soil, and Aggregate Layers in Road Construction... 3
Linus Dep, Robert E. Troxler, and William F. Troxler Jr.

A Study into the Benefit and Cost-Effectiveness of Using State-of-the-Art Technology for Road Network Level Condition Assessment... 19
Herman Visser, Simon Tetley, Tony Lewis, Kaslyn Naidoo, Tasneem Moolla, and Justin Pillay

Asphalt Pavement Innovations

Paths to Successful Asphalt Road Structures Performance Based Specifications for Hot Mix Asphalt 35
Bernhard Hofko

Study on Rejuvenators and Their Effect on Performance Characteristics of an Asphalt Concrete Containing 30% Reclaimed Asphalt 46
Pavla Vacková, Majda Belhaj, Jan Valentin, and Liang He

Sustainable Use of Cool Pavement and Reclaimed Asphalt in the City of Phoenix 57
Ryan Stevens

Optimizing the Adhesion Quality of Asphalt Binders Based on Energy Parameters Derived from Surface Free Energy 72
Allex E. Alvarez and Lady D. Vega

Estimating Asphalt Film Thickness in Asphalt Mixtures Using Microscopy to Further Enhance the Performance of UAE Roadways 87
Alaa Sukkari, Ghazi Al Khateeb, Waleed Ziada, and Helal Ezzat

Effect of Crumb Rubber and Polymer Modifiers on Performance Related Properties of Asphalt Binders 97
Joseph Dib, Nariman J. Khalil, and Edwina Saroufim

Road Safety Leadership

How Do Best Performing Countries in Road Safety Save Lives on the Roads? Lessons Learned from Case Studies in Singapore 117
Alina Florentina Burlacu and Emily Tan

Data for Road Incident Visualization, Evaluation, and Reporting (DRIVER): Case Studies of Cebu, Philippines and the Asia-Pacific Road Safety Observatory (APRSO) 134
Florentina Alina Burlacu, Miguel Enrico III Cabag Paala, Veronica Ines Raffo, and Juan Miguel Velasquez Torres

Road Financing Strategies

Is a Further Increase in Fuel Levy in Kenya Justified? 151
Evans Omondi Ochola and Jennaro Boniface Odoki

Bridge and Structural Management

Aerial Robotic System for Complete Bridge Inspections 171
Antidio Viguria, Rafael Caballero, Ángel Petrus, Francisco Javier Pérez-Grau, and Miguel Ángel Trujillo

Damage Inspection, Structural Evaluation and Rehabilitation of a Balanced Cantilever Bridge with Center Hinges 187
Chawalit Tipagornwong, Koonnamas Punthutaecha, Peerapat Phutantikul, Setthaphong Thongprapha, and Kridayuth Chompooming

Reinforcement of Scoured Pile Group Bridge Foundations with Spun Micro Piles 202
Apichai Wachiraprakarnpong, Juti Kraikuan, Jirasak Watcharakornyotin, and Thanwa Wiboonsarun

Repair of Settled Pile Bent Bridge Foundations with Spun Micro Piles 217
Apichai Wachiraprakarnpong, Juti Kraikuan, Wichai Yu, and Pawin Ritthiruth

Overview of Integral Abutment Bridge Applications in the United States 232
Amin Akhnoukh, Rajprabhu Thungappa, and Rudolf Seracino

Case Studies in ITS Development

Pacemaker Lighting Application to Prevent Traffic Congestion in a V-Shaped Tunnel and Provide Sustainable Operation: A Case Study: Eurasia Tunnel 243
Aşkın Kaan Kaptan and Murat Gücüyener

Real Time Multi Object Detection & Tracking on Urban Cameras 257
Rifkat Minnikhanov, Maria Dagaeva, Timur Aslyamov, Tikhon Bolshakov, and Emil Faizrakhmanov

Effective Advanced Warning for Connected Safety Applications - Supplementing Automated Driving Systems for Improved Vehicle Reaction 269
Gregory M. Baumgardner, Rama Krishna Boyapati, and Amudha Varshini Kamaraj

Traffic Planning and Forecasting - 1

A Case Study of Rural Freight Transport – Two Regions in North Carolina 287
Daniel J. Findley, Steven A. Bert, George List, Peter Coclanis, and Dana Magliola

Traffic Delay Evaluation and Simulating Sustainable Solutions for Adjacent Signalized Roundabouts 303
Mohamad Yaman Fares and Muamer Abuzwidah

Safe Roads by Design - 1

A Comparative Evaluation of the Safety Performance of Median Barriers on Rural Highways; A Case-Study 317
Victoria Gitelman and Etti Doveh

Ascendi's Safety Barriers Upgrading Program 335
Telma Silva and João Neves

Effectiveness of Cable Median Barriers in Preventing Cross Median Crashes and Related Casualties in the United States - A Systematic Review 345
Baraah Qawasmeh and Deogratias Eustace

Integrated Transport Planning

Transforming Infrastructure Projects Using Agile 357
Nihal Erian and Brendan Halleman

Savings Potential in Highway Planning, Construction and Maintenance Using BIM - German Experience with PPP 365
Veit Appelt

Demonstrating Connectivity and Exchange of Data Between BIM and Asset Management Systems in Road Infrastructure Asset Management... 379
Sukalpa Biswas, John Proust, Tadas Andriejauskas, Alex Wright, Carl Van Geem, Darko Kokot, António Antunes, Vânia Marecos, José Barateiro, Shubham Bhusari, and Jelena Petrović

Greening Road Projects

Road Construction Using Locally Available Materials 395
Robert D. Friedman and Ahmed F. Abdelkader

Practical Applications of Big Data Science

Road Roughness Estimation Using Acceleration Data from Smartphones 407
Arak Montha, Attaphon Huytook, Tatree Rakmark, and Ponlathep Lertworawanich

Improving Traffic Safety by Using Waze User Reports 416
Raitis Steinbergs and Maris Kligis

A Glimpse into the Near Future – Digital Twins and the Internet of Things 425
Howard Shotz and James Birdsall

Long Term Pavement and Asset Performance - 1

Field Monitoring of Road Pavement Responses and Their Performance in Thailand 441
Auckpath Sawangsuriya

Supplementary Cementitious Materials in Concrete Industry – A New Horizon 450
Amin Akhnoukh and Tejan Ekhande

Long Term Anti Corrosion Measures for Magosaki Viaduct by Cover Plate Method 458
Hiromasa Kobayashi, Yukio Usuda, Yuki Kishi, and Yukio Nagao

Degradation of Friction Performance Indicator Over the Time in Highways Using Linear Mixed Models 474
Adriana Santos, Elisabete Freitas, Susana Faria, Joel Oliveira, and Ana Maria A. C. Rocha

Road Safety Risk Diagnosis

Application of an Innovative Network Wide Road Safety Assessment Procedure Based on Human Factors 493
Andrea Paliotto, Monica Meocci, and Valentina Branzi

A Review of the Spatial Analysis Techniques for the Identification of Road Accident Black Spots and It's Application in Context to India 511
Shawon Aziz and Sewa Ram

Road Network Safety Screening of County Wide Road Network. The Case of the Province of Brescia (Northern Italy) 525
Michela Bonera, Benedetto Barabino, and Giulio Maternini

An iRAP Based Risk Impact Analysis at National Highway-1 for a Proposed Route Connecting Coastal Areas of Bangladesh 542
Armana Sabiha Huq

Maintenance Strategies - 1

Improving Pavement Condition at an Accelerated Pace: The City of Phoenix Accelerated Pavement Maintenance Program 559
Ryan Stevens

Smart Infrastructure Asset Management System on Metropolitan Expressway in Japan 575
Hirotaka Nakashima, Taishi Nakamura, Yusuke Hosoi, Koji Konno, and Hinari Kawamura

Decarbonizing Road Transport

Evaluation of the CO_2 Reduction Effect of Low Rolling Resistance Asphalt Pavement Using the Fuel Consumption Simulation Method ... 589
Yu Shirai, Atsushi Kawakami, Masaru Terada, and Kenji Himeno

Green Energy Sources Based on Thermo-Electrochemical Cells for Electricity Generating from Transport, Engineering Buildings and Environment Waste Heat 602
Igor N. Burmistrov, Nikolay V. Kiselev, Elena A. Boychenko, Nikolay V. Gorshkov, Evgeny A. Kolesnikov, and Stanislav L. Mamulat

Long Term Pavement and Asset Performance - 2

An Approach to Estimate Pavement's Friction Correlation Between PCI, IRI & Skid Number ... 611
Carlos J. Obando, Jose R. Medina, and Kamil E. Kaloush

Effect of Different Aggregate Gradations on Rutting Performance of Asphalt Mixtures for UAE Roadways ... 622
Khalil Almbaidheen, Ghazi Al-Khateeb, Waleed Ziada, and Myasar Abulkhair

Evaluation of AASHTO Mechanistic Empirical Design Guide Inputs to the Performance of Tennessee Pavements ... 634
Onyango Mbakisya, Msechu Kelvin, Udeh Sampson, and Owino Joseph

Influence of Dynamic Analysis on Estimation of Rutting Performance Using the Fixed Vehicle Approach ... 647
Gauri R. Mahajan, Radhika Bayya, and Krishna Prapoorna Biligiri

Street and Highway Designs for CAVs - 1

A Review on Benefits and Security Concerns for Self Driving Vehicles ... 667
Gozde Bakioglu and Ali Osman Atahan

Contrast Ratio of Road Markings in Poland - Evaluation for Machine Vision Applications Based on Naturalistic Driving Study ... 676
Tomasz E. Burghardt and Anton Pashkevich

San Diego Bus Rapid Transit Using Connected and Autonomous Vehicle Technology ... 691
Dan Lukasik and Dmitri Khijniak

Maintenance Strategies - 2

Introduction to the RoadMark System Low Cost, Fast and Flexible Data Collection for Rural Roads ... 707
Mark J. Thriscutt

Similarity Between an Optimal Budget in Pavement Management & an Equilibrium Quantity of Demand-Supply Analysis in Economics ... 719
Ponlathep Lertworawanich

Integrating Flexible Pavement Surface Macrotexture to Pavement Management System to Optimize Pavement Preservation Treatment Recommendation Strategy ... 735
Seng Hkawn N-Sang, Jose Medina, and Kamil Kaloush

Safe Roads by Design - 2

MASH TL-3 Development and Evaluation of the Thrie-Beam Bullnose Attenuator 755
Robert Bielenberg, Ron Faller, and Cody Stolle

Safety Performance Evaluation of Modified Thrie-Beam Guardrail 772
Robert Bielenberg, Ron Faller, and Karla Lechtenberg

In-Service Performance Evaluation of Iowa's Sloped End Treatments 789
Cody Stolle, Jessica Lingenfelter, Khyle Clute, and Robert Bielenberg

A Synthesis of 787-mm Tall, Non-proprietary, Strong-Post, W-beam Guardrail Systems 805
Scott Rosenbaugh, Robert Bielenberg, and Ronald Faller

A Synthesis of MASH Crashworthy, Non-proprietary, Weak-Post, W-beam Guardrail Systems 821
Scott Rosenbaugh, Robert Bielenberg, and Ronald Faller

The Safety Highway Geometry Based on Unbalanced Centripetal Acceleration 837
Creso de Franco Peixoto and Maria Teresa Françoso

Stiffening Guidance for Temporary Concrete Barrier Systems in Work Zone and Construction Situations 850
Karla Lechtenberg, Chen Fang, and Ronald Faller

Safety Performance Evaluation of a Non-proprietary Type III Barricade for Use in Work Zones 868
Karla Lechtenberg, Ronald Faller, Jennifer Rasmussen, and Mojdeh Asadollahi Pajouh

ITS Design and Implementation Strategies

Unified ITS Environment in the Republic of Tatarstan 881
Rifkat Minnikhanov, Maria Dagaeva, Sofya Kildeeva, and Alisa Makhmutova

Road Pricing and Tolling

Assessment of the Potential Implementation of High-Occupancy Toll Lanes on the Major Freeways in the United Arab Emirates 897
Ahmed Shabib, Mahmoud Khalil, and Muamer Abuzwidah

Driver Behavior Strategies

Posted Road Speed Limits in Abu Dhabi: Are They Too High? Should They Have Been Raised? Evidence Based Answers 915
Francisco Daniel B. Albuquerque

Transport Responses to the Pandemic

The Impact of the COVID-19 Pandemic on Mobility Behavior in Istanbul After One Year of Pandemic 933
Ali Atahan and Lina Alhelo

The Evaluation of the Impacts on Traffic of the Countermeasures on Pandemic in Istanbul 950
Mahmut Esad Ergin, Halit Ozen, and Mustafa Ilıcalı

New Approaches to Performance Delivery

From Reactive to Proactive Maintenance in Road Asset Management 963
Timo Saarenketo and Vesa Männistö

Observing How Influence of Nature Phenomena Against Inside Tunnel by Air Pressure Information 975
Kensaku Kawauchi, Yumi Watanabe, Yuichi Mizushima, and Takeo Hosokai

Road Asset Management: Innovative Approaches 986
Soughah Salem Al-Samahi and Fernando Varela Soto

Traffic Planning and Forecasting – 2

Evaluating the Efficiency of Constructability Review Meetings for Highway Department Projects 997
Amin K. Akhnoukh, Minerva Bonilla, Nicolas Norboge, Daniel Findley, William Rasdorf, and Clare Fullerton

A Novel Method for Aggregate Tour-Based Modeling with Empirical Evidence 1007
Yanling Xiang, Shiying She, Meng Zheng, Heng Liu, and Huanyu Lei

Multilayer Perceptron Modelling of Travelers Towards Park-and-Ride Service in Karachi 1026
Irfan Ahmed Memon, Ubedullah Soomro, Sabeen Qureshi, Imtiaz Ahmed Chandio, Mir Aftab Hussain Talpur, and Madzlan Napiah

Congestion on Canada's Busiest Highway, 401 Problems, Causes, and Mitigation Strategies 1039
Abdul Basith Siddiqui

Multi-Stakeholder Transportation Strategies

A Holistic Approach for the Road Sector in Sub-Saharan Africa 1053
Tim Lukas Kornprobst, Ulrich Thüer, and Yana Tumakova

Innovations in Road Materials

The Introduction of Micro - & Nanodispersed Fillers into the Bitumen Binders for the Effective Microwave Absorption (for the Road, Airfield & Bridge Pavements) 1071
Stanislav Mamulat, Igor Burmistrov, Yuriy Mamulat, Dmitry Metlenkin, and Svetlana Shekhovtsova

Rheological Properties of Rubber Modified Asphalt Binder in the UAE 1083
Mohammed Ismail, Waleed A. Zeiada, Ghazi Al-Khateeb, and Helal Ezzat

Recycling Waste Rubber Tires in Pervious Concrete Evaluation of Hydrological and Strength Characteristics 1098
Sahil Surehali, Avishreshth Singh, and Krishna Prapoorna Biligiri

Incorporation of CFRP and GFRP Composite Wastes in Pervious Concrete Pavements 1112
Akhil Charak, Avishreshth Singh, Krishna Prapoorna Biligiri, and Venkataraman Pandurangan

Asphalt Modified with Recycled Waste Plastic in South Africa Encouraging Results of Trial Section Performance 1125
Simon Tetley, Tony Lewis, Waynand Nortje, Deane Koekemoer, and Herman Visser

Climate Resilient Road Design – 1

GIS Aided Vulnerability Assessment for Roads 1139
Berna Çalışkan, Ali Osman Atahan, and Ali Sercan Kesten

Investigation of Historical and Future Air Temperature Changes in the UAE 1148
Reem N. Hassan, Waleed A. Zeiada, Muamer Abuzwidah, Sham M. Mirou, and Ayat G. Ashour

Climate Teleconnections Contribution to Seasonal Precipitation Forecasts Using Hybrid Intelligent Model 1167
Rim Ouachani, Zoubeida Bargaoui, and Taha Ouarda

Development of Pavement Temperature Prediction Models for Tropical Regions Incorporation into Flexible Pavement Design Framework 1181
Chaitanya Gubbala, Krishna Prapoorna Biligiri, and Amarendra Kumar Sandra

Impacts of Transport Investments

Experiences of High Capacity Transport in Finland 1197
Vesa Männistö

**Impacts of Transportation Infrastructure Investments and Options
for Sustainable Funding** 1207
Daniel J. Findley, Steven A. Bert, Weston Head, Nicolas Norboge,
and Kelly Fuller

Climate Resilient Road Design – 2

Climate Resilient Urban Mobility by Non-motorized Transport 1225
Kigozi Joseph

Author Index 1237

Long Term Pavement and Asset Performance - 2

An Approach to Estimate Pavement's Friction Correlation Between PCI, IRI & Skid Number

Carlos J. Obando[✉], Jose R. Medina, and Kamil E. Kaloush

Arizona State University, Tempe, AZ, USA
cobandog@asu.edu

Abstract. Friction in pavements is an especial phenomenon that varies upon many variables such as weather, pavement's age, material properties, and tire-surface contact. Furthermore, it is a very important factor since it is related to safety. To measure friction there are different procedures and methodologies. Depending on the method, and even the equipment used to measure skid resistance, output data can vary. In addition, friction varies with time. Some studies state that friction has a decreasing linear behavior, and others say behavior dependents on temperature and the level of exposure of aggregates' surface, so friction in pavements is a subject still on development and having consistent measurements of skid resistance is complicated. Based on the Long-Term Pavement Performance (LTPP) data base, this study found that friction can be related to the pavement condition index (PCI) and roughness (IRI). These findings can support alternative approaches to estimate skid resistance on a more suitable way and take proactive actions to maintain adequate safety ranges on roads.

Keywords: Skid number · Texture · Pavement · Pavement condition index (PCI) · International Roughness Index (IRI)

1 Introduction

According to the HMA Pavement Evaluation and Rehabilitation Reference Manual, friction is a functional factor that affects Hot Mix Asphalt (HMA) pavement's performance (U.S. Department of Transportation 2001). The concept of friction or skid resistance is very important in pavements because this is related to safety. Friction involves two bodies, tire and pavement surface, more specifically, the pavement's microtexture and macrotexture.

The measure of friction is made by an abstract quantity known as coefficient of friction (µ) which is calculated by dividing the motion frictional resistance by the load acting perpendicular to the interface (Carmon and Ben-Dor 2018). Other way to talk about friction is in terms of skid number (SN), which is equal to multiply µ by 100. In addition, a standard methodology called International Friction Index (IFI) that measures friction at a standard velocity of 60 km/h, and under wet conditions, has been adopted by the Permanent International Association of Road Congress (PIARC) (U.S. Department of Transportation 1999). This method is used to standardize friction data that was estimated using methods as the Locked Wheel Trailer (LWST) and the

Sideway-Force Coefficient Routine Investigatory Machine (SCRI) (Barrantes Quiros 2017). However, according to a research in the University of South Florida, "The IFI concept is based on the assumption that the friction value of a given surface depends on the slip speed at which measurements are taken, the texture properties of the pavement surface (both micro and macrotexture) and characteristics of the device used to obtain the measurements" (Fuentes 2009). Until today, measurements and concepts about skid number are still on debate.

Historical friction data from the Long-Term Pavement Performance InfoPave database has been analyzed by others and interesting differences were found. A study by Carmon & Ben-Dor found that new constructed pavements have high skid number, while older pavements can experience the effects of aging, producing friction loss (Carmon and Ben-Dor 2018). From this perspective, friction value decreases along the time once the pavement has been built. Other studies have found that skid resistance values commonly rise around the first two years just after construction because the roadway is damaged away by traffic loads and coarse aggregate surfaces turn out to be exposed, then decreases along the lasting pavement life due to aggregates become more polished. In addition, friction experience seasonal changes with higher friction during the Winter and Fall and lower during the Summer and Spring, and care must be taken when analyzing friction data (Jayawickrama and Thomas 1998). According to this view, friction's variation on pavements is based on pavement's micro and macrotexture and has a beginning early service life stage an increase and after a decrease until the end of pavement's service life. In sum, then, the issue is whether the variation of friction is always in a decreasing way, or this fluctuation has different patterns deepening on internal and/or external factors.

This paper discusses that even though there are studies and hypothesis about friction fluctuation values along pavement's life, it is possible to explain those variation from a different view based on real collected data. In this study a relationship between friction, distresses and the international roughness index (IRI) are examined to better understand how friction fluctuates and what are involved. Since currently, estimating friction is complicated, having a suitable way to estimate friction could proactively advice about pavement's safety.

2 Method

This document is based on the LTPP InfoPave database. Friction, distresses, and roughness data was downloaded for more than 460 LTPP test sections, from 41 States, including all climatic regions (dry-freeze, dry-non-freeze, wet-freeze and wet-nonfreeze). The dataset also included more than 2300 data points. Table 1 shows a summary of the input data.

Table 1. General information about chosen sections

No.	Row labels	Data points	No. sections	No.	Row labels	Data points	No. sections
1	Alabama	15	6	22	Montana	31	16
2	Arizona	108	36	23	Nebraska	14	6
3	Arkansas	1	1	24	Nevada	1	1
4	California	91	32	25	New Hampshire	6	1
5	Colorado	35	18	26	New Jersey	35	5
6	Connecticut	45	9	27	New Mexico	15	11
7	Delaware	9	3	28	New York	38	4
8	Florida	26	5	29	North Carolina	75	10
9	Georgia	39	4	30	Ohio	66	9
10	Idaho	5	5	31	Oklahoma	163	28
11	Illinois	74	23	32	Oregon	16	3
12	Indiana	245	32	33	Pennsylvania	44	10
13	Iowa	193	21	34	Rhode Island	6	1
14	Kansas	163	16	35	South Dakota	2	2
15	Kentucky	8	1	36	Tennessee	52	14
16	Maine	42	10	37	Texas	245	41
17	Maryland	202	11	38	Utah	6	3
18	Michigan	19	8	39	Vermont	16	4
19	Minnesota	5	5	40	Virginia	68	17
20	Mississippi	8	4	41	Washington	52	6
21	Missouri	99	24		Total	2383	466

Information from InfoPave was downloaded using the "Data Selection and Download" function located in the tab "Data" from the website https://infopave.fhwa.dot.gov. 476 sections from different states of U.S were chosen. Type of sections sorts from urban to rural, and from minor to principal arterial. The age of roads in this study ranges between 8 and 41 years. It was selected information related to Pavement Distresses for Asphalt Concrete (AC, all the 3 alternatives), Surface Characteristics (Longitudinal Profile, Section Level IRI, and Friction).

Friction, distresses and IRI data were compared considering same year, same construction number (CN), and same identification for each section. To avoid variations in friction values due to resurfacing, reconstruction, or any maintenance to improve friction, the concept of CN was implied.

The severity and density of the distresses to calculate the pavement condition index (PCI) for each section was done based on the downloaded distresses data and involved an Excel tool developed at Arizona State University to directly import distress data from InfoPave and estimates PCI based on ASTM D6433-18 (Wu 2015).

To ensure all data corresponds to the same section it was needed to create a unique identification code, which was made merging the code of the state, the SHRP code, the CN, and the year of the test/visit date of each section. Note so, that the data corresponds to the same section at different years and different CNs, then one section can be taken into the analysis more than once. Graphs comparing roughness (in terms of IRI), friction, alligator cracking, transversal cracking, and finally PCI were plotted to find possible correlations.

LTPP data related with skid resistance was collected using a method called locked wheel. This method is the most common in U.S (Schnebele et al. 2015) and uses a typical lock-wheel skid measurement system that consists in a vehicle with one or more test wheels incorporated or as part of a pulled trailer. Measurement of friction is done in a testing speed of 64 km/hr (40 mph) and water is applied ahead of the test tire to generate a wetted road surface. This equipment reports friction as a Friction Number (FN), or Skid Number (SN). Alligator cracking is reported in square meters (m2) per section, and IRI is reported in meter per kilometer (m/km).

3 Results

Collected data showed that IRI always increases as the pavement gets older. As a previous study stated that there is a significant relationship between IRI and cracking (Mubaraki 2016), so the higher amount of alligator cracking (fatigue), the higher IRI. Another result of the pavement's wear is the appearance of transverse cracks. According to the data analyzed a trend exists as well. The higher number of transverse cracks, the higher IRI, however, the relationship between IR and transverse cracks has a lower correspondence. Figure 1 shows a linear relationship between IRI and alligator cracks, and the relationship between IRI and transversal cracks.

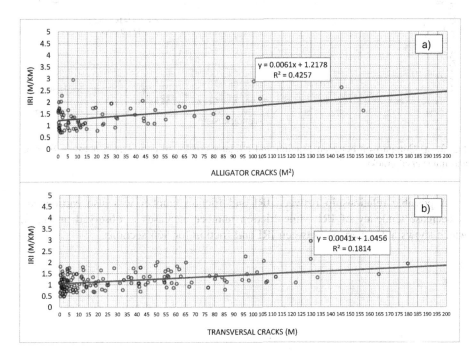

Fig. 1. a) Comparison between alligator cracks and IRI, b) Comparison between transverse cracks and IRI

Additionally, data shows that friction fluctuates with time. The older pavement, the lower friction number. It is common to find an increase of cracks on the pavement over time, then, realize what happened with cracks and friction along the time is important to understand friction fluctuation. Figure 2 shows a comparison between friction number and alligator cracks with an important power type correspondence. The higher alligator cracks, the value of friction decreases, while transverse cracks do not show an important relationship with friction.

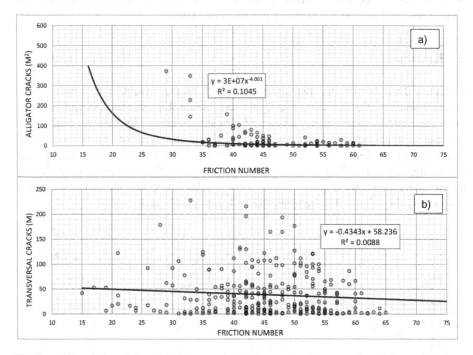

Fig. 2. a) Comparison between Friction and alligator cracks, b) comparison between Friction and transversal cracks

A possible relationship between IRI and friction number was also carried out. A decrease tendency of IRI while friction number increases is observed in Fig. 3, which shows a comparison between IRI and friction for 466 sections.

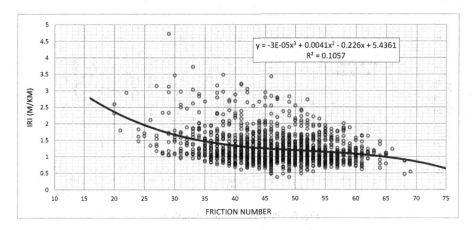

Fig. 3. Friction vs. IRI

Since relationships between IRI and Friction vs. distresses like cracking were found, further analysis was made to get establish the relationship between IRI and Friction Number to the pavement condition index (PCI), which involves all pavement distresses. As mentioned earlier, the estimation of PCI was done using the tool developed at Arizona State University, in which PCI is based on ASTM D6433-03.

Figures 4 and 5 shows comparisons between IRI and Friction vs. PCI respectively, in which certainly IRI decreases as the PCI increases, and the value of Friction increases as the value of PCI is higher.

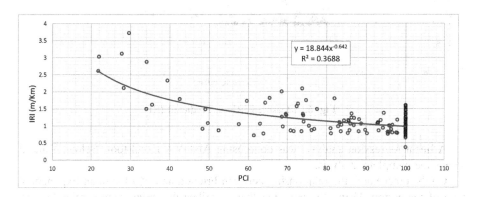

Fig. 4. PCI vs. IRI

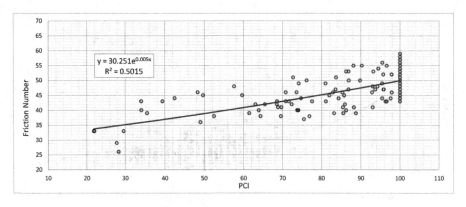

Fig. 5. PCI vs. Friction Number

The relationship between IRI and PCI fits in a Power type of regression. Power equation that describes this relationship (refer Fig. 4) was used to calculate a predicted IRI based on PCI. Figure 6 shows a comparison between the predicted IRI and the measured IRI values.

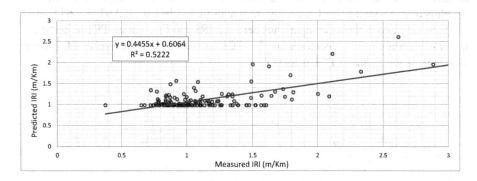

Fig. 6. Measured IRI vs. Predicted IRI

A linear regression analysis was done using ANOVA Excel tool. Below, Table 2 shows the complete analysis.

On the other hand, the relationship between Friction and PCI fits in an Exponential type of regression. Using the equation that describes this relationship (refer Fig. 5), predicted Friction was estimated based on PCI. Figure 7 shows a comparison between the predicted Friction and the measured Friction values.

Table 2. Summary ANOVA's output for IRI

Regression Statistics	
Multiple R	0.7226
R Square	0.5221
Adjusted R Square	0.5191
Standard Error	0.3306
Observations	157

ANOVA

	df	SS	MS	F	Significance F
Regression	1	18.5244	18.5244	169.4023	1.22E-26
Residual	155	16.9495	0.1093		
Total	156	35.4739			

	Coefficients	Standard Error	t Stat	P-value	Lower 95%	Upper 95%	Lower 95.0%	Upper 95.0%
Intercept	-0.1504	0.1050	-1.4324	0.1540	-0.3579	0.0570	-0.3579	0.0570
Predicted IRI Poly	1.1720	0.0900	13.0154	1.22E-26	0.9942	1.3499	0.9942	1.3499

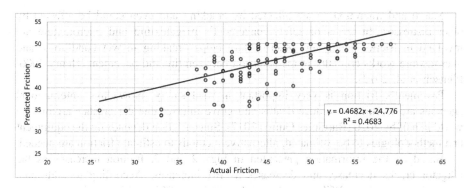

Fig. 7. Measured Friction vs. Predicted Friction

4 Discussion

A linear regression analysis was done using ANOVA Excel tool. Below, Table 3 shows the complete analysis.

Based on the graphs and statistics analysis before presented, IRI is related to distresses such us alligator cracks and at the same time, friction is related to those distresses, then friction and IRI could be associated (refer Fig. 3). The higher IRI, the lower friction number. The higher alligator, the value of friction decreases. These initial findings took the research to estimate PCI for each section and do the comparisons of PCI with IRI and Friction. Friction in pavements besides to be affected by temperature changes, seasonal variations, method, and equipment used to measure it, as it is shown in this study, also is affected by distresses, which means that friction changes depending on the PCI.

Table 3. Summary ANOVA's output for Friction

Regression Statistics	
Multiple R	0.6842
R Square	0.4682
Adjusted R Square	0.4648
Standard Error	4.5117
Observations	157

ANOVA

	df	SS	MS	F	Significance F
Regression	1	2778.426	2778.426	136.4935	5.125E-23
Residual	155	3155.140	20.355		
Total	156	5933.566			

	Coefficients	Standard Error	t Stat	P-value	Lower 95%	Upper 95%	Lower 95.0%	Upper 95.0%
Intercept	0.2655	4.0245	0.0659	0.94747	-7.6844	8.2155	-7.6844	8.2155
Predicted Fric Poly	1.0000	0.0855	11.6830	5.125E-23	0.8309	1.1691	0.8309	1.1691

Macro and microtexture are affected by cracking and even could be affected by other distresses. In terms of safety, pavement's megatexture and microtexture are involved in friction changing. Macrotexture helps breaking in wet conditions, and microtexture works crucially in the adhesion between tire and pavement (Henry 2000; Flintsch et al. 2012).

Pavement's friction is very important in road safety. Nowadays, international agencies are working to reduce deaths and grave injuries due car crashes around the world in half by 2020 (Barrantes Quiros 2017). Since friction depends on the equipment and methodologies to be estimated, alternative methods to estimate friction are needed.

Based on the information presented in Figs. 2 and 3, friction increases as the alligator cracks decrease, and decreases as IRI increases. According to additional comparisons involving PCI, it was found that the better PCI the higher friction number (refer Fig. 5).

The authors of this research propose the Eq. 1 shown in Fig. 4 to estimate IRI based on PCI to after correlate IRI and Friction, and Eq. 2, presented in Fig. 5, to estimate friction based on PCI.

$$IRI = 18.844 PCI^{-0.642} \qquad (1)$$

Where:
IRI = International Roughness Index (m/km)
PCI = Pavement Condition Index (0–100)

$$Fn = 30.251 e^{0.005 PCI} \qquad (2)$$

Where:
Fn = Friction Number
PCI = Pavement Condition Index (0–100)

Table 4. Thresholds for Friction and IRI based on PCI

Friction number (from Eq. 2)		PCI		IRI m/km (from Eq. 1)	
Fn < 36	Not Acceptable	PCI < 20	Failed	IRI > 2.75	Very Poor
		20 ≤ PCI < 35	Very Poor	2.75 ≥ IRI > 1.53	Poor
36.0 ≤ Fn < 38.8	Regular	35 ≤ PCI < 50	Poor	1.53 ≥ IRI > 1.18	Regular
38.8 ≤ Fn < 47.4	Good	50 ≤ PCI < 75	Fair	1.18 ≥ IRI > 1.05	Good
		75 ≤ PCI < 90	Very Good		
47.4 ≤ Fn	Very Good	90 ≤ PCI	Excellent	1.05 ≥ IRI	Very Good

Table 4 shows a summary of the general thresholds using Eqs. 1 and 2 to easy report and understand friction and IRI. The thresholds go accordingly with previous studies done by the Federal Highway Administration (FHWA) (Lavaud 2010), and the Transportation Research Board (NCHRP 2009).

Then, for values of PCI below to 35 and values of IRI above 2.75 m/km, friction number could be below 36, which is a not acceptable value in terms of safety. Very good friction values are so when PCI is equal or higher than 90 and IRI equal or lower than 1.05. Note that to correlate IRI with Friction, Eq. 1 could be applied in function of PCI to later use Table 2, however, based on IRI data found directly in the field, Friction can be estimated going directly to Table 2.

According to the results presented in this study, friction could be estimated based on PCI data and associated to IRI. This propose presents new views to consider friction in pavement towards safety and would help to develop suitable pavement management plans to control safety on roads.

5 Conclusions

There is a relationship between IRI and Friction vs. cracking, in which data showed a good arrangement. Further analysis revealed that similarly, Fn and PCI can be associated by an exponential behavior with R^2 of 0.47 using Eq. 2. To correlate IRI with Friction, Eq. 1 could be applied in function of PCI with a power behaviour with R^2 of 0.52, however, based on IRI data found directly in the field, friction can be estimated as well using Table 4.

The threshold values for Fn were established based on PCI and IRI, which go accordingly with previous studies done by the Federal Highway Administration (FHWA), and the Transportation Research Board.

Friction fluctuates along the time, but contrary to some hypothesis about friction behavior, skid resistance is not linear and not depends only on the time. Friction increases or decreases depending on the level of interaction between tire and pavement which at its time depends on pavements surface condition. To avoid dealing with difficulties related to different methodologies, calibration procedures and equipment to measure friction, it could be estimated based on PCI and/or IRI data.

Acknowledgements. The authors would like to acknowledge the invaluable support provided by the Colombian Program Colombia Cientifica and the Scholarship Fulbright - Pasaporte a la Ciencia. Based on the focuses/challenges of the Program Colombia Cientifica, this study becomes a tool for the Sustainable Construction and a more efficient Transportation development. Also, the authors would like to thank The National Center of Excellence for SMART Innovations at ASU.

Futures Research. Since IRI and PCI can be associated to friction, studies about how vehicles can measure and monitor PCI and IRI can be helpful to determine pavement friction in real time operation. Data collection could be done even by mobile applications, then, analysis done by management systems of road safety (MSRS) could make possible to notified in real time about safety conditions for a specific road and to take care about hazards related to friction. Therefore, further research about data collection of distresses and/or IRI to systematically feed management systems could be helpful to estimate friction to act proactively and ensure road safety.

References

ASTM: D6433-18 Standard Practice for Roads and Parking Lots Pavement Condition Index Surveys. s.l.:ASTM International 2018

Barrantes Quiros, S.: Practical method for locked-wheel and CFME friction measurment intervonversion. Virginia Polytechnic Institute and State University, Blacksburg (Virginia) (2017)

Carmon, N., Ben-Dor, E.B.-D.: Mapping Asphaltic Roads' Skid Resistance Using. Remote Sensoring, 10 March 2018

Flintsch, G.W., McGhee, K.K., de León Izeppi, E., Najafi, S.: The Little Book of Tire Pavement Friction. s.l.:s.n (2012)

Fuentes, L.G.: Investigation of the Factors Influencing Skid Resistance and the International Friction Index (2009). http://scholarcommons.usf.edu/etd/3920

Henry, J.: Evaluation of pavement friction characteristics. Transportation Research Board (2000)

Jayawickrama, P., Thomas, B.: Correction of field skid measurements for seasonal variations in Texas. Transp. Res. Rec. **1639**, 147–154 (1998). TRB, National Research Council

Lavaud, P.: Importance of Surface Regularity (IRI) in the Construction of Hot Asphalt Pavements, s.l.: Roadtec Inc. (2010)

Mubaraki, M.: Study the relationship between pavement surface distress and roughness data. In: MATEC Web of Conferences (2016)

NCHRP: Guide for Pavement Friction, Champaign, IL: s.n. (2009)

Schnebele, E., Tanyu, B.F., Cervone, G., Waters, N.: Review of remote sensing methodologies for pavement management and assessment. Eur. Transp. Res. Rev. 7(2), 1–19 (2015). https://doi.org/10.1007/s12544-015-0156-6

U.S. Department of Transportation: Assessment of LTPP Friction Data. McLean (Virginia): s.n (1999)

U.S. Department of Transportation: HMA Pavement Evaluation and Rehabilitation, Reference Manual. Champaign (Illinois): s.n (2001)

U.S. Department of Transportation (2018). https://infopave.fhwa.dot.gov/. Accessed Oct 2018

Wu, K.: Development of PCI-based pavement performance model for management of road infrastructure system. Arizona State University, Tempe (2015)

Effect of Different Aggregate Gradations on Rutting Performance of Asphalt Mixtures for UAE Roadways

Khalil Almbaidheen[1(✉)], Ghazi Al-Khateeb[1,2], Waleed Ziada[1,3], and Myasar Abulkhair[1]

[1] University of Sarjah, Sharjah, United Arab Emirates
u17200710@sharjah.ac.ae
[2] Jordan University of Science and Technology, Irbid, Jordan
[3] Mansoura University, Mansoura, Egypt

Abstract. Rutting or Permanent deformation is one of the primary distresses that occur in asphalt pavements. Many researchers showed that about 55 to 80% of rutting occurred in the surface layer of asphalt pavement. Since this layer contains 90–95% aggregates by weight, studying the properties of the aggregate component becomes very crucial. Gradation is one of the important characteristics of the aggregates affecting the performance of hot-mix asphalt (HMA) particularly against rutting. The skeleton of the internal aggregate structure is highly affected by gradation, which directly impacts the rutting resistance of the asphalt mixture. The main objectives of this research study are to evaluate the effect of aggregate gradation type including GAP-graded, FHWA gradation, UAE gradation, and gradation by Bailey method on the rutting performance of HMA for the UAE's hot climate. A total of 48 Superpave gyratory-compacted specimens of different aggregate gradations were prepared and tested in the mixture design stage to determine the design asphalt binder content for each mixture/gradation type. Another set of Superpave gyratory-compacted (SGC) specimens were also prepared using the design asphalt binder content obtained for each asphalt mixture/gradation for the performance testing phase. The test results and analysis provided a full picture of the performance and behavior of asphalt mixtures using different aggregate gradations. The results shows that the GAP and the Bailey gradations have the highest rutting resistance in both Dynamic modulus and flow number tests.

Keywords: Aggregate gradation · GAP gradation · Rutting · Dynamic modulus · Flow number · Bailey method · Mixture design · Superpave gyratory compactor

1 Introduction

Aggregate gradation and its effect on the performance of the Hot-Mix Asphalt (HMA) has been studied for many years. Various studies, e.g. (Kandhal et al. 1998; Nukunya et al. 2001; Birgisson et al. 2004) have shown that there is a clear relationship between the aggregate gradation and the ability of the HMA to resist the distresses.

Moreover, the behavior of the main Components of the HMA (Aggregate, Asphalt and Air voids) is playing an important role to give the idea of how the mixture will act during the service life of the project (Campen et al. 1959; Kumar and Goetz 1977).

Any mixture will act as per its microstructure, the asphalt mixture microstructure is formed mostly by aggregate particles which it will be affected by the main properties of the aggregates such as size, density, shape and distribution. The asphalt binder will flow through the particles making a film on the surface of the aggregate particles and start binding them. Understanding this mechanistic will help to predict the type of action the mixture will act under the load.

Due to the high repeating load on the roads specially from the trucks in UAE, rutting considered as one of the main distressed that make a challenge for the road authorities. The percentage and depth of rutting be subject to external and internal factors. External factors include load and volume of truck traffic, tire pressure, temperature, and construction practices. Internal factors include thickness of pavement, asphalt binder, aggregate, and mixture properties (Zaniewski and Srinivasan 2004).

Button et al. (1990) studied the rutting and the factors that causing the rutting, but he found that the aggregate properties and its distribution is the main factor that effect on rutting cause and resistance. Stakston and Bahia (2003) also have pointed that the resistance of rutting is mainly dependent on the gradation of the aggregate in the mixture, and if the mixtures produced with the best high quality raw materials it would fail if the gradation were not properly selected.

Many research and numerical modelling show that the external energy to the asphalt mixture caused by vehicle tire loading will case the permanent deformations. There are three different mechanisms caused these deformations. Reducing the friction between the aggregates which coated with asphalt binder is considered the first mechanism. Friction resistance in these aggregates like all granular materials is related to their mineral components, roughness, and asphalt binder properties. The second mechanism is reducing the interlock between aggregates that it pushes the aggregates away from each other. Increasing air-void in asphalt mixture is the result of this kind of dilatory behavior. The expansion depends on three main properties the gradation, angularity, and the shape of the aggregates. Losing the adhesion between the aggregates particle and the asphalt binder is considered the third mechanism. It is expected that the effect of each mechanism depends on mixture design and pavement layer thickness. For a thin pavement layer, stone-on-stone friction and interlock between aggregates are the principal mechanisms against rutting. Increasing the thickness of pavement layer would reduce the effect of friction and interlock between aggregate and therefore binder deformation properties have the most effect on the resistance against failure in cohesion and continuity between binder and aggregates (Jung and Young 1998).

Therefore, this research will focus on the effect of aggregate gradation on the rutting resistance of HMA asphalt pavements for the UAE's hot climate and investigate the best gradation that have the maximum rutting resistance.

2 Aggregate Gradation

Many researchers studied the aggregate gradation until the year of publishing the Superpave design method in 1993. Researchers after that focused on the new Superpave design method and the tests related to it. From the beginning of the 2000s, the understanding of Superpave method was clear and the researchers start again doing some research on the aggregate gradation since they noticed that there are some problems they are facing on site and the proposed gradation from the Superpave was not satisfying the need to solve these problems.

The above-mentioned gradation theories and analysis methods cannot directly guide the process of mixture design. There are several design methods developed based on the theories above, such as Superpave method, Bailey method, Coarse Aggregate Void-filling (CAVF) method (Zhang et al. 2001), Aggregate Gradation, Design and Test method for Stone Asphalt Concrete. For different design conditions, the specific requirements of gradation are varied. As for asphalt mixture design, currently, there are two main gradation types categorized by aggregate morphology, namely continuous gradation, and gap gradation, or three types by air voids, namely dense gradation, open gradation, and semi-open gradation.

Continuous gradation is the representation of all the standard particle sizes in a certain proportion, in which the voids between larger particles is effectively filled by smaller particles to produce a well-packed structure. GAP gradation is a kind of gradation which lacks one or more intermediate size.

The main procedure is how to control the gradation. There are several terminologies commonly used in gradation design of asphalt mixture, such as sieve screening, fine gradation, coarse gradation, fine aggregate, coarse aggregate, mineral filler, mineral dust, control points and so on. The first challenge on working with asphalt mixture is to differentiate those terminologies. Sieve test is the most effective way to do analysis of aggregate gradation. From the test, the percentage retained and the percentage passing by mass can be calculated. For the fine and coarse gradations, empirically and based on visual investigation, they are conceptually separated depending on whether the gradation curve is passing below or above the maximum density 0.45 power curve.

3 Mixture Design and Test Methods

Basically, the asphalt mixture is composed of two parts: aggregate and asphalt binder; identifying each one and the blending percentage of each one is the main process. After the combination of the two parts, a new product is obtained with new physical properties. The first concept of the HMA design has two components: the mixture design as simulation and the weight-volume terms. The simulation is a very important step to address the big difference between the field and the laboratory conditions. The weight-volume terms are also important since the asphalt mixture design in its nature is a volumetric-based process. And hence, understanding the volume, specific gravity, and voids of the phasing system of the asphalt mixture is a major step in the mixture design process.

The second one is the mixture design variables including the aggregates and the main aggregate properties such as gradation, size, toughness, shape and texture, the asphalt binder, and the ratio between the aggregate and the asphalt binder. Understanding the objective of the mixture design is the third and the last one. The objective of the mixture design is to obtain a mixture with optimum properties that can provide the best possible performance under traffic and climatic conditions. Mixture design procedures include Hveem, Marshall, and Superpave. Superpave mixture design is the latest procedure that is known to provide better HMA performance than other methods, which was used in this study to evaluate the gradations.

The design asphalt binder content can be found from the relationship between asphalt binder content and the air voids content (V_a). The asphalt binder content at the air voids content of 4% is considered the design asphalt binder content that can be used in the mixture. As shown in Fig. 1.

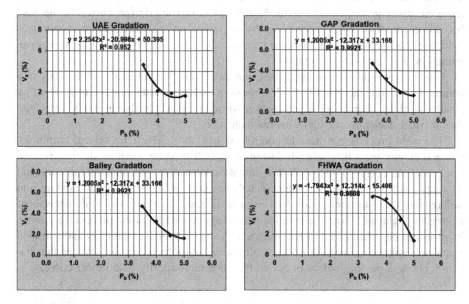

Fig. 1. The relationship between air voids and asphalt binder content for the different aggregate gradations

3.1 Dynamic Modulus, |E*| Test

There are many mechanical properties that can describe the Viscoelastic behavior of the Asphalt mixture. These properties can be obtained from advanced mechanical tests to find the viscoelastic response. The main properties are the complex modulus, relaxation modulus, and creep compliance. As per AASHTO TP 62-07 (2007), to determine the dynamic modulus (|E*|) the samples should be tested at five temperatures: −10, 4.4, 21.1, 37.8, and 54.4 °C and using six loading frequencies: 25, 10, 5, 1, 0.5, and 0.1 Hz. The level of the stress varies with the loading frequency to make sure that the specimen response is within the linear viscoelastic limits.

3.2 Flow Number Test

The flow number test (FN) is used to determine the permanent deformation characteristics of paving materials The prepared sample should be kept into the environmental chamber until it reaches the required degree (54.4 °C), the of why this temperature is chosen is to represent the real rutting may occur, for sample is expose to a confining pressure and a repeated haversine load pulse is applied for a period of 0.1 s. The sample then rests for 0.9 s during which time the recovered strains are monitored. In this project the FN will be tested on four different gradations with three cylindrical specimens of 100 mm in diameter and 150 mm in height replicates of each. The samples were manufactured with an air voids target 7 ± 0.5%.

4 Materials

4.1 Asphalt Binder

The asphalt binder used in this research study is a 60/70-penetration grade asphalt binder. Penetration test, softening point test, and rotational viscosity test were conducted on the asphalt binder as summarized in Table 1.

Table 1. Test results of 60/70 asphalt binder

Parameter	Methods	Unit	Results
Penetration @25 °C (100 gm) 5 s	ASTM D5/ DSM - 13	dmm	62
Softening point	ASTM D36/D36M-14e 1	°C	49
Rotational viscosity at 135 °C	ASTM D4402/D4402M - 15	cP	381.5

4.2 Aggregate Properties and Gradations

The type of the aggregate which have been used is Gabbro locally sources, this kind of aggregate is a commonly used type in the road construction projects in UAE. The main physical and chemical tests have been conducted on the aggregate to make sure it is in line with the local authorities' specification limits. The results can be found in Tables 2 and 3.

Table 2. Specific gravity of aggregate.

Aggregate size	G_{SA} (GM/CC)	G_{SB} (GM/CC)	G_{SE} (GM/CC)
3/4″	2.986	2.890	2.637
3/8″	2.926	2.801	1.0319
3/16″	2.971	2.818	4.6
Filler	2.962	2.962	95.4

Four gradations have been used in this study: GAP-gradation, FHWA recommended gradation for hot areas, UAE common gradation, and gradation by Bailey method as shown in Tables 4 and 5. The selection of the of the GAP, FHWA, Bailey gradations was based on the international recommendations to enhance the rutting resistance performance in similar hot weather as the UAE.

Table 3. Aggregate properties

Description	Spec. Standard	Spec.limit Coarse aggregate	Fine	Test results 20 mm aggregate	10 mm aggregate	5 mm aggregate
Soundness test (magnesium sulphate)	ASTM C-88	10% Max,	10% Max,	2.4	2.93	3.80
Loss by abrasion	ASTM C-131/C535	25% Max	–	12	10	–
Aggregate crushing value (acv)	BS 812 P-110	20% Max	–	13	11	–
Flakiness index	BS 812, P 105.1	25% Max	–	11	20	–
Elongation index	BS 812, P 105.2	25% Max	–	18	26	–
Acid soluble chloride	BS 812, P 117	0.1% Max	0.1% Max	0.01	0.01	0.01
Acid soluble sulphates	BS EN 1744-1:1998	0.5% Max	0.5% Max	0.04	0.04	0.06
Water absorption	ASTM C-127 &128	2.0% max	2.3% Max	0.4	0.7	1.4

5 Asphalt Mixture Test Results

The dynamic modulus (E*) and flow number (FN) performance tests were conducted on all asphalt mixtures used in this study using a Universal Testing Machine (UTM) with 30-kN load capacity. The following section presents the results of these tests.

5.1 Dynamic Modulus, |E*| Test Results

As discussed, the tests include the testing of four different mixtures and three replicates for each. The dimensions of the cylindrical specimens are 100 mm in diameter and

Table 4. Aggregate gradations

Sieve size (mm)	UAE gradation (%)	FHWA gradation (%)	GAP gradation (%)
25.0	100	100	100
19.0	75.4	95.9	98
12.5	68.4	74.5	83
9.5	59	54.9	68
4.75	37.1	31.8	42
2.360	24.3	23.9	22
1.18	17	17.5	14
0.6	11.8	12.6	9
0.3	8	7.4	5
0.15	6.3	4.3	3
0.075	4.9	3.2	1.9

Table 5. Bailey method gradation

Stockpiles	Aggregate (%)
20MM	18
10MM	51.4
5MM	28
Filler	2.6

150 mm in height. All $|E^*|$ test specimens were prepared according to the AASHTO T 342 protocol. The maximum specific gravity (G_{mm}), the bulk specific gravity (G_{mb}), and air voids were found for each. The results are summarized in Table 6.

Table 6. No. of gyrations, G_{mb}, G_{mm}, and air voids of the $|E^*|$ test specimens.

Mixture type	Specimen ID	No. of gyrations	G_{mb}	G_{mm}	Air voids (%)
UAE gradation	UAE-1	36	2.502	2.81	7.1
	UAE-2	34	2.502		7.2
	UAE-3	29	2.502		7.4
FHWA gradation	FHWA-1	26	2.464	2.735	7.6
	FHWA-2	29	2.464		7.2
	FHWA-3	42	2.519		6.7
Bailey gradation	BG-1	132	2.401	2.76	6.6
	BG-2	115	2.401		6.9
	BG-3	109	2.401		7.3
GAP gradation	GAP-1	110	2.448	2.690	7.0
	GAP-2	91	2.421		7.3
	GAP-3	98	2.421		7.2

5.1.1 Construction and Comparison of |E*| Master Curves

One of the main outputs of the dynamic modulus test is the master curve, which is constructed at a reference temperature of 21.1 °C. The main principle of the curve is based on the time-temperature superposition. The data at various temperatures are shifted with respect to time until the curves merge into a single smooth function (Stephen and Yatish 2007). Figure 2 show the |E*| master curves for the four tested asphalt mixtures.

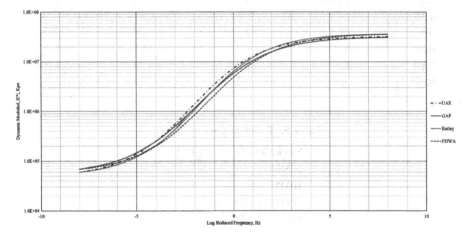

Fig. 2. |E*| master curve of the four gradations mixture

The sigmoidal and shifting parameters for the different asphalt mixtures are summarized in Table 7.

Table 7. |E*| sigmoidal and shifting parameters of tested asphalt mixtures.

Parameter	Mixture type			
	UAE	Bailey	FHWA	GAP
☐	3.8431	3.8627	3.9292	3.9036
☐	2.8308	2.8648	2.7983	2.7648
☐	1.2274	1.0237	0.7794	1.0082
☐	0.5735	0.5791	0.5563	0.5428
a	0.0000	0.0000	0.0000	0.0001
b	−0.0712	−0.0708	−0.0704	−0.0714
c	4.7456	4.7489	4.7343	4.7289

As shown in the previous figures, the asphalt mixtures with the Bailey and the GAP gradations have higher |E*| values than those of the asphalt mixtures with the FHWA and the UAE gradations. Another comparison as shown in Fig. 3 is between the E* values for each aggregate gradation and each loading frequency at the reference temperature 21.1 °C, the asphalt mixtures with the Bailey gradation or GAP gradation have the highest E*; either ranked 1st or 2nd. At low loading frequencies (0.1, 0.5, and 1 Hz), the asphalt mixture with the GAP gradation has the highest E* value.

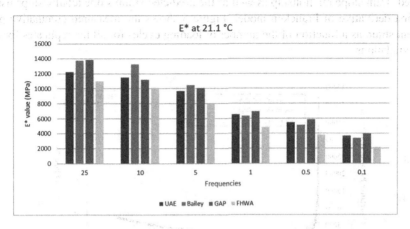

Fig. 3. Comparison of |E*| values for the different mixtures at 21.1 °C

5.2 Flow Number Test Results

5.2.1 The Tested Specimens

In this research study, the FN was tested on the asphalt mixtures with the four different gradations using three replicates (100 mm in diameter × 150 mm in height) for each mixture. The specimens were manufactured with an air voids content of 7 ± 0.5%. Table 8 shows the number of gyrations, bulk specific gravity, and maximum specific gravity, and the air voids of the specimens for each asphalt mixture.

Table 8. No. of gyration, Bulk Sp. Gr., Maximum Sp. Gr., and saturated surface dry air voids of the |E*| test specimens

Mixture type	Specimen ID	No. of gyration	G_{mb}	G_{mm}	Air voids (%)
UAE gradation	UAE-1	26	2.502	2.81	7.35
	UAE-2	29	2.502		7.41
	UAE-3	31	2.502		6.98
FHWA gradation	FHWA-1	26	2.464	2.735	7.39
	FHWA-2	29	2.464		7.41
	FHWA-3	31	2.519		6.98
Bailey gradation	BG-1	112	2.401	2.76	7.0
	BG-2	108	2.401		7.4
	BG-3	120	2.401		7.2
GAP gradation	GAP-1	101	2.448	2.690	7.5
	GAP-2	91	2.421		7.5
	GAP-3	98	2.421		7.4
	GAP-3				

5.2.2 Flow Number Test Results and Analysis

Figure 4 represent a sample of the repeated load/strain plots of the asphalt mixtures with the different gradations. This figure includes the measured cumulative permanent strain as a function of the number of loading cycles. The figure also shows the calculated strain slope relationship as well as the predicted strain slope relationships using the first derivative of Francken model. Figure 5 shows the measured cumulative permanent strain as a function of the number of loading cycles for all the replicates for the four gradations.

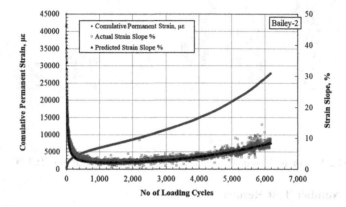

Fig. 4. Repeated load/flow number results for bailey gradation specimens (sample figure)

The Statistica Software was used to analyze the test results. The used model was Francken and the model coefficients were found by non-linear regression analysis.

The tables include a summary of results of the FN test: resilient deformation (ε_R), the standard deviation, and the coefficient of variation (Table 9).

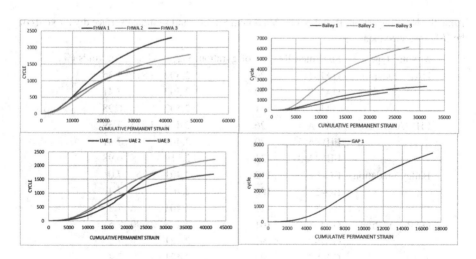

Fig. 5. Cumulative permanent strain µε for the replicates of the four gradations

Table 9. Summary of flow number test results at failure for all gradations

Mixture type	AVG flow number (cycles)	AVG permanent deformation, ε_P ($\mu\varepsilon$)
UAE gradation	644.6	13366
Bailey gradation	1247	8157
FHWA gradation	594	11423
GAP gradation	1777	8304.6

It is noticed that the Bailey and the GAP asphalt mixtures have higher FN values than the UAE and the FHWA asphalt mixtures. This means that the asphalt mixtures with the Bailey and the GAP gradations have the potential to resist permanent deformation (rutting) better than the other asphalt mixtures. The asphalt mixture with the Bailey gradation has a FN value that is almost double the values of the asphalt mixtures with the UAE and the FHWA gradations. In addition, the asphalt mixture with GAP gradation has a FN value of approximately three time the value of the asphalt mixtures with the UAE and the FHWA gradations.

Overall, the GAP and the Bailey mixtures show similar permanent deformation at failure. They showed permanent deformation values at failure lower than those of the UAE and FHWA mixtures; about 40% and 30% less than the UAE and FHWA mixtures, respectively.

6 Summary of Findings

Based on the analysis and results of this study, the following major findings were drawn:

It was found that the asphalt mixtures with the GAP and the Bailey gradations have higher $|E^*|$ values at temperature 21.1 °C and on the main frequencies (25,10,1, and 0.5 Hz) on average by 40% and 45% than those with the UAE and the FHWA gradations.

At the reference temperature of 21.1 °C, the analysis showed that the $|E^*|$ values of the GAP and Bailey asphalt mixtures were about 15% higher than those of UAE and FHWA mixtures.

For the rutting resistance, both the GAP and Bailey gradations showed FN values with approximately three and two times those of the UAE and FHWA gradations.

The GAP and Bailey gradations showed about 40% and 30% lower permanent deformation values than those of the UAE and FHWA gradations, respectively.

More studies on different gradations by using the local UAE raw materials need to be considered.

Implement new methods to design the gradation of the UAE roads such as Bailey methods is recommended to enhance the roads quality and reduce the rutting formation as shown in this study.

References

AASHTO: Standard method of test for determining dynamic modulus of hot-mix asphalt (HMA). AASHTO TP 62-07, Washington, DC (2007)

Birgisson, B., Darko, D., Roque, R., Page, G.C.: The Need for inducing Shear Instability to obtain Relevant Parameters for HMA Rut Resistance (2004)

Button, J.W., Perdomo, D., Lytton, R.L.: Influence of Aggregate on Rutting in Asphalt Concrete Pavements, Transportation Research Board (1990)

Campen, J.F., Smith, J.R., Erickson, L.G., Mertz, L.R.: The relationships between voids, surface area, film thickness and stability in bituminous paving mixtures. In: Proceedings, AAPT, vol. 28 (1959)

Abulkhair, M., Zeiada, W., Al-Khateeb, G., Shanableh, A., Dabous, S.A.: Stiffness and rutting assessment of asphalt mixtures using steel slag aggregates. In: The 5th World Congress on Civil, Structural, and Environmental Engineering (2020). https://doi.org/10.11159/icgre20.145

Cross, S.A., Jakatimath, Y.: Determination of Dynamic Modulus Master Curves for Oklahoma HMA Mixtures, FHWA/OK 07 (05), The Oklahoma Department of Transportation, Stillwater, Oklahoma (2007)

Kandhal, P.S., Foo, K.Y., Mallick, R.B.: A Critical Review of VMA Requirements in Superpave. NCAT Report No. 98-1, National Center for Asphalt Technology, Auburn, Al (1998)

Kandhal, P.S., Mallick, R.B.: Effect of mix gradation on rutting potential of dense-graded asphalt mixtures. Transp. Res. Rec. **1767**, 146–157 (2001)

Khosla, N.P., Sadasivam, S., Malpass, G.: Performance Evaluation of Fine Graded Superpave Mixtures for Surface Courses. North Carolina State University (2001)

Kumar, A., Goetz, W.H.: Asphalt hardening as affected by film thickness, voids and permeability in asphaltic mixtures. In: Proceedings, AAPT, vol. 46, USA (1977)

Jung, D.H., Young, K.N.: Relationship between Asphalt Binder Viscosity and Pavement Rutting, Transportation Research Board, ID: 00-0112 (1998)

Nukunya, B., Roque, R., Tia, M., Birgisson, B.: Evaluation of VMA and other volumetric properties as criteria for the design and acceptance of superpave mixtures. Presented at Annual Meeting of the Association of Asphalt Paving Technologists, Clearwater Beach, Fla. (2001)

Stakston, A.D., Bahia, H.: The Effect of Fine Aggregate Angularity, Asphalt Content and Performance Graded Asphalts on Hot-Mix Asphalt Performance, University of Wisconsin – Madison, Department of Civil and Environmental Engineering, Submitted to Wisconsin Department of Transportation, Division of Transportation Infrastructure Development, Research Coordination Section, WisDOT Highway Research Study 0092-45-98 (2003)

Zaniewski, J.P., Srinivasan, G.: Evaluation of Indirect Tensile Strength to Identify Asphalt Concrete Rutting Potential, Asphalt Technology Program, Department of Civil and Environmental Engineering, West Virginia University, Performed in Cooperation with the U.S. Department of Transportation - Federal Highway Administration (2004)

Evaluation of AASHTO Mechanistic Empirical Design Guide Inputs to the Performance of Tennessee Pavements

Onyango Mbakisya[1(✉)], Msechu Kelvin[1], Udeh Sampson[2], and Owino Joseph[1]

[1] University of Tennessee at Chattanooga, Chattanooga, USA
Mbakisya-Onyango@utc.edu
[2] Tennessee Department of Transportation, Nashville, USA

Abstract. Tennessee Department of Transportation (TDOT) is among the States Departments of Transportation, taking measures to implement the Pavement Mechanistic-Empirical Design (PMED) approach and analysis from the AASHTO 1993 pavement design procedure. TDOT has funded the calibration of Tennessee distress models and traffic inputs for its interstate and state routes. In this study, local and national calibrated models are used to evaluate the performance of Long-Term Pavement Performance (LTPP) sites in Tennessee using AASHTOWare Pavement Mechanistic-Empirical Design v2.5.5 (PMED v2.5.5). Traffic inputs include level 1, LTPP traffic volume adjustment factors; level 2, Tennessee statewide traffic volume adjustment factors; and level 3 the default national traffic volume adjustment factors. The study also considers LTPP sites that are relatively close to climatic stations, Modern Era Retrospective-analysis for Research and Application (MERRA) and North American Regional Reanalysis (NARR), in distance and elevation. The distances and elevation differences were considered to eliminate the need for creating visual weather stations. The statistical analysis comparing measured and predicted distresses showed that local calibrated distress models with level 2 traffic inputs predicted relatively close to the measured distresses compared to traffic levels 1 and 3. For all traffic levels and local calibration, MERRA predicted better on bottom up cracking and permanent deformation, while NARR predicted better on terminal IRI. For national calibrations, MERRA predicted better on bottom up cracking, Jointed Plain Concrete Pavement (JPCP) transverse cracking, and total transverse cracking, while NARR predicted better on permanent deformation and terminal IRI.

Keywords: MEPDG · PMED · Pavement analysis · Performance evaluation · Calibration · Climate data inputs

1 Introduction

The Pavement Mechanistic-Empirical Design (PMED) is the state-of-the-art pavement design method developed to replace the existing empirical design method, AASHTO 1993 Guide for Pavement Design. Great efforts have been invested to its development,

© The Author(s), under exclusive license to Springer Nature Switzerland AG 2022
A. Akhnoukh et al. (Eds.): IRF 2021, SUCI, pp. 634–646, 2022.
https://doi.org/10.1007/978-3-030-79801-7_46

and equally great effort is required for its implementation. The main advantage of the PMED approach is the consideration of various design factors that affect the performance and life of pavements. Both AASHTO 1993 and PMED considers traffic, materials properties, and climatic conditions as design inputs. The difference is that the PMED inputs for pavement design and analysis, are more detailed and in large amounts, hence the use of a design software is warranted, currently, AASHTOWare Pavement ME Design v2.5.5 (PMED v2.5.5). The amount of data needed for design and analysis has brought about some concerns on its quality, on which each state must invest on obtaining quality input data.

The PMED design approach has a three-level hierarchy of data for all its inputs (ARA 2004), based on the quality or means of which the data is collected;

- Level 1 – refers to the data that is obtained from actual pavement section that is to be analyzed or designed. Level 1 data has the best quality as they offer a more realistic behavior of the measured inputs.
- Level 2 – refers to data that is obtained from local calibration practices based on regression analysis of common trends and behaviors of existing measured quantities. They are commonly referred to as regional/local data and are less accurate and less costly compared to Level 1 data.
- Level 3 – this level of data is commonly referred to as national data. Level 3 values are formulated based on a general trend from existing data, hence having the least quality and less costly compared to the two levels above.

Materials inputs in the PMED are classified into three categories. First, materials inputs that are required in the computation/prediction of pavement distresses (stress, strain and displacement), these properties are dynamic/elastic modulus and Poisson's ratio. Second, materials inputs that are directly used in the distress/smoothness models/transfer functions, such as strength, plasticity and erodibility. Third, is the materials properties that are linked to the climatic models. Material properties for climatic modelling help with the determination of moisture profile and temperature in the pavement cross-section. Material properties for climatic modeling include gradation parameters, thermal properties and plasticity index (ARA 2004).

Traffic inputs represents the magnitude and number of load repetition throughout the pavement design life. The PMED requires traffic inputs for any pavement type to include; base year truck traffic volume, vehicle class distribution (VCD), hourly, daily and monthly factors, axle load distribution factors, truck-traffic directional distribution and lane distribution factors, axle and wheel base configurations, vehicle operational speed, truck lateral distribution factor, truck growth factors, tire characteristics and inflation pressure. The axle load spectra (ALS) accounts for the magnitude of truck traffic, this eliminates the need of equivalent single axle load (ESAL) used in the AASHTO 1993 design approach (Administration 2001; ARA 2004).

Climatic effects to the pavement are considered in the PMED using the Enhanced Integrated Climatic Model (EICM). The EICM consists of three models; Climatic-Materials-Structural Model (CMS Model) (Dempsey et al. 1985), Frost Heave and Thaw Settlement Model (Guymon et al. 1993) and Infiltration and Drainage Model (ID Model) (Lytton et al. 1993). With these models the EICM computes and predicts resilience modulus adjustment factor, frost and thaw depths, temperature, pore water pressure, drainage performance and frost heave throughout the entire pavement profile. The implementation of the PMED's EICM Model requires the following hourly climatic data (HCD): precipitation, air temperature, wind speed, percent sunshine and relative humidity (ARA 2004).

2 Background

State Departments of Transportation (State DOTs) are working on PMED implementation to shift from the empirical based AASHTO 1993. The shift requires local calibration of design inputs and validations based on the conditions that exist in the respective state. The Tennessee Department of Transportation (TDOT) is among the state DOTs that are working on PMED implementation. TDOT has so far calibrated most of its flexible pavement distress prediction models for interstate and state routes, traffic volume adjustment factors for all routes, and currently, TDOT is working on the calibration of climate data inputs.

The local calibrated distress models for TDOT include rutting, bottom-up cracking, top-down cracking, and roughness (Gong et al. 2017). Other states that have performed similar calibrations include; Minnesota (Velasquez et al. 2009), Washington state (Li et al. 2009), North Carolina (Muthadi 2007), Utah and Ohio (Darter et al. 2009).

The local calibrated traffic adjustment factors for Tennessee are vehicle class distribution, monthly adjustment factors and number of axles per truck. These practices have also been conducted in other states like Georgia (Selezneva and Von Quintus 2014), North Carolina (Stone et al. 2011), Virginia (Smith and Brian 2009), Arizona (Darter et al. 2013) and Alabama (Turochy et al. 2015).

2.1 MEPDG Evaluation Practices

Pavement management system (PMS) data was used in the state of Louisiana in the effort to evaluate MEPDG flexible pavement design. A total of 39 Louisiana roads sections were evaluated based on these criteria; new or full depth rehabilitated flexible pavement, minimum length of 0.5 mile and having no less than 5 years of service life. Level 3 data was used for the evaluation of the Louisiana flexible pavements. To assess the bias of MEPDG predictions, goodness-of-fit statistics and hypothesis tests were used to compare the predicted and measured mean pavement distresses. IRI predicted and measured values showed close agreements, rutting and fatigue cracking models were dependent on the pavement type analyzed (Wu et al. 2013).

Using goodness-of-fit statistics and hypothesis tests, the state of Iowa conducted a verification of MEPDG national calibrations on sixteen of its pavements. The predicted performance was compared to the measured performance from the Pavement Management Information System (PMIS). From the statistical analysis it was concluded that IRI had a close agreement to the PMIS values, while the rutting and faulting prediction models showed a systematic difference (Sunghwan et al. 2010).

For the validation and recalibration of the national calibrated models in Ohio, LTPP sections and data were used. Both statistical and non-statistical methods were used for analysis. Diagnostic statistical methods were used when the measured performances were above zero. Non-statistical methods were only used in situations where the measured distresses were zero or close to zero. For these cases of measured distresses having zero values or values approaching to zero, the diagnostic statistics are meaning less and or impossible to implement. From these evaluations the MEPDG rutting model predictions had a significant bias to the measure values. MEPDG over predicted IRI values when the measured IRI were low and under predicted the IRI values when the measured values were high. Hot Mix Asphalt (HMA) transverse cracking was diagnosed with a non-statistical method due to lack of enough data. HMA transverse cracking measured and predicted values had a close agreement, hence the national calibrations were seeing fit for this model. Identical conclusion was reached from the analysis of the JPCP transverse cracking model. JPCP transverse faulting and JPCP IRI models had a good correlation with the measured values. All model with bad correlation between measured and MEPDG predicted distresses were recalibrated (Mallela et al. 2009).

Like other DOT, Tennessee has put in place PMED implementation plan and is working on developing local calibrations and other inputs for the successful design implementation. This paper outlines the evaluation climate data sources in relation to pavement performance.

3 Methodology

3.1 Study Objectives

The objective of this study is to compare the predicted and measured pavement distresses on LTPP sites in Tennessee using PMED v2.5.5, MERRA and NARR weather stations, local and national calibrated distress models along with different levels of traffic input data. The comparison seeks to increase the understanding of pavement design inputs and their impacts on pavement performance.

3.2 PMED Design Inputs

For distress prediction models, this study used local calibration coefficients for alligator cracking, longitudinal cracking and rutting (Gong et al. 2017; Zhou et al. 2013) and national calibrated distress model. Material inputs utilized in this study were those from the LTPP Infopave database for all pavement sections, as well as the default national values found in the PMED v2.5.5.

Traffic inputs were divided into three levels;

- Level 1 traffic inputs in were obtained from LTPP sites to include annual average daily truck traffic (AADTT), number of lanes, vehicle class distribution (VCD) and growth, axle distributions, monthly adjustment factors (MAF), and number of axles per truck.
- Level 2 traffic inputs were obtained from Tennessee local calibration (Onyango et al. 2019) to include month adjustment factors, vehicle class distribution and growth, and axle per truck. Axle distribution for Level 2 are those from Level 1 LTPP site.

- Level 3 traffic inputs are the national or global traffic default values that are found in the PMED v2.5.5.

The analysis utilized two climate data sources, Modern Era Retrospective-analysis for Research and Application (MERRA) and North American Regional Reanalysis (NARR). However, the site condition annual water table depth values for the PMED v2.5.5 climatic module, were obtained from the United States Geological Survey (USGS) database.

3.3 LTPP Site Selection

The selected Long-Term Pavement Performance (LTPP) sites comprised of complete input data needed for the analysis. Moreover, the selection considered the vicinity of the weather stations to the LTPP sites. Since this study involves the comparison of the performance of pavement using two sources of climatic data (MERRA and NARR), LTPP sites were selected based on the distance between the LTPP sites and the climatic data station of these two sources. This consideration was made to eliminate the need of interpolation amongst weather stations, which may lead to less representative data if; the terrain is not similar, elevation difference is high and, when there is a large distance between the stations used to create the visual weather stations (VWS) (Schwartz 2015). On a validation of VWS in Canada it was observed that, VWS increased the AC rutting predictions by 1.6 times compared to the actual station (Saha et al. 2014). In another study, to assure the quality of climatic data interpolation between station was avoided (Khazanovich et al. 2013). Comparison of VWS to the actual weather station was observed to have significant effects in a study to validate VWS for MEPDG using LTPP data (Li et al. 2010; Msechu 2021).

The climate data stations selected for hourly climate data had a proximity between the weather stations of the two climate data sources (MERRA and NARR). This consideration enables an assumption of a similar climatic conditions for the two climate stations considered. Hence, a closer and better comparison of the distress prediction with respect to climate inputs is expected. Therefore, the LTPP sites selected on this study are close to the MERRA and NARR weather stations, among other things, (Table 1) the elevation difference was also checked between the weather stations (MERRA and NARR), and the elevation difference between the weather stations (MERRA and NARR) to the respective LTPP sites (Table 1).

Based on the criteria above, a total of twenty LTPP sites were identified. Only ten (10) sites had complete LTPP site input parameters and distress data (Table 1). The selected sites included rigid pavements (LTPP ID: 601,602 and 605), one rehabilitated rigid pavement (LTPP ID: 604), two flexible pavements (LTPP ID: C320 and C330) and 4 rehabilitated flexible pavements (LTPP ID: C311, 1023, 2008 and 1029) (ARA 2004).

Table 1. Climatic stations and LTPP sites relationships

MERRA station ID	NARR station ID	MERRA-NARR distance (mi.)	MERRA – NARR elevation difference (ft.)	LTPP site ID	MERRA-site distance (mi.)	NARR-site distance (mi.)	Site-MERRA elevation difference (ft.)	Site-NARR elevation difference (ft.)
138963	3811	11.38	107.92	601	16.26	17.84	30.39	138.31
				602	16.16	17.71	33.38	141.3
				604	15.55	16.81	19.9	127.82
				605	15.94	17.39	28.4	136.32
				2008	24.81	20.69	−51.86	56.06
138392	13882	24.16	24.16	1029	3.91	24.11	−162.37	−49.45
139546	53868	8.10	8.1	C311	19.96	13.64	141.43	186.19
				C320	20.03	13.63	169.55	214.31
				C330	20.10	13.62	154.29	199.05
				1023	20.12	13.59	124.9	169.66

3.4 Distress Prediction

Distress predictions of the ten (10) LTPP sites were performed using AASHTOWare PMED v2.5.5 software. The prediction considered LTPP sites pavement cross-sections and material data, two weather stations (NARR and MERRA), three traffic levels 1, 2, and 3, and local and national calibrated distress models. Each level of traffic was analyzed with local and national distress models, and each of the distress models were analyzed with NARR and MERRA climatic data making a total of four outputs for each of the respective three traffic levels.

The pavement analysis year was selected based on the available measured distress data on LTPP site. In occasions where different distress surveys were conducted in different years for the LTPP site, multiple pavement analyses were conducted in the PMED v2.5.5 to match the year of LTPP distress survey data.

Based on the availability of distress data on the LTPP database, this study considered terminal IRI, permanent deformation, Asphalt Concrete (AC) bottom up fatigue cracking, total transverse cracking and JPCP transverse cracking as the pavement distresses for the comparison. AC top down cracking was not included due to the previous reports of the model inadequacy (Schwartz et al. 2015; Ziedan et al. 2019).

4 Results and Analysis

4.1 Distress Prediction Results

Results of the distress predictions from PMED v2.5.5 show NARR (N) and MERRA (M) weather stations, traffic levels (Level 1–3) and the distress model calibrations (local and national) are presented in Tables 2, 3, 4, 5 and 6. All the ten LTPP sites were considered for terminal IRI prediction, since they had measured IRI values on all pavement types. The analysis of rigid pavements used national calibrated distress model because the local distress model calibration for rigid pavements is currently lacking. From the IRI prediction results (Table 2), all the three traffic input levels with national calibrated distress models predicted higher distresses than with local calibrated distress models. For all three traffic input levels and both distress models, the terminal IRI predictions with MERRA (M) hourly climatic data were relatively higher than that of NARR (N). Values are measured as inches per mile (in/mile). This could be attributed to the slightly closer proximity to the site of the MERRA station compared to NARR station.

Table 2. Terminal IRI

Traffic	Level 1				Level 2				Level 3				LTPP site data
Calibration	Local		National		Local		National		Local		National		
Site ID	N	M	N	M	N	M	N	M	N	M	N	M	
601	–	–	169	174	–	–	204	210	–	–	191	200	215
602	–	–	171	176	–	–	205	212	–	–	195	212	104
605	–	–	161	165	–	–	197	201	–	–	183	188	128
604	105	105	106	107	109	108	109	111	106	107	108	109	114
C320	100	101	102	106	103	104	112	106	103	104	106	111	62
C330	95	97	97	101	98	100	101	107	97	99	100	106	107
1023	120	121	123	127	121	122	124	129	121	121	124	128	128
C311	112	113	116	118	117	129	123	132	117	122	123	129	80
2008	124	123	126	126	124	122	126	125	123	122	125	124	119
1029	140	144	152	158	139	141	151	156	137	137	149	152	107

Total pavement permanent deformation prediction was done for all flexible and rehabilitated flexible LTPP pavement sections. The distress prediction results (Table 3) shows predictions using national calibrated distress model, being relatively higher than which used local calibrated distress models, through all traffic levels. MERRA (M) stations predictions are relatively higher than the NARR (N) predictions. Total pavement permanent deformation predictions are recorded in inches (in.)

Table 3. Permanent deformation – total pavement

Traffic	Level 1				Level 2				Level 3				LTPP site data
Calibration	Local		National		Local		National		Local		National		
Site ID	N	M	N	M	N	M	N	M	N	M	N	M	
C320	0.12	0.15	0.17	0.28	0.17	0.23	0.26	0.43	0.18	0.23	0.26	0.42	0.35
C330	0.13	0.17	0.19	0.32	0.19	0.26	0.29	0.5	0.19	0.26	0.29	0.49	0.31
1023	0.13	0.17	0.19	0.29	0.14	0.18	0.21	0.32	0.14	0.18	0.2	0.31	0.2
C311	0.16	0.19	0.21	0.28	0.26	0.31	0.35	0.46	0.25	0.31	0.34	0.45	0.35
2008	0.1	0.11	0.13	0.15	0.1	0.1	0.13	0.14	0.09	0.09	0.12	0.13	0.12
1029	0.25	0.33	0.32	0.49	0.26	0.34	0.33	0.5	0.25	0.33	0.32	0.49	0.24

For AC bottom up fatigue cracking, only two flexible pavements and one rehabilitated rigid pavement were considered due to lack available measured distress data

(Table 4). The local calibrated distress models for all traffic levels predicts relatively higher distress than the national calibrated distress model predictions (Table 4). Level 2 traffic inputs have the highest prediction than Level 1 and Level 3, while MERRA (M) predicted more than NARR (N) on some instances. AC bottom up cracking values are recorded as percent lane area (% lane area).

Table 4. AC bottom up fatigue cracking

Traffic	Level 1				Level 2				Level 3				LTPP site data
Calibration	Local		National		Local		National		Local		National		
Site ID	N	M	N	M	N	M	N	M	N	M	N	M	
604	1.86	1.86	1.86	1.86	1.86	1.86	1.86	1.86	1.86	1.86	1.86	1.86	2
C320	1.86	1.88	1.86	1.86	2.17	6.09	1.94	3.09	2	3.89	1.9	2.41	21
C330	1.86	1.89	1.87	1.98	3.13	15.86	2.2	6.97	2.49	9.09	2.02	4.18	20

Total transverse cracking predictions were considered for five (5) rehabilitated pavement sections (Table 5). Identical distress predictions were observed in three of the five (5) sites, at all traffic levels and transfer function coefficients. Only one site had completely different predictions at all traffic levels and calibrated transfer function coefficients (ARA,2004). On this site, Level 2 traffic had higher distress predictions than both Level 1 and Level 3. Total transverse cracking values are recorded in foot per mile (ft/mile).

Three LTPP rigid pavement sections where analyzed for JPCP transverse cracking (Table 5) which was measured as the percent of slab cracked per section (% slab). From the prediction results, it can be noted that Level 3 traffic with both climatic data produced higher distresses followed by Level 2 and Level 1 having the least predictions. In all cases the MERRA (M) climatic station produced higher distresses than NARR (N) station.

Table 5. Transverse cracking – flexible and rigid pavements

Total transverse cracking flexible							
Traffic	Level 1		Level 2		Level 3		LTPP site data
Calibration	National		National		National		
Site ID	N	M	N	M	N	M	
604	4085.6	4054.75	4282.94	4289.25	4282.94	4289.25	5924
1023	797.74	242.26	797.74	242.26	797.74	242.26	527
C311	814.85	242.26	814.85	242.26	814.85	242.26	191
2008	2052.93	2265.15	2052.93	2265.15	2052.93	2265.15	2006
1029	151.9	158.43	205.53	205.53	148.5	152.12	106.95
JPCP transverse cracking rigid							
Traffic	Level 1		Level 2		Level 3		LTPP site data
Calibration	National		National		National		
Site ID	N	M	N	M	N	M	
601	1.23	2.79	1.23	3.66	3.22	8.43	5
602	3.38	7.18	4.27	9.84	11.19	28.48	17
605	1.23	1.23	1.23	1.23	1.23	3.22	2

4.2 Statistical Analysis of Results

Statistical analysis was conducted to compare LTPP sites distress prediction results against measured distresses. The statistical methods used included t-test and the non-parametric Wilcoxon rank sum test. Both tests had a null hypothesis (H_0): No difference between the means of predicted results and measured distresses on LTPP sites, and the alternative hypothesis (H_1): there exists a difference between the predicted and measured distresses on LTPP sites. Both the tests were performed at a 95% confidence level.

T-test was performed on the distress predictions that were normally distributed; AC bottom up fatigue cracking and permanent deformation – total pavement, while the Wilcoxon rank sum test was performed on the distresses that were not normally distributed (Hollander et al. 1999); terminal IRI, JPCP transverse cracking and total transverse cracking.

Both the t-test and Wilcoxon rank sum tests where done based on the traffic level, weather station source and calibration coefficients used. The t-test p-values from the analysis of AC bottom up cracking and permanent deformation can be observed on Table 6, and the Wilcoxon rank sum test p-values for terminal IRI and JPCP transverse cracking can also be observed on Table 6.

Identical Wilcoxon rank sum test p-values of 0.625 for NARR weather station and 0.8125 for MERRA weather station were obtained for all the three traffic levels for total transverse cracking due to this, these values are not reported on Table 6.

Table 6. P-Values for PMED predicted distresses

	Bottom up fatigue cracking				Permanent deformation – total pavement			
	Local calibration		National calibration		Local calibration		National calibration	
Traffic	NARR	MERRA	NARR	MERRA	NARR	MERRA	NARR	MERRA
Level 1	0.1807	0.1811	0.1808	0.1816	0.0310	0.1544	0.2223	0.0740
Level 2	0.1922	0.4446	0.1838	0.2312	0.1364	0.6404	0.9580	0.0911
Level 3	0.1863	0.2643	0.1822	0.2018	0.1192	0.5989	0.8974	0.1147
	Terminal IRI				JPCP transverse cracking	
	Local calibration		National calibration		National calibration	
Traffic	NARR	MERRA	NARR	MERRA	NARR	MERRA
Level 1	0.62	0.62	0.3527	0.2475	0.184	0.7
Level 2	0.8125	0.5362	0.0431	0.0431	0.184	0.7
Level 3	0.535	0.4557	0.1309	0.0753	0.7	0.7

NOTE: All p-values are > 0.05 signifying failure to reject the null hypothesis; therefore, the difference between measured and predicted values is statistically insignificant

4.3 Discussion of the Results

From the statistical analysis results (Table 6) distresses predicted by both the MERRA and NARR climate data show no significant difference from the LTPP site measured distresses. The MERRA climate data show higher p-values for its predictions indicating a better agreement to the null hypothesis compared to NARR, hence closer to the measured values.

For the AC bottom up fatigue cracking (Table 6), it is noted that Level 1 traffic p-values with respect to national calibrated distress model gave a slightly better agreement to the null hypothesis than the local calibrated distress models, whereas in both cases the MERRA climate data gave a better agreement to the null hypothesis. Level 2 traffic with local calibrated distress models gave a better agreement to the null hypothesis than with default (national) transfer functions, whereas in both cases MERRA climate data gave a better agreement to the null hypothesis than the NARR station. Level 3 traffic prediction were much more like Level 2 predictions favoring local calibrated distress models and MERRA climate data. In overall it can be stated clearly that among the three traffic levels Level 2 gave the best agreement to the null hypothesis.

Permanent deformation based on the p-values (Table 6) show that NARR climate data with the national calibrated distress model gave a better agreement to the null hypothesis than all other conditions. However, Level 2 traffic with MERRA climate data and local calibrated distress models gave a slight better agreement to the null hypothesis than the NARR climate data. In overall Level 2 traffic gave a better agreement to the null hypothesis with respect to local calibrated distress models having MERRA climate data as the best agreement to the null hypothesis. Level 3 traffic gave a better agreement to the null hypothesis with the national calibrated distress models and having the best agreements with the NARR climatic station.

The terminal IRI model used in the state of Tennessee has adopted the default national values for its calibration coefficients, however changes in the terminal IRI predictions were observed with change of other distress calibration models (cracking or permanent deformation) (Table 6). For this reason, the IRI distresses recorded in Table 6 for national and local calibration models are different. From Table 6, all levels of traffic gave a better agreement to the null hypothesis with local calibrated distress model than with national calibrated distress model for both MERRA and NARR climate data. Level 2 traffic gave relatively the best agreement to the null hypothesis with local calibrations but the least with national calibrations. NARR climate data gave a slightly better agreement to the null hypothesis than MERRA in all scenarios.

Like the terminal IRI model, JPCP transverse cracking model is not calibrated for the state of Tennessee. From Table 6, the p-values of the national calibrated JPCP transverse model shows that MERRA data has resulted to a better agreement to the null hypothesis with all traffic levels than NARR.

Total transverse cracking being only nationally calibrated, produced identical p-values for all levels of traffic. P-values with respect to NARR was 0.625 and 0.8125 for MERRA, this phenomenon of equal p-values for all the traffic levels can be related to transverse cracking not being a load related distress, but dependent to climatic conditions.

5 Conclusions

From the evaluation carried out in this study the following can be concluded;

- For the calibrated distress prediction models, local calibrated distress model predictions were much closer to the measured values compared to national calibrated distress models.
- When considering local calibrated distress models, Level 2 traffic (local calibrated) predicted distresses that were much closer to the measured values than all the traffic levels whereas the national calibrated distress models mostly favored level 3 traffic inputs.
- Furthermore, on the local calibrated distress models, MERRA climatic stations distress predictions were much closer to the measured values on bottom up cracking and permanent deformation while NARR predicted values were much closer to the measured values with terminal IRI predictions.
- On national calibrated distress models, MERRA climate data distress predictions were much closer to the measured values for both JPCP transverse cracking and total transverse cracking while, NARR distress predictions were much closer to the measured values with bottom up cracking, terminal IRI and permanent deformation.
- From Table 1, it can also be observed that the MERRA stations elevation differences to the LTPP sites are less compared to the NARR stations to the LTPP sites while, the difference in distances to the sites are relatively close for both weather station.

From these observations the state of Tennessee can adopt the local calibrated distress model, Level 2 traffic and MERRA climatic stations for pavement analysis and design. MERRA climatic stations is more advantageous than NARR weather stations since they cover larger geographic area (Ziedan et al. 2019). Almost all performances had no significant difference to the LTPP site measured data. P-values for most predictions were not very high despite been above 0.05. Higher p-values imply to a stronger agreement with the null hypothesis. A stronger agreement with the null hypothesis informs on how close the predicted values are from the measured values. Distress predictions with low p-values can be attributed to the use of national default values in cases where the LTPP database did not have the required inputs.

This evaluation involved only ten (10) LTPP station from the twenty candidate LTPP sites, due to the lack of measured pavement distress data (ARA 2004). For better comparison of inputs, more distress data from Tennessee roads should be utilized to validate and improve the local calibrated distress models and factors.

Acknowledgements. Authors appreciate and thank the cooperation and research funding provided by the Tennessee Department of Transportation for development and implementation of the PMED inputs for the state of Tennessee. Likewise, thanks are due to LTPP Infopave for available data that can be applied for pavement studies and improvements.

References

Administration, F.H.: Guide to LTPP traffic data collection and processing. In: Federal Highway Administration, Office of Infrastructure Research (2001)

Ara, I.: Guide for mechanistic-empirical design of new and rehabilitated pavement structures. In: Transportation Research Board of the National Academies. Washington, DC (2004)

Darter, M.I., Titus-Glover, L., Von Quintus, H.L.: Draft user's guide for UDOT mechanistic-empirical pavement design (2009)

Darter, M.I., Titus-Glover, L., Wolf, D.J.: Development of a traffic data input system in Arizona for the MEPDG (2013)

Dempsey, B., Herlach, W., Patel, A.: The climatic-material-structural pavement analysis program, Final report. Washington DC: Federal Highway Administration (1985)

Gong, H., Huang, B., Shu, X., Udeh, S.: Local calibration of the fatigue cracking models in the mechanistic-empirical pavement design guide for Tennessee. Road Mater. Pavement Des. **18**(sup3), 130–138 (2017)

Guymon, G.L., Berg, R.L., Hromadka, T.V.: Mathematical model of frost heave and thaw settlement in pavements (1993)

Hollander, M., Wolfe, D.A., Chicken, E.: Nonparametric Statistical Methods. John Wiley & Sons, New York (1999)

Khazanovich, L., et al.: Design and construction guidelines for thermally insulated concrete pavements (2013)

Li, J., Pierce, L.M., Uhlmeyer, J.: Calibration of flexible pavement in mechanistic–empirical pavement design guide for Washington state. Transp. Res. Rec. **2095**(1), 73–83 (2009)

Li, Q., Wang, K.C., Hall, K.D.: Verification of virtual climatic data in MEPDG using the LTPP database. Int. J. Pav. Res. Technology **3**(1), 10–15 (2010)

Lytton, R., Pufahl, D., Michalak, C., Liang, H., Dempsey, B.: An integrated model of the climatic effects on pavements (1993)

Mallela, J., et al.: Guidelines for Implementing NCHRP 1–37A ME design procedures in ohio: Volume 1–Summary of Findings, Implementation Plan, and Next Steps (2009)

Msechu, K.J.: Improvement of pavement mechanistic-empirical design (PMED) virtual weather station interpolation model using radial basis function-Tennessee case study. Thesis University of Tennessee at Chattanooga (2021)

Muthadi, N.R.: Local calibration of the MEPDG for flexible pavement design (2007)

Onyango, M., Owino, J., Wu, W., Fomunung, I.: Traffic Data Input for Mechanistic Empirical Pavement Design Guide (MEPDG) for Tennessee (2019). https://www.tn.gov/content/dam/tn/tdot/documents/RES2016-22.pdf

Saha, J., Nassiri, S., Bayat, A., Soleymani, H.: Evaluation of the effects of Canadian climate conditions on the MEPDG predictions for flexible pavement performance. Int. J. Pavement Eng. **15**(5), 392–401 (2014)

Schwartz, C.W.: Evaluation of LTPP climatic data for use in mechanistic-empirical pavement design guide (MEPDG) Calibration and Other Pavement Analysis. US Department of Transportation, Federal Highway Administration, Research (2015)

Schwartz, C.W., Forman, B.A., Leininger, C.W.: Alternative source of climate data for mechanistic–empirical pavement performance prediction. Transp. Res. Rec. **2524**(1), 83–91 (2015)

Selezneva, O., Von Quintus, H.: Traffic load spectra for implementing and using the mechanistic-empirical pavement design guide in Georgia (2014)

Smith, B.C., Brian, K.: Diefenderfer. Development of Truck Equivalent Single-Axle Load (ESAL) Factors Based on Weigh-in-Motion Data for Pavement Design in Virginia. VTRC, 09-R18 (2009)

Stone, J.R., et al.: Development of traffic data input resources for the mechanistic empirical pavement design process (2011)

Sunghwan, K., Ceylan, H., Gopalakrishnan, K., Smadi, O., Brakke, C., Behnami, F.: Verification of mechanistic-empirical pavement design guide (MEPDG) performance predictions using pavement management information system (PMIS). Transp. Res. Rec. **2395** (2010)

Turochy, R.E., Timm, D.H., Mai, D.: Development of Alabama traffic factors for use in mechanistic-empirical pavement design (2015)

Velasquez, R., et al.: Implementation of the MEPDG for new and rehabilitated pavement structures for design of concrete and asphalt pavements in Minnesota (2009)

Wu, Z., Yang, X., Zhang, Z.: Evaluation of MEPDG flexible pavement design using pavement management system data: Louisiana experience. Int. J. Pavement Eng. **14**(7), 674–685 (2013)

Zhou, C., Huang, B., Shu, X., Dong, Q.: Validating MEPDG with Tennessee pavement performance data. J. Transp. Eng. **139**(3), 306–312 (2013)

Ziedan, A., Onyango, M., Wu, W., Udeh, S., Owino, J., Fomunung, I.: Comparative analysis between modern-era retrospective analysis for research and applications and updated mechanistic-empirical pavement design guide climate database in the state of Tennessee. Transp. Res. Rec. **2673**(6), 279–287 (2019)

Influence of Dynamic Analysis on Estimation of Rutting Performance Using the Fixed Vehicle Approach

Gauri R. Mahajan, Radhika Bayya, and Krishna Prapoorna Biligiri[(✉)]

Indian Institute of Technology Tirupati, Tirupati, India
bkp@iittp.ac.in

Abstract. Vehicles traversing on asphalt pavements are of diverse types comprising different axle configurations. Globally, the flexible pavement design guidelines are based on fixed vehicle approach and the pavement analysis is carried out by adopting static or quasi-static loads ascribed to the lack of provision for dynamic load representation and intricacies in the analyses. However, it has been found that the dynamic component significantly contributes to the time-varying nature of vehicular loads that affects the asphalt pavement performance. On this note, this paper focused on estimating the rutting performance of asphalt pavements under the influence of dynamic vehicular loads. The dynamic analysis was performed on a pavement system modeled in ABAQUS® software by considering traffic loads due to four vehicle types with different axle configurations. The surface wearing course was represented by the power-law creep model to account for viscoelastic nature of the asphalt. The time-varying loads were represented by a lumped parameter model for the vehicle accounting for the inertia, stiffness, and damping characteristics in addition to the self-weight. Further, the vertical strains obtained from the dynamic analysis were compared with the estimates of static analysis, thereby, examining the differences in the strain ratios responsible for rutting failure due to the two analyses. Additionally, calibration factors for strain ratios were proposed, which will exemplify the realistic contribution of dynamic loads in rutting performance estimation. Thus, this study will help advance the existing pavement design framework through the incorporation of the vehicle dynamics effects in rational asphalt pavement performance evaluation.

Keywords: Asphalt pavements · Fixed vehicle approach · Static and dynamic analyses · Vehicle model · Power-law creep model · Rutting · Calibration factors

1 Introduction

The various vehicles traversing on asphalt pavements cause distresses such as rutting, fatigue cracking, potholes, raveling, and many more. The vehicles comprise combinations of different axle configurations such as single, tandem, tridem, and quad, and the number of repetitions of axles deteriorate the pavements further. The vehicular loads consist of static and dynamic loading components, which correspond to the self-weight of the vehicle and time-varying nature of the vehicles, respectively. Further, the surface roughness increases vehicular vibrations, which then transfers the static and dynamic loads together onto the asphalt pavements leading to occurrence of distresses.

In traditional pavement design approach, rutting and fatigue cracking are evaluated under the influence of static and/or quasi-static (moving) loads and do not consider the effect of dynamic time-varying nature of traffic loads (AASHTO 1993; AUSTROADS 2004; European Commission 1999; IRC:37 2018; MEPDG 2008). Additionally, researchers have incorporated the static and quasi-static loads for pavement analysis and design through various software such as KENPAVE™, VESYS, VISCO-ROUTE, MICHPAVE, ILLIPAVE®, etc. (Chen et al. 1995; Chabot et al. 2006; Huang 1973; Kinder 1986; Thompson and Elliott 1985). It is noteworthy that the software programs evaluate pavement performance under the influence of static or moving loads but do not account for the time-varying component of vehicular loads.

In the late 1980s, due to the advancements in vehicle design with multiple axle configurations, researchers recognized the importance of dynamic aspects of vehicular loading for pavement performance evaluation (Brademeyer et al. 1986; Hardy and Cebon 1993; Markow et al. 1988). This led to the consideration of dynamic vehicle load component in the pavement design process by incorporating materials properties through the use of dynamic/stiffness modulus. However, the process still does not account for the vehicular characteristics such as inertia, damping, and stiffness, which play a significant role in transferring the vehicular load to the pavement systems. The vehicle characteristics are widely captured in vehicle dynamics using dynamic vehicle system models such as quarter-car, half-car, and full-car models (Barbosa 2011; Hamersma and Els 2015; Jiao 2013; Li et al. 2009; Zhu and Ishitobi 2006). The vehicle models comprise spring and dashpot elements that correspond to the stiffness and damping properties of the vehicles, respectively. Notwithstanding, the vehicle models have not been utilized to evaluate and quantify asphalt pavement distresses (Mahajan et al. 2020).

Despite the availability of vehicle system models, the existing flexible pavement design guidelines use the fixed vehicle approach with static component of vehicular loads as an important input parameter. Essentially, the analysis is performed using a linear elastic pavement system subjected to a stationary single or multi-axle load with a magnitude of 80 kN (AASHTO 1993; AUSTROADS 2004; European Commission 1999; IRC:37 2018; MEPDG 2008; SAPEM 2013). As part of the design process, the vertical compressive and/or horizontal tensile strains obtained as key outputs are used to quantify rutting and fatigue performance criteria. However, these performance criteria are evaluated only under the influence of static vehicular loads, which may underestimate or overestimate the final pavement thickness due to the absence of dynamic loads. Hence, it is necessary to incorporate the dynamic aspects of vehicular loads to replicate the actual traffic loading scenario for performance criteria evaluation.

Thus, the objective of this study was to examine the influence of dynamic loads on rutting performance evaluation in asphalt pavement systems, which was accomplished by using a dynamic vehicle model traversing on a single-layered viscoelastic asphalt pavement system modeled in ABAQUS® software. The pavement model was subjected to the dynamic loads due to four vehicle types with different axle configurations. Further, calibration factors were proposed based on the accumulated strain for each vehicle type to quantify the contribution of dynamic and static loads in assessing rutting performance.

2 Study Methodology

2.1 Data Collection and Representation of Traffic Loads

The traffic survey and weigh-in motion (WiM) data were collected from Andhra Pradesh Road Development Corporation (APRDC) for the Ongole-Nandyala State highway in Prakasam district of Andhra Pradesh, India. The traffic survey was carried out for 24 h for 3 consecutive days and the data was recorded at an interval of 15 min for each vehicle group. Based on the traffic count, there were four vehicle types with varied axle configurations. The information about the various axle configurations for the four vehicle types are given in Table 1. Also, based on the collected WiM data, the individual axle loads from each vehicle type were converted into an equivalent standard axle load (ESAL) of 80 kN using the fixed vehicle approach.

Table 1. Information about axle configurations for vehicle types

Vehicle type	Axle configuration	
	Front axle	Rear axle
Type 1 (2-axle truck)	Single axle with single wheel on each side	Single axle with dual wheels on each side
Type 2 (3-axle truck)		Tandem axle
Type 3 (Bus)		Single axle with dual wheels on each side
Type 4 (Multi-axle truck)		Tandem axle + Tridem axle

The ESAL is represented by a standard axle configuration of a single axle with single wheel (front axle) and a single axle with dual wheels (rear axle) according to several guidelines (AASHTO 1993; AUSTROADS 2004; European Commission 1999; IRC:37 2018; MEPDG 2008; South African National Roads Agency Ltd. 2013). In this study, the axle loads of four vehicle types were converted into 80 kN ESAL using the truck factor (TF) designated in the USA-based Pavement-ME approach (MEPDG 2008). For this, the WiM data was used to derive the number of axles for each axle load category, which was further multiplied with the equivalent axle load factors (EALF) based on the axle configurations to determine the ESAL of the particular vehicle type. The computation of ESALs for each vehicle type is given by Eq. (1):

$$ESAL = Number\ of\ axles\ for\ each\ load\ category \times EALF \quad (1)$$

Then, the summation of ESALs of each axle load category was divided by the number of trucks or vehicles of the particular type to determine the TF, represented by Eq. (2) (MEPDG 2008):

$$TF = \frac{Total\ 80\ \text{kN}\ ESALs}{N} \quad (2)$$

where N is the number of trucks.

The predicted number of vehicular truck traffic for the four vehicle types used in this study was determined using Eq. (3) (MEPDG 2008):

$$TT = AADTT_X \times MDF \times HDF \times DDF \times LDF \times No.\ of\ days \quad (3)$$

where TT is the total number of trucks predicted for the considered pavement design period; $AADTT_X$ is the average annual daily truck traffic at age X; MDF is the monthly distribution factor; HDF is the hourly distribution factor; DDF is the directional distribution factor; LDF is the lane distribution factor. The $AADTT_X$ was determined according to Eq. (4) (MEPDG 2008):

$$AADTT_X = AADTT_{BY} \times GR^{AGE} \quad (4)$$

where $AADTT_{BY}$ is the base year annual average daily truck traffic; GR is the traffic growth rate (in %); AGE is the design period in years. The computed TT for each vehicle type helped determine the number of standard axles of 80 kN required to be used in the asphalt pavement model for both static and dynamic analyses.

In this study, the dynamic aspects of the vehicular axle loads were based on two-degrees of freedom (2-DOF) model with spring, mass, and dashpot arrangement as representative of vehicle body (sprung mass) and tire (unsprung mass) to determine the vertical strain in asphalt pavements. The vehicle model was assumed to have an axle configuration similar to that of an 80-kN standard load configuration. The 2-DOF vehicle system model is mathematically represented in Eqs. (5) and (6):

$$m_v \ddot{x}_v + c_v(\dot{x}_v - \dot{x}_t) + k_v(x_v - x_t) = 0 \quad (5)$$

$$m_t \ddot{x}_t + c_v(\dot{x}_t - \dot{x}_v) + k_v(x_t - x_v) + c_t \dot{x}_t + k_t x_t = 0 \quad (6)$$

where m_v and m_t are the masses of vehicle and tire, respectively in kg; c_v and c_t are the vehicle and tire damping coefficient, respectively in Ns/m^2; k_v and k_t are the linear stiffness of vehicle and tire, respectively in N/m; x_v and x_t are the vehicle body and tire displacements, respectively in m; $(c_t \dot{x}_t + k_t x_t)$ represents the force transferred through a tire in N.

For the standard axle configuration with a load of 80 kN (40 kN on each axle), the stiffness and damping values for the vehicle model were determined based on the natural frequency of vehicle vibration. Additionally, the axle spacing and vehicle speed were used based on the Indian Roads Congress guidelines and the speed limits proposed for State Highways in India (IRC: 6 2016; Motor Vehicle Act 1988). Further, the time duration (TD) for each vehicle system model's load application was computed based on the axle spacing and vehicle speed. In this study, the dynamic analysis using the 2-DOF vehicle model was performed in MATLAB® to generate the time-varying pressure history of the axle loads. The pressure to be applied on the pavement model was determined based on the dynamic time-varying load and the tire pavement contact area. Likewise, the static loads based on ESALs of 80 kN axle load for four vehicle types were applied on the pavement model as a constant stationary load of 40 kN each for front and rear axles. The TD for static load application was similar to the aforementioned dynamic analysis. The details about the viscoelastic asphalt pavement model used for the static and dynamic loads application is discussed next.

2.2 Modeling of Asphalt Pavement System

The asphalt pavement model was developed in ABAQUS® and subjected to dynamic and static loads for determining the rutting response. The geometry of the developed pavement model consisted of a 1.75 m wide and 3 m long pavement section. The assumed width was taken due to symmetry and is half of one lane width of 3.5 m (IRC: SP-73 2018). The length of the pavement model corresponded to the extent of the stress distribution occurring due to the largest magnitude of axle load from WiM data. The pavement model consisted of a circular tire-pavement contact area, which is commonly used due to its simplicity for application of dynamic and static loads (Huang 2004). Note that static and dynamic analyses were performed using a viscoelastic pavement material model to study the influence of dynamic loads on rutting response in asphalt pavements. Further, static analysis was also performed using a linear elastic pavement material model, available within the USA-based Pavement-ME guidelines (MEPDG 2008).

The viscoelastic properties were assigned to the developed pavement model to capture the behavior of the pavement material under the influence of traffic loads. The viscoelastic nature of the asphalt surface course was described using the power-law creep model, which has been used by researchers for studying the rutting performance in asphalt pavements (Imaninasab and Bakhshi 2017; Uzarowski 2006; White 2002). The mathematical equation of the classic power-law creep model is given in Eq. (7):

$$\dot{\varepsilon}_c = A\sigma^n t^m \qquad (7)$$

where $\dot{\varepsilon}_c$ is the creep strain rate; σ is the uniaxial equivalent deviatoric stress in N/mm^2; t is the time in s; A, n and m are the material constants, which depend on the asphalt mix and temperature.

The viscoelastic and linear elastic asphalt material properties were extracted from the experimental results of repeated load permanent deformation (RLPD) test, as explained later. Further, the vertical strains were determined based on static and dynamic analyses using viscoelastic and linear elastic pavement model, as discussed next.

2.3 Determination of Vertical Strains from Static and Dynamic Analyses

The verical strain in the asphalt pavements is the governing parameter required to quantify rutting. In this study, the vertical strains generated due to the four vehicle types in the viscoelastic and linear elastic asphalt pavement model were computed based on the static and dynamic analyses. Pavement-ME uses Jacob Uzan Layered Elastic Analysis (JULEA), an elastic layer theory program to estimate vertical strains due to applied load and other materials properties. Further, the accumulated permanent axial strain was determined using Eq. (8) based on static analysis using the linear elastic pavement model (MEPDG 2008):

$$\varepsilon_p = \frac{\beta_{1r} \times k_z \times \varepsilon_{r(HMA)} \times 10^{k_{1r}} \times n^{k_{2r}\beta_{2r}} \times T^{k_{3r}\beta_{3r}}}{h_{(HMA)}} \qquad (8)$$

where, ε_p is accumulated permanent axial strain in the HMA layer; $\varepsilon_{r(HMA)}$ is the resilient or elastic strain at the mid-depth of HMA layer; n is the number of axle load repetitions; T is the mix or pavement temperature in °F; $k_{1r,2r,3r}$ are the global field calibration parameters; $\beta_{1r,2r,3r}$ are the local or mixture field calibration constants for the global calibration, set to 1; $h_{(HMA)}$ is the thickness of the HMA layer in inches. k_z is depth confinement factor represented by Eq. (9):

$$k_z = (C_1 + C_2 D) \times 0.328196^D \qquad (9)$$

where, D is the depth below the surface in inches, and C_1, C_2 based on Eqs. (10) and (11):

$$C_1 = -0.1039(H_{HMA})^2 + (2.4868 \times H_{HMA}) - 17.342 \qquad (10)$$

$$C_2 = 0.0172(H_{HMA})^2 - (1.7331 \times H_{HMA}) + 27.428 \qquad (11)$$

Note that Eq. (8) fundamentally is the field-calibrated mathematical form obtained based on the laboratory results of the RLPD test indicative of viscoelastic material properties in asphalt pavements. Furthermore, this study incorporated the viscoelastic material behavior of asphalt concrete using power-law creep model, and the resultant accumulated vertical creep and elastic strains obtained from viscoelastic pavement model during static and dynamic analyses in ABAQUS® were in turn utilized to derive the ratio of plastic to elastic strain, called strain ratio $\left(\frac{\varepsilon_p}{\varepsilon_r}\right)$. Thus, the strain ratios obtained from static and dynamic analyses of the viscoelastic pavement model were subsequently compared with the strain ratio of the static analysis based on linear elastic model, actually implemented in the existing pavement design guidelines. Therefore, the comparison of strain ratios between the different methods helped demonstrate the plausible susceptibility of asphalt pavements to rutting, which assisted in proposing calibration factors due to static and dynamic viscoelastic pavement model analyses.

3 Analyses

3.1 Determination of Truck Traffic Based on Fixed Vehicle Approach

The calculation of truck traffic requires TF, $AADTT_X$, MDF, HDF, DDF, and LDF in Eq. (3), which was utilized to determine the number of vehicles required for performing static and dynamic analyses. The individual axle loads from each vehicle type obtained from WiM data were converted into an ESAL of 80 kN using EALF, as discussed in Sect. 2.1. The TF of the vehicle types were calculated using Eq. (1) and $AADTT_{BY}$ was obtained from the traffic count data assembled in the available APRDC database. Additionally, a design period of 4 months (1 summer season in India) was considered to determine the number of vehicles that will run on the pavement model developed in ABAQUS® for static and dynamic analyses. Due to non-availability of actual MDF, HDF, DDF, and LDF values for the conditions in India, the magnitudes were assumed to be 1, 4.18, 0.5, and 1, respectively, as proposed in Pavement-ME for two-lane highways (MEPDG 2008). The annual traffic growth rate was taken as 10.11%, as recommended for traffic conditions in India (MORTH Annual Report 2019-2020). Further, the TT for different vehicles was computed, and reported as in Table 2. For the obtained TT; static and dynamic analyses were performed, as discussed in the next section.

Table 2. Computed TF, $AADTT_{BY}$, and TT for different vehicles

Vehicle type	TF	$AADTT_{BY}$	TT
Type 1 (2-axle truck)	0.0196	504	125
Type 2 (3-axle truck)	0.1418	955	237
Type 3 (Bus)	0.0068	674	167
Type 4 (Multi-axle truck)	0.1237	171	42

3.2 Dynamic and Static Load Analyses

The dynamic analysis was performed using the 2-DOF vehicle model, as mentioned in Sect. 2.1, which was utilized for analysis under time-varying load application on viscoelastic asphalt surface wearing course. In this study, a constant damping value of 5% was assumed, and the k_v and k_t were derived based on the standard axle load of 80 kN by assuming the natural frequency of the vehicle free vibration as 1.5 Hz (Jazar 2008). Hence, for a standard axle load of 80 kN, the magnitude of the axle loads was taken as 40 kN each for the front and rear axles. The 2-DOF vehicle system model parameters adopted in this study are shown in Table 3.

Table 3. Vehicle system model parameters based on the standard axle load of 80 kN

Axle	m_v (kN)	k_v (N/m)	k_t (N/m)
Front	40	354945.6	3549456
Rear	40	354945.6	3549456

The analysis of the vehicle model was initiated with an assumed external displacement of 0.2 m, which is indicative of the roughness of the pavement to vibrate the vehicle system at the first time instant to generate the tire forces. Further, the longitudinal spacing between the front and rear axles was taken as 3.2 m for the standard axle configuration conforming to IRC:6 2016. The speed for the vehicle system model was 65 kmph, which is the maximum speed limit of the vehicle configuration used on State highways in Andhra Pradesh, India (Motor Vehicle Act 1988). Based on the given axle spacing and speed, the TD was computed as 0.1772 s, which was further divided into three categories to account for the approaching, full load, and leaving effects of each axle. From the computed TD, 10% of the time was assumed for full load application of the axle load, while the remaining 90% was distributed between the approaching and leaving effects. Since this study focused on rutting evaluation in asphalt pavements, the assumed 10% and 90% apportionment of TD represented loading and unloading patterns followed in the RLPD (flow number) test, which characterizes rutting in the laboratory (AASHTO TP-79 2010; Witczak 2007).

On the other hand, a time headway of 2 s was considered between the successive vehicles: rear axle full load application of the first vehicle and the beginning of the full load of the front axle of the following vehicle (SWOV Fact sheet 2003). Further, based on the TD and time headway, Eqs. (5) and (6) pertaining to the 2-DOF vehicle model were solved analytically to compute the force transferred through the tire. The computed tire force was subsequently combined with the contact area to determine the pressure magnitude that was applied for approaching, full load, and leaving effects. Based on the TD and computed TT, the total time duration required for dynamic and static analyses for different vehicles was determined and summarized in Table 4.

Table 4. Total time duration for dynamic and static analyses

Vehicle type	Total time duration (seconds)
Type 1 (2-axle truck)	273
Type 2 (3-axle truck)	519
Type 3 (Bus)	365
Type 4 (Multi-axle)	91

In this way, the pavement analyses were performed based on the time-varying pressure history generated using the vehicle system model and static load of 40 kN each for the front and rear axles for the computed total time durations. From the obtained values of total time duration, it was observed that each vehicle type incurred different computation time for analysis based on the number of vehicles that used the facility or the computed TT for the respective vehicle type. The numerical modeling of asphalt pavement for dynamic and static load applications is discussed in the next section.

3.3 Asphalt Pavement Numerical Model

In this study, the asphalt pavement dynamic and static response analysis was performed using a 3-D model geometry developed for the Ongole-Nandyala State Highway, State of Andhra Pradesh, India. Since this study focused on proposing the calibration factors based on dynamic and static analyses, a simple viscoelastic single-layered asphalt surface course with a thickness of 0.12 m was developed in ABAQUS®. The thickness was equal to the depth of the asphalt surface wearing course of the Ongole-Nandyala roadway section. The radius of contact area was estimated to be 0.1 m based on the assumed tire pressure of 689.476 kPa and the maximum axle load obtained from the WiM data.

The material constants for the considered power-law creep model were extracted from the results of the RLPD (flow number) test performed at 54 °C on a bituminous concrete grade 1 (BC-grade 1) that was blended with a viscosity grade (VG30) binder with an asphalt content of 5.5% (Babitha 2015; MoRTH 2013). The extracted material parameters for the viscoelastic asphalt surface course model are displayed in Table 5. With the mentioned geometry and material inputs, the boundary conditions at the bottom of the asphalt surface course were assumed to be fixed. On the contrary, the boundary nodes along the pavement edges were horizontally constrained but allowed to move in the vertical direction to capture the vertical compressive strains. Figure 1 shows the load and boundary conditions of the asphalt pavement model developed in ABAQUS®.

Table 5. Material parameters used for asphalt surface course in ABAQUS®

Mix type	Instantaneous modulus (MPa)	Poisson's ratio	A	n	m
BC-grade 1 with VG30 binder	65	0.4	0.0002473	0.7561	−0.3975

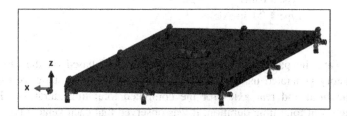

Fig. 1. Load and boundary conditions of asphalt pavement model

The eight-node linear brick reduced integration element (C3D8R) from the ABAQUS® solid element library was selected for analysis. The selection of the given element type was to achieve a reliable solution from finite element analysis, as mentioned in the literature (Bussler and Ramesh 1993). Further, the analysis of the viscoelastic pavement model subjected to dynamic time-varying pressure history and static load of 40 kN was implemented using VISCO option in the step module of ABAQUS® and the Quasi-Newton algorithm, which minimizes the computation time in updating the stiffness matrix for each finite element analysis step. Also, a linear elastic pavement model was developed using the same geometry and elastic property, as mentioned earlier in this section. The static analysis using linear elastic pavement model was performed using STATIC GENERAL option in the step module and Newton algorithm was adopted owing to the linearity of the model. The aforesaid numerical viscoelastic and linear elastic pavement models were used for performing respectively, dynamic and static analyses of the axle loads to obtain the resultant vertical accumulated strains.

4 Results and Discussion

Table 6 summarizes the details of the vehicle model parameters for two vehicles obtained during the analyses. The dynamic time-varying pressure history obtained due to the 2-DOF vehicle model is shown in Fig. 2. This was applied on the viscoelastic asphalt surface model to obtain the vertical accumulated strains, which are representative of the rutting mechanism in asphalt pavements.

Table 6. Vehicle model parameters

Vehicle number	Axle	Mass of the vehicle body (kg)	Stiffness of vehicle body (1×10^4 N/m)	Mass of the tire (kg)	Stiffness of the tire (1×10^5 N/m)	Time duration (seconds)
1	Front axle	4078.86	35.495	500	35.495	0
1	Rear axle	4078.86	35.495	500	35.495	0.1772
2	Front axle	4078.86	35.495	500	35.495	2.0000
2	Rear axle	4078.86	35.495	500	35.495	0.1772

The time-varying pressure history (similar to Fig. 2) was generated for the computed TT for each vehicle type. Based on the generated time-varying history, the accumulated strains were obtained from the dynamic analysis of the viscoelastic pavement model. The elastic strain obtained using static analysis of the linear elastic pavement model were subsequently utilized to obtain the accumulated strains using Eq. (8). The summary of the strain values obtained from the static and dynamic analyses of the pavement model is displayed in Table 7. Further, a typical plot of the resultant vertical strain for a viscoelastic pavement model analysis is shown in Fig. 3.

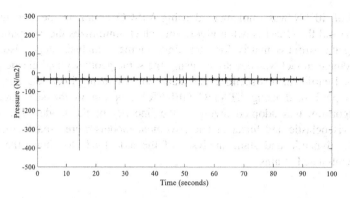

Fig. 2. A typical time-varying pressure history generated from 2-DOF vehicle model

Table 7. Vertical strains obtained from dynamic and static analyses of pavement model

Vehicle type	Dynamic analysis (viscoelastic)		Static analysis (viscoelastic)		Static analysis (linear elastic)	
	Elastic strain (1×10^{-7})	Creep strain (1×10^{-4})	Elastic strain	Creep strain	Elastic strain	Creep strain
Type 1 (2-axle truck)	3.15	4.92	0.0037	1.2689	0.0037	0.0030
Type 2 (3-axle truck)	3.39	4.92	0.0037	1.2713	0.0037	0.0041
Type 3 (Bus)	2.89	4.92	0.0037	1.2702	0.0037	0.0035
Type 4 (Multi-axle truck)	3.54	4.92	0.0037	1.2664	0.0037	0.0018

As observed, the accumulated strains based on the static analysis using linear elastic model are lower than that of the static analysis using the viscoelastic model. The difference in the strain values was mainly due to the absence of viscoelastic effects in the linear elastic model. Therefore, the static linear elastic analysis may underestimate the accumulated strain, and possibly rutting response magnitudes in asphalt pavements. Further, the static analysis using linear elastic as well as viscoelastic pavement models overestimate the accumulated strains compared to the dynamic viscoelastic analysis. This may lead to overdesign of the asphalt pavements, wastage of materials and may turn out to be expensive. The overestimation of the accumulated strains in the static analysis could be due to the absence of the time-varying vehicular loading as well as the vehicle speed effects, which are very much considered in the dynamic analysis. Additionally, the dynamic analysis also accounts for the approaching, full load, and leaving load effects that are observed in the real-field scenario.

From the results of the dynamic analysis, it was observed that there seems to be a negligible change between the creep strain values within a particular vehicle analysis type. The negligible change may perhaps be due to the similar loading pattern in a particular analysis type with a single axle load magnitude of 80 kN for each vehicle. Also, due to the consideration of a single load magnitude, no significant difference in the generated tire force was observed, and also the pressure values at every time instant simulated using the analysis of the vehicle model. Based on both dynamic and static analyses using viscoelastic pavement material model, the maximum strains occurring in the considered pavement were 492 and 1,271,300 microstrains ($\mu\varepsilon$), respectively. The order of magnitudes of the accumulated strains for static viscoelastic analysis was closer to the static linear elastic analysis adopted in the existing pavement design procedures. But, it is necessary to design a pavement system based on the anticipated strains obtained from viscoelastic material models to capture the realistic asphalt concrete material behavior. Therefore, the dynamic viscoelastic analysis needs to be adopted to account for the dynamic vehicular loading as well as to simulate the material behavior of asphalt concrete observed in the field.

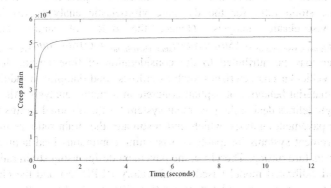

Fig. 3. A typical creep strain plot obtained from viscoelastic pavement model analysis

Further, the ratio of the accumulated strain to elastic strain $\left(\frac{\varepsilon_p}{\varepsilon_r}\right)$ was determined for the static linear elastic analysis using Eq. (8) to predict the rutting response. For the computation of strain ratio based on the linear elastic analysis, elastic strains and the computed TT were utilized. Similarly, the strain ratio was also computed using elastic and accumulated strain values obtained from dynamic and static analyses using the viscoelastic pavement model. These ratios were compared with the linear elastic analysis strain ratio to understand the significance of dynamic analysis, as demonstrated in Table 8.

Table 8. Strain ratio $\left(\frac{\varepsilon_p}{\varepsilon_r}\right)$ for dynamic and static analyses

Vehicle type	Static analysis (linear elastic)	Dynamic analysis (viscoelastic)	Static analysis (viscoelastic)
Type 1 (2-axle truck)	0.81	1561.90	342.95
Type 2 (3-axle truck)	1.11	1451.33	343.59
Type 3 (Bus)	0.94	1702.42	343.30
Type 4 (Multi-axle truck)	0.49	1389.83	342.28

As observed, the strain ratio for the static analysis using linear elastic pavement model was much lower than that of the viscoelastic pavement model analyses. However, the strain ratio for the dynamic viscoelastic analysis was larger than the static viscoelastic analysis. Hence, the order of strain ratios were: $(Strain\ ratio)_{dynamic\ viscoelastic} > (Strain\ ratio)_{static\ viscoelastic} > (Strain\ ratio)_{static\ linear\ elastic}$.

The order may be attributed to the consideration of time-varying loading and inclusion of vehicular characteristics such as stiffness and damping in addition to the viscoelastic material behavior of asphalt concrete in dynamic analysis. Also, note that the existing guidelines design the pavement system for the obtained strains from static linear elastic pavement analysis, which underestimates the strain ratio, resulting in the designed pavement systems incapable of sustaining continuous load applications and leading to premature rutting. Therefore, the pavement design using strain ratio obtained from the field calibrated model based on laboratory RLPD test and the elastic strain incorporated to compute accumulated strains (Eq. (8)) may lead to inefficient design of the asphalt pavement systems. Hence, it would be rational to use dynamic vehicular loads to obtain realistic values of strains. In this direction, calibration factors for strain ratios (CFSR) for each vehicle type and model analysis were proposed, as illustrated in Eqs. (12) and (13), and presented in Table 9.

$$(CFSR)_{static\ viscoelastic} = \frac{(Strain\ ratio)_{static\ viscoelastic}}{(Strain\ ratio)_{static\ linear\ elastic}} \quad (12)$$

$$(CFSR)_{dynamic\ viscoelastic} = \frac{(Strain\ ratio)_{dynamic\ viscoelastic}}{(Strain\ ratio)_{static\ linear\ elastic}} \quad (13)$$

Table 9. Calibration factors based on strain ratios of static and dynamic analyses

Vehicle type	$(CFSR)_{static\ viscoelastic}$	$(CFSR)_{dynamic\ viscoelastic}$
Type 1 (2-axle truck)	418	1906
Type 2 (3-axle truck)	309	1303
Type 3 (Bus)	365	1808
Type 4 (Multi-axle truck)	704	2859

$(CFSR)_{dynamic\,viscoelastic}$ was four times greater than $(CFSR)_{static\,viscoelastic}$ due to the incorporation of the time-varying nature of vehicular loads using dynamic 2-DOF vehicle model. Hence, it is necessary to design the pavement based on the dynamic viscoelastic analysis. Further, the CFSR obtained from both static and dynamic analyses for the Type 4 vehicle comprising single axle, tandem axle, and tridem axle configurations is maximum amongst all the types. Therefore, the pavement design needs to be based on the maximum CFSR obtained due to Type 4 in this study. But, the design should account for the actual traffic loading scenario and material behavior that are observed in the field and this aspect was accounted for in the dynamic viscoelastic pavement analysis.

Overall, this study provided a methodology for rational pavement design based on the accumulated strains and strain ratios obtained from static and dynamic analyses. However, it is to be noted that the validation for the strain ratios and CFSR Obtained in this study due to the three analyses, namely, static linear elastic, static viscoelastic and dynamic viscoelastic, needs to be undertaken through field studies.

5 Conclusions and Recommendations

The main objective of this study was to propose calibration factors for various vehicle types based on dynamic analysis of asphalt pavement system and underscore the contribution of dynamic time-varying nature of vehicular loads required for rational pavement design. The following enlist the significant conclusions and recommendations drawn from this study:

- The 2-DOF quarter car model has potential to incorporate time-varying dynamic vehicular loads, and hence can be utilized in addition to the prevailing static load consideration in the pavement design procedures.
- Damping and stiffness characteristics of the vehicle have significant influence on the pressure transferred to the pavements at every time instant affecting the accumulated strains in the pavement system.
- The fixed vehicle approach based on ESAL of 80 kN did not show variations in the resultant accumulated strains obtained due to the effect of various vehicles since the approach accounted only for the fixed vehicle configuration and single load magnitude.
- The vertical accumulated strains computed using dynamic analysis were significantly lower compared to the vertical strains obtained from static analysis ascribed to the absence of vehicular speed and time headway effect adopted in static analysis.
- The strain ratio obtained based on linear elastic static analysis may lead to conservative design or rather an inefficient design of the asphalt pavements leading to premature rutting due to the absence of viscoelastic material properties and time-varying vehicular loading inputs.
- Recommendations: a major contribution of this study was to understand the effect of dynamic vehicular loading on the rutting response of asphalt pavements. However, there is still scope to advance the research as follows.

- o The traffic growth rate of 4% adopted in the mechanistic-empirical pavement design may be reconsidered along with the actual *MDF* and *HDF* values to determine the accumulated strain and strain ratio for the static and dynamic analyses. This may provide performance prediction in asphalt pavements based on the actual traffic and environmental conditions prevailing on field.
- o Calibration factors for static and dynamic viscoelastic analyses could be proposed based on the strain response using field instrumentation.
- o The half or full vehicle system could be considered for analysis using advanced dynamic vehicle system models such as half-car or full-car to account for the transverse axle spacing between different vehicle configurations.
- o The single-layered viscoelastic asphalt pavement model could be extended to include additional pavement layers for determining the rutting in unbound layers. Also, for accurate vertical strains prediction, advanced pavement materials models such as two-stage (viscoelastic-viscoplastic, elasto-viscoplastic) and three-stage (viscoelastic-viscoplastic-viscodamage) models could be implemented for dynamic analysis.

References

American Association of State Highway and Transportation Officials. AASHTO guide for design of pavement Structures. American Association of State Highway and Transportation Officials, USA (1993)

AUSTROADS. Pavement design - A guide to the structural design of road pavements. Sydney, Australia (2004)

AASHTO TP 79-10. Determining the dynamic modulus and flow number for hot mix asphalt (HMA) using the asphalt mixture performance tester (AMPT). American Association of State Highway and Transportation Officials, USA (2010)

Brademeyer, B.D., Delatte, N.J., Markow, M.J.: Analyses of moving dynamic loads on highway pavements. Part II: pavement response. Transp. Res. Rec., 381–393 (1986)

Bussler, M.L., Ramesh, A.: The eight-node hexahedral element in FEA of part designs. Found. Manag. Technol. **121**(11), 26–28 (1993)

Barbosa, R.S.: Vehicle dynamic response due to pavement roughness. J. Braz. Soc. Mech. Sci. Eng. **33**(3), 302–307 (2011)

Babitha, K.N.: Evaluation of Rutting Characteristics of Asphalt Mixtures Based on E*/Sinφ Stiffness Parameter. M.Tech Thesis, Indian Institute of Technology Kharagpur, Kharagpur, West Bengal, India (2015)

Chen, D.H., Zaman, M., Laguros, J., Soltani, A.: Assessment of computer programs for analysis of flexible pavement structure. Transp. Res. Rec. **1482**, 123–133 (1995)

Chabot, A., Tamagny, P., Poché, D., Duhamel, D.: Visco-elastic modelling for asphalt pavements- software viscoRoute. In: Proceedings of the 10th International Conference on Asphalt Pavements, Quebec, Canada, vol. 2, pp. 5–14 (2006)

European Commission. COST 333: Development of new bituminous pavement design method: Final report of the action. European Commission, Belgium (1999)

Huang, Y.H.: Stresses and strains in viscoelastic multilayer systems subjected to moving loads. Highway Res. Rec. **457**, 60–71 (1973)

Hardy, M.S.A., Cebon, D.: Response of continuous pavements to moving dynamic loads. J. Eng. Mech. **119**(9), 1762–1780 (1993)

Huang, Y.H.: Pavement Analysis and Design, 2nd edn. Pearson Education Inc., Prentice Hall, Upper Saddle River (2004)

Hamersma, H.A., Els, S.: A comparison of quarter, half and full vehicle models with experimental ride comfort data. In: International design Engineering Technical Conferences and Computers and Information in Engineering Conference, 2–5 August 2015. American Society of Mechanical Engineers, Boston (2015

Indian Roads Congress. IRC: 6 - Standard Specifications and Code of Practice for Road Bridges, Section II – Loads and Load Combinations (Seventh Revision). Indian Roads Congress, New Delhi, India (2016)

Imaninasab, R., Bakhshi, B.: Rutting analysis of modified asphalt concrete pavements. Proc. Inst. Civil Eng.-Constr. Mater. **170**(4), 166–177 (2017)

Indian Roads Congress. IRC: 37- guidelines for the design of flexible pavements (Fourth Revision). Indian Roads Congress, New Delhi, India (2018)

Indian Roads Congress. IRC: SP: 73 - Manual of Specifications and Standards for Two Laning of Highways with Paved Shoulder (Second Revision). Indian Roads Congress, New Delhi, India (2018)

Jazar, R.N.: Quarter car. In: Vehicle Dynamics: Theory and Application, pp. 931–975, Springer, Boston (2008)

Jiao, L.: Vehicle model for tyre-ground contact force evaluation. MS Thesis, KTH Royal Institute of Technology, Stockholm, Sweden (2013)

Kinder, D.F.: VESYS and the structural analysis of road pavements. In: Asphalt: Road to 2000. 6th International Asphalt Conference, Sydney, Australia, 27–30 January 1986 (1986)

Li, H., Yang, S., Li, S.: Influence of vehicle parameters on the dynamics of pavement structure due to vehicle-road interaction. In: International Conference on Measuring Technology and Mechatronics Automation, China, vol. 3, pp. 522–525 (2009)

Markow, M.J., Hedrick, J.K., Brademeyer, B.D., Abbo, E.: Analyzing the interactions between dynamic vehicle loads and highway pavements. Transp. Res. Rec. **1196**, 161–169 (1988)

Motor Vehicles Act. Table of Maximum Speed Limit at a Glance (Extra., Pt. II, Sec. 3(ii), dated 9th June, 1989) published in the Gazette of India (1988)

Mechanistic Empirical Pavement Design Guide: A Manual Practice (MEPDG). American Association of State Highway and Transportation Officials, Washington, DC (2008)

Mahajan, G.R., Radhika, B., Biligiri, K.P.: A critical review of vehicle-pavement interaction mechanism in evaluating flexible pavement performance characteristics. Road Mater. Pavement Des., 1–35 (2020)

Ministry of Road Transport and Highways (MORTH). Specifications for Roads and Bridge Works (Fifth revision). Ministry of Road Transport and Highways, New Delhi, India (2013)

Ministry of Road Transport and Highways (MORTH), Annual Report, Bharatmala – Road to Prosperity. Ministry of Road Transport and Highways, New Delhi, India (2020)

SWOV Fact sheet, Headway times and road safety. SWOV Institute for Road Safety Research, The Netherlands Technical Report, Netherlands (2003)

South African National Roads Agency Ltd. South African pavement engineering manual (SAPEM) – Chapter 1: Introduction (First Revision). South Africa (2013)

Thompson, M.R., Elliott, R.P.: ILLI-PAVE based response algorithms for design of conventional flexible pavements. Transp. Res. Rec. **1043**, 50–57 (1985)

Uzarowski, L.: The development of asphalt mix creep parameters and finite element modeling of asphalt rutting. Ph.D. Thesis, University of Waterloo, Canada (2006)

White, T.D., Haddock, J.F., Hand, A.J.T., Fang, H.: NCHRP report 468: Contributions of Pavement Structural Layers to Rutting of Hot-Mix Asphalt Pavements. Transportation Research Board, Washington, D.C (2002)

Witczak, M.W.: Specification criteria for simple performance tests for rutting. volume I: Dynamic modulus (E^*) and Volume II: Flow number and flow time (NCHRP Report 580). National Cooperative Highway Research Program, Transportation Research Record, National Research Council, Washington, D.C (2007)

Zhu, Q., Ishitobi, M.: Chaotic vibration of a nonlinear full-vehicle model. Int. J. Solids Struct. **43** (3–4), 747–759 (2006)

Street and Highway Designs for CAVs - 1

A Review on Benefits and Security Concerns for Self Driving Vehicles

Gozde Bakioglu[✉] and Ali Osman Atahan

Department of Civil Engineering, Faculty of Civil Engineering,
Istanbul Technical University, 34469 Maslak, Istanbul, Turkey
`bakioglugo@itu.edu.tr`

Abstract. Self-driving vehicle technologies are receiving attention and there is a promising trend in market for future development. Based on their advanced features, self-driving vehicles are expected to increase the mobility for those with disabilities, reduce the amount and severity of accidents, enhance the utility of time on travel and reduce the air pollution. On the other hand, this technology brings safety and security challenges, which need to be addressed before their mass implementation in roads. The purpose of this study is to discuss the advantages and disadvantages of autonomous vehicle technology through providing comparative analysis. This paper also presents an overview of recent research on self-driving vehicle safety issues and the shortcomings of this technology in a broader sense. As a result of the study, suggestions about the future of self-driving vehicle will be provided for planners and policy-makers.

Keywords: Self-driving vehicle · Safety issues · Security challenges

1 Introduction

With the enhancement in vehicle automation technology, future of transportation systems will involve more intelligent and automated systems. Autonomous (or self-driving, driverless) vehicles (AV) are crucial to combine digital and physical worlds, which are expected to increase the mobility for those with disabilities, reduce the severity of accidents, enhance the utility of time on travel and reduce air pollution. The embrace of AVs in smart cities is associated with numerous potential advantages.

On the other hand, this technology brings safety and security challenges, which need to be addressed before their mass implementation on roads. The main concerns comprise potentially problematic interaction between users and automated technology (Sivak and Schoettle 2014). Automation level of autonomous vehicles becomes higher when human intervention is less. Different automation levels have been proposed with respect to human engagement and technological aspects. Table 1 indicates the all levels of automation ranging from no automation to full automation provided by the NHTSA (2021).

© The Author(s), under exclusive license to Springer Nature Switzerland AG 2022
A. Akhnoukh et al. (Eds.): IRF 2021, SUCI, pp. 667–675, 2022.
https://doi.org/10.1007/978-3-030-79801-7_48

Table 1. Levels of automation

Levels of automation		Functions
Level 0	No automation	Full human engagement
Level 1	Driver assistance	Human engagement with driving assistance
Level 2	Partial automation	Human engagement with driving assistance
Level 3	Conditional automation	Human engagement if needed
Level 4	High automation	No human engagement
Level 5	Full automation	No human engagement

It is assumed that AVs will provide effective solution to issues about transportation, such as mobility (less travel time, less cost, less traffic congestion), environment (lower vehicle emission and energy consumption), parking and efficiency. Automated vehicles may also avoid driver related mistakes including tiredness, distraction, and dangerous driving. Furthermore, AVs have potential to gather information related to infrastructure and environment more efficiently than human through Vehicle-to-Vehicle (V2V) and Vehicle-to-Infrastructure (V2I) technology and sensors (Howard and Dai 2014; Bansal and Kockelman 2016).

The automation technology comes with disadvantages, concerns and risks. It is reported that 49 AVs incidents occurred in California during 2018, more than half of which were due to rear-end collisions, and most of rest were brought about by non-hardware and software reasons, such as faulty planning (California DMV 2018). It is also asserted that automated vehicles can be vulnerable to hacking and can be the target of attack on the vehicle's hardware and software. AVs could also be exposed to system failure, such as malfunction of sensors (Bakioglu and Atahan 2020). Another risk that can be linked to AVs is miscommunication which can occur while using manual and automated modes at the same time.

Numerous researchers have investigated the potential effects of AVs on many aspects of people's lives. Perceptions of that vehicles have also been analyzed in the recent past years. However, obstacles and barriers need to be addressed to attain these benefits. Most research is being conducted on understanding and handling different kinds of obstacles associated with the embrace of AVs. Some research focuses only on the small area of autonomous vehicles' risks. Although there exists a number of works addressing the potential benefits and barriers of AVs, there is no study that has reviewed all risks involved with AVs in one review. The identified knowledge gap of the pre-literature review is that there is not any comparative study which takes into account all kinds of risks in self-driving vehicles. This review attempts to integrate all pros and cons of AVs, and safety issues and risks in that vehicles discussed by the previous studies. Thus, the purpose of this study is to discuss the advantages and disadvantages of autonomous vehicle technology through providing comparative analysis. This paper also presents an overview of recent research on self-driving vehicle safety issues and the shortcomings of this technology in a broader sense. As a result of the study, suggestions about the future of self-driving vehicle will be provided for planners and policy-makers.

2 Benefits and Advantages of Using Self-driving Vehicles

This section explores the advantages of autonomous vehicle usage that were documented in the past studies. Table 2 summarizes the benefits and advantages of AVs which were included in the aforementioned research. It is anticipated that AVs will reduce the traffic stress level, parking cost, insurance cost, and better fuel economy. Also, self-driving vehicle will increase mobility for those having disabilities and the elderly, and increase productivity due to multitasking. Besides, automated vehicles will enhance urban tourism, develop new software and hardware companies and emerge new markets. Driverless vehicles are expected to provide better freight delivery, lower insurance rates, and better energy efficiency, energy conservation and lower vehicle emission.

Table 2. Summary of advantages and benefits of using AVs

Advantages and benefits of AVs	References
Traffic flow control, decreased needs for right-of-way	Casley et al. (2013), Bakioglu and Dogru (2018), Liu et al. (2019)
Reduction of traffic stress level	Begg (2014), Shabanpour et al. (2018)
Reduction in parking cost, insurance cost, and better fuel economy	Wadud (2017), Begg (2014)
Increased mobility for those having disabilities and the elderly	Howard and Dai (2014), Bansal and Kockelman (2016), König and Neumayr (2017)
Better energy efficiency, energy conservation and lower vehicle emission	Begg (2014), Vahidi and Sciarretta (2018), Silberg et al. (2013)
Reduced congestions, increased accessibility, travel speed and shorter travel times	Casley et al. (2013), Silberg et al. (2013), Schoettle and Sivak (2014), Bansal and Kockelman (2016), Kröger et al. (2018)
Increased productivity due to multitasking	Howard and Dai (2014), Bansal and Kockelman (2016), König and Neumayr (2017)
Improvement in safety, reduction in accidents and severity of crashes, and increased safety	Silberg et al. (2013), Howard and Dai (2014), Bansal and Kockelman (2016), Shabanpour et al.(2018)
Better freight delivery	Alessandrini et al. (2015)
Enhanced urban tourism	Cohen and Hopkins (2019)
More environmental friendly	Howard and Dai (2014); Bakioglu and Karaman (2018)
Developed new software and hardware companies and emerge new markets	Bamonte (2013)
Decrease in need for parking	Bansal and Kockelman (2016)
Lower insurance rates	Schoettle and Sivak (2014), Shabanpour et al. (2018)

3 Security Concerns and Disadvantages of Using Self-driving Vehicles

As autonomous vehicle have various benefits, it could also be associated with different barriers and concerns about adoption of AVs. It is anticipated that AVs will increase safety concerns, cost of the system (Casley et al. 2013), and cause some issues, such as legal and liability problems for drivers or owners. Exclusive lane should be dedicated to operate this system well, which can be seen a problem. Also, there is always a possibility of vehicle malfunction, such as equipment or system failure, inaccurate positioning. Another issue that can be linked to AVs is cyber security problems. The digitization of the transportation network can be vulnerable to hacking (Alfonso et al. 2018). Vehicle-to-Vehicle (V2V) and Vehicle-to-Infrastructure communications will gain importance to operate this system. Interaction with conventional cars can give rise to communication issues, and it is still unclear the extent to which AVs' performance under different traffic conditions. Furthermore, it is worth to investigate the extent to which people will accept and use automated vehicles, and feel nervous while riding in AVs. Table 3 shows the summary of disadvantages and security concerns of using AVs.

Table 3. Summary of disadvantages and security concerns of using AVs

Disadvantages and security concerns of AVs	References
Increased safety concerns	Casley et al. (2013), Howard and Dai (2014), König and Neumayr (2017)
Legal and liability issues for drivers or owners, morality and ethical considerations	Casley et al. (2013), Begg (2014), Bansal et al. (2016), Chen et al. (2017)
Cyber security issues, data privacy problems, hacking computer systems of vehicle	Begg (2014), Schoettle and Sivak (2014), Xu and Duan (2019)
Accurate positioning and mapping	Signifredi et al. (2015), Kala and Warwick (2013)
Environmental concerns, interactions with conventional vehicles	Bansal et al. (2016), Haboucha et al. (2017)
Increased cost of the system, high expected purchase price of AVs, affordability of AVs	Casley et al. (2013), Shabanpour et al. (2018)
Equipment or system failure	Bansal et al. (2016), Schoettle and Sivak (2014)
User acceptance and reaction, possibility of feeling nervous while riding in AVs	Fagnant and Kockelman (2015), Kyriakidis et al. (2015)
Communication Systems, interactions with conventional vehicles	Bansal et al. (2016), Zhou et al. (2018)
Imperfect performance under different traffic conditions, lack of control	Shabanpour et al. (2018), König and Neumayr (2017)
Existence of exclusive lane	Shabanpour et al. (2018)

3.1 Risk Landscape of Autonomous Vehicles

Self-driving vehicles are considered to be transformative and to assure the safety of transportation systems. Nevertheless, automated cars have some potential safety and security concerns. Most research focuses on the hardware and software failures (Checkoway et al. 2011). Complex electronic systems are vulnerable to performance design malfunction. A minor in-vehicle system failure, such as inaccurate signal, non-functioning sensor, or software malfunction can directly compromise traffic safety. The accurate functionality of the Global Position System (GPS) sensors is crucial to self-driving vehicles, serving self-localization with sensitive precision (Parkinson et al. 2017). Thus, failed GPS data affects the localization of self-driving vehicles bring about accident hazard.

Even more so, vehicle operating system failure can be exploited by those with malicious intentions, resulting in main security concern. Automated driving technologies are sensitive to manipulation by the advanced communication system. Zaidi and Raharajan (2015) presented that remote attacks are the most dangerous one for the private autonomous vehicle which threatens the security of the vehicle. Thus, safety and security are of decisive importance to self-driving vehicles. Any safety risks, failures, and attacks may give rise to major safety losses, which need to be addressed. Safety and security issues in automated driving technology may make difficult the adoption of self-driving vehicle usage.

In this section, risks associated with automated vehicle will be discussed in detail. Bakioglu and Atahan (2021) addressed and prioritized these risks by performing hybrid Multicriteria Decision Making (MCDM) methods. Risk identification will help autonomous vehicle developers to make provision against non-anticipated traffic conditions. Risks involved with autonomous vehicles are listed as below (Bakioglu and Atahan 2021):

- Electronic Infrastructure Risk: This arises due to lack of sophisticated technology elements, such as vehicle-to-vehicle and vehicle-to-infrastructure communication on the road and the vehicle.
- Cyber Attack Risk: Risk involved with deliberate exploitation of automated vehicles system by unauthorized entities. The target of the attack can vary, ranging from the attack on software to manage the system, or physical attack on the vehicle's hardware.
- Environmental Adaptation Risk: It refers to adaptation of external environment (e.g., dense fog, slick road) while en-route.
- Reputational risk: Risk associated with the failure of an autonomous which occur during production of AVs phase, such as problem with vehicle design, AV sensors, testing and performance.
- Internet Outage Risk: Risks that arises from limitation of internet access while en route. The journey with AVs depends completely on the availability of the internet in order to update the real-time traffic information acquired from navigation system. Loss of internet connection, for example, because of congestion of users could lead to potential pre-crash events.

- Behavioural adaptation Risk: This arises due to lack of adjusting driver behaviour to the new technology. The behavioural adaptation is more likely when drivers are aware of a change rather than not being aware of a change.
- Road Infrastructure Risk: Risk associated with road infrastructures such as obstructed signs, bridge collapse.
- Disruption/catastrophic Risk: It refers to the natural and man-made disasters, including earthquakes, flood, terrorist attacks, and political challenges.

3.2 Traffic Accident Aspect of Autonomous Vehicle

Autonomous vehicles are expected to have a substantial impact on traffic in the future. Self-driving vehicle technology improves the road safety, where driver error is predicted to bring about 94% of total accidents (NHTSA 2015).

Although traffic accident sample regarding AVs is limited, some researchers have investigated traffic crash reports involving self-driving vehicles. Preliminary analysis of AVs' accidents in 2015 demonstrated that number of accidents is positively correlated with autonomous miles traveled, and average reaction time was found 0.83 s in this study (Dixit et al. 2016).

When it comes to type of collisions, it is found that the most frequent type of collision is rear-end – front bumper of conventional vehicle and rear bumper of autonomous vehicle. The vehicle speed was also found as less than 10 mph in the most cases (Favarò et al. 2017). Simulation analyses show the same results. Tibljaš et al. (2018) simulated the introduction of AVs (10%–50% in traffic flow) on 4 roundabouts recorded an increase of rear-end type crashes. Thus, it can be emphasized that AVs are prone to get in rear-end traffic accident both in real life and simulation packages.

Petrović et al. (2020) determined the specific features of traffic accidents with AVs. They expressed that the introduction of AVs reduce the share of types of collision "broadside" and "pedestrian", and they are capable of compensating error of the driver conventional vehicle "right of way violation".

4 Discussion and Conclusion

This paper reviews the literature about self-driving vehicles. This review integrates all pros and cons of AVs, and safety issues and risks in the vehicles discussed by the past studies. Risks associated with AVs are also addressed in this paper, which does not exist in the literature. It is anticipated that AVs could have a number of advantages which will ease the adoption of this new technology. However, understanding and identifying the various barriers and concerns that restrain the embrace of these technologies will enable decision-makers to tackle these disadvantages and will help them to define required equipment and infrastructure that ensure the smooth operation and safety of AVs.

As a result of reviewing literature, the most prominent advantages of having AVs is increased mobility for those having disabilities and the elderly. There are also several traffic related benefits, such as reduced traffic congestion and number of accidents, improved traffic flow and enhanced capacity. However, some accidents were reported during riding in AVs so that people are still susceptible to adopting self-driving vehicles. Malicious hacking and cyber security issues are found to be the most affecting disadvantages for embrace of automated vehicles. Legal and liability issues for drivers or owners, morality and ethical considerations can also be assumed to be outstanding barrier to use this technology.

Within the frame of risk landscape, cyber-attack risk is determined as the most important risk and the subsequent one is reputational risk. The mechanical failure, such as problem with vehicle design gives rise to concerns about safety in AVs. Driverless vehicle can be prone to cybercrimes and mechanical malfunction. Government and the automotive industry pertaining to autonomous vehicles should ensure the safety of these vehicles. They may develop robust software to protect those vehicle from malicious hacking. Also, testing all parts of that vehicle during manufacturing process is needed for all vehicle before releasing them into market.

It seems that there is still a long way to overcome the disadvantages of AVs and let those vehicles to release to automotive industry.

There are numerous possible research directions for future studies as a follow up of this study. Following valuable directions may be the extension of this study:

- There is still much room for improvement in the accuracy and positioning so that new algorithm can be proposed to solve positioning problems and try to work Global Position System (GPS) better even without internet access.
- Formulate a system to ensure that AVs under non-anticipated traffic conditions still perform well.
- Develop V2V and V2I communication systems to provide AVs to interact with convectional vehicles in a smooth way.
- Define legal responsibilities for drivers and passengers of AVs while en-route.

References

Alessandrini, A., Campagna, A., Site, P.D., Filippi, F., Persia, L.: Automated vehicles and the rethinking of mobility and cities. Transp. Res. Procedia 5, 145–160 (2015)

Alfonso, J., Naranjo, J.E., Menéndez, J.M., Alonso, A.: Vehicular communications. In: Intelligent Vehicles, pp. 103–139. Elsevier (2018)

Bakioglu, G., Atahan, A.O.: Evaluating the influencing factors on adoption of self-driving vehicles by using interval-valued pythagorean fuzzy AHP. In: Kahraman, C., Onar, S.C., Oztaysi, B., Sari, I.U., Selcuk Cebi, A., Tolga, Cagri (eds.) INFUS 2020. AISC, vol. 1197, pp. 503–511. Springer, Cham (2021). https://doi.org/10.1007/978-3-030-51156-2_58

Bakioglu, G., Atahan, A.O.: AHP integrated TOPSIS and VIKOR methods with Pythagorean fuzzy sets to prioritize risks in self-driving vehicles. Appl. Soft Comput. 99, 106948 (2021)

Bakioglu, G., Dogru, A.: GIS-Based Visualization for Estimating Level of Service (2018)

Bakioğlu, G., Karaman, H.: Accessibility of medical services following an earthquake: a case study of traffic and economic aspects affecting the Istanbul roadway. Int. J. Disast. Risk Reduct. 31, 403–418 (2018)

Bansal, P., Kockelman, K.M.: Forecasting Americans' long-term adoption of connected and autonomous vehicle technologies. In: Paper Presented at the 95th Annual Meeting of the Transportation Research Board (2016)

Begg, D.: A 2050 vision for London: What are the implications of driverless transport, transport times, London, UK (2014)

California Department of Motor Vehicles. Autonomous Vehicle Disengagement Reports (2018)

Casley, S.V., Jardim, A.S., Quartulli, A.M.: A study of public acceptance of autonomous cars, interactive qualifying project, Worcester Polytechnic Institute (2013)

Checkoway S., et al.: Comprehensive experimental analyses of automotive attack surfaces. In: Proceedings of USENIX Security (2011)

Chen, D., Ahn, S., Chitturi, M., Noyce, D.A.: Towards vehicle automation: roadway capacity formulation for traffic mixed with regular and automated vehicles. Transp. Res. Part B Methodol. **100**, 196–221 (2017)

Cohen, S.A., Hopkins, D.: Autonomous vehicles and the future of urban tourism. Ann. Tour. Res. **74**, 33–42 (2019)

Da Xu, L., Duan, L.: Big data for cyber physical systems in industry 4.0: a survey. Enterp. Inf. Syst. **13**, 148–169 (2019)

Dixit, V.V., Chand, S., Nair, D.J.: Autonomous vehicles: disengagements, accidents and reaction times. PLoS one **11**(12), e0168054 (2016)

Fagnant, D.J., Kockelman, K.: Preparing a nation for autonomous vehicles: opportunities, barriers and policy recommendations. Transp. Res. Part A Policy Pract. **77**, 167–181 (2015)

Favarò, F.M., Nader, N., Eurich, S.O., Tripp, M., Varadaraju, N.: Examining accident reports involving autonomous vehicles in California. PLoS One **12**, e0184952 (2017). https://doi.org/10.1371/journal.pone.0184952

Howard, D., Dai, D.: Public perceptions of self-driving cars: the case of Berkeley, California. In: Paper Presented at the 93rd Annual Meeting of the Transportation Research Board, Washington D.C (2014)

Kala, W.K.: Motion planning of autonomous vehicles in a non-autonomous vehicle environment without speed lanes. Eng. Appl. Artif. Intell. **26**, 1588–1601 (2013)

König, M., Neumayr, L.: Users' resistance towards radical innovations: the case of the self-driving car. Transp. Res. Part F **44**, 42–52 (2017)

Kröger, L., Kuhnimhof, T., Trommer, S.: Does context matter? a comparative study modelling autonomous vehicle impact on travel behaviour for Germany and the USA. Transp. Res. Part A Policy Pract. **122**, 146–161 (2018)

Kyriakidis, M., Happee, R., Winter, J.C.F.: Public opinion on automated driving: results of an international questionnaire among 5000 respondents. Transp. Res. Part F Traffic Psychol. Behav. **32**, 127–140 (2015)

Liu, B., Shi, Q., Song, Z., El Kamel, A.: Trajectory planning for autonomous intersection management of connected vehicles. Simul. Model. Pract Theory **90**, 16–30 (2019)

NHTSA (2021). https://www.nhtsa.gov/technology-innovation/automated-vehicles-safety

Parkinson, S., Ward, P., Wilson, K., Miller, J.: Cyber threats facing autonomous and connected vehicles: future challenges. IEEE Trans. Intell. Transp. Syst. **18**(11), 2898–2915 (2017)

Petrović, Đ., Mijailović, R., Pešić, D.: Traffic accidents with autonomous vehicles: type of collisions, manoeuvres and errors of conventional vehicles' drivers. Transp. Res. Procedia **45**, 161–168 (2020)

Schoettle, B., Sivak, M.: A survey of public opinion about autonomous and self-driving vehicles in the U.S., the U.K., and Australia, Technical Report, The University of Michigan Transportation Research Institute (2014)

Shabanpour, R., Golshani, N., Shamshiripour, A., Mohammadian, A.K.: Eliciting preferences for adoption of fully automated vehicles using best-worst analysis. Transp. Res. Part C: Emerg. Technol. **93**, 463–478 (2018)

Signifredi, A., Luca, B., Coati, A., Medina, J.S., Molinari, D.: A general purpose approach for global and local path planning combination. In: 2015 IEEE 18th International Conference Intelligent Transportation System, pp. 996–1001. IEEE (2015)

Silberg, G., et al.: Self-Driving Cars: Are we Ready?, Technical Report, KPMG (2013)

Singh S, for NHTSA. Critical reasons for crashes investigated in the national motor vehicle crash causation survey -DOT HS 812 115, US Department of Transportation. https://crashstats.nhtsa.dot.gov/Api/Public/ViewPublication/812115

Tibljaš, A.D., Giuffrè, T., Surdonja, S., Trubia, S.: Introduction of autonomous vehicles: roundabouts design and safety performance evaluation. Sustainability **10**, 1–14 (2018). https://doi.org/10.3390/su10041060

Vahidi, A., Sciarretta, A.: Energy saving potentials of connected and auto- mated vehicles. Transp. Res. Part C Emerg. Technol. **95**, 822–843 (2018)

Wadud, Z.: Fully automated vehicles: a cost of ownership analysis to in- form early adoption. Transp. Res. Part A Policy Pract. **101**, 163–176 (2017)

Zaidi, K., Rajarajan, M.: Vehicular Internet: security & privacy challenges and opportunities. Future Internet **7**(3), 257–275 (2015)

Zhou, Y., Li, H., Shi, C., Lu, N., Cheng, N.: A fuzzy-rule based data delivery scheme in VANETs with intelligent speed prediction and relay selection. Wirel. Commun. Mob. Comput., 1–15 (2018)

Contrast Ratio of Road Markings in Poland - Evaluation for Machine Vision Applications Based on Naturalistic Driving Study

Tomasz E. Burghardt[1] and Anton Pashkevich[2(✉)]

[1] M. Swarovski GmbH, Amstetten, Austria
tomasz.burghardt@swarco.com
[2] Politechnika Krakowska, Kraków, Poland
apashkevich@pk.edu.pl

ABSTRACT. Road markings are critical road safety features for both human drivers and for machine vision technology used in advanced driver assistance systems and in the emerging technology of automated vehicles. Amongst the parameters, contrast ratio is necessary for appropriate recognition of road markings. To assess the contrast ratio of road markings at various roads, still images of roadway were obtained from a dashboard camcorder used for a naturalistic driving study in Poland. Road markings and neighbouring roadway surface were measured for luminance and Weber contrast was calculated. At the studied representative roads average contrast ratio was 0.8 under daytime illumination and 2.0 at night; enhancement of the contrast through digital image manipulation resulted in increases to 2.3 and 6.8, respectively. Under poor visibility daytime conditions (interference from glare or rain), average contrast ratio dropped to 0.5 (enhanced 1.4); in the worst case it was below 0.1. Consequently, the current machine vision technology could fail under some poor visibility circumstances. The image enhancement indicated that both the initial and digitally enhanced contrast ratios were important.

Keywords: Road visibility · Dashboard camcorder · Driver support systems · Automated vehicles · Luminance · Image enhancement

1 Introduction

Road markings (RM), ubiquitous features present at majority of roads, are considered their necessary basic safety feature. While RM are quite inexpensive and easy to install, they bring enormous economic and social benefits through channelling of vehicular traffic and thus improving road safety (Miller 1992); they help drivers to remain within traffic lane (Calvi 2015, Steyvers and De Waard 2000). There is no currently feasible alternate technology (Mosböck and Burghardt 2018). For human drivers, the recognition of RM depends on visibility level, which relies on the contrast against the neighbouring surface (Brémond 2020, Brémond et al. 2013). Comparison of luminance of RM to the luminance of the neighbouring road surface provides a measurable value of contrast ratio (CR). Researchers working in the transportation field with luminance and CR as bases for visibility have directed their considerable work toward road signs

(He et al. 2021, Schnell et al. 2004), but, very surprisingly, reports concerning CR of horizontal road signalisation could not be found.

RM are necessary not only for human drivers, but also for the advanced driver assistance systems and the emerging technology of automated vehicles (AV) (Burghardt et al. 2020, Mosböck and Burghardt 2018, Najeh et al. 2020). All camera-based machine vision (MV) equipment relies on software that identifies RM, which are distinguished from the roadway surface through CR. Research about MV concentrates on two main aspects: the equipment itself (Goelles et al. 2020, Marti et al. 2019, Rosique et al. 2019), and the algorithms to correctly recognise images and their features (Ranft & Stiller 2016). One of the first reports that presented the method to process a video recording for MV needs was published three decades ago (Dickmanns et al. 1990); since then, advances in both the technology and the software were rapid (Bengler et al. 2014). Currently, essentially all AV use vehicle localisation within roadway based on the data incoming from MV sensing units that include cameras (Xing et al. 2018). Reliance on high-resolution three-dimensional road maps also primarily depends on RM (Liu et al. 2020, Poggenhans et al. 2018); indeed, literature reports of a successful broader use of road mapping navigation without RM could not be found. Meanwhile, other concepts are being considered and discussed without much implementation (Kuutti et al. 2018).

Taking into account the importance of RM for MV, they are surprisingly rarely considered separately from the road itself and the correlation between their quality and MV efficiency is often ignored (Carreras et al. 2018, Kunze et al. 2018). However, because successful MV depends on pattern recognition, clear distinction between the RM and the surrounding pavement is critical (Narote et al. 2018). Research reports clearly pointed out that inadequate visibility of RM was causing faulty operation of MV (Hadi and Sinha 2011, Hadi et al. 2007, Matowicki et al. 2016); poor quality of RM was recently pointed out as contributing to crashes of AV (Calvert et al. 2020). Contrariwise, high quality RM and their proper recognition enabled the success of the historic automotive journey done in a fully autonomous manner (Ziegler et al. 2014). It was reported that for a successful MV recognition CR of RM should be higher than 2.0, but values above 3.0 were preferred (Carlson 2017, Carlson and Poorsartep 2017). Particularly negative effect on CR has glare – it is a serious issue affecting road safety for human drivers (Theeuwes et al. 2002) and being equally detrimental for MV (Carlson 2017).

Given the absence of reports related to CR of RM and the importance of this practical topic, a research was initiated to identify possible issues and to start filling the knowledge gap. It was envisaged that the assessment would be done based on data obtained from a naturalistic driving study – to best reflect the real situations to which drivers and MV sensors are exposed, and only readily available equipment and software would be utilised – to eliminate complex transformations that could demand unrealistic computing power. Hence, there was a dual purpose of this effort: firstly – evaluation of the methodology based on image analysis from a dashboard camcorder, and secondly – measurement of CR at selected roads in Poland under various lighting conditions as the demonstration of the methodology. Despite the broad availability and low cost of dashboard camcorders, research of their use for this purpose was so far done with accent on development of image processing algorithms (Mathibela et al. 2015, Poggenhans et al. 2015). The image analysis from various road stretches for the purpose of establishing typical contrast ratios was not reported. Because image enhancement is a readily available and well-known procedure (Otsu 1979), its example was utilised to verify whether improved CR values could be readily obtained in all of the cases.

2 Methodology

Data subjected to the analysis was collected from a camcorder MiVue 752 WIFI Dual (Mitac Europe Ltd., Crawley, United Kingdom) that was positioned, without any obstruction of the driver's view, next to the rear-view mirror in a passenger car (i.e. at the height of approximately 1.2 m above ground). The video was recorded at 30 Hz, with resolution 1080 × 1920 pixels, at field of view 140°, using high dynamic range recording parameters. The films were collected for an ongoing naturalistic driving study aiming at evaluation of vertical road signage, driving speeds, and drivers' behaviour in Poland (Pashkevich et al. 2021). All required privacy protection rules were followed. For the analysis presented herein, the recordings were reviewed and representative scenes were captured as still images (example is given in Fig. 1). Selected were images, in which other vehicles did not obscure the visibility; it could be considered as a shortcoming, but during driving there would be a continuity of recording and preview.

Fig. 1. Original representative image (1080 × 1920 pixels), a motorway.

The images were analysed with an open-source software (GNU Image Manipulation Program, version 2.10.24) that was rendering them according to build-in sRGB profile. Firstly, an area 750 × 1600 pixels was selected for analysis: cropped out were 220 pixels from the top and the bottom and 160 pixels from each edge, i.e. the vehicle's hood, majority of the sky, and extreme sides, while any street lighting was included (example of cropped image from Fig. 1 is given as Fig. 2). Next, using *colour picker* tool set to 2 pixel radius, obtained were measurements of perceptual brightness (L*) according to CIE L*a*b* colour space. In each case, measured were L* of the RM and their nearest neighbouring road surface backgrounds. For each region, analysis was performed at three predefined distances within each image: (1) near region, representing area about 2–5 m ahead of the vehicle – pixels 680–700 from the top,

(2) middle region, representing the preview area approximately 7–9 m ahead of vehicle – pixels 600–620 from the top, and (3) far region, representing the end of usable view for such analysis from images obtained from the utilised equipment, distance about 30–45 m ahead of the vehicle – pixels 550–570 from the top. The given distances assume that driving was done on a flat surface. These measurement areas are marked in Fig. 2.

The influence of any artefacts caused by objects reflecting in the windshield, dirt, bugs, etc. became part of the analysis, with an assumption that they would be a representation of the expected imperfections from image capture by MV equipment. In all of the cases, averages from these three distances are given, but one must note that in some situations the values between the locations could vary considerably or the markings could be missing for the analysed regions. For data presentation clarity, individual regions shall not be discussed herein.

Fig. 2. Image after cropping (750 × 1600 pixels). Horizontal lines represent regions used for analyses.

For the analysis, measurements of luminance and contrast ratio were done according to the methodology that was previously used for assessment of road signs (Schnell et al. 2004), but with the added enhancement step. As was mentioned above, the enhancement step is a well-known protocol (Otsu 1979), with a plethora of variations designed for particular needs. Usually, such procedure is done on greyscale images; however, for the purpose of this work the prior extraction of L* component was not done because it was not bringing any additional advantage. For the purpose of this work, the images were enhanced using tool *posterise* at level 6 (i.e. the number of colours in each RGB channel was lowered from 256 to 6) in *overlay* mode – this resulted in images like shown in Fig. 3. The theoretical equation for the transformation, obtained from the software documentation, is shown in Eq. 1: for each pixel, E is the resulting pixel colour, I is the original pixel colour, and M is the pixel colour after reduction of the colour numbers (level 6). After enhancement of the image, measurements were taken again, in the same locations.

$$E = \frac{I}{255} \cdot (I + \frac{2M}{255} \cdot (255 - I)) \qquad (1)$$

Fig. 3. Enhanced cropped image – *posterise* (level 6), *overlay*.

For each case, calculation of contrast ratio CR (Weber contrast) was done according to Eq. 2, in which, L_m is the luminance of the road marking and L_b is the luminance of the neighbouring background.

$$CR = \frac{L_m - L_b}{L_b} \qquad (2)$$

The selection of locations was done arbitrarily, to cover a wide range of conditions and configurations of RM; this aspect is not worth any further discussion at present. The evaluation areas included the following road classes: (1) motorways (and expressways) – limited access divided highways with at least two lanes in each direction, speed limit 140 km/h (120 km/h at expressways), (2) rural roads – two lane bidirectional roads, speed limit 90 km/h, (3) city highways – divided limited access highways through city with two or more lanes in each direction, speed limit 70 km/h or 80 km/h, and (4) city streets – two lane bidirectional roads passing through residential areas of cities or towns, speed limit 50 km/h. The analysis did not include city thoroughfares (broad streets within cities, difficult to cross except for traffic controlled intersections, with speed limit 50–70 km/h), lighted regions at motorways or expressways, ramps, etc. As a clarification, the authors define 'city highways' as roads within city that cannot be crossed by pedestrians – such roads are forming physical barriers and breaking the uniformity of development, thus causing decline of the cities and suburbanisation (Baum-Snow 2007).

The width of RM in Poland varies depending on line location and road class: the middle line is always 12 cm wide, while the edge lines at all roads with broad hard shoulders are 24 cm wide (at roads with dirt shoulders – 12 cm). At all single carriageway bidirectional roads, double centre line (2 × 12 cm) is used in regions where overtaking is forbidden (from one or both sides) and single line (1 × 12 cm) where overtaking is allowed. The analysed edge lines at city streets were examples of lines 24 cm wide separating the roadway from bicycle paths or delimiting bus stops. The middle line is defined as marking dividing traffic lanes in the same direction, while centre line(s) divide traffic in opposite direction. In all of the cases white RM were analysed; yellow is reserved for temporary signage.

In addition to the different locations, analysis was done under different weather and lighting conditions. Daytime illumination was compared with night time (without illumination other than vehicle headlights) as examples of good visibility conditions. CR was also assessed under daytime illumination but with poor visibility conditions caused by rain, drizzle, or glare.

3 Results

To establish a baseline, evaluation was firstly done at various roads when the lighting conditions were good. The results for daylight conditions, shown in Table 1, revealed that the average CR was rather poor, only 0.8 (range 0.4–1.5) but upon image enhancement rose to 2.3 (range 0.5–7.0). Analysis from night time conditions (Table 2), resulted in average measured CR higher than in the presence of daytime illumination: CR of 2.0 (range 0.4–3.8) increased after image enhancement to 6.8 (range 0.6–14.2). Higher CR at night was expected based on professional knowledge: it is furnished by both the vehicle headlights illuminating the road ahead and retroreflectivity of the RM. Under poor visibility daytime conditions a drop in CR by an average 38% as compared to the values obtained under favourable visibility was measured (Table 3): average CR decreased to 0.5 (range 0.1–1.3) and upon digital image enhancement only a modest increase to 1.4 (range 0.1–5.3) could be achieved. Hence, insufficient CR could be improved through digital manipulation of the images only in some cases; in situations with really low CR no meaningful increase in CR could be achieved. These results are charted in Fig. 4.

A comparison between various road classes done during daytime under good and poor visibility conditions revealed major differences, except for the case of motorways; the outcome is shown graphically in Fig. 5. The almost imperceptible drop of CR at motorways under poor visibility as compared to the good visibility remains unexplained at present; it cannot be positively associated with width of the RM.

Whereas the CR values given above were averages for several arbitrarily selected roads, one must consider whether the differences measured under dissimilar weather conditions would be representative for one particular location. The results presented in Table 4 from one location at a rural road confirm the validity of the measured mean values. For illustration, selected analysed images are provided. Good lighting conditions, when low initial CR could be easily increased through digital image manipulation, are shown in Fig. 6 and Fig. 7, correspondingly before and after enhancement. Poor lighting conditions, with sun glare, are illustrated as Fig. 8 and Fig. 9 (again, before and after enhancement) – low initial CR could be enhanced only slightly. One must observe that in these cases, despite low CR, line markings were clearly visible. In such situation, RM recognition by human drivers would be adequate but MV equipment could fail; hence, the critical importance of high CR for MV must be emphasised. When comparing between the centre and edge lines in such situations, one must consider the location of the interfering light source as compared to the camera position.

Very poor visibility conditions at city streets, with sunshine glare interference, are illustrated Fig. 10 (original image) and Fig. 11 (enhanced image). In these cases (note that they are not associated with data given in Table 4), RM are not visible because of sunshine positioned directly ahead; CR was negligible and could not be increased.

Table 1. Contrast ratio under good lighting conditions: daytime.[a]

Road type	Motorways and expressways			Rural roads		City highways		City streets	
Line marking location	Right edge	Middle	Left edge	Right edge	Centre	Middle	Right edge	Centre
Number of analysed locations	8	8	8	8	8	6	8	8
Luminance (marking)	64 (39–87)	70 (36–94)	69 (37–98)	69 (37–100)	70 (44–98)	55 (36–83)	61 (46–88)	63 (42–97)
Luminance (background)	35 (16–50)	36 (19–50)	35 (13–53)	38 (16–62)	39 (17–63)	31 (13–39)	36 (25–46)	37 (22–49)
Contrast ratio	0.9 (0.4–1.5)	1.0 (0.6–1.5)	1.0 (0.4–1.9)	0.9 (0.3–1.6)	0.9 (0.3–2.2)	0.8 (0.4–2.0)	0.7 (0.0–1.6)	0.7 (0.2–1.2)
Luminance (marking), enhanced	68 (34–94)	77 (31–95)	75 (34–100)	72 (30–100)	73 (35–100)	54 (29–93)	65 (38–95)	65 (35–100)
Luminance (background), enhanced	23 (6–40)	26 (6–40)	24 (4–41)	31 (4–69)	31 (5–69)	19 (4–32)	26 (10–37)	27 (7–41)
Contrast ratio, enhanced	2.6 (1.0–5.6)	2.4 (1.3–5.0)	3.1 (0.8–8.8)	2.2 (0.4–8.3)	2.2 (0.4–10.4)	2.7 (0.5–9.1)	2.1 (0.1–5.7)	1.8 (0.3–4.6)

[a] Given are average values for the three distances at each location, ranges (minimum – maximum) are in parentheses.

Table 2. Contrast ratio under good lighting conditions: night time.[a]

Road type	Motorways and expressways			Rural roads		City highways		City streets	
Line marking location	Right edge	Middle	Left edge	Right edge	Centre	Middle	Right edge	Centre
Number of analysed locations	9	9	9	10	10	7	6	6
Luminance (marking)	81 (33–99)	96 (78–100)	86 (43–100)	88 (56–100)	91 (65–100)	62 (39–92)	84 (60–97)	76 (42–100)
Luminance (background)	27 (6–44)	40 (19–62)	30 (4–66)	35 (13–48)	31 (8–47)	30 (14–39)	37 (36–39)	36 (13–59)
Contrast ratio	2.6 (0.7–5.1)	1.6 (0.6–3.4)	2.9 (0.5–10.0)	1.8 (0.6–5.2)	2.3 (0.9–9.5)	1.1 (0.4–2.4)	1.3 (0.6–1.7)	1.4 (0.3–4.0)
Luminance (marking), enhanced	83 (23–100)	99 (92–100)	87 (28–100)	93 (65–100)	95 (73–100)	63 (31–92)	91 (76–100)	78 (34–100)
Luminance (background), enhanced	15 (1–37)	33 (7–69)	19 (1–69)	25 (2–39)	20 (3–40)	16 (5–31)	30 (29–31)	26 (3–68)
Contrast ratio, enhanced	9.8 (1.2–25.1)	3.9 (0.5–13.9)	11.4 (0.5–36.0)	6.2 (1.2–37.4)	8.1 (1.5–34.7)	4.3 (0.4–11.0)	2.0 (1.6–2.2)	4.5 (0.3–18.8)

[a] Given are average values for the three distances at each location, ranges (minimum – maximum) are in parentheses.

Table 3. Contrast ratio under poor daytime visibility conditions (rain, drizzle, glare).[a]

Road type	Motorways and expressways			Rural roads		City highways	City streets	
Line marking location	Right edge	Middle	Left edge	Right edge	Centre	Middle	Right edge	Centre
Number of analysed locations	8	8	8	7	7	17	7	7
Luminance (marking)	39 (19–61)	50 (30–74)	55 (17–77)	39 (12–70)	57 (18–98)	42 (18–96)	36 (12–55)	47 (15–84)
Luminance (background)	24 (8–38)	30 (13–43)	31 (5–62)	31 (7–54)	42 (9–64)	32 (8–88)	27 (9–35)	42 (18–90)
Contrast ratio	0.7 (0.2–1.4)	0.7 (0.3–1.6)	0.9 (0.1–2.3)	0.4 (0.0–1.3)	0.4 (0.1–1.1)	0.4 (0.0–1.8)	0.3 (0.0–0.8)	0.5 (0.1–2.1)
Luminance (marking), enhanced	30 (5–69)	46 (20–89)	58 (5–92)	33 (4–76)	56 (8–85)	34 (5–100)	28 (3–64)	42 (4–96)
Luminance (background), enhanced	11 (1–30)	19 (4–43)	22 (1–51)	21 (2–59)	36 (2–76)	21 (2–96)	16 (2–28)	35 (5–94)
Contrast ratio, enhanced	2.3 (0.3–5.9)	2.2 (0.2–8.5)	2.9 (0.1–9.3)	0.8 (0.0–2.2)	1.1 (0.0–4.2)	1.1 (0.0–6.7)	0.9 (0.0–5.1)	1.2 (0.0–3.6)

[a] Given are average values for the three distances at each location, ranges (minimum – maximum) are in parentheses.

Table 4. Contrast ratio under different daytime lighting conditions at one location (rural highway).[a]

Time of day	Morning		Morning		Morning		Dawn		Afternoon		Afternoon		Early evening		Early evening	
Sky	Cloudy		Cloudy		Clear		Clear		Clear		Cloudy		Clear		Clear	
Road surface	Dry		Moist		Dry		Dry		Dry		Dry		Dry		Dry	
Interference	None		Wet road surface		Sun glare		Sun glare		None		None		None		Headlights glare	
Line marking location	Right edge	Centre	Right edge	Centre	Right edge	Centre	Right edge	Centre	Right edge	Centre	Right edge	Centre	Right edge	Centre	Right edge	Centre
Luminance (marking)	50	57	57	66	44	62	38	50	73	92	56	62	45	50	54	71
Luminance (background)	35	33	39	40	37	49	35	30	45	56	39	38	31	28	37	41
Contrast ratio	0.4	0.7	0.5	0.6	0.2	0.3	0.1	0.6	0.6	0.6	0.4	0.6	0.5	0.8	0.5	0.7
Luminance (marking), enhanced	47	55	55	67	41	68	26	43	77	98	57	63	37	49	54	68
Luminance (background), enhanced	23	22	30	31	25	40	23	20	42	55	30	30	21	15	25	33
Contrast ratio, enhanced	1.3	1.7	0.8	1.1	0.9	0.7	0.2	1.6	0.8	0.8	0.9	1.1	1.7	3.2	1.5	1.0

[a] Given are average values for the three analysed distances, one measurement per conditions.

684 T. E. Burghardt and A. Pashkevich

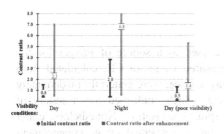

Fig. 4. Average contrast ratios under different conditions (all roads).

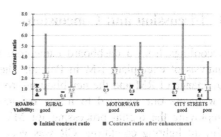

Fig. 5. Average contrast ratios during daytime: comparison of roads and conditions.

Fig. 6. Rural road, good lighting conditions; original cropped camcorder view. CR 0.8 (centre lines) and 0.5 (right edge line).

Fig. 7. Rural road, good lighting conditions; image after digital enhancement. CR 3.2 (centre lines) and 1.7 (right edge line).

Fig. 8. Rural road, poor lighting conditions, sun glare interference; original cropped camcorder view. CR 0.3 (centre lines) and 0.2 (right edge line).

Fig. 9. Rural road, poor lighting conditions, sun glare interference; image after digital enhancement. CR 0.7 (centre lines) and 0.9 (right edge line).

Fig. 10. City street, extreme sun glare interference; original cropped camcorder view. CR 0.1 (centre lines) and 0.2 (middle line).

Fig. 11. City street, extreme sun glare interference; image after digital enhancement. CR 0.1 (centre lines) and 0.5 (middle line).

4 Discussion and Conclusions

Recordings from a dashboard camcorder were used as a source of images for the analysis of exemplary RM in Poland. The measured values of CR, calculated from luminance of RM and the neighbouring roadway surface, were generally inadequately low unless one resorted to image enhancement, which confirmed the necessity of employing the digital processing for MV equipment utilisation. Nonetheless, under difficult lighting conditions CR was very low and the selected image enhancement option was not sufficient: in an example provided in Fig. 10 and Fig. 11, as well as in a few additional cases, CR was essentially nil and could not be increased. Whereas the enhancement parameters (*posterise* level 6, *overlay*) were selected based on analysis of representative images as giving the best outcome, they definitely appear to be insufficient under some lighting conditions. Because varying the enhancement options was not in scope of this work, it is possible that adjustment of the parameters by selection of different modes could improve the outcome.

While digital image enhancement applies only to MV, but one must consider the human drivers as well (Mosböck and Burghardt 2018, Burghardt et al. 2020). Because perception and understanding of RM by human drivers relies on a multitude of cues, CR becomes only one of numerous inputs, some of which are not accessible to MV. Indeed, the finest contemporary MV equipment cannot even approach the broad comprehension capability of human drivers. Consequently, for accurate steering, MV must receive unambiguous signals and, for the success of AV, it must be more reliable than human drivers; if this is not achieved, one of the main goals of AV – decrease in the number of accidents – would not be reached and acceptance of the novel technology would be low (Becker and Axhausen 2017). Whereas most researchers repeat that human errors were responsible for over 90% of collisions and thus the introduction of AV would proportionally increase road safety, they seem to forget about 'ironies of automation' (Noy et al. 2018) and a plethora of other factors (Shinar 2019). To minimise these special situations, high CR is necessary.

This exploratory work has not addressed many very important visibility issues, which could be considered as weakness of this report; however, the plethora of such considerations is so broad and diverse that their inclusion would depart from the first purpose: demonstration of the feasibility of the utilised methodology. Amongst missing visibility issues are those associated with the precipitation and the presence of snow on the roadway, severely limited visibility due to a dense fog or heavy rain, the presence of phantom markings and work zone signalisation, etc. Such conditions remain a meaningful challenge for both human drivers and MV (Yoneda et al. 2019). A study of individual cases – road configurations and lighting conditions—and their effect on CR was beyond the scope of this effort as well. The effect of retroreflectivity of RM, such as a comparison between spring and autumn, is also a separate research area. Because these are special topics that require additional extensive analyses and separate description, they must be treated independently; appropriate material is being constantly collected and subjected to similar assessment and the results shall be reported appropriately.

It was reported that increase in retroreflectivity of RM at least partially helps in overcoming the obstacle of low CR under any conditions (Burghardt et al. 2021b). Such increase was not only positively perceived by drivers (Diamandouros and Gatscha 2016, Horberry et al. 2006, Pashkevich et al. 2017), but also correlated with decreased number of accidents (Avelar and Carlson 2014). To accomplish it, one could to reach for novel, high-end RM systems – combination of high-end paints and 'premium' glass beads were described by us several times, in various configurations: in all of the cases the performance was enhanced in terms of retroreflectivity and in addition prolonged service life was achieved (Burghardt 2018, Burghardt et al. 2016, 2017, 2019a, 2021c, 2021d, Pashkevich et al. 2020). The length of service life was calculated to be a major environmental factor, which should be an additional advantage (Burghardt and Pashkevich 2020, 2021, Burghardt et al. 2021a). RM are being observed by drivers, as was verified using an eye tracker, even though the counted number of gazes was low in comparison with visual attention given to other road features (Pashkevich et al. 2019). Interestingly, it was found that retroreflectivity of RM affected the number and distribution of gazes during night time driving; so far, only preliminary results were presented (Burghardt et al. 2019b).

To summarise, average CR of the analysed RM was 0.8 under daylight and 2.0 under night time conditions, without meaningful variations between road classes. Expectedly, under poor visibility conditions, such as glare or rain, average CR during daytime dropped by approximately 38%, to 0.5. Digital image enhancement through applying filters reducing the number of colours resulted in increase of average CR to 2.3, 6.8, and 1.4, correspondingly for daytime, night time, and daytime poor visibility conditions. As such, it is deemed sufficient for MV equipment in most situations. Nonetheless, in some cases, particularly in the presence of glare, CR was very low and its enhancement did not increase it adequately, which could cause failures of MV.

This methodology of using dashboard camcorder to evaluate CR of RM is providing valuable knowledge that can be used by road administrators and researchers. Even though herein were presented only preliminary analyses from representative still images, it is envisaged that evaluation of the entire recording can be accomplished with appropriate software. Thus, a simple and inexpensive test for CR of RM could be done as a method of proofing their quality for the use by AV, but also for improved perception by human drivers—for the ultimate goal of increasing road safety.

References

Avelar, R.E., Carlson, P.J.: Link between pavement marking retroreflectivity and night crashes on Michigan two-lane highways. Transp. Res. Rec. J. Transp. Res. Board **2404**, 59–67 (2014). https://doi.org/10.3141/2404-07

Baum-Snow, N.: Did highways cause suburbanization? Q. J. Econ. **122**(2), 775–805 (2007). https://doi.org/10.1162/qjec.122.2.775

Becker, F., Axhausen, K.: Literature review on surveys investigating the acceptance of automated vehicles. Transportation **44**(6), 1293–1306 (2017). https://doi.org/10.1007/s11116-017-9808-9

Bengler, K., Dietmayer, K., Farber, B., Maurer, M., Stiller, C., Winner, H.: Three decades of driver assistance systems: review and future perspectives. IEEE Intell. Transp. Syst. Mag. **6** (4), 6–22 (2014). https://doi.org/10.1109/MITS.2014.2336271

Brémond, R.: Visual performance models in road lighting: a historical perspective. LEUKOS **17**, 212–241 (2020). https://doi.org/10.1080/15502724.2019.1708204

Brémond, R., Bodard, V., Dumont, E., Nouailles-Mayeur, A.: Target visibility level and detection distance on a driving simulator. Light. Res. Technol. **45**(1), 76–89 (2013). https://doi.org/10.1177/1477153511433782

Burghardt, T.E.: High durability – high retroreflectivity solution for a structured road marking system. In: Proceedings of International Conference on Traffic and Transport Engineering; Belgrade, Serbia, 27–28 September 2018, pp. 1096–1102 (2018)

Burghardt, T.E., Pashkevich, A.: Materials selection for structured horizontal road markings: financial and environmental case studies. Eur. Transp. Res. Rev. **12**(1), 1 (2020). https://doi.org/10.1186/s12544-020-0397-x

Burghardt, T.E., Pashkevich, A.: Green public procurement for road marking materials from insiders' perspective. J. Cleaner Product. **298**, 126521 (2021). https://doi.org/10.1016/j.jclepro.2021.126521

Burghardt, T., Pashkevich, A., Żakowska, L.: Influence of volatile organic compounds emissions from road marking paints on ground-level ozone formation: case study of Kraków, Poland. Transp. Res. Procedia **14**, 714–723 (2016). https://doi.org/10.1016/j.trpro.2016.05.338

Burghardt, T.E., Ščukanec, A., Babić, D., Babić, D.: Durability of waterborne road marking systems with various glass beads. In: Proceedings of International Conference on Traffic Development, Logistics and Sustainable Transport; Opatija, Croatia, 1–2 June 2017, pp. 51–58 (2017)

Burghardt, T., Pashkevich, A., Mosböck, H.: Yellow pedestrian crossings: from innovative technology for glass beads to a new retroreflectivity regulation. Case Stud. Transp. Policy **7** (4), 862–870 (2019a). https://doi.org/10.1016/j.cstp.2019.07.007

Burghardt, T.E., Pashkevich, A., Bairamov, E.: Eye tracker study of retroreflectivity perception by drivers. In: IRF International Symposium on Traffic Signs and Pavement Markings; Zagreb, Croatia, 3–4 October 2019 (2019a)

Burghardt, T.E., Mosböck, H., Pashkevich, A., Fiolić, M.: Horizontal road markings for human and machine vision. Transp. Res. Procedia **48**, 3622–3633 (2020). https://doi.org/10.1016/j.trpro.2020.08.089

Burghardt, T.E., Babić, D., Pashkevich, A.: Performance and environmental assessment of prefabricated retroreflective spots for road marking. Case Stud. Construct. Mater. **15**, e00555 (2021a). https://doi.org/10.1016/j.cscm.2021.e00555

Burghardt, T.E., et al.: Visibility of various road markings for machine vision. Case Stud. Constr. Mater. **15**, e00579 (2021b). https://doi.org/10.1016/j.cscm.2021.e00579

Burghardt, T.E., Maki, E., Pashkevich, A.: Yellow thermoplastic road markings with high retroreflectivity: demonstration study in Texas. Case Stud. Constr. Mater. **14**, e00539 (2021c). https://doi.org/10.1016/j.cscm.2021.e00539

Burghardt, T.E., Pashkevich, A., Bartusiak, J.: Solution for a two-year renewal cycle of structured road markings. Roads Bridges **20**, 5–18 (2021d). https://doi.org/10.7409/rabdim.021.001

Calvert, S.C., Mecacci, G., van Arem, B., de Sio, F.S., Heikoop, D.D., Hagenzieker, M.: Gaps in the control of automated vehicles on roads. IEEE Intell. Transp. Syst. Mag. **13**, 146–153 (2020). https://doi.org/10.1109/mits.2019.2926278

Calvi, A.: A study on driving performance along horizontal curves of rural roads. J. Transp. Saf. Secur. **7**(3), 243–267 (2015). https://doi.org/10.1080/19439962.2014.952468

Carlson, P.: Pavement markings for machine vision systems. In: ITS World Congress 2017, Montréal, Quebec, Canada, 29 October–2 November 2017 (2017)

Carlson, P.J., Poorsartep, M.: Enhancing the roadway physical infrastructure for advanced vehicle technologies: a case study in pavement markings for machine vision and a road map toward a better understanding. In: Proceedings of Transportation Research Board 96th Annual Meeting, Washington, District of Columbia, United States, 8–12 January 2017, paper 17-06250 (2017)

Carreras, A., Daura, X., Erhart, J., Ruehrup, S.: Road infrastructure support levels for automated driving. In: Proceedings of the 25th ITS World Congress, Copenhagen, Denmark, 17–21 September 2018, pp. 17–21 (2018)

Diamandouros, K., Gatscha, M.: Rainvision: the impact of road markings on driver behaviour-wet night visibility. Transp. Res. Procedia **14**, 4344–4353 (2016). https://doi.org/10.1016/j.trpro.2016.05.356

Dickmanns, E.D., Mysliwetz, B., Christians, T.: An integrated spatio-temporal approach to automatic visual guidance of autonomous vehicles. IEEE Trans. Syst. Man Cybern. **20**(6), 1273–1284 (1990). https://doi.org/10.1109/21.61200

Goelles, T., Schlager, B., Muckenhuber, S.: Fault detection, isolation, identification and recovery (FDIIR) methods for automotive perception sensors including a detailed literature survey for LiDAR. Sensors **20**(13), 3662 (2020). https://doi.org/10.3390/s20133662

Hadi, M., Sinha, P.: Effect of pavement marking retroreflectivity on the performance of vision-based lane departure warning systems. J. Intell. Transp. Syst. **15**(1), 42–51 (2011). https://doi.org/10.1080/15472450.2011.544587

Hadi, M., Sinha, P., Easterling, J.: Effect of environmental conditions on performance of image recognition-based lane departure warning system. Transp. Res. Rec. J. Transp. Res. Board **2000**(1), 114–120 (2007). https://doi.org/10.3141/2000-14

He, Y., Li, Y., Xing, L., Qiu, Z., Zhang, X.: Influence of text luminance, text colour and background luminance of variable-message signs on legibility in urban areas at night. Light. Res. Technol. **53**(3), 263–279 (2021). https://doi.org/10.1177/1477153520958466

Horberry, T., Anderson, J., Regan, M.A.: The possible safety benefits of enhanced road markings: a driving simulator evaluation. Transp. Res. Part F Traff. Psychol. Behav. **9**(1), 77–87 (2006). https://doi.org/10.1016/j.trf.2005.09.002

Kunze, L., Bruls, T., Suleymanov, T., Newman, P.: Reading between the lanes: road layout reconstruction from partially segmented scenes. In: Proceedings of 21st International Conference on Intelligent Transportation Systems, Maui, Hawaii, United States, 4–7 November 2018, pp. 401–408 (2018). https://doi.org/10.1109/ITSC.2018.8569270

Kuutti, S., Fallah, S., Katsaros, K., Dianati, M., Mccullough, F., Mouzakitis, A.: A survey of the state-of-the-art localization techniques and their potentials for autonomous vehicle applications. IEEE Internet Things J. **5**(2), 829–846 (2018). https://doi.org/10.1109/JIOT.2018.2812300

Liu, R., Wang, J., Zhang, B.: High definition map for automated driving: overview and analysis. J. Navig. **73**(2), 324–341 (2020). https://doi.org/10.1017/S0373463319000638

Marti, E., Angel, M., de Miguel, F., Garcia, J.: A Review of sensor technologies for perception in automated driving. IEEE Intell. Transp. Syst. Mag. **11**(4), 94–108 (2019). https://doi.org/10.1109/MITS.2019.2907630

Mathibela, B., Newman, P., Posner, I.: Reading the road: road marking classification and interpretation. IEEE Trans. Intell. Transp. Syst. **16**(4), 2072–2081 (2015). https://doi.org/10.1109/TITS.2015.2393715

Matowicki, M., Přibyl, O., Přibyl, P.: Analysis of possibility to utilize road marking for the needs of autonomous vehicles. In: Smart Cities Symposium Prague, Prague, Czech Republic, 26–27 May 2016 (2016). https://doi.org/10.1109/SCSP.2016.7501026

Miller, T.R.: Benefit–cost analysis of lane marking. Transp. Res. Rec. J. Transp. Res. Board **1334**, 38–45 (1992)

Mosböck, H., Burghardt, T.E.: Horizontal road markings and autonomous driving – back from the future. In: Proceedings of 37th Annual Southern African Transport Conference, Pretoria, South Africa, 9–12 July 2018, pp. 557–568 (2018). http://hdl.handle.net/2263/69562

Najeh, I., Bouillaut, L., Daucher, D., Redondin, M.: Maintenance strategy for the road infrastructure for the autonomous vehicle. In: ESREL 2020-PSAM 15, 30th European Safety and Reliability Conference and the 15th Probabilistic Safety Assessment and Management Conference, Venice, Italy, 1–5 November 2020 (2020)

Narote, S.P., Bhujbal, P.N., Narote, A.S., Dhane, D.M.: A review of recent advances in lane detection and departure warning system. Pattern Recogn. **73**, 216–234 (2018). https://doi.org/10.1016/j.patcog.2017.08.014

Noy, I.Y., Shinar, D., Horrey, W.J.: Automated driving: safety blind spots. Saf. Sci. **102**, 68–78 (2018). https://doi.org/10.1016/j.ssci.2017.07.018

Otsu, N.: A threshold selection method from gray-level histograms. IEEE Trans. Syst. Man Cybern. **9**(1), 62–66 (1979)

Pashkevich, A., Bartusiak, J., Żakowska, L., Burghardt, T.E.: Durable waterborne horizontal road markings for improvement of air quality. Transp. Res. Procedia **45**, 530–538 (2020a). https://doi.org/10.1016/j.trpro.2020.03.060

Pashkevich, A., Bartusiak, J., Burghardt, T.E., Šucha, M.: Naturalistic driving study: methodological aspects and exemplary analysis of a long roadwork zone. In: Macioszek, E., Sierpiński, G. (eds.) Research Methods in Modern Urban Transportation Systems and Networks. LNNS, vol. 207, pp. 165–183. Springer, Cham (2021). https://doi.org/10.1007/978-3-030-71708-7_11

Pashkevich, A., Burghardt, T.E., Shubenkova, K., Makarova, I.: Analysis of drivers' eye movements to observe horizontal road markings ahead of intersections. In: Varhelyi, A., Žuraulis, V., Prentkovskis, O. (eds.) VISZERO 2018. LNITI, pp. 1–10. Springer, Cham (2019). https://doi.org/10.1007/978-3-030-22375-5_1

Pashkevich, A., Burghardt, T.E., Żakowska, L., Nowak, M., Koterbicki, M., Piegza, M.: Highly retroreflective horizontal road markings: drivers' perception. In: Proceedings of International Conference on Traffic Development, Logistics and Sustainable Transport, Opatija, Croatia, 1–2 June 2017, pp. 277–287 (2017)

Poggenhans, F., Salscheider, N.O., Stiller, C.: Precise localization in high-definition road maps for urban regions. In: IEEE/RSJ International Conference on Intelligent Robots and Systems (IROS). Madrid, Spain, 1–5 October 2018, pp. 2167–2174 (2018). https://doi.org/10.1109/IROS.2018.8594414

Poggenhans, F., Schreiber, M., Stiller, C.: A universal approach to detect and classify road surface markings. In: IEEE 18th International Conference on Intelligent Transportation Systems, Las Palmas de Gran Canaria, Spain, 15–18 September 2015, pp. 1915–1921 (2015). https://doi.org/10.1109/ITSC.2018.8569270

Ranft, B., Stiller, C.: The role of machine vision for intelligent vehicles. IEEE Trans. Intell. Veh. **1**(1), 8–19 (2016). https://doi.org/10.1109/TIV.2016.2551553

Rosique, F., Navarro, P., Fernández, C., Padilla, A.: A systematic review of perception system and simulators for autonomous vehicles research. Sensors **19**(3), 648 (2019). https://doi.org/10.3390/s19030648

Schnell, T., Aktan, F., Li, C.: Traffic sign luminance requirements of nighttime drivers for symbolic signs. Transp. Res. Rec. J. Transp. Res. Board **1862**(1), 24–35 (2004). https://doi.org/10.3141/1862-04

Shinar, D.: Crash causes, countermeasures, and safety policy implications. Accid. Anal. Prev. **125**, 224–231 (2019). https://doi.org/10.1016/j.aap.2019.02.015

Steyvers, F.J., De Waard, D.: Road-edge delineation in rural areas: effects on driving behaviour. Ergonomics **43**(2), 223–238 (2000). https://doi.org/10.1080/001401300184576

Theeuwes, J., Alferdinck, J.W., Perel, M.: Relation between glare and driving performance. Hum. Factors **44**, 95–107 (2002). https://doi.org/10.1518/0018720024494775

Xing, Y., et al.: Advances in vision-based lane detection: algorithms, integration, assessment, and perspectives on ACP-based parallel vision. IEEE/CAA J. Automatica Sinica **5**(3), 645–661 (2018). https://doi.org/10.1109/JAS.2018.7511063

Yoneda, K., Suganuma, N., Yanase, R., Aldibaja, M.: Automated driving recognition technologies for adverse weather conditions. IATSS Res. **43**(4), 253–262 (2019). https://doi.org/10.1016/j.iatssr.2019.11.005

Ziegler, J., et al.: Making Bertha drive—an autonomous journey on a historic route. IEEE Intell. Transp. Syst. Mag. **6**(2), 8–20 (2014). https://doi.org/10.1109/MITS.2014.2306552

San Diego Bus Rapid Transit Using Connected and Autonomous Vehicle Technology

Dan Lukasik[✉] and Dmitri Khijniak

Parsons, New York City, USA
Daniel.lukasik@parsons.com

Abstract. Parsons has been contracted by the San Diego Association of Governments to deliver an innovative pilot demonstration of Bus-On-Shoulder (BOS) operations on a limited access freeway using state-of-the-art technology for driver assistance. Through the deployment of Vehicle to Infrastructure Connected Vehicle (V2I CV), Lane Departure Warning (LDW), Blind Spot Warning (BSW), and Forward Collision Warning (FCW) technologies, this BOS pilot project will demonstrate that bus performance can be improved by allowing buses to drive on the freeway transit-only shoulder lane with minimal changes to the roadway. The pilot project will extend along Interstate 805 (I-805) from Chula Vista/South Bay to State Route 94 (SR 94) into downtown San Diego in both directions of travel. Upon completion of the pilot, all improvements to the freeway that were installed for the purposes of this pilot project will be removed.

Parsons is developing, integrating, and providing the following technology for this project:

- LDW system alerting bus drivers that they are drifting into adjacent lanes.
- BSW system alerting bus drivers of bus blind spots.
- FCW system alerting bus drivers of forward obstructions.
- Ramp metering Transit Signal Priority system holding vehicles at the ramps, as well as displaying. messages on Changeable Messages Signs at the ramps warning drivers of the buses

Keywords: Smart mobility · Intelligent transportation systems · Connected vehicle · Autonomous vehicle · Transit · Bus rapid transit

1 Introduction

The Bus-on-Shoulder (BOS) is a pilot demonstration of the integrated state-of-the-art driver assist technology, video detection machine vision technology and vehicle-to-roadside communication technology aiming to enable buses to utilize freeway shoulder in a safe and effective manner. The BOS system enables authorized South Bay Bus Rapid Transit (BRT) buses to travel on the shoulder of a limited access freeway in San Diego during peak congestion hours with minimal changes to the roadway. The project is deployed along Interstate 805 (I-805) and State Route 94 (SR-94) from Chula Vista to San Diego downtown in both direction of travel.

The project was launched in 2019 when Parson was selected as a system integration for the BOS enabling technologies. The project is conducted in partnership with the San Diego Association of Governments (SANDAG), Metropolitan Transit System (MTS), the California Department of Transportation (Caltrans), the California Highway Patrol (CHP) and the U.S. Department of Transportation (USDOT). The project is going into full deployment in late 2021 and will continue in operation for the next 3 years.

This paper provides an architecture and design overview of the project highlighting main components and reasons for their selection. It also shares lessons learned during the design and development of the project.

2 Overview

Like in any large city, travel on freeways in San Diego shows patterns of congestion during morning and afternoon peak traffic hours. The congestions affect travelers driving their personal vehicles to work as well as those which choose to take transit buses. The concept behind the BOS offers an alternative solution to conventional congestion mitigation methods by giving advantage to certain buses to utilize additional space on the freeway which is reserved as the freeway shoulder. The shoulder is not used similar to a conventional lane on a freeway. Rather it is reserved for an authorized transit buses to utilize during rush hours to bypass the congested lanes.

The BOS project allows only certain buses to use the freeway shoulder on the seven-mile segment of the freeway in the designated project area. The buses are equipped with the Advanced Driver Assistance System (ADAS) which enhances safety and bus operations along the road corridor. Use of ADAS was selected specifically because it can assist drivers when buses manoeuvre between regular freeway lanes and the shoulder lane, as well as when buses traverse on-ramp areas by traveling on the shoulder.

The BOS will provide more reliable travel during peak hours with minimal changes to the roadway by utilizing roadway capacity already available. Some of the metrics which will be monitored during the project implementation are bus schedule adherence and travel reliability during the hours of peak congestion as well as utilization of advance driver assistance technologies to ensure safe bus operation while driving on the freeway shoulder.

Concept of Operation

The typical traffic pattern displays congestion in the northbound direction along I-805 and along SR-94 westbound in the morning hours between 6 AM and 9 AM when most of the commuter traffic is heading into the San Diego downtown. During this time, traffic often backs up which reduces speed below 30 miles per hour (mph) across all lanes. The congestion reverses in the late afternoon when traffic heads out along SR-94 Eastbound and then I-805 Southbound. The BRT bus fleet shown on Fig. 1 is comprised of 60-foot articulated buses traveling between a station near the Otay Mesa border crossing and into San Diego downtown. The buses operate in Express mode (i.e., without stops) when traveling along the freeway.

The Fig. 2 depicts the area of the freeway where driving on the shoulder is permitted. The Blue line highlights the area where outside shoulder is available while the Orange colour shades the areas where the inside shoulder can be used. The Bus-on-Shoulder is set to operate between 6 AM and 9 AM on weekdays in the northbound direction and between 4 PM to 7 PM in the southbound direction along the designated freeways. During those hours buses continue to drive normally on the freeway. In addition, during BOS operating hours, buses can additionally utilize outer shoulders on I-805 and inside shoulder along SR-94 as an additional lane to bypass zones of congestion.

Fig. 1. 60-foot articulated BRT transit bus used for the BOS project

There are several stipulations which apply to how the Bus-on-Shoulder must operate. Those stipulations were developed early during the project planning phase and agreed by the project stakeholders:

- Buses can utilize the shoulder only in the designated portion of the I-805 and SR-94 freeways
- Only BRT buses equipped with the ADAS and V2X technologies are authorized use the shoulder lanes
- Buses can enter the shoulder if speed on general-purpose lanes of the freeway drops below 35 mph
- Buses cannot travel on the shoulder faster than 15 mph over the speed of the adjacent freeway lane
- Buses cannot use the shoulder if the pavement is wet, which occurs quite infrequently in San Diego
- Buses must exit the shoulder if the shoulder is blocked by a large debris or used by CHP vehicle to conduct their business (e.g., clear the accident)

While driving on the shoulder, bus drivers expect to conform to the general driving safety principles as well as maintain maximum speed as discussed above. Thus, a special training for bus drivers is also one of the preconditions for starting the BOS project. Bus drivers are expected make decision whether to use the shoulder or continue using general-purpose lanes based on the driving conditions several other factors outlined above.

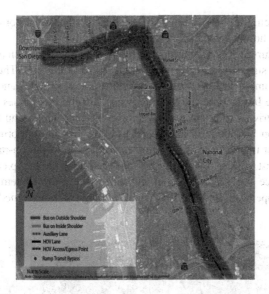

Fig. 2. BOS project corridor in San Diego along I-805 and SR-94 freeways.

The project called for several technological enhancements:

- The buses were required to be equipped with the Advanced Driver Safety Assist technology. The technology was required to assist drivers when they maneuver along the shoulder and be used for the rest of the traveling route.
- Buses must include communication technology which allow them to bypass ramp merge points while holding off the ramp traffic.
- The roadside must include a video detection technology which serves as a backup detection to notify ramp system about approaching bus.
- The system must include mechanism for monitoring system wide operating conditions and if critical failure is detected in one of the system components, inform drivers that driving on the shoulder is not available.

One of the innovative concepts the BOS project introduced was the use of transit signal priority (TSP) by the bus to get a priority when bypassing the on-ramp merge point. As shown on the Fig. 3, the bus traveling on the outer shoulder, will arrive to the on-ramp merge zone where it may conflict with other vehicles arriving from the on-ramp. More so, vehicles coming from the on-ramp will generally accelerate to arrive to the merge point with speeds comparable those of vehicles on the freeway. If a bus is traveling along the shoulder, vehicles drivers may respond differently with some slowing down and some accelerating trying to overcome the bus.

The TSP system was proposed to avoid unsafe situations at the merge point between on-ramp and the shoulder lane. As the bus approaches the merge point, the bus will use V2X radio to communicate to the ramp system. The bus V2X radio will transmit bus position, speed and heading at a frequency of 10 messages per second. The ramp system will use bus position to activate a priority request in anticipation of bus transition. When, the priority request is activated, signals for both ramp lanes will be set to red, thus holding vehicles at the ramp limit line.

Additionally, during TSP activation, the ramp system will turn on an additional Bus Blank-Out (BBO) signs to notify drivers about approaching bus. The TSP activation must occur with some time ahead of bus arrival to allow vehicles ahead of the bus to merge. Also, some buffer time must be allowed after the bus passes the ramp merge point, to prevent drivers to attempt to overcome the bus. This ramp holding must be calculated based on bus location and speed to allow bus the opportunity to pass the ramp with the least interference from vehicles coming from the on-ramp.

The TSP request will automatically be cancelled after the bus passed the on-ramp merging area. Once the TSP request is cancelled, the ramp system will resume its normal metering operation and the Bus Blank-out sign will be turned off.

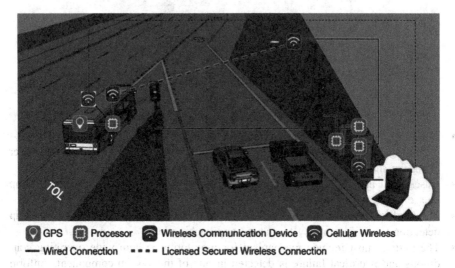

Fig. 3. Bus communication with the ramp system while traversing freeway on-ramp merge point

3 Design

The System Architecture diagram on.

Figure 4 provides an design overview of the BOS system.

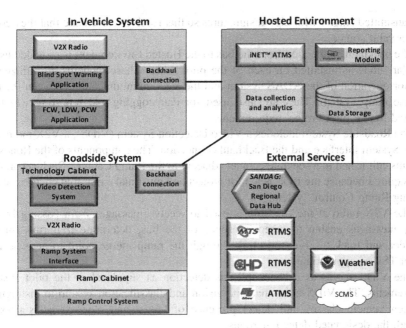

Fig. 4. BOS system architecture

The BOS added enhancements to the four major areas:

- In-Vehicle Systems
- Roadside Systems
- Hosted Environment
- Interfaces to External Services

The In-Vehicle System enhancements include installation of sensors and processing units which enable applications such as Lane Departure Warning (LDW), Forward Collision Warning (FCW), Blind Spot Warning (BSW) and Pedestrian Collision Warning (PCW). The ADAS system is used to detect situations which could precipitate unsafe driving conditions and activates in-cabin audio-visual warnings assisting the driver. For example, LDW detects when bus starts to cross the lane markings in an adjacent lane without activation of the corresponding turn signal. The FCW detects when a bus moves closer than set limit to a vehicle in front of it. The BSW warns the driver when a vehicle in adjacent lane moving side-by-side with a bus may not be seen in the mirror. The driver alert is activated upon activation of the turn signal indicating a maneuver in the direction of the vehicle detected in the adjacent lane. Lastly, the PCW detects pedestrian and bicyclists which are moving in proximity of the bus and being in danger to cause a collision.

The bus is equipped with a Vehicle-to-Everything (V2X) radio and a backhaul data connection. The V2X radio is used to communication to the roadside ramp controller. The communication includes transmission of the bus position, speed and heading using Basic Safety Messages (BSM) at a frequency of 10 messages per second. The messages are transmitted singed with digital signatures so that receiver can ensure that they come from a valid source.

The backhaul connection from the bus to the Hosted Environment is provided using Cellular Modems installed on each of the pilot buses. These modems are utilized to monitor operation of the ADAS system and the V2X communication between the bus and the ramp system. They are also used for data logging and reporting which is required by the pilot project requirements.

The Roadside System includes a Video Detection System (ViDS), a V2X radio, the Ramp System Interface and the Backhaul connection. The components of the Roadside Systems collocated in a separate cabinet adjacent to the Ramp Cabinet. Such separation was required because the BOS is a pilot project and must make minimal changes to the existing Ramp Control System.

The V2X radio at the roadside is used to receive messages from passing buses. These messages enable to detect presence of the bus, determine if it runs on the shoulder and track the bus trajectory through the ramp merge point to activate and cancel TSP as appropriate.

The ViDS serves as a "backup" bus detection as stipulated in the pilot system requirements. The ViDS users machine vision and classification system to distinguish between vehicle classes and detect presence of buses, and other vehicles passing through the designated detection zones.

The ViDS has two primary uses in the BOS. The ViDS can confirm presence of the bus in the shoulder. If the bus V2X radio doesn't work, the TSP can be triggered based on the ViDS detection of bus presence in the shoulder. In addition, the ViDS is used to detect any other vehicles which may use the shoulder but are not allowed to do it. The ViDS is not used for enforcement rather it is a monitoring tool to determine frequency of shoulder misuse by non-authorized vehicles thus helping to assess if any additional enforcement is required.

The Ramp System Interface is used to send control commands to the Ramp Control System in the Caltrans cabinet. The interface also allows the BOS Roadside Interface to monitor activation of the BBO signs which confirms that the TSP request was successfully executed. By monitoring the activation of the BBO sign, the Roadside System can detect a possible malfunction, e.g., when TSP request is placed but the BBO sign is not activated. If such malfunction is detected, it will be recorded and used to alert BOS operation.

The Roadside System cabinets are equipped with Cellular Modems to provide backhaul connection into the Hosted Environment. Similar to the buses, the backhaul is used to monitor the roadside system operation, collect data from V2X radios and maintain the roadside equipment remotely.

The Hosted Environment (HE) is the "heart and brain" of the BOS system. The HE runs "Data Collection and Analytics" which is a set of processes to collect, process and archive data from buses and roadside systems. The data processing is used primarily to ensure that system is operating within specified guidelines. The data analytics also provide performance analysis which monitors and reports on the key system metrics. It is also used for the detection and reporting of abnormal events in the system.

The Data Collection and Analytics uses the Data Storage which is broken into non-relational storage for high frequency and high-volume data in real time coming from buses and roadside systems. This data primarily includes data logged from V2X radios, system events and process logs. A set of HE processes use the "raw" data and transform it into the data analytics which is stored in a separate relational database. The latter is queried for generating analytics and performance reports.

The Advanced Traffic Management Systems (ATMS) is used as a command-and-control system and a user interface for the BOS. Using ATMS, a user can monitor location of buses, various systems events and control system operating mode. The ATMS is also linked with the Reporting Module such as based on the Microsoft PowerBI to generate various user and performance reports.

The Hosted Environment interfaces to external systems to retrieve data from the several agencies including Caltrans, MTS and CHP. The HE uses interface to the SANDAG Regional Data Hub which provides an interagency data "data clearing house" and collect information about road and traffic including incidents, road closures, ramp operation status, etc. from several regional systems. This information is vital to the BOS operation since it determines availability of the shoulder for safe use by buses. The HE use this data to monitor for incidents, lane closures, construction events and automatically sends alerts to the buses if any of those conditions would indicate a conflict or unavailability of the shoulder for the BOS. The HE retrieves weather data from local weather stations to detect precipitation conditions. Also, the HE interfaces with the Security Credential Management System (SCMS) to facilitate delivery of digital certificates to the V2X roadside units.

4 Building Blocks

In this section, further details for the foundational components will be provided including:

- Advanced Driver Assistance System
- Connected Vehicle Technology
- Transit Signal Priority - Caltrans Ramp Controllers
- Backup bus detection utilizing machine vision
- iNET™ Monitoring and Analytics

Advanced Driver Assistance System
The Fig. 5 shows various components and detections zones for a bus equipped with an ADAS system.

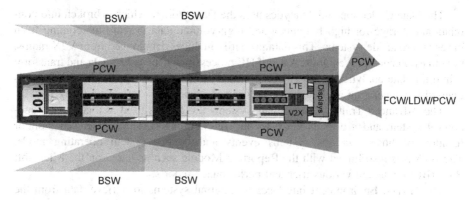

Fig. 5. ADAS detection zones for a bus used in the BOS

The BOS ADAS system consists of two subsystems installed and integrated together. The MobilEye Shield + system uses video sensor and provides driver notifications for the Lane Departure Warning, Forward Collision Warning, and Pedestrian Departure Warning applications. The system uses two sets of cameras on each side of the bus and two front facing cameras. It also includes driver notification displays to provide audio-visual warnings. Blue triangles on Fig. 5 shows general zones of detection for vehicles and pedestrians using the Shield + cameras.

The Blind Spot Warning application is provided by Hexagon's system utilizes sidefire radars also installed on both sides of the bus. The BSW detects vehicles traveling on adjacent to the bus lanes. The BSW is used to assist drivers when performing lane change maneuvers alerting the driver about a vehicle traveling along the bus. The BSW sends warnings to the same displays as the Shield + system.

This integration of the ADAS systems was selected because they can assist drivers in the moving scenarios when a bus travels at high-speed along the freeway and when bus performs manoeuvres to use the freeway shoulder. The bus driver can rely on LDW, FCW and BSW to maintain safe distance and perform lane changes and merges with other vehicles. The PCW is effective when bus manoeuvres in the densely populated streets with heavy pedestrian and bicyclists traffic as typical in the city downtown.

Connected Vehicle Technology

The project utilizes the latest generation of the connected vehicle technology based on cellular V2X (C-V2X). The C-V2X is a standard based high-speed wireless communication which can enable fast communication vehicle-to-vehicle and vehicle-to-infrastructure. For the BOS project, C-V2X radios are installed on each bus and at each ramp. Approaching the ramp, each bus will communicate its position, speed and heading to the roadside radio. The Roadside system will use the C-V2X roadside radio to receive messages from the bus and track the bus location as it traverses the ramp with very low latency thus allowing to send request for TSP when the bus is about to approach the on-ramp merge area.

Transit Signal Priority - Caltrans Ramp Controllers

The BOS sends Transit Signal Priority requests to the Caltrans Ramp System which runs "Universal Ramp Metering Software" (URMS) on the ramp controllers. The Controller software manages all aspects of the ramp taking into account ramp schedule, freeway speeds and vehicle passing through the ramp lane detectors. For this project, the URMS software was enhanced to support TSP calls from a "virtual lane" designated for the bus traveling the shoulder. The primary interface to the BOS was using contact closure triggered by the BOS Roadside System. The Ramp Controller used modified logic to allow TSP request to temporary stop ramp metering for a short period of time (i.e., up to 15 s) thus allowing the bus to pass through the on-ramp merge point unimpeded.

Backup Bus Detection Utilizing Machine Vision

Machine Vision based ViDS system from MioVision is used to detect vehicles and recognize buses as a separate class of vehicles. The ViDS system was configured to detect buses traveling on the shoulder and serve as a "backup" to send TSP request to the Ramp Controller. The ViDS system also was setup to detect any other vehicles (except buses) that might use the shoulder lane. The latter is important to ensure that the shoulder is not used by other vehicles as means to bypass the congested traffic areas.

Use of iNET™ for Monitoring and Analytics

Parsons iNET™ Advanced Traffic Management System (ATMS) provides comprehensive set of tools to monitor and control traffic system from a central user interface as shown on Fig. 6. The iNET™ ATMS was selected to manage the BOS. The ATMS software already included interface to the SANDAG Regional Data Hub and provided a visual interface to monitor system performance, road conditions and respond to alerts. The iNET™ was enhanced to integrate with the BOS specific components designed into the roadside and in-vehicle systems. Using iNET™ ATMS, an operator can observe buses using the shoulder, monitor traffic conditions and incident reports related to the project corridor. The operator can also send notification to the bus built-in ADAS screen that shoulder use is not available.

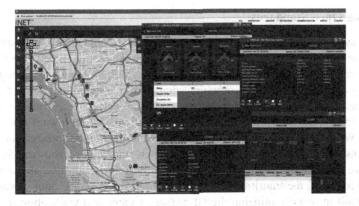

Fig. 6. Parsons iNET™ ATMS user portal

ATMS also includes integration with the Microsoft PowerBI reporting platform. A user can used PowerBI to create comprehensive reports using BOS analytical data. These reports highlight the system status, provide reports on critical events and overall benefits of the system including travel time savings, number of completed routes and use of the shoulders.

5 Lessons Learned

5.1 Building Complex Systems Out from Existing Technologies

The BOS uses existing technologies including ADAS, ViDS, V2X, ATMS and other. Leveraging technologies reduces project risk and project investments. However, there are some downsides where reuse of the established applications comes with a reduced flexibility compared to the development of the brand new technology from scratch. Conversely, established applications comes with thorough validation. While adding new features enhances their functionality but also may introduce "bug". This brings a two-prong approach where planning and design phases can identify reusable technologies and need for their further enhancements. Additionally, system integration and validation needs to focus on those areas where changes were introduced to perform thorough testing and validation.

5.2 User Interfaces and Driver Distraction

As much as it is interesting to showcase BOS technologies in the buses by adding new user interfaces, one of the early project decisions was minimize driver distractions. Thus, ADAS indicators were permitted to be added into buses. However, V2X communication between ramp and approaching bus was decided must happen in the background without any distinct signal to the driver. This way, the driver can continue to operate the buses following operating and safety guidelines established by MTS, while V2X technology requesting TSP in the background provides improved driving conditions for transitioning through the on-ramp merging zone.

5.3 Using Open Standards

The BOS system takes advantage of many off-the-shelf components. This still leaves the need for interoperability between buses and the roadside. Fortunately, the last 15 or so years brought into reality vehicle-to-everything (V2X) wireless technology. This technology enables short range communication at a high-speed and low latency utilizing licensed 5.9 Gigahertz (GHz) spectrum. More recently, the technology has been migrating from Dedicated Short Range Communication (DSRC) toward cellular V2X (C-V2X). When launched, the project was planned around DSRC project but recently started transition to the C-V2X. Utilizing the open standards such as the Society of Automotive Engineering (SAE) J2735 for encoding information exchanged between buses and roadside, the transition was quite easy. Also, the project took advantage of the V2X built-in security utilizing digital message signing and verification secured by certificates issued by the Security Credential Management System (SCMS). Use of the open standard also helped to establish data schemas for collect and archived data which can be used for the project performance analysis as well as the historical data for future analysis.

5.4 Building on Project Stakeholder Consensus

Many ITS projects put emphasis on the technology and its benefits. This is valid because in many cases proving that technology works and provide expected benefits is one of the main project objectives. This project being no exception benefited tremendously from the preliminary work to organize project stakeholders and discuss project details with them. The project was conceived as solving a problem of congestion utilizing advanced safety and sensor technologies. However, for the solution to work, it had to be deployed within jurisdiction of several public agencies. Freeway and ramp systems are operated and maintained by Caltrans. The buses are operated by MTS. The SANDAG runs the Regional Data Hub and became the technical project manager. System engineering principles were followed in the project development starting from preparation of the Concept of Operation and System Requirements. The CHP played considerable role in evaluating project safety. Under SANDAG sponsorship, multiple face-to-face meetings between agencies resulted in development of project approach and decision documents which codified operational aspects and supporting activities including driver training, public outreach, project coordination etc.

6 Conclusions

The paper provided an overview of the Bus on Shoulder project carried out in San Diego. The BOS integrates advanced technologies including ADAS, ViDS and V2X to enable bus to travel on the freeway shoulder with improved safety. With minimum changes to the freeway infrastructure, buses can achieve improved schedule performance and travel reliability.

An overview of the system architecture is discussed including component technologies which comprised the BOS system. The system spans upgrades to the buses, additional components to the roadside ramp system and the Hosted Environment which collect system data and runs analysis to inform system operators of the operating conditions, exceptions and performance.

Lastly, several lessons learned are shared about utilizing of-the-shelf components, use of open standards for interoperability and importance of the consensus development among stakeholders to implement the BOS system.

Acknowledgements. This project is a culmination of the dedicate work of several organizations and individuals. The financial support came from the US DOT. The SANDAG provided project leadership and coordination among various stakeholders that were essential to ensure success of this endeavour. We would like to acknowledge support from MTS, Caltrans, and CHP as instrumentals in supporting project implementation.

References

SAE J2735. Society of Automotive Engineers V2X Communications Message Set Dictionary, Revision 2016-03 (2016)

SAE J2945/1. On-Board System Requirements for V2V Safety Communications, Revision 2016-03 (2016)

SANDAG. SANDAG Bus on Shoulder Concept of Operations; SANDAG internal document, Sang Diego (2016)

Maintenance Strategies - 2

Introduction to the RoadMark System Low Cost, Fast and Flexible Data Collection for Rural Roads

Mark J. Thriscutt[✉]

Penhallow Limited, Littlehampton, UK
Mark@Penhallow.ltd.uk

Abstract. Rural roads are often the poor cousins in the road sub-sector, despite comprising the majority of the public road network, representing a considerable public investment. They have a profound impact on the quality of life for the significant rural population and economy; and directly impact urban citizens too, influencing the availability and price of many basic commodities. Rural roads have important economic, social, political, environmental and equity consequences that no successful country can ignore.

However, attracting and retaining the quality of professional staff required is challenging and funding is frequently woefully inadequate, exacerbated by weak planning due to inadequate data about these rural road assets.

Based on projects undertaken in Myanmar from 2016 and in Cambodia in 2018–9, the RoadMark system was developed to address the shortcomings common in managing rural roads, which were inadequately addressed by other existing systems. RoadMark provides:

- The ability to capture basic inventory and condition data for all the key types of road assets, using only a GPS- enabled Smartphones, thereby avoiding the requirement for specialised and expensive equipment and its ongoing servicing;
- Fast survey speeds (important for large rural networks);
- Minimal training, (avoiding reliance on specialised or expensive experts);
- No requirement for any network referencing system to be established beforehand;
- Ability to monitor surveys remotely;
- Surveys that require no internet or mobile signals, (often problematical in rural areas);
- Ability to map flexibly, all road assets;
- Permanent availability of data and use in other applications;
- Cost-effective, inexpensive and flexible licensing options.

Keywords: Rural roads · Data collection · RoadMark · Planning · Budgeting

1 Introduction

This paper describes the significant benefits possible to road managers, from using a simple, low cost data collection system (RoadMark), to collect basic inventory and condition data about poorly defined road assets. This paper describes:

- How the system has been designed to accommodate the institutional and organizational weaknesses commonly found in the management of rural roads in many developing countries;
- The results from the system's use on rural roads in Myanmar and Cambodia;
- Examples of the much improved strategic network plans and budgeting developed using the data collected; and
- Examples of the reports and mapping that can be produced using the data, greatly assisting road network managers and funding agencies to better assess their network's needs and for network monitoring.

2 Background

2.1 The Importance of Rural Roads

Rural roads are frequently viewed as the poor cousin in the roads sub-sector, receiving far less attention, funding and resources than National roads and even urban roads, where political pressures for improved road infrastructure are often greater. However, rural roads usually represent the majority of the public road network in any country by length; and they act as essential lifelines connecting the large rural population to cities, markets and government services.

Even where governments claim to understand their importance, their focus is frequently on the expansion of the rural road networks, with inadequate support to their subsequent maintenance and management. As any competent road manager knows, this represents a poor use the substantial amounts of public resources invested in these road assets. Rural roads (and especially their maintenance) are politically unattractive: decision makers and funding agencies are often more focused on other more urgent and higher profile causes instead. Nevertheless, the underlying realities of rural roads and their needs remains.

2.2 The Challenges Facing the Management of Rural Roads in Many Countries

Unfortunately, those institutions responsible for managing rural roads do not help themselves. Too often, their requests for funding are poorly justified and/or are insufficiently based on objective criteria. Similarly, their accountability is equally weak. In turn, this automatically presents potential funding agencies (e.g. Ministries of Finance) with a problem, as they too are unable to justify their decisions and have inadequate assurances about how their funds are being used. It is therefore easier for them to allocate funds elsewhere. Meanwhile, the state of the rural road assets continues to deteriorate from a lack of planning, inadequate funding and a poor allocation of the resources that are available.

Objective data about the rural road assets is the very basis for providing any sufficiently justified budget request; and for providing sufficient justification for any rural road expenditures. And yet too often, rural road agencies do not even have definitive inventory information, such as the extent of their road assets (e.g. how many kilometers of paved and unpaved roads they have), nor have any reliable details about their bridges or culverts; let alone any objective information about their current condition. Very often, these road assets are not formally gazetted and are unreferenced.

These problems are compounded by weak technical capacity, especially in the more remote, local offices (where any knowledge about the rural roads should be better than in regional or national offices). Attracting the skilled staff necessary to manage these rural road assets is a constant problem and competent staff are often attracted to better opportunities elsewhere. Access to working, specialized equipment (and knowledge about how to use it) is also often very limited and its maintenance is frequently not adequate, due to a lack of funds and organizational capacity.

Addressing these challenges effectively takes a long time, sustained commitment, funding and effort. Meanwhile, the local road assets continue to deteriorate from inadequate maintenance.

3 Requirements for Any Data Collection System

For road managers, the starting point is to clarify the size of the current maintenance needs, based on the actual situation on the road network. This is the very basis for their competent management and provides a strong justification for any budget requests to meet these needs. Achieving this requires collecting and analysing data about these assets on a regular basis. Unfortunately, rural road networks are often extensive. With the often limited local technical, organisational and financial capacity available, even rudimentary data collection in a systematic and consistent standard (and at the necessary frequency) is often not undertaken as extensively as required. This significantly undermines the credibility of any plans and budget estimates made.

To maximise the likelihood that this necessary task will actually be carried out to necessary standards and frequencies, several conditions need to be in place:

- Data collection needs to be kept as simple as possible, so that it can be undertaken by local staff, who are frequently not specialised nor trained road engineers nor technicians.
- Where staff from local administrations are not available, external companies could be used instead, but consistent standards and survey monitoring is required to ensure acceptable delivery.
- Data collection also needs to be as fast as possible, to minimise the logistics and cost involved in undertaking the surveys, especially on such extensive networks.
- The need for specialised or expensive equipment or highly skilled or trained staff must be minimised, as these are simply unaffordable, unsustainable and unrealistic,

given local constraints. The availability, reliability and cost of high capacity or high speed internet or mobile phone signals is often limited or non-existent in rural areas.
- Road managers need to know about <u>all</u> the relevant assets associated with their rural roads, including bridges, culverts, drainage and (perhaps) street furniture, local land use and so on. All these features about their rural roads are relevant to the cost of providing safe rural road access.

Traditionally, data has been collected manually, using paper based forms. However, this is relatively slow, prone to errors (and falsification) and difficult to verify or monitor. It also requires transferring the data into a computer system for subsequent analysis, which is both very time-consuming and prone to transcription errors.

Modern Smartphones offer a much more flexible, robust and readily available way to capture data about the road assets, to a consistent standard, which can be more easily managed, monitored and analysed. The RoadMark system which has been continually developed since 2015, was originally developed to address these specific challenges common on rural roads. It has subsequently been trialled in Myanmar and Cambodia.

4 Approach Adopted

Road management and the data needed to support it, can be immensely complex and expensive. This is frequently beyond both the financial and technical capacity of many rural road agencies. RoadMark has been developed in an attempt to provide a sustainable and affordable approach, achieving many of the most important benefits with minimal effort, (known as the 80:20 rule or Pareto principle).

The primary purpose of the RoadMark data collection software is to provide rural road managers with a quick, network-wide assessment of the extent and basic condition of their road assets, to a level sufficient to develop medium term strategic plans and budget estimates. This includes other non-carriageway road assets (avoiding the need for separate surveys), giving road agencies a comprehensive assessment of the overall requirements for the network. These can then be reflected in the plans and budgets developed.

The level and accuracy of data collected by the RoadMark system is not therefore generally adequate for detailed designs or for Bills of Quantities to be developed: this level of accuracy and detail would be onerous for the extensive size of most rural road assets. However, it does help to locate specific projects for inclusion in annual Works Programmes and thus, specific areas where more detailed assessments might be necessary for these works. This approach helps scarce resources to be better focused and utilised by road managers.

The system also provides managers, non-technical decision makers (and the general public) with a wide range of reports and mapping options, giving a far greater understanding about the current status of the rural road network than is often currently available. These reports and maps can incorporate any or all of the road features surveyed.

When analysing and interpreting the data collected, the overall approach is to keep this as simple, as flexible and as transparent as possible, allowing local road managers to see and understand how the data they collect is reflected in their plans, budget and annual work programmes. Only the database management functions are protected (within RoadMark Analysis, RMA), as it is the most critical area to avoid accidental data corruption and loss. Nevertheless, all the data stored within the RMA database management system is fully available to local road managers in perpetuity.

Although it could be used for other road types, such as urban or national roads, RoadMark was primarily designed to be used on rural roads, where motorised traffic volumes are generally low (and are therefore of lesser importance in determining what actions to take).

4.1 Condition Ratings

Basic condition categories can be recorded for each type of road asset surveyed. These are deliberately kept as simple as possible to facilitate survey speeds and understanding, whilst minimising subjectivity. In Cambodia and Myanmar, each condition category (e.g. Good, Fair, Poor or Bad) was based on the estimated severity (i.e. cost) of any actions that would be required within the next 12–18 months, to restore the road asset into a "Good" condition. This approach was fast and relatively easy for non-engineering surveyors to understand and use. It significantly simplifies subsequent analysis and allows an estimate of the assets' current values to be made (and compared to their "As new" value), providing important and worthwhile indicators for the overall condition of the rural road assets.

4.2 Asset Referencing and Road Sectioning

In many countries, the rural road assets often lack any consistent or widespread referencing system, providing unique reference numbering for each road asset. This is a major problem for storing data about these road assets, where a common referencing system is needed to cross-reference the assets and to compare data over time. Unfortunately however, establishing and implementing any systematic and widely used referencing system for such an extensive network is a major task, requiring long-term, high level and consistent commitment to implement across all types of rural road assets.

RoadMark does not require any referencing system, allowing surveyors to travel to any part of the network and start recording data straightaway. The system assigns unique references internally and all non-carriageway data is associated with the road along which the data is collected.

The carriageway is segmented using a combination of fixed and flexible criteria, providing maximum flexibility in dividing up the carriageways into homogeneous sections:

- Fixed criteria (which are flagged by the surveyor during the surveys), are those points along a road where certain rarely changing conditions occur, such as a change in road class, geometry or surface type, the presence of an administrative boundary, intersections or changes in adjacent land use.
- Flexible criteria (which are again indicated by the surveyors onsite) occur when there is a change in the condition of the carriageway.

Unfortunately however, the lack of a common referencing system (particularly for the carriageways) means that it is currently not possible to cross-reference specific parts of the network, from one year to the next. (An overall assessment of how the network's asset groups change over time can be made however.) Consequently, it is currently necessary to recollect all the relevant data each year. However, because the surveys are kept as simple (and as fast) as possible, this additional requirement is not significant, given that condition data needs to be re-collected on a regular basis anyway. See also comments below (under future developments).

4.3 Recording Lengths and Dimensions

The RoadMark system calculates road lengths automatically, using the GPS coordinates that are continually recorded along each road. This is accurate to within about 3 m and avoids the need for surveyors to measure any road lengths manually.

Based on feedback from pilot surveys, road widths are recorded as the number or traffic lanes possible (each being 3 metres wide), as this was deemed to be easier to estimate onsite.

Bridge dimensions (e.g. span length, footpath and deck widths) need to be measured (or estimated).

4.4 Scalability and Organisation of Data Collection

RoadMark is only a tool to allow road agencies to carry out their primary tasks more efficiently and effectively. It was therefore designed to mirror and accommodate how the road agency wishes to manage its functions. For example, road data could be collected locally or by external consultants, managed regionally and/or analysed centrally.

One weakness common in many road management systems has been that those expected to collect road data differ from those using this data. Those collecting the data therefore have no interest in the data they are expected to collect. RoadMark can however, be used at the most local of levels if desired, minimising this risk and promoting decentralisation.

Over time, the amount of data collected and stored is likely to grow significantly. RoadMark has been designed to be scalable and to manage multiple databases, in order to reflect these changing needs of road managers.

5 Key Features of the Roadmark System

Figure 1 shows the basic structure of the RoadMark system and how it can integrate with external applications.

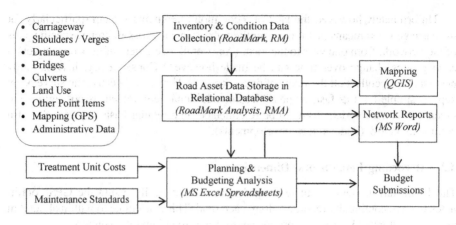

Fig. 1. Basic structure of roadmark system & interactions with external applications

- The **RoadMark smartphone application**, which allows users to collect inventory and basic condition data about a wide range of road assets:

  o **Carriageway** (pavement type, width, administrative data, condition, dual or single);

  o **Shoulders** (type, width, condition); **Culverts** (Type, size and condition); **Longitudinal drainage** (type, condition);

  o **Bridges** (including type, dimensions and condition for the key components: substructure, superstructure, parapets, carriageway and footpaths);

  o **Other point items** including signs (type and condition), trees, access points (driveways), intersection, schools, markets, major structures and river crossings;

  o **Adjacent land use** (for each side) including any corridor restrictions (e.g. buildings and water features), and earthwork details.

- The **RoadMark Analysis (RMA) Windows10 application**, into which collected data is stored securely.

6 Pilot Surveys

In surveys undertaken on rural roads in Myanmar and Cambodia (covering a total of approximately 1,000 km), the data stored in the RMA application was exported and analysed in Excel, to develop simple strategic network plans and budget estimates. This data was also mapped using the freely available, cross-platform QGIS software, (the data can also be imported into other common GIS applications).

These projects have provided valuable feedback that has guided subsequent developments of the system. The main findings from the surveys were:

- Training (which was provided onsite – any nearby roads can be used), only required a few hours before local surveyors (many of whom spoke limited English) were sufficiently familiar with the main features of the Android Smartphone App.

- The application was considered to be very easy to understand and use.
- Surveyors only required a GPS-enabled Android Smartphone and a vehicle/motorcycle, both of which are widely available, even in remote rural areas.
- Survey teams usually consisted of one surveyor and one vehicle operator only.
- Survey speeds varied (depending on the operator and the number of road features to be recorded on the network) but averaged between 15–20 kph.
- No local internet or mobile phone signal was required to use the RoadMark application, only a GPS signal.
- The application worked well, even if the data collection surveys were interrupted (for example from an empty battery or incoming call). The application was reliable and stable, storing the recorded data safely on the Smartphone (for later transfer).
- Data requirements were very small (about 10-15kb per kilometre surveyed), making the transfer of data from site to remote head office (as email attachments), simple and fast, even when internet connectivity was poor.
- All the data was time and date-stamped and geolocated, avoiding any falsification of the data. This also allowed survey managers to easily monitor and manage the data collection surveys. (RMA provides a report summarising the surveys carried out).
- The importation of survey data into the database application was fast and simple. This worked well, even when located in a different country to where the surveys were undertaken. (A batch loading option has been added subsequently.)
- A progressively larger number of reports and data outputs have been included from RMA, to increase the range of data analysis options available.
- Database integrity worked well and was scalable. (An ability for the user to switch between different databases has been provided subsequently.)

7 Analyis of Data

The data collected during the surveys was exported from the RMA database and used to develop:

- Needs-based network maintenance plans (by asset class, administrative area, type of works, etc.);
- Budget and maintenance backlog requirements (using the same categories above);
- Development of various key performance indicators (KPIs), including an estimate of the network's current (and optimum) values and various condition indices;
- Sensitivity testing, to identify the impact of changing treatment standards and unit costs on the overall budget and works volumes (Figs. 2, 3 and 4).

Introduction to the RoadMark System Low Cost 715

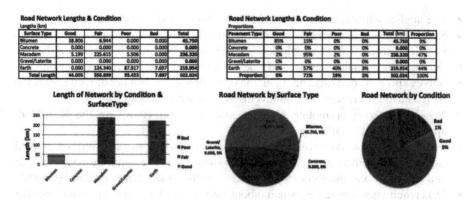

Fig. 2. Example of current surveyed network summaries

Fig. 3. Development of scheduled and backlog maintenance requirements

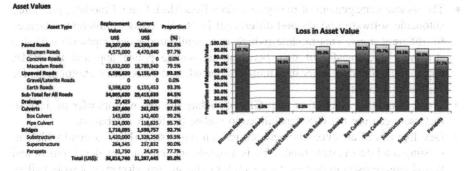

Fig. 4. Example of asset values reporting possible from RoadMark system

Extensive and flexible reports and maps were also developed, providing road managers with a comprehensive assessment and reporting capability for their rural road assets. These all represented a considerable improvement on their previous planning and reporting capabilities (Fig. 5).

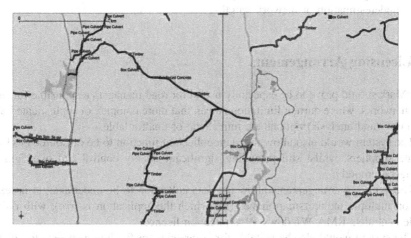

Fig. 5. Example of mapping of rural road assets using RoadMark data

8 Further System Developments

RoadMark was designed to be as simple and as robust as possible, reflecting the realities of many rural environments in many countries. Nevertheless, continuing advances in mobile technologies, together with changes in communication and computers capabilities mean that the software requires continual development to keep up with these changes and make use of the new opportunities they present. The system also continues to undergo developments as resources allow, to extend its usability and capabilities, in response to user comments. At present, the following areas are under consideration or development:

- The remote management of the system via a RoadMark Central module, providing automatic software updates and differential levels of access to different databases for different users. This module (which is currently under development) will also streamline the transfer of data between the mobile phone app and the RMA application, allowing larger data transfers (including the storage of photographs for various road assets).
- Tighter/smoother integration with GIS (including the two-way transfer of data, so that users can use the GIS capabilities to update the RMA databases).
- Development of an automatic network referencing system (NRS) around which all existing road data is structured. This is a significant and complex development and would require users to provide various rules to be applied. However, it would allow data on the network to be cross-referenced over time and allow previous survey data to be downloaded to the mobile application, further reducing future data entry requirements.
- Other developments include proposals to
  o Provide the ability to quickly define the network, at normal driving speeds;
  o Overcome the currently limited accuracy in the vertical (altitude) reading of the GPS system, which in more hilly or mountainous areas can result in a noticeable underestimation of network lengths.

9 Licensing Arrangements

RoadMark should prove to be especially useful for road managers responsible for rural road networks, where current limitations mean that more complex or sophisticated data collection (and analysis) options are unrealistic or unaffordable.

The system would also allow this data collection function to be outsourced to local private suppliers, whilst still providing significant quality control and oversight by managers remotely.

The licensing arrangements allow the data collection to be undertaken at no extra cost on multiple mobile smartphones (for which the application is free), with only a single centralise RMA Windows 10 application licence.

The system would also be suitable for companies (or funding agencies) to quickly assess or check the current condition and extent of the various road assets. This can be useful for funding and monitoring purposes.

Licensing can be provided for permanent use or for a limited time at lower cost. (The user will however, always have access to any data that they have already collected and stored in the RMA database). Multiple licensing discounts are available for larger users. These arrangement are designed to be as flexible as possible, to meet the specific needs of different clients.

10 Conclusions

The RoadMark system was developed because no other cost-effective data collection systems appeared to effectively address the most common institutional, organisational, funding and technical weaknesses that are endemic in managing rural road assets in many countries. A trade-off is always necessary between data collection requirements and its accuracy and usefulness. RoadMark aims to maximise the use of any road data, given what can realistically be collected in the situation prevailing on many rural road networks. Its simplicity and fast survey speeds, together with avoidance of specialised equipment, excessive training or the need for any network referencing system to be established beforehand, means that road managers should be able to collect a considerable amount of necessary data about their road assets, allowing far more robust and justified, needs-based strategic plans and budget estimates to be developed quickly and simply.

Following use in Cambodia and Myanmar, the system continues to undergo further developments, in order to provide additional flexibility and usability for road managers, without increasing its complexity or data demands for users.

Similarity Between an Optimal Budget in Pavement Management & an Equilibrium Quantity of Demand-Supply Analysis in Economics

Ponlathep Lertworawanich[✉]

Department of Highways, Bangkok, Thailand
ponla.le@doh.go.th

Abstract. The Department of Highways (DOH) is a state agency under the Ministry of Transport, Thailand. The DOH has developed a pavement management system called the Thailand Pavement Management System where the International Roughness Index (IRI) is a serviceability measure index for highway maintenance performance evaluation. The DOH sets the thresholds of IRI at 3.50 m/km for the regular highway network with Annual Average Daily Traffic (AADT) less than 8,000 and at 3.00 m/km for the major trunk highway network with AADT of 8000 or above. Contrary to the traditional approach that specifies the output state, for instance the average IRI, of the network and then back-calculates the required budget to attain that state, this study presents a new method based on the so-called cost-benefit analysis of economics that optimizes both the required budget and the output state of the highway network to achieve the maximum net benefit of highway maintenance. The optimal budget is shown to have similar characteristics to the equilibrium quantity of goods of demand-supply analysis in economics. This study also proposes a new model to optimally allocate maintenance budgets between two highway networks to maximize the total net benefit for a given budget constraint. It is found that at the optimum the maintenance budget should be allocated in such a way that the marginal net benefit divided by the marginal cost should be constant for both networks. The proposed methodology has demonstrated itself to be a potential model for budget planning of the pavement management system.

Keywords: Optimal budget · Cost-benefit analysis · Demand-supply analysis · Highway maintenance

1 Introduction

The Department of Highways (DOH) is a state agency under the Ministry of Transport, Thailand. It is responsible for planning, designing, constructing, and maintaining national highways throughout Thailand. Efficient and adequate highway systems are an important driving force for economic and social development. Roads open up more areas and stimulate economic and social development. On the other hand, rapid economic growth has led to a state of economic dependence and deterioration of natural resources. For Thailand, the 1997 economic crisis served as a costly lesson of

© The Author(s), under exclusive license to Springer Nature Switzerland AG 2022
A. Akhnoukh et al. (Eds.): IRF 2021, SUCI, pp. 719–734, 2022.
https://doi.org/10.1007/978-3-030-79801-7_52

unbalanced and unstable growth, partly due to the improper economic and social development process. In some countries, due to the economic crisis, the trend of budgetary pressures on highway agencies is increasing. At the same time, road users are increasingly demanding in terms of highway quality, efficiency and safety. Several highway maintenance and rehabilitation projects have been delayed because of budget constraints. The economic crisis has also stimulated a wider debate about the state of the road network infrastructure and the consequences of past large investment in new construction and under-investment in projects. During the late 1990s economic crisis in Thailand, the construction of new highways almost ceased and the scarce funds available were used essentially for maintenance and rehabilitation of existing highways.

Pavements gradually deteriorate as they are opened to traffic. Highway authorities must maintain proper serviceability of highways to ensure efficiency and safety for public use. However, highway maintenance management with restricted budgets is very challenging. At present, the national network of Thailand under the DOH consists of 58,000 km (actual length) of highways. To maintain these highways, there are more than 130 highway districts performing maintenance tasks on a daily basis. These districts are located in all regions of Thailand. All highways deteriorate at different rates and exhibit different distressed conditions depending on their utilization and environmental, weather, and geographical conditions as well as other factors such as natural disasters. Therefore, maintenance activities to restore highways to a good serviceable condition must be systematically planned and managed so that every baht (Thai currency) spent on them must economically return benefits to the nation.

As a result of restricted budget, there is an urgent need for an efficient pavement management system that can provide cost-effective solutions to the required highway maintenance activities. Haas et al. (1994) mentioned that good pavement management is not business as usual. Pavement management, in its broadest sense, includes all the activities involved in the planning and programming, design, construction, maintenance, and rehabilitation of the pavement portion of a public work program. It is a set of tools or methods that assist decision makers in finding optimal strategies for providing and maintaining pavements in a serviceable condition over a given period of time. Cook (1984) applied goal programming technique in highway rehabilitation from the capital budgeting approach. Shafizadeh and Mannering (2003) investigated the public attitude toward acceptable levels of road roughness on urban highways. Their findings were compatible with the FHWA IRI guideline of 2.70 m/km for the acceptable ride quality. Khan et al. (2016) employed the HDM-4 model to establish road maintenance plans for the Queensland region by setting the acceptable IRI at 4.0 m/km. Hu et al. (2017) presented a mathematical relationship between IRI and driving workload of drivers. They found that standard IRI values for pavement maintenance in China were beyond the comfort and safety thresholds for both car and truck drivers. Denysiuk et al. (2018) presented an optimization model for road asset management by incorporating degradation by searching an optimal maintenance scheduling that minimizes both the asset degradation and maintenance cost. Lee and Lee (2018) examined two asset valuation methods by comparing the Government Accounting and Standard Board Statement 34 (GASB34) and the depreciation replacement cost (DRC) methods. They found that the method based on DRC provided a higher current value than does the modified GASB34 method. Mensah et al. (2019) investigated social equity in highway maintenance by considering the Gini coefficient and the Theil index.

In the 1980s, the DOH started using a pavement management system based on the World Bank system called the Highway Development and Management model (HDM). Along the course of the three decades, the DOH has developed its own pavement management model called the Thailand Pavement Management System (TPMS). The core element of the TPMS is based on the HDM model but modified to suit Thailand environments and requirements. The TPMS is capable of budget analysis and budget allocation according to pavement distressed conditions and appropriate engineering measures. The TPMS has an embedded algorithm with an objective to maximize benefits of the selected maintenance activities on each highway portion under budget constraints. The system can be used to analyze both operational plans (one-year plans) and strategic plans (multi-year plans) for highway networks. The analysis provides information on optimal maintenance activities on individual highway sections associated with the relevant maintenance costs as well as the resulting international roughness index (IRI) values. The corresponding benefits to road users are calculated for each selected maintenance treatment and each highway section. Generally, benefits of road maintenance consist of both road-user and non road-user benefits. Road-user benefits occur as a decrease in the cost of vehicle operations, a decrease in time spent on the road, a decrease in accident rates, or a decrease in the strain and discomfort of nonuniform driving. Non road-user benefits can accrue as an increase in land values and stimulation of economic growth along a better road. Most studies consider only the road-user benefits. The benefits are estimated by comparing the amount of travel time, vehicle operating cost and expected number of road accidents for the alternative to the base case. Similar to any other transportation investment projects, for road maintenance projects, the benefits come from three components in monetary units, 1) vehicle operating cost saving, 2) travel time saving, and 3) safety benefits, due to better road conditions or lower IRI. However, the TPMS includes only the first two components because currently there is no accurate accident prediction model available in Thailand.

The primary factor influencing serviceability of roads is the roughness of the road surface. Roughness evaluation is crucial to the pavement management process as it provides highway authorities with a direct measurement influencing the public's perception of the quality of service provided by the pavement. A universal roughness standard has been the subject of extensive discussion and the International Roughness Index (IRI) is widely used internationally. It is the accepted choice of the DOH and is used by many other road agencies in Thailand. In the metric system, IRI is expressed in m/km. Users of the measure must still make decisions as to what is an acceptable level of roughness. Mucka (2017) summarized IRI limit values for new, reconstructed, or rehabilitated roads; for existing roads; and road classification schemes used around the world where limit values are a function of road surface type, road functional category, road speed limit, road construction type, or Annual Average Daily Traffic (AADT). Initially, the DOH specified the level of IRI at 3.50 m/km as a threshold to distinguish between acceptable and unacceptable pavement conditions. Different countries set different IRI thresholds. In case of Thailand, the threshold is set according to the study by the DOH to maintain the balance between the maintenance budget the DOH typically acquires and the serviceability the DOH can deliver to public. The IRI threshold at 3.5 m/km is only used for the national highway network under responsibility of the DOH. Different IRI thresholds are set for other roads such as local roads or rural roads under different road authorities. Until recently the DOH has started using different IRI thresholds for different functional classification of highways based on AADT. The threshold for regular highways with AADT smaller than 8,000 is 3.50 m/km while that for major trunk highways with AADT of 8000 or above is 3.00 m/km. There are 31,400 km (actual length) of regular highways and 26,400 km (actual length) of major trunk highways in Thailand. Setting a stricter IRI threshold would surely result in a better road condition, nonetheless, with more pressure on budget requirements. If there had been no budget restriction, the DOH could have maintained the entire national highway network within the acceptable IRI thresholds or even better. But, as a matter of fact, Thailand is a developing country whose budget must be allocated to a multitude of development activities that are deemed important to society as a whole. Therefore, the question to be answered is how much of the highway network should be maintained to meet their specified IRI thresholds so that the state welfare or net benefit of maintenance will be at optimum.

For a given budget as an input, the TPMS has the objective to maximize the benefit from road user cost reduction due to highway maintenance activities. With the given budget input, the TPMS provides a one-to-one relationship between the resulting benefit and the resulting state of the highway network. This resulting state is the percentage of the highway network length with IRI no greater than the specified thresholds. However, the TPMS cannot directly give a compromising solution that provides a balance between the required budget and the gained economic benefit. In fact, the budget itself is a variable of interest for road authorities. By specifying the resulting state of the highway network, then back-calculating the required budget will not lead to an optimal condition as long as we do not know the optimal economic impact of road maintenance.

In this study, we attempt to find an optimal budget that produces an optimal economic benefit of highway maintenance that leads to an optimal maintained state of the highway network. It provides information on the optimal percentage of the regular and major trunk highway network lengths that should be maintained to meet their IRI thresholds. Please note that it is not necessary to have all highways meet their IRI thresholds to produce the maximum net economic benefit. For instance, when a marginal cost exceeds a marginal benefit, further maintenance will not produce a higher net economic benefit. The optimal maintenance budget is shown to have similar characteristics to the equilibrium quantity of goods of demand-supply analysis in economics. The concept of social surplus in economics is applied to define an economic effect of the scarcity of maintenance budget in a monetary unit. This study also examines an optimal allocation of maintenance budget between the regular and the major trunk highway network when an available budget is less than the required optimum. A thorough literature search finds no existing system proposes the similar concept to the one proposed in this study. This paper presents a new way of looking at highway maintenance that balances economic and engineering perspectives of pavement management. With this notion, the objectives of this study are:

- To perform analysis on highway maintenance activities and required budgets with the TPMS,
- To propose a new method that gives optimal budgets and optimal percentage of the highway network length to be maintained to meet their specified IRI thresholds based on an economics concept,
- To propose a new optimal maintenance budget allocation method between the regular and the major trunk highway networks that produces the maximum net benefit based on an economics concept.

In the remainder of the paper, the model development is presented where an optimal highway maintenance budget model is formulated using an economics concept of cost-benefit analysis. Next, an optimal maintenance budget allocation model between the regular and the major trunk highway networks is proposed. Then, the following section is results and discussions where a case study with Thailand data is examined, followed by conclusions and future research recommendations.

2 Model Development

Boardman et al. (2011) mentions that cost-benefit analysis is a policy assessment method that quantifies in monetary terms the value of all consequences of a policy to all members of society. The objective is to have more efficient allocation of society's resources. In economics, the quantity demanded is the amount of a good that consumers are willing to buy at a given price during a specified period, holding constant the other factors that influence purchases. The quantity demanded of a good or service can exceed the quantity actually sold. The vertical axis (labeled Price as shown in Fig. 1) of the demand curve can be interpreted as the highest price someone is willing to pay for an additional unit of the good. A standard assumption in economics is that demand curves slope downward which is based on the principle of diminishing marginal utility where each additional unit of the good is valued slightly less by each consumer than the preceding unit. Thus, the area under the demand curve measures the total benefits society would receive from consuming X^* units of good X. In a competitive market, consumers pay the market price, P^*. Thus, consumers spend P^*X^* to consume X^* units. The net benefit to consumers equals the total benefits minus consumers' actual expenditures, (P^*X^*). This area below the demand curve but above the price line, is called consumer surplus (CS) as shown in Fig. 1. For the supply curve, the quantity supplied is the amount of a good that firms want to sell during a given period at a given price, holding constant other factors that influence supply decisions, such as costs and government actions. The supply curve indicates the marginal cost of each additional unit of the good produced. Thus, the area under the marginal cost curve represents the total variable cost of producing a given amount of good X, say X^*. Suppose that the market price of a good is P^* and, consequently, firms supply X^* units. Their revenue in dollars would be P^*X^*. Producer surplus (PS) measures the benefit going to firms. It equals the difference between actual revenues and the minimum total revenue that firms must receive before they would be willing to produce X^* units at a price of P^*, the area under the supply curve as shown in Fig. 1.

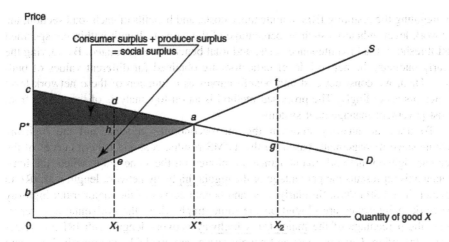

Fig. 1. Social surplus at equilibrium between demand and supply

The sum of consumer surplus and producer surplus is called social surplus (SS). Social surplus is illustrated in Fig. 1, which depicts both a demand curve and a supply curve in the same graph. Equilibrium occurs at a price of P* and a quantity of X*. Consumer surplus is the area caP*, producer surplus is the area P*ab, and social surplus is the sum of these areas, cab. Now, net social benefits equal the difference between total consumer benefits and total producer costs. Total consumer benefits equal the area under the demand curve, caX*0, while total costs equal total variable costs, the area under the supply curve, baX*0. The difference is the area, cab. This makes it clear that social surplus equals net social benefits. Remembering that the demand curve reflects marginal benefits and the supply curve reflects marginal cost, at the competitive equilibrium demand equals supply and marginal benefits equals marginal cost. Therefore, net social benefits are maximized. Either positive or negative deviations from the optimal quantity, X*, will reduce social surplus.

2.1 Optimal Maintenance Budget Estimation

Regarding the highway maintenance management, with the outlined microeconomics concept above, to construct cost and benefit curves of the highway maintenance, this study uses the TPMS. It requires inputs such as road network inventories, distressed conditions (IRI, rutting, and cracking), traffic loading conditions, units of costs of each maintenance treatment, and yearly maintenance budgets. Different maintenance treatments have different impacts of the life extension of pavements. The life extension of pavements also has impacts on cost and benefit of the maintenance. The process of selecting treatments is performed at road-section level of the optimization process in the TPMS. The TPMS outputs for *each road section*: maintenance costs, resulting IRIs, maintenance treatments, and benefits from savings of road user cost due to the reduction of travel times and vehicle operation costs. Like most pavement management systems, there are deterioration models of IRI, cracking, and rutting in the TPMS. By aggregating the resulting IRIs, maintenance costs, and benefits of each road section, the network-level indicators such as percentage of the network length within the specified IRI threshold, total maintenance costs, and total benefits are calculated. By varying the yearly budgets, the network-level indicators are obtained for different values of budgets. Then, we construct cost and benefit curves as a function of these network-level indicators accordingly. The proposed method is an on-top analysis of the results from most pavement management systems.

To draw an analogy between the microeconomics concept and the highway maintenance management, based on the TPMS results, cost and benefit curves of the regular highway network maintenance are plotted on the same graph where the horizontal axis represents the percentage of the regular highway network length with IRI no greater than 3.50 m/km. Similarly, cost and benefit curves of the major trunk highway network maintenance are plotted on the same graph where the horizontal axis represents the percentage of the major trunk highway network length with IRI no greater than 3.00 m/km. Once the cost and benefit curves are available, the marginal cost and benefit curves are calculated as the derivative of each. The optimum is where the marginal cost and the marginal benefit curves intersect as shown in Fig. 2.

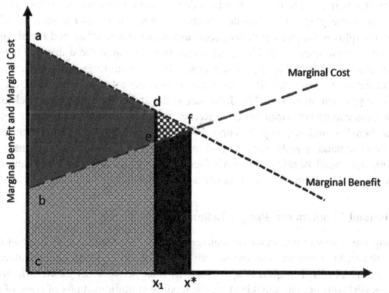

Fig. 2. Marginal benefit and marginal cost of highway maintenance

Let us consider Fig. 2. Suppose we maintain a highway network length to meet the IRI threshold up to x* percent. The benefit of the highway maintenance is the area of the trapezoid(afx*c) under the marginal benefit curve while the cost of the highway maintenance is the area of the trapezoid(cbfx*) under the marginal cost curve. The net benefit (net worth) of highway maintenance is the difference between the benefit and the cost or the area of △abf. If we maintain x_1 percentage of the highway network length with IRI no greater than the specified threshold, then the net benefit equals the area of trapezoid(abed). In fact, we can mathematically derive an optimal condition as follows.

$$\text{Net Benefit (NB)} = \text{Benefit} - \text{Cost}$$
$$= \int_0^x MB(z) - MC(z)dz \qquad (1)$$
$$\frac{d(NB)}{dx} = MB(x) - MC(x) = 0 \rightarrow MB(x) = MC(x)$$

Where $MB(x)$ = the marginal benefit function of the highway maintenance
$MC(x)$ = the marginal cost function of the highway maintenance

Based on this conceptual analysis, the optimum or the maximum net benefit occurs when the percentage of the highway network length with IRI no greater than the specified threshold is x* or the intersection point of the marginal benefit and the marginal cost curves. Any deviations from the optimal point will reduce the net benefit. For example, if we maintain x_1 percentage of highway network length with IRI no greater than the specified threshold, then the net benefit equals the area of the trapezoid(abed). The area of △def is defined as a loss due to a deviation from the optimal net benefit, similar to the dead weight loss (DWL) defined in welfare economics. Since there are two types of the highway networks, the regular highway network (AADT < 8,000) and the major trunk highway network (AADT ≥ 8000), the optimal maintenance budgets for each network can be individually estimated using the proposed method. The total optimal budget is the sum of two optimal budgets from each type of the highway networks. Note that in Thailand road maintenance budget is planned and approved on a yearly basis; therefore, this study focuses on one-year budget analysis. However, the proposed method will be further extended to multi-year analysis in future research.

2.2 Optimal Budget Allocation Between Two Highway Networks

When an available budget is less than the sum of the optimal budgets from two types of highway networks, the budget should be allocated to each network in a way that the sum of the net benefits is maximized. In other words, the sum of the dead weight losses, as defined in Fig. 2, is minimized given that the sum of maintenance costs of each network does not exceed the total available budget. This budget allocation problem can be mathematically expressed as shown below.

$$Z = MIN \left\{ \int_{x_1}^{x^*} (MB_1(x) - MC_1(x))dx + \int_{x_2}^{x^{**}} (MB_2(x) - MC_2(x))dx \right\} \qquad (2)$$

$$\text{Subject to } C_1(x_1) + C_2(x_2) = B \qquad (3)$$

Where $MC_1(x)$ = Marginal cost function of the regular highway network
$MC_2(x)$ = Marginal cost function of the major trunk highway network
$MB_1(x)$ = Marginal benefit function of the regular highway network
$MB_2(x)$ = Marginal benefit function of the major trunk highway network
$C_1(x)$= Cost function of the regular highway network
$C_2(x)$= Cost function of the major trunk highway network
B = Available maintenance budget
x_1 = Percentage of the regular highway network length with IRI no greater than 3.5 m/km
x_2 = Percentage of the major trunk highway network length with IRI no greater than 3.0 m/km

Equation (2) represents the objective function of minimizing the sum of dead weight losses due to deviations from individual optimal budgets of the regular and major trunk highway networks. Equation (3) is the budget constraint indicating that the sum of maintenance costs associated to each network equals the available budget. The set of Eqs. (2) and (3) can readily be solved using Lagrange Multiplier technique as follows.

$$\mathcal{L}(x_1, x_2, \lambda) = \int_{x_1}^{x^*} (MB_1(x) - MC_1(x))dx + \int_{x_2}^{x^{**}} (MB_2(x) - MC_2(x))dx + \lambda(B - C_1(x_1) - C_2(x_2)) \quad (4)$$

By differentiating the Lagrangian function in Eq. (4) respect to x_i and λ, and equating each to zero, we obtain:

$$\frac{\partial \mathcal{L}}{\partial x_i} = MB_i(x_i) - MC_i(x_i) - \lambda \frac{\partial C_i(x_i)}{\partial x_i} = 0; \forall i = 1, 2 \quad (5)$$

$$\frac{\partial \mathcal{L}}{\partial \lambda} = B - C_1(x_1) - C_2(x_x) = 0 \quad (6)$$

$$\lambda = \frac{MB_i(x_i) - MC_i(x_i)}{MC_i(x_i)} \quad (7)$$

Taking a closer look at the resulting Eqs. (6) and (7), it means that at the optimal allocation the maintenance budget should be allocated in a way that the marginal net benefit divided by the marginal cost of each highway network type should be the same value for both the regular and the major trunk highway networks. Otherwise, the budget should be transferred to the network with the higher ratio of the marginal net benefit over the marginal cost because it will produce a higher net benefit. In the next section, an application of the proposed model is presented with Thailand highway maintenance data.

3 Results and Discussions

The proposed methodology is applied to Thailand highway maintenance data to analyze the required maintenance budget in 2020 fiscal year. Optimal maintenance budgets and their corresponding optimal percentage of the regular and the major trunk highway network lengths with IRI no greater than their corresponding thresholds are calculated when there is no budget constraint. Then, we will investigate different scenarios when available budgets are less than the optimal one. Initial conditions in 2019 are as follows. There are two highway networks under the DOH responsibility classified by AADT, namely 1) the regular highway network of 31,400 km (highways of this type have AADT < 8000), and 2) the major trunk highway network of 26,400 km (highways of this type have AADT $\geq$ 8000). For existing conditions, 81.452% of the regular highway network length and 72.865% of the major trunk highway network length meet their IRI serviceability thresholds. The regular highway network carries 44,447.71 × 10^6 veh-km while the major trunk highway network carries 214,610.19 × 10^6 veh-km.

3.1 Results of Optimal Maintenance Budgets

Cost, benefit, and net benefit curves of maintenance for the regular highway network from the TPMS analysis are plotted in Fig. 3. The net benefit curve is concaved with a unique maximum. To obtain the marginal cost and the marginal benefit curves, regression lines that provide the best-fits to the cost and benefit curves are calculated and their derivatives are computed as shown in Fig. 4. The intersection point of the marginal cost and the marginal benefit curves for the regular highway network is at 85.75% of the regular highway network length with IRI no greater than 3.50 m/km. The optimal maintenance budget is equal to 21.27 billion bahts and the optimal benefit is equal to 40.04 billion bahts. The maximum net benefit is the area between the marginal benefit and the marginal cost curve or A_1 in Fig. 4 and is equivalent to 18.77 billion bahts. The benefit-to-cost (B/C) ratio at the optimal condition of the regular highway network is 1.88. Similarly, we apply the proposed methodology to the major trunk highway network. Cost, benefit, and net benefit curves are plotted in Fig. 5. It is clear that there is a unique maximum net benefit. To obtain marginal cost and marginal benefit curves, regression lines that provide the best-fits to the cost and benefit curves are calculated and their derivatives are computed as shown in Fig. 6.

The intersection point of the marginal cost and the marginal benefit curves is at 76.36% of the major trunk highway network length with IRI no greater than 3.00 m/km. The optimal budget is equal to 50.18 billion bahts and the optimal benefit is equal to 221.89 billion bahts. The maximum net benefit is the area between the marginal benefit and the marginal cost curve or A_2 in Fig. 6 and is equivalent to 171.7145 billion bahts. The optimal maintenance budget for the major trunk highway network is greater than that for the regular highway network. This is because initially the major trunk highway network has less percentage of the network length within the acceptable IRI threshold than the regular highway network does. In addition, higher traffic (214,610.19 × 10^6 veh-km) on the major trunk highway network will likely induce higher benefit of maintenance when compared to the regular highway network (44,447.71 × 10^6 veh-km). The benefit-to-cost (B/C) ratio at the optimal condition of

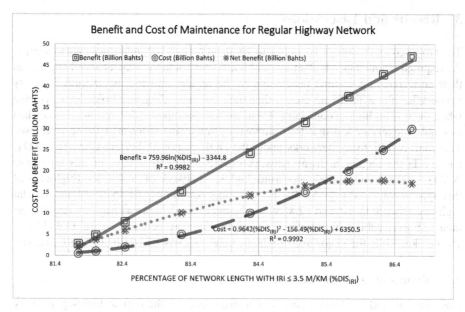

Fig. 3. Benefit and cost of maintenance for the regular highway network

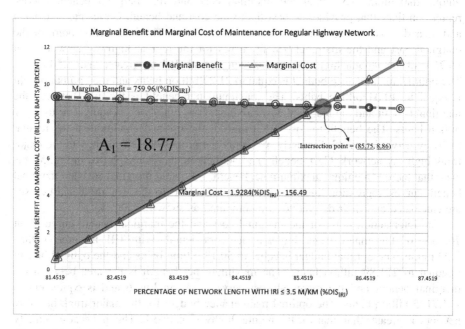

Fig. 4. Marginal benefit and marginal cost of maintenance for the regular highway network

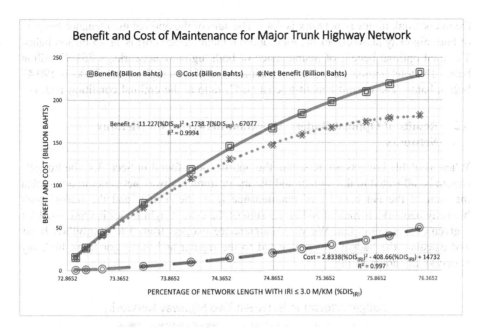

Fig. 5. Benefit and cost of maintenance for the major trunk highway network

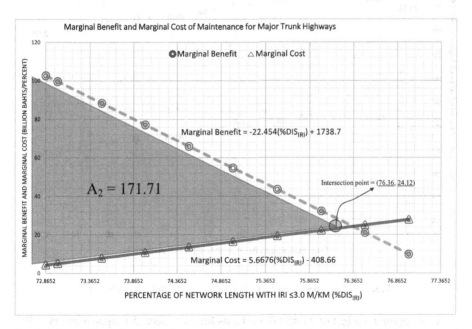

Fig. 6. Marginal benefit and marginal cost of maintenance for the major trunk highway network

the major trunk highway network is 4.42. The sum of the optimal maintenance budgets of both highway networks or the total optimal maintenance cost is 71.45 billion bahts. The total optimal benefit of maintenance of both highway networks is 261.93 billion bahts. The total optimal net benefit of maintenance of both highway networks is 190.48 billion bahts. The overall benefit-to-cost (B/C) ratio at the optimal condition is 3.67.

3.2 Results of Optimal Budget Allocation Between Two Highway Networks

When available budgets are less than the optimal ones found in Sect. 3.1, the budgets should be allocated to the regular and the major trunk highway networks in a way that the sum of the net benefits of maintenance from the regular and the major trunk highway networks is maximized or the sum of the dead weight losses is minimized for a given budget constraint. We vary maintenance budgets from 5 to 100 billion bahts and calculate the budget splits allocated to the regular and the major trunk highway networks as shown in Fig. 7.

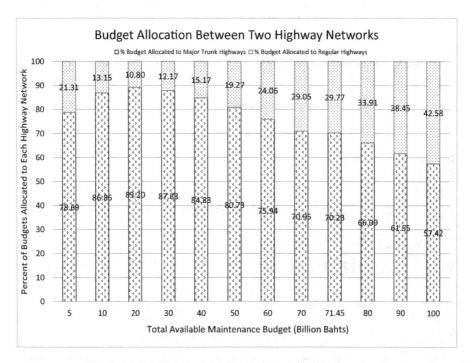

Fig. 7. Maintenance budget allocated to each highway network

At the optimal budget of 71.45 billion bahts found in Sect. 3.1, approximately 70% of the total budget is allocated to the major trunk highway network while roughly 30% is allocated to the regular highway network. This produces the maximum net benefit of 190.48 billion bahts. For different total available budgets, the percentage of budgets allocated to the regular highway network varies from 10 to 40% of the total available budget and the rest goes to the major trunk highway network. It is clear that the net benefits of maintenance of the major trunk highway network are much higher than those of the regular highway network. This is because the major trunk highways accommodate higher traffic volumes which generate higher benefits than the regular highways do. We also investigate benefit-to-cost (B/C) ratios for different total available budgets as shown in Fig. 8.

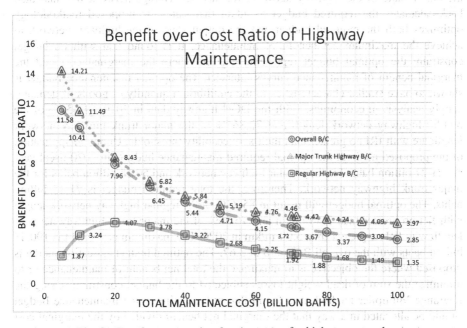

Fig. 8. Benefit-to-cost ratio of maintenance for highway networks

The B/C ratios are always greater than one within the budget range from 5 to 100 billion bahts. The overall B/C ratios and the B/C ratios of the major trunk highway network are decreasing functions of the total available budgets within the budget range of interest. However, the B/C ratios of the regular highway network are increasing with the total available budget up to 20 billion bahts. When the total available budget is greater than 20 billion bahts, the B/C ratios of the regular highway network are decreasing with the total available budget. Please note that if the B/C ratios of each highway network type are plotted against its own budget (not total available budget), they become decreasing functions. This is because the marginal benefit is a decreasing function while the marginal cost is an increasing function of its own budget as shown in Figs. 3 and 5.

4 Conclusions and Future Research

In this study, a new methodology to identify an optimal highway maintenance budget is proposed. The methodology is based on an economics concept of cost-benefit analysis. The proposed model attempts to find an optimal maintenance budget that maximizes net benefit of highway maintenance. Contrary to the traditional approach that specifies the output state of the highway network, for instance average network IRI, and then back-calculates the required budget to attain that state, the proposed methodology optimizes both the required budget and the output state of the highway network to achieve the maximum net benefit of maintenance. It is found that with no budget constraint the optimal budget represents a point where the marginal cost and the marginal benefit of maintenance curves intersect. The optimal maintenance budget is shown to have similar characteristics to the equilibrium quantity of goods of demand-supply analysis in economics. With the DOH highway data in 2019, 81.452% of the regular highway network length and 72.865% of the major trunk highway network length are with IRI no greater than their serviceability thresholds. Upon the application of the proposed method, the optimal required maintenance budgets in 2020 fiscal year are 21.27 billion bahts for the regular highway network and 50.18 billion bahts for the major trunk highway network. Therefore, the total required budget is 71.45 billion bahts. The optimal budget will result in 85.75% of the regular highway network length with IRI no greater than 3.50 m/km and 76.36% of the major trunk highway network length with IRI no greater than 3.00 m/km in 2020. The maximum net benefit is 190.48 billion bahts with the overall B/C ratio of 3.67. The optimal budget allocation model is proposed where the objective is to maximize the total net benefit of maintenance or to minimize the sum of dead weight losses subjected to the budget constraint. Using the Lagrange Multiplier method, it is found that at the optimum the maintenance budget should be allocated in a way that the marginal <u>net</u> benefit divided by the marginal cost of each highway network type should be the same value for both the regular and the major trunk highway networks. The B/C ratios for different budget scenarios are calculated. The resulting overall B/C ratios for the total budget range of 5 to 100 billion bahts are well above one and a decreasing function of the total budget. The proposed model lends an economics concept of cost-benefit analysis to answer questions of optimal budgets and optimal budget allocation of highway maintenance without specifying the output state of the highway network. Instead, the output state is treated as a variable to be found. In future, we plan to use the proposed method for optimal budget planning of the next Thailand fiscal year. Based on the findings of this paper, some road maintenance policy implications are identified as follows.

- It is not necessary to maintain all highways to their best serviceable conditions because it does not produce the maximum net benefit from an economics point of view.
- There exists an optimal state of the highway network that produces the maximum net benefit of maintenance and it should be the target for maintenance budget planning.
- The required budget to achieve the optimal state is the optimal budget which can theoretically be found using the proposed cost-benefit analysis method. The optimal

budget is shown to have similar characteristics to the equilibrium quantity of goods of demand-supply analysis in economics
- The optimal budget found is the upper bound for budget consideration because the higher budget will not result in the higher net benefit.
- Similar to social welfare economics, budget decision-makers can weigh economic effects of the scarcity of road maintenance budgets in monetary units using the proposed method.

In addition, we find some research gaps that should be addressed in future as follows.

- Majority of the maintenance budgets are allocated to the major trunk highway network because of its high traffic usages which produce higher benefits. Small portions of the maintenance budgets are allocated to the regular highway network. There will be some small roads, with small traffic volumes but with severe damages, which are not selected for maintenance due to lower benefits. This situation should be taken into consideration. Therefore, equity of maintenance between the major trunk highway and the regular highway networks should be systematically addressed in future research.
- Extend the proposed method to multi-year budget planning.

References

Boardman, A.E., Greenberg, D.H., Vining, A.R., Weimer, D.L.: Cost-Benefit Analysis: Concepts and Practice. Prentice Hall, Upper Saddle River (2011)
Cook, W.D.: Goal programming and financial planning models for highway rehabilitation. J. Oper. Res. Soc. **35**(3), 217–223 (1984)
Denysiuk, R., Fernandes, J., Matos, J., Neves, L.C., Berardinelli, U.: A computational framework for infrastructure asset maintenance scheduling. Struct. Eng. Int. **26**(2), 94–104 (2018)
Haas, R., Hudson, W.R., Zaniewshi, J.: Modern Pavement Management. Krieger Publishing Company, Malabar (1994)
Hu, J., Gao, X., Wang, R., Sun, S.: Research on comfort and safety threshold of pavement roughness. Transp. Res. Rec. **2641**, 187–193 (2017)
Khan, M.U., Mesbah, M., Ferreira, L., Williams, D.J.: Development of optimum pavement maintenance strategies for a road network. Aust. J. Civ. Eng. **16**(2), 85–96 (2016)
Lee, D.Y., Lee, M.J.: A study of the asset valuation method for efficient road facility maintenance. J. Asian Arch. Build. Eng. **13**(2), 279–286 (2018)
Mensah, J.F., Kothari, C., O'Brien, W.J., Jiao, J.: Integrating social equity in highway maintenance and rehabilitation programming: a quantitative approach. Sustain. Cities Soc. **48**, 1–12 (2019)
Mucka, P.: International roughness index specifications around the world. Road Mater. Pav. Des. **18**(4), 929–965 (2017)
Shafizadeh, K., Mannering, F.: Acceptability of pavement roughness on urban highways by driving public. Transp. Res. Rec. **1860**, 187–193 (2003)

Integrating Flexible Pavement Surface Macrotexture to Pavement Management System to Optimize Pavement Preservation Treatment Recommendation Strategy

Seng Hkawn N-Sang[✉], Jose Medina, and Kamil Kaloush

Arizona State University, Tempe, AZ, USA
snsang@asu.edu

Abstract. Transportation agencies implement pavement preservation program to keep the good road in good condition for an extended period. This paper focuses on addressing pavement surface texture issues such as bleeding and raveling by investigating pavement surface macrotexture characteristics, developing surface macrotexture performance models for hot-mix dense graded overlay and chip seal, and establishing thresholds to trigger surface texture treatments. The surface texture evaluation of 20 pavement sections capturing nine (9) different types of pavement surfaces in the City of Phoenix were performed using automated and manual survey methods. Both FHWA Long-Term Pavement Performance (LTPP) and City of Phoenix pavement performance datasets were used for this study. Macrotexture is characterized by the surface texture between individual aggregates and thus is influenced by aggregates composition and binder type of the wearing course. A high macrotexture depth indicated by Mean Profile Depth (MPD) or Macrotexture Depth (MTD) represents a rough textured pavement surface. MPD above 1.2 mm exhibits a rougher pavement texture. The rate of change of MPD greater than 0.11 mm/yr. for thin hot-mix dense graded asphalt overlay and 0 mm/yr., for chip seal can be established as the raveling limits. The rate of change of MPD less than 0 mm/yr. for thin hot-mix dense graded asphalt overlay and 0.15 mm/yr., for chip seal can be established as the bleeding limits. Therefore, pavement macrotexture can be integrated to the pavement management system to optimize preservation treatment recommendation strategy.

Keywords: Pavement Management System (PMS) · Pavement performance · Flexible pavement preservation treatments · Pavement surface distress · Macrotexture · Mean Profile Depth (MPD) · Macrotexture Depth (MTD)

1 Introduction

The International Organization for Standardization (ISO) defined pavement texture as the deviation of a pavement surface from a true planar surface, with a texture wavelength less than 0.5 m. Pavement surface texture can be grouped into four different categories based on their wavelengths: microtexture, macrotexture, megatexture, and roughness or unevenness (ISO 13473-1 2019). This study is limited in scope to macrotexture with its wavelength ranging from 0.5 mm to 50.0 mm and is motivated by raveling observed on young pavements in the City of Phoenix (COP) and improper assessment of new chip seal as raveling. Based on a review of 10% of 7,800 km of COP's flexible pavement street network, approximately 37% had a rougher pavement surface texture than slurry seal and approximately 3% has a rougher pavement surface texture than chip seal. The use of surface macrotexture to assess and monitor asphalt concrete pavements surface texture roughness and deficiencies such as bleeding and raveling is proposed. Bleeding caused by excessive binder content and low voids results in a partial or complete immersion of the aggregate into the bituminous binder. Raveling is a disintegration of pavement surface by loss of binder and both fine and coarse aggregates. Raveling results from the separation of the bituminous film from the aggregates due to water, chemical, and mechanical actions, or poor material or construction qualities. Unless severely raveled and deeply pitted, raveling may not affect the pavement structure but both raveling and bleeding can have a substantial effect on the ride quality and safety.

Asphalt pavement macrotexture is affected by the maximum aggregate dimensions, fine and coarse aggregate types, mix gradations, binder types, asphalt mixtures, and surface finishing or texturing methods (Hall et al. 2009). Macrotexture is associated with the coarseness of pavement surface and is attributed to exciting shock absorbers in vehicle suspension systems, deforming tire sidewalls of a moving vehicle, affecting energy dissipation, wasting heat, and rolling resistance by vehicles (Praticò and Vaiana 2015). Macrotexture has been known to affect the high-speed, wet skid resistance and an increased in macrotexture reduced accidents at lower speeds (Henry 2000). Since it is desirable for pavement surface to exhibit a balance of friction for safety and roughness for ride quality and driver comfort, pavement wearing surface has been designed to provide a good friction, low levels of roughness, and low level of noise (Ahammed and Tighe 2010; Flintsch et al. 2003). Flintsch et al. (2003) applied macrotexture to detect and monitor frictional properties and segregation in hot mix asphalt pavement to assist in drafting the construction quality control specification. Although numerous studies had been conducted to gain a better understanding of surface macrotexture, none had explored its long-term performance nor its applications to preservation treatment recommendation strategy. Therefore, the objectives of this study are to evaluate surface macrotexture performance over time and to develop a pavement preservation treatment recommendation strategy to mitigate surface texture deficiencies.

The research methodology encompasses the following:

1. Analyse the historical pavement condition, site condition, traffic condition, and original construction and subsequent maintenance history data.
2. Perform a rigorous data scrubbing to remove incomplete, erroneous, inaccurate, or unusual pavement macrotexture data.
3. Field investigate surface macrotexture for different types of flexible pavement surfaces using Automated Road Analyzer (ARAN) and volumetric method in accordance with ASTM specification E965 using glass sphere and establish the relationship between automated and manual survey methods.

4. Develop pavement surface macrotexture indicator, mean profile depth (MPD), based performance models for thin hot-mix dense graded asphalt overlay and chip seal.
5. Establish an optimal timing for the preservation treatment.
6. Develop a decision tree for preservation treatment selection.

2 Pavement Surface Macrotexture Characteristics

The commonly used indicators for pavement surface macrotextures are macrotexture depth (MTD) and mean profile depth (MPD). The classic three-dimensional measure of pavement macrotexture is a volumetric method that can be performed in the laboratory and in the field. In this study, the volumetric method was used as a referenced macrotexture measurement or a benchmark. The test was performed as specified in ASTM standard E965–15 (2019) and the average of four equally spaced diameters was recorded and the macrotexture depth (MTD) of the test pavement surface was calculated using Eq. 1:

$$MTD(in\ mm) = \frac{Vol\ of\ Material\ (in\ mm^3)}{\pi * Average\ of\ Diamter\ of\ Area\ Covered\ by\ Material(in\ mm^2)} \quad (1)$$

As per ISO 13473-1 (2019), MPD is determined by dividing the measured profile into segments of 100 mm in length in the direction of travel and computed using the Eq. 2:

$$MPD(in\ mm) = \frac{Peak\ Level(1st) + Peak\ Level(2nd)}{2} - Average\ Level \quad (2)$$

The COP utilized ARAN, a fully automated pavement data collection van equipped with Laser Crack Measurement System (LCMS), and customized distress rating and analysis to a single 3.048 m lane. The survey cross section is divided into five zones: left exterior, left wheel path, center, right wheel path, and right exterior. The wheel paths and the center between the wheel paths are 0.9144 mm and the space outside of the wheel paths to left or the right edge of pavement detection zone are 0.1524 m. Although ARAN can extract macrotexture measurements the full width, only the pavement surface texture in the left wheel path and right wheel path are analyzed by the COP. The algorithm to compute macrotexture measure, MPD, is based on a "digital sand patch method" which computes the air void-content volume between a three-dimensional rendering of pavement surface and the road surface itself.

The surface macrotexture or surface texture roughness assessment of 20 pavement sections in the COP was performed in the field by the volumetric method and the automated method using ARAN. These 20 sections captured eight different types of preservation treatments the COP currently utilizes. At least one macrotexture test per pavement type were performed to generate a full range of macrotexture depth. Fog seal is a light application of emulsified asphalt with or without rejuvenator that is sprayed so thin that it does not correct pavement surface texture at macro level. Seal coat is a high-performance fiber/mineral reinforced asphalt emulsion blended with polymers and specially graded fine aggregate. Seal coats such as LIQUIDROAD replenish fine aggregates to the existing pavement. Micro seal and slurry seal are carefully designed mixture of asphalt emulsion (which may be polymer-modified or latex polymer emulsified), virgin or reclaimed asphalt pavement (RAP) aggregate, mineral filler, water, and additives and uniformly spread over a properly prepared surface at a single stone thickness. Micro seal with nominal maximum aggregate size (NMAS) of 9.53 mm is applied on high-speed roads and slurry seal with NMAS of 4.75 mm is applied on low-speed roads. Chip seal with precoated chips is called Fractured Aggregate Surface Treatment (FAST). FAST with differing chip gradation can be specified as low volume FAST or high volume FAST. For low volume FAST, the largest sieve opening the aggregate passes is 13 mm and 19 mm sieve opening for the high volume FAST. Thin overlay consists of a mixture of aggregates and terminal blend polymer-modified asphalt rubber binder graded at PG 76–22 TR +. The specified binder content is 6.0% on non-residential streets and 6.2% on residential streets. The asphalt concrete mix uses 13 mm dense graded hot-mix asphalt concrete. The thin overlay pavement thickness is 25 mm on residential street and 32 mm on non-residential street.

Field macrotexture measurements in MPD and MTD are reported in (Table 1). A high MPD or MTD indicates a rough pavement surface texture and a low MPD or MTD indicates a smooth pavement surface texture. MPD ranges from 0.61 mm for fiber slurry seal to 2.97 mm for rough pavement surface. MTD ranges from 0.67 mm for thin overlay to 6.93 mm for rough textured pavement surface. The lowest MPD is fiber slurry sealed pavement while the lowest MTD is thin overlay pavement. There are three main factors that can be attributed to the differences in average macrotexture depth values. Firstly, MTD is a three-dimensional measure while MPD is a two-dimensional measure. Secondly, MTD is performed manually and MPD is captured using ARAN. Thirdly, MPD is the average macrotexture depth measured along the left and right wheel paths on a given test section while MTD is the average macrotexture depth measured at a random spot anywhere within an evaluation section as specified in ASTM E965. The COP section ID 7320 is the rough pavement section with the highest MPD and MTD values. As shown in (Fig. 1), a strong correlation between MPD and MTD was observed and represented by R^2 of 0.83. A linear relationship established between field MPD and MTD measurements for 20 pavement sections is represented by Eq. 3:

$$MTD(in\,mm) = 2.04 * MPD(in\,mm) - 0.03 \tag{3}$$

The effectiveness of preservation treatment in mitigating surface texture defects and the surface macrotexture performance before and after preservation treatments were explored. There were 270 road sections which received a specific preservation treatment less than 18 months ago that were evaluated. They included 49 fog sealed sections, 43 seal coat applied sections, 24 slurry sealed sections, 14 micro sealed sections, 3 high volume FAST applied sections, 6 low volume FAST applied sections, and 131 thin overlay sections. (Fig. 2) shows a boxplot of the macrotexture measurements

Table 1. Field measured MPD and MTD data

COP section ID	MPD (mm)	MTD (mm)	Pavement surface type
373210	0.97	0.67	Thin overlay
7880	0.61	0.94	Fiber slurry seal
133901	1.18	1.47	Slurry seal
39835	0.88	1.08	Micro seal
43340	1.05	2.56	Micro seal
10190	1.55	1.32	Seal coat
40120	0.68	0.86	Seal coat
10920	1.22	0.99	Fog seal
334409	0.92	1.33	Fog seal
82605	1.41	2.16	Seal coat over low volume FAST
73001	2.52	4.33	High volume FAST
353114	2.51	4.64	High volume FAST
502001	2.15	2.69	High volume FAST
7320	2.97	6.93	Rough surface texture
7325	2.35	4.66	Rough surface texture
15365	2.34	3.69	Rough surface texture
40310	2.76	4.19	Rough surface texture
40315	2.44	3.57	Rough surface texture
42190	2.02	2.87	Rough surface texture
42195	1.98	3.21	Rough surface texture

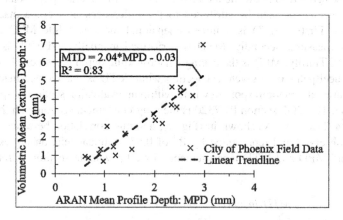

Fig. 1. Correlation between macrotexture using ARAN and volumetric method.

captured by ARAN before and after the preservation treatment. The mean MPD values indicated by the "x" on the boxplot are the average MPD values of all sections receiving a specific treatment type and range from MPD of 0.79 mm after thin overlay to MPD of 2.30 mm after high volume FAST. A comparison of the mean MPD before and after treatment indicates an increase in macrotexture after fog seal and FAST applications. While fog seal does not correct pavement surface texture, the aging and oxidation effects occurring during the 7-month to 28-month time elapsed between the before and after treatment condition surveys cause pavement macrotexture depth to increase. Except for the high-volume FAST which was applied over the chip seal section, the remaining FAST sections were applied on existing overlay or slurry seal pavement sections. Since FAST aggregates are larger in size and more uniform than slurry seal or hot mix dense graded asphalt pavement, an increased in MPD by 63% for high volume FAST and 37% for low volume FAST after treatments were observed. On the other hand, the greatest reduction in pavement macrotexture depth occurred after the slurry seal treatment with a reduction in MPD by 37%. Reductions in MPD were also observed for seal coat, micro seal, and a thin overlay. The smaller section of the boxplot indicating a smaller spread of average MPD data is observed for slurry seal, micro seal, low volume FAST, and thin overlay.

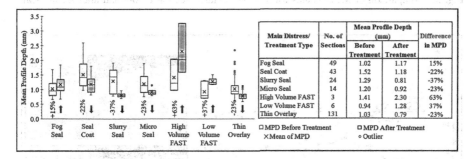

Fig. 2. A comparison of MPD before and after treatment.

The effectiveness of cape seal and seal coat in mitigating rough texture on the wearing course were also explored. (Fig. 3) shows a rough and pitted pavement with MPD of 2.76 mm which was treated with a cape seal. Cape seal is an application of micro seal over FAST. A reduction in MPD by 64%, from 2.76 mm before treatment to 1.0 mm, was observed after cape seal. Cape seal corrected the pitted pavement surface by filling and leveling the pavement surface. Although the road was programmed to receive slurry seal three years later, to address the overwhelming requests from residents to correct the pavement rough texture, COP's inhouse maintenance crew performed a double seal coat application. The double seal coat reduced macrotexture depth by about 31%; from MPD of 2.23 mm to 1.54 mm as shown in (Fig. 4). A specially formulated seal coat, LIQUIDROAD, composed of small and large silica and limestone, and clay ball, is expected to replenish fine aggregates to the existing aged pavement and provide a smoother pavement surface texture. A double seal coat application on a rough pavement improves the pavement texture but does not fill the undulation in the deeply pitted pavement that MPD remains high and adversely mud cracks were observed unexpectedly. (Fig. 5) shows field pictures taken from the top and from the side views of the pavement that had a very rough surface texture and deep pits and after a single coating of LIQUIDROAD and after double seal coat.

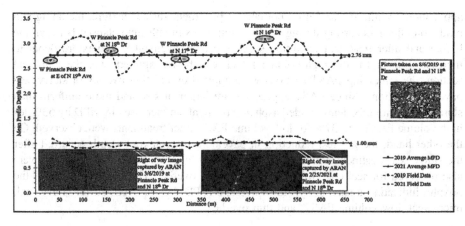

Fig. 3. Comparison of MPD values for a rough textured section before and after a cape seal.

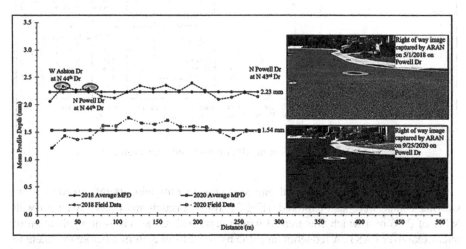

Fig. 4. Comparison of MPD values for a rough textured section before and after a seal coat application.

Fig. 5. Pavement surface before and after the first coat of seal coat application (left and center images) and the final surface after the double seal coat (right image).

The pavement surface macrotexture for raveled, rough, and fatigue pavements and the change in macrotexture between two consecutive condition surveys were explored. The raveled section shown in (Fig. 6) was treated with Asphalt-Rubber Asphalt

Concrete (ARAC) overlay consisting of a mixture of aggregate, mineral admixture, and asphalt-rubber binder (ARB) in 2014. The specified target binder content was 8.5% and effective void range was 3% ± 1 for the 75-blow Marshall ARAC mix. Crumb rubber in ARB was at least 18% by weight of total binder and the maximum size of the crumb rubber was 2 mm. Raveling at the intersections was noticeable a year after ARAC overlay. The average MPD for the raveled section was 1.29 mm in 2019 and 1.59 mm in 2021 respectively. The average rate of change of MPD within the 23-month period was about 0.3 mm/yr. The rate of change of MPD increased considerably more at the intersection of N 44th St and E Redfield Rd with its rate at 0.7 mm/yr. The average MPD increased for the raveled section was 23% but increased to 48% at the intersection of N 44th St and E Redfield Rd. MPD spikes at the intersections and at the speed hump can be attributed to the stop and go or turning maneuvers aggravating the pavement surface texture.

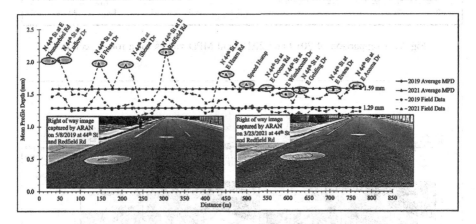

Fig. 6. Comparison of 2019 and 2021 field MPD values for a raveled section.

The rough textured and polished aggregate section shown in (Fig. 7) was constructed in 1998 and received no subsequent preservation treatment. The average MPD for the rough section was 2.77 mm in 2019 and 2.90 mm in 2021 respectively. The average rate of change of MPD within the 26-month period was 0.06 mm/yr. This aged and pitted pavement surface had already lost binder, fines, and coarse aggregates that macrotexture depth was increased by 5% between the two survey periods.

The pavement section shown in (Fig. 8) was widened in 1998 with no additional maintenance activity performed on the existing pavement. The macrotexture condition assessments were performed on the fatigue section and the average MPD was 1.23 mm in 2019 and 1.24 mm in 2021 respectively. The change in pavement macrotexture depth is negligible, less than 1%, for the fatigue section. The average rate of change of MPD within the 23-month period is about 0.01 mm/yr.

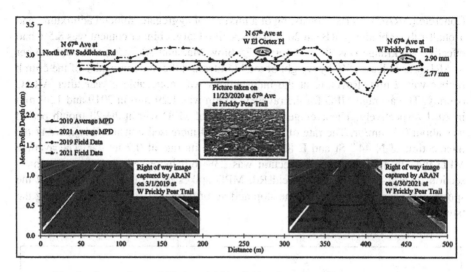

Fig. 7. Comparison of 2019 and 2021 field MPD values for a rough section.

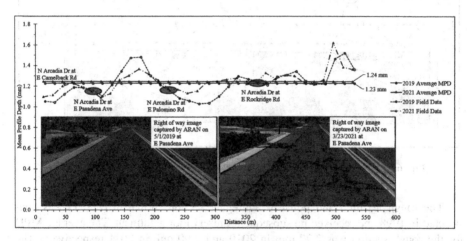

Fig. 8. Comparison of 2019 and 2021 field MPD values for a fatigue section.

3 Pavement Surface Macrotexture Performance Model

The objectives of the pavement performance model are to allow agencies to predict pavement condition in the future, to develop annual pavement maintenance program, and to compare the performance of different preservation treatments. Thus, the pavement performance model needs to be reliable, accurate, and up to date. Iuele (2016) mentioned the difficulty in estimating the texture deterioration rate. Her study was based on four different dense graded friction courses laid on four road sections in Cosenza, Italy. Pavement surface texture for the test sections were measured and monitored immediately after construction and every 6 months for up to 18 months after the construction. The City of Phoenix data which included 13 thin overlay sections that had record of at least two surveys performed in less than eighteen months also supported Iuele's (2016) finding. As shown in (Fig. 9), MPD increased right after an overlay and then decreased after 0.16 yr. Then another increased in MPD after 0.2 yr. was followed by another decrease in MPD.

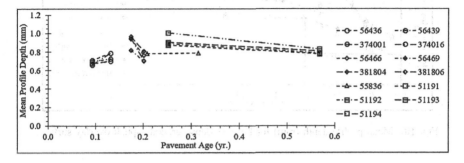

Fig. 9. Mean profile depth after thin overlay.

Since the City of Phoenix acquired ARAN 9000 model in November 2017, the macrotexture data was limited. Therefore, Long-Term Pavement Performance (LTPP) database was used to investigate the macrotexture performance right after overlay and the performance over time after overlay and chip seal. There were 441 sections out of 1,807 total LTPP asphalt concrete test sections for which macrotexture survey were performed between April 21, 2013 and November 14, 2019. Macrotexture database went through a rigorous data cleaning procedure where MPD was evaluated and sections with only one MPD survey record were removed. As shown in (Fig. 10), a similar variation in macrotexture performance was observed for thick hot-mix dense graded asphalt overlay sections and a gradual increase in macrotexture was observed 18 months after the overlay. Research performed by Iuele (2016) also noted a similar surface texture characteristic on four test sites during the first 18 months after dense graded friction courses were applied. She attributed the initial increase in macrotexture depth to the loss of bitumen film from the aggregate surface. The bond between the aggregate and the binder depended on the asphalt binder in the asphalt mixture and the nature of the aggregate. Instead of vehicular traffic load smoothing and kneading the asphalt binder, because of a lower stripping resistance aggregate, the bitumen film was removed from the aggregate. It was then followed by a substantial decrease in the macrotexture depth due to the smoothing and polishing affect from the traffic loading over time as well as dust and oil buildups. With its compaction like action, the traffic loading induced binder smearing and binder migration which filled the voids and thereby decreased the macrotexture depth. Another significant increase in macrotexture a year after construction was attributed to the removal of migrated binder from trafficking. With such great variation in macrotexture depth during the first several months after construction, estimating the rate of macrotexture deterioration was indeed difficult. Therefore, data points for which the rate of change of MPD rapidly decreased or increased after preservation treatments were removed from consideration and macrotexture data collected 12 months after overlay and sections with increasing MPD over time were considered for the study.

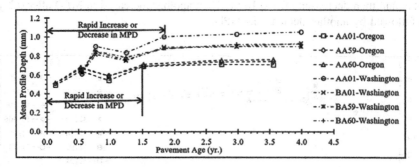

Fig. 10. Mean profile depth (mm) for hot-mix dense graded asphalt overlay surface.

The macrotexture depth for hot-mix dense graded asphalt pavement increases over time while the macrotexture depth for chip seal decreases over time. The environment alone can cause the asphalt pavement to age and oxidize over time. As the pavement ages and oxidizes, the asphalt surface start losing binder and then fine aggregates. Traffic loading also affects the pavement surface texture by causing bitumen film to wear off, fine aggregates to lose, and coarse aggregates to polish. The extent and rate of aggregate loss is dependent on the asphalt pavement mixture. Beside pavement surface macrotexture contributions to roadway safety and noise, its characteristic is indicative of pavement surface wear overtime or premature pavement defects. Therefore, it is critical to periodically survey pavement surface texture in the field at the flow of the traffic using a reliable and practical pavement surface texture assessment method, so pavement surface texture can be monitored, and surface texture issues can be detected and addressed appropriately.

Due to the data availability, MPD performance model was developed using 34 hot-mix dense graded thin overlay and 37 chip seal sections from LTPP database. The surface texture depth for the thin overlay were assessed approximately between 14 months to 20 years after overlay. The field texture surveys for the chip seal were performed approximately between 11 months to 19 years after the treatment. The field measured MPD ranges from 0.50 mm to 1.50 mm for hot-mix dense graded thin overlay and 0.85 mm to 2.30 mm for chip seal. Gradual increase in pavement surface texture depth indicator, MPD, over time for hot-mix dense graded asphalt pavement and the gradual decrease in MPD over time for chip seal can be captured by the power function. Uz and Gökalp's (2017) concluded that the macrotexture depth for chip seal increased as the chip size increased. The traffic load working and kneading to achieve the desired chip embedment, MPD, is expected to decrease over time for chip seal. The expression for MPD and the rate of change of MPD (MPD') are represented by Eq. 4 and Eq. 5:

$$MPD\ (in\ mm) = \alpha + \beta * T^\gamma \qquad (4)$$

$$MPD\ (in\ mm) = \beta * \gamma * T^{(\gamma-1)} \qquad (5)$$

where: T is adjusted time (in years) since the last major construction or preservation treatment to best fit the power function and α, β and γ are shape parameters of the power function. The shape parameters for thin overlay are α = 0.51, β = 0.05, and γ = 1.15 while the shape parameters for chip seal are α = −16.56, β = 18.78, and γ = −0.03 respectively. The pavement surface texture-based performance models based on the historical MPD data for the thin hot-mix dense graded asphalt overlay and chip seal are shown in (Fig. 11). The field macrotexture depths were represented by the triangle markers for overlay and diamond markers for chip seal. Each MPD survey data was connected by the dashed line to illustrate the steepness of the slope which represented the rate of change of MPD. Comparing the standard error ratio for the pavement surface texture models, the model for overlay pavement was slightly more accurate with S_e/S_y of 0.26 and its corresponding R^2 of 0.96 than the model for chip seal with a slightly higher standard error ratio with S_e/S_y of 0.31 and its corresponding R^2 of 0.94. Since R^2 denoting a measure of an accuracy of the pavement macrotexture performance model are high, the models are highly correlated to the field MPD data.

A high pavement texture depth is not suggestive of texture issue but a higher-than-normal rate of change of MPD is indicative of raveling, and a lower-than-normal rate of change of MPD is indicative of bleeding. A normal thin overlay or chip seal is expected to follow their respective rate of change of MPD trend which was derived from the pavement surface texture performance models and shown in (Fig. 12). (Fig. 12) also displayed 25 data points for each unique LTPP sections. Each data point represented the rate of change of MPD and was determined by dividing the difference in MPD measurements of a section by the time difference between the two MPD survey dates. These raveled and bleeding sections were gathered from the comments on the LTPP field distress survey forms and verified with the distress images since raveling or bleeding were noted but not always rated. Since there was only one thin overlay section that was bleeding and no section that was raveling, a few thick overlay sections were considered in addition. This does not mean that thin overlay pavements do not ravel or bleed but rather indicates limitations on data availability. There were ten thick overlay sections that were bleeding and another nine sections that were raveling. As shown in (Fig. 12), the rate of change of MPD for all eleven bleeding sections fell below the thin overlay rate of change of MPD performance curve and all nine raveled sections fell above the thin overlay rate of change of MPD performance curve. Similarly, four filled diamond points, which fell above the rate of change of MPD performance curve, denoted three raveled chip seal sections. The rate of change of MPD for a chip seal section that was bleeding was represented by the unfilled diamond marker and fell just below the pavement surface texture performance curve.

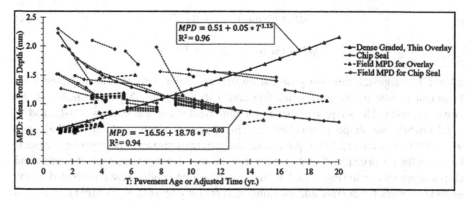

Fig. 11. MPD based pavement surface texture performance curves.

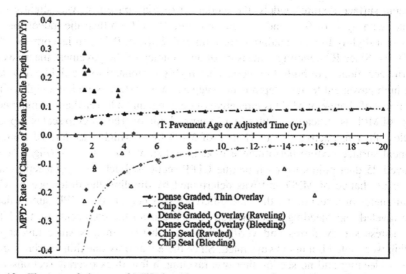

Fig. 12. The rate of change of MPD based on pavement surface texture performance curves.

4 Pavement Preservation Treatment Strategy

It is recommended that new pavements or pavements that received preservation treatment undergo a full year of summer-winter cycle even before considering for a light preservation treatment such as fog seal. The threshold values to trigger a specific preservation treatment were not determined based on the pavement age but instead by evaluating the macrotexture measure of an existing pavement surface. The pavement surface macrotexture indicator, MPD was investigated to be integrated to the pavement surface texture treatment recommendation strategy so pavement with surface texture issues can be addressed accordingly. There were 43 hot-mix dense graded overlay sections investigated of which seven were thick overlay. There were 33 sections that did not ravel nor bleed but four of them had high MPD. There were three raveled and seven bleeding sections. As shown in (Fig. 13), the solid line with triangle markers represented MPD performance curve for thin overlay and the dash, dot, dot line with the triangle markers represented the rate of change of MPD. The MPD for hot-mix dense graded thin overlay pavement surface texture ranged from 0.5 mm to 1.5 mm. LTPP sections with MPD of approximately 1.2 mm or higher were represented by the lines with the circle markers. Although MPD of 1.2 mm and higher exhibited a rougher surface texture, the rate of change of MPD, represented by the circle makers, were below the rate of change of MPD performance curve and were still above the rate of change of MPD of zero. Therefore, besides exhibiting rougher surface texture, no raveling or bleeding were observed on these sections. Since pavement surface macrotexture is dictated by the asphalt binder and aggregate composition in the overlay mixture, a higher MPD or rougher surface texture does not necessarily indicate raveling and a lower MPD or smoother surface texture does not necessarily indicate bleeding.

The rate of change of MPD is a better measure to use in establishing bleeding and raveling limits. The MPD for a bleeding overlay section represented by the dashed line with unfilled triangle markers decreased over time and the rate of change of MPD at approximately, -0.23 mm/yr. fell below zero. The rate of change of MPD for dense graded overlay without texture issues gradually increased over time and thus the rate of change of MPD falling below zero represented a reduction in macrotexture depth over time and denoted bleeding. The hot-mix dense graded thick overlay section was added to (Fig. 13), to illustrate MPD and the rate of change of MPD for a bleeding section. Even with the limited data, using engineering judgement and available historical data, the rate of change of MPD less than 0 mm/yr. can be established as the threshold for bleeding. Since there was only one raveled thin overlay section, six thick overlay sections that were noted in the field as raveling were also evaluated to determine the threshold for raveling. The MPD for the seven raveled hot-mix dense graded overlay LTPP pavement sections were not significantly higher than the rest of the overlay pavement section. However, the lowest rate of change of MPD for the raveled sections was 0.11 mm/yr. and fell above the rate of change of MPD performance curve. There were 16 chip seal treated LTPP sections evaluated as shown in (Fig. 14). There were eleven sections without texture issue, two raveled sections, and three bleeding sections. MPD for chip seal sections ranged from 0.67 mm to 2.1 mm. Generally, MPD is expected to decrease over time for chip seal but MPD was observed to increase for the raveled chip seal. Therefore, the rate of change of MPD greater than zero was set as the threshold for raveling. Based on the field MPD data and a review of distress survey, the bleeding limit was established as the rate of change of MPD lower than -0.15 mm/yr.

Pavement maintenance strategies are designed to always keep the pavement above a target acceptable level and to prevent pavement from requiring rehabilitation (Peterson 1981). The preservation treatment selection decision tree shown on (Fig. 15) was developed from analyzing a meaningful performance measure, surface macrotexture. This macrotexture performance measure can provide quantifiable benefits associated with preservation treatment applications and trackable pavement performance over time. It can also be used to trigger surface treatment. On a structurally sound pavement that is raveling, low-cost surface treatments such as seal coat, slurry seal and micro seal can be used to restore existing asphalt pavement surface texture and retard further disintegration or dislodging of fine and coarse aggregates. When raveling

Integrating Flexible Pavement Surface Macrotexture to Pavement Management System 749

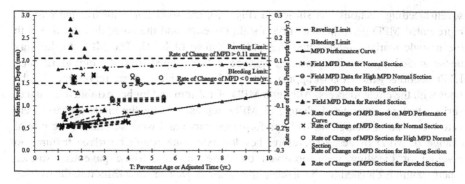

Fig. 13. Pavement surface texture thresholds for hot-mix dense graded overlay pavement.

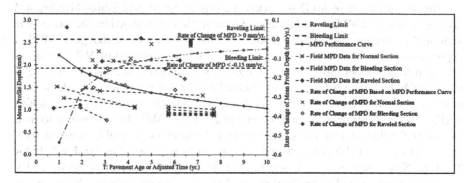

Fig. 14. Pavement surface texture thresholds for chip seal.

is severe and affects pavement structure, a more extensive treatment is required to restore pavement structure. A mild bleeding on a structurally sound pavement can be addressed by spreading the blotting material such as sand to soak up excess binder or by slurry seal for a more sever bleeding.

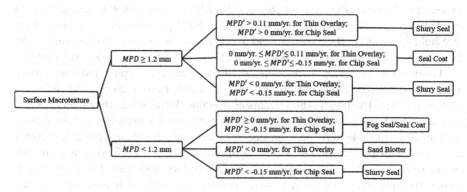

Fig. 15. Preservation treatment selection decision tree.

Li et al. (2017) performed the effects of pavement macrotexture on bicycle ride quality and based on the survey responses from 155 bicycle club members, slurry seal on city streets were highly acceptable. Therefore, MPD of 1.2 mm for slurry seal was selected to delineate the rough pavement surface texture. For a pavement section that has rough surface texture and is raveled, MPD is greater than or equal to 1.2 mm, and the rate of change of MPD is greater than 0.11 mm/yr. for thin hot-mix dense graded asphalt overlay and 0 mm/yr. for chip seal. Slurry seal is recommended to mitigate surface texture roughness and raveling. An existing oxidized pavement with MPD greater than or equal to 1.2 mm, and rate of change of MPD between 0 and 0.11 mm/yr. for overlay or between 0 and −0.15 mm/yr. for chip seal is considered rough but has not raveled or bled yet that seal coat is recommended to retard raveling potential in the future. A pavement section with MPD greater than or equal to 1.2 mm, and the rate of change of MPD less than 0 mm/yr. for thin overlay or −0.15 mm/yr. for chip seal is considered bleeding and missing aggregates that slurry seal is recommended. A pavement with MPD less than 1.2 mm, and the rate of change of MPD greater than or equal to 0 mm/yr. for overlay or −0.15 mm/yr. for chip seal may have started raveling but has not significantly lost aggregates yet that fog seal or seal coat is recommended to lock aggregate in place and prevent further raveling. A hot-mix dense graded asphalt overlay section that is bleeding with MPD less than 1.2 mm and the rate of change of MPD less than 0 mm/yr. can be treated using sand blotter. A chip seal treated pavement section with MPD less than 1.2 mm and the rate of change of MPD less than −0.15 mm/yr. is considered bleeding and has lost chips considerably that slurry seal is recommended to provide a new wearing course. The macrotexture indicator, MPD was used to objectively detect raveling and bleeding as well as trigger preservation treatments to mitigate pavement surface texture issues. However, agencies can potentially use a different macrotexture indicator such as MTD and adjust the macrotexture and the rate of change of macrotexture depth limits to reflect the local network performance and to tailor to the budget scenario. Therefore, the pavement surface macrotexture can be integrated to pavement management system and used in optimizing pavement preservation treatment recommendation strategy.

5 Summary and Conclusions

The primary purpose of a pavement preservation program is to keep the good road in good condition for an extended period of time. A successful implementation of the pavement preservation program allows an agency to provide guidance in the selection and timing of the preservation treatments for its street network considering the budget, to objectively monitor the performance a specific treatment over its service life, and to measure the cost-effectiveness of the preservation treatment. A common practice for the pavement preservation treatment program has been applying the right preservation treatment on the right road at the right time. Preservation treatments are known to extend the service life of the pavement by protecting the good pavement from rapid deterioration at a lower cost compared to the cost of rehabilitation for the life of the pavement. There are several preservation treatments available and each with its unique intended purpose and benefits.

This study investigated pavement surface macrotexture characteristics, developing surface macrotexture performance models for hot-mix dense graded overlay and chip seal, and establishing thresholds to trigger surface texture treatments. The major findings from this research work are highlighted below:

- Pavement surface macrotexture for various pavement surface types was different: macrotexture depth for fiber slurry seal was the lowest with MPD of 0.61 mm and the highest with MPD of 2.97 mm for the pitted and rough textured pavement.
- Pavement surface macrotexture indicators, MTD, measured using the manual method, and MPD, measured using the automated method, were strongly correlated.
- An analysis of before and after preservation treatments demonstrated a 37% reduction in MPD for slurry seal and 63% increase in MPD after FAST.
- In approximately two-year period, an increased in MPD for a ravelled pavement was 23% and was less than 1% for the fatigue pavement. The increase in MPD for a pitted pavement which had already lost binder and fine aggregate was about 5%.
- Based on a review of the pavement surface macrotexture performance after an overlay, during the first 18 months, pavement surface texture may go through a rapid increase and rapid decrease in macrotexture depth due to the traffic, aggregate gradation, asphalt type, and mixture composition.
- The pavement surface macrotexture performance models developed from the historical MPD data from LTPP database showed a gradual increase in MPD over time for a hot-mix dense graded asphalt overlay and a gradual decrease in MPD over time for chip seal.
- The rate of change of MPD, derived from the MPD performance model, can be used to objectively rate ravelling and bleeding.
- Since pavement surface macrotexture greater than MPD of 1.2 mm measured using LCMS is rougher than slurry seal surface, MPD greater than 1.2 mm is defined as a rough pavement surface texture that may or may not be ravelled.
- Based on pavement surface macrotexture performance measures, a preservation treatment decision tree to address pavement surface texture issues can be developed and incorporated to the preservation treatment program.

This paper presented a methodology from analysing the existing pavement surface macrotexture performance measure to developing pavement performance model and establishing a preservation treatment selection strategy to address texture related distresses. Although, the automated macrotexture detection method which can be performed at the flow of the traffic is not inexpensive currently, in the near future with improvements in Light Detection And Ranging (LiDAR) and camera, an assessment of pavement macrotexture can be performed using a smartphone. Data limitation only allowed this study to presented performance models for thin hot-mix dense graded asphalt overlay and chip seal. Therefore, a periodic assessment of pavement macrotexture for at least three rounds of survey can be used to develop pavement macrotexture performance models for other types of pavement. Further research, using additional data, is necessary in this area.

References

Ahammed, M.A., Tighe, S.T.: Pavement surface friction and noise: integration into the pavement management system. Can. J. Civ. Eng. **37**(10), 1331–1340 (2010)

ASTM E965-15. Standard test method for measuring pavement macrotexture depth using a volumetric technique. West Conshohocken: ASTM International (2019). https://doi.org/10.1520/E0965-15R19

Flintsch, G.W., de León, E., McGhee, K.K., Al-Qadi, I.L.: Pavement surface macrotexture measurement and applications. Transp. Res. Rec. **1860**, 168–177 (2003)

Hall, J.W., Smith, K.L., Titus-Glover, L., Wambold, J.C., Yager, T.J., Rado, Z.: NCHRP Web-Only Document 108: Guide for pavement friction, Final Report for NCHRP Project 01-43 (2009)

Henry, J.J.: NCHRP synthesis 291: evaluation of pavement friction characteristics. Transp. Res. Board (2000). http://onlinepubs.trb.org/onlinepubs/nchrp/nchrp_syn_291.pdf

International Organization for Standardization, Characterization of pavement texture by use of surface profiles - Part 1: Determination of mean profile depth. ISO 13473–1:2019. ISO, Geneva, Switzerland (2019)

Iuele, T.:. Road surface micro- and macrotexture evolution in relation to asphalt mix composition. In: Advanced Material and Structural Engineering, pp. 433–436. ISBN 978-1-138-02786-2 (2016)

Li, H., et al.: Development of recommended guidelines for preservation treatments for bicycle routes. University of California Pavement Research Center, Davis and Berkeley. UCPRC-RR-2013-0UCPRC-RR-2016-02 (2017)

Peterson, D.E.: NCHRP synthesis of highway practice report 77: evaluation of pavement maintenance strategies. Transp. Res. Board (1981)

Praticò, F.G., Vaiana, R.: A study on the relationship between mean texture depth and mean profile depth of asphalt pavements. Constr. Build. Mater. **101**, 72–79 (2015)

Uz, V.E., Gökalp, I.: Comparative laboratory evaluation of macro texture depth of surface coatings with standard volumetric test methods. Constr. Build. Mater. (2017). https://doi.org/10.1016/j.conbuildmat.2017.02.059

Safe Roads by Design - 2

MASH TL-3 Development and Evaluation of the Thrie-Beam Bullnose Attenuator

Robert Bielenberg[✉], Ron Faller, and Cody Stolle

Midwest Roadside Safety Facility, University of Nebraska, Lincoln, NE, USA
rbielenberg2@unl.edu

Abstract. The thrie-beam bullnose system is a non-gating, redirective crash cushion that provides a safe, cost effective, non-proprietary option for shielding median piers and other median hazards. It consists of a guardrail envelope comprised of thrie-beam panels mounted at a height of 803 mm and supported by breakaway steel posts. The nose rail section and the first two adjacent rail sections are radiused and slotted, and the fourth and fifth rail sections are slotted. The combination of the radiused and slotted rail segments along with the breakaway posts allow the system to safely capture and decelerate vehicles impacting near the end or nose of the system, while redirecting vehicles impacting along the side. The barrier system was originally developed and evaluated under National Cooperative Highway Research Program (NCHRP) Report 350.

In 2009, the American Association of State Highway and Transportation Officials (AASHTO) implemented an updated standard for the evaluation of roadside hardware. The new standard, the Manual for Assessing Safety Hardware (MASH), improved the criteria for evaluating roadside hardware beyond the previous NCHRP Report 350 standard through updates to test vehicles, test matrices, and impact conditions. A research effort was undertaken to evaluate the thrie-beam bullnose to MASH Test Level 3 through full-scale crash testing and modify the design as needed to meet the revised safety criteria. This paper details the evaluation of the thrie-beam bullnose through MASH TL-3 full-scale crash tests, modification of the original design to meet MASH criteria, and recommendations for implementation of the system.

Keywords: Highway safety · Crash test · Roadside appurtenances · Breakaway posts · Compliance test · MASH 2016 · Bullnose · Thrie beam

1 Introduction

The use of the divided highway separated by a median area has been a valuable safety feature in modern roadway design. The median allows a safe recovery area for errant vehicles to come to rest without impeding upon oncoming traffic. However, many roadway structures are built in the median, such as bridge supports, drainage structures, and large sign supports. These structures present hazards to errant vehicles traveling in the median area.

Enclosed guardrail envelopes, commonly called bullnose systems, are a common solution for shielding errant vehicles from hazards in highway medians. A bullnose

© The Author(s), under exclusive license to Springer Nature Switzerland AG 2022
A. Akhnoukh et al. (Eds.): IRF 2021, SUCI, pp. 755–771, 2022.
https://doi.org/10.1007/978-3-030-79801-7_54

guardrail system involves wrapping a semi-rigid guardrail system completely around the hazard. When a vehicle impacts the radiused nose of a bullnose, the barrier captures the vehicle, collapses inward, and dissipates energy to safely decelerate the vehicle. Impacts along the sides of the barrier are redirected similar to standard guardrail systems. From 1997 through 2000, the Midwest Roadside Safety Facility (MwRSF) developed a thrie-beam bullnose guardrail system for shielding median hazards found between divided highways (Bielenberg et al. 1998a, Bielenberg et al. 1998b, Bielenberg et al. 2000, Bielenberg et al. 2001). The new, non-proprietary bullnose guardrail system was successfully full-scale crash tested and evaluated according to the Test Level 3 (TL-3) safety performance evaluation criteria provided in National Cooperative Highway Research Program (NCHRP) Report 350 (Ross et al. 1993).

Controlled Release Terminal (CRT) wood posts were used in the original bullnose guardrail system, as shown in Fig. 1. Although the CRT posts adequately met the TL-3 safety requirements, these wood posts have several drawbacks. First, the properties and performance of wood posts are highly variable due to knots, checks, and splits, thus requiring grading and inspection of posts. Second, two holes are drilled into the CRT posts at grade and below that allow them to break away upon impact. These holes expose the interior of the wood to the environment, which can accelerate deterioration. Further, chemical preservatives used to treat the wood posts have been identified as harmful to the environment by some government agencies. Thus, the treated wood posts may require special consideration during disposal. Due to these concerns, a breakaway steel post option was developed for use in the thrie-beam bullnose guardrail system.

A Universal Breakaway Steel Post (UBSP) was developed to replace the CRT posts in the thrie-beam bullnose median barrier system (Arens et al. 2009, Schmidt et al. 2010a, Schmidt et al. 2010b), as shown in Fig. 1. The UBSP post was based on a fracturing-bolt concept and was designed to match the cantilevered bending capacities of existing wood CRT posts about their strong and weak axes, as well as for a biaxial loading condition. The embedded portion of the UBSP post had a similar cross section as compared to the CRT post, which provided comparable rotational resistance in the soil. The mass, general geometry, and the breakaway characteristics of the upper UBSP post section were also similar to the CRT wood post. The lower portion of the UBSP post consisted of a foundation tube with the lower base plate. The upper portion of the UBSP post consisted of a post with the upper base plate. The bullnose system utilized Breakaway Cable Terminal (BCT) posts for the first two posts as well as for the last two anchorage posts on each side of the barrier. Post nos. 3 through 8 were UBSP posts, and the remaining posts were standard thrie-beam guardrail steel posts. The bullnose with breakaway steel posts was subjected to test designation nos. 3–30, 3–31, and 3–38 of NCHRP Report No. 350 to determine if it met the TL-3 safety performance criteria and to ensure the UBSP post was providing similar performance to the original CRT posts. In all three full-scale crash tests, the vehicle was safely contained and decelerated, and the barrier did not cause vehicle instability.

In 2009, the American Association of State Highway and Transportation Officials (AASHTO) implemented an updated standard for the evaluation of roadside hardware. The new standard, entitled the *Manual for Assessing Safety Hardware* (MASH) (AASHTO 2009), improved the criteria for evaluating roadside hardware beyond the

a) CRT post   b) UBSP post

Fig. 1. Bullnose system breakaway posts

previous NCHRP Report No. 350 standard through updates to test vehicles, test matrices, and impact conditions. In an effort to encourage state departments of transportation and hardware developers to advance hardware designs, the Federal Highway Administration (FHWA) and AASHTO collaborated to develop a MASH implementation policy that included sunset dates for various roadside hardware categories. Further, the 2009 MASH safety criteria were updated in 2016, thus resulting in the MASH 2016 document (AASHTO 2016). The new policy requires that safety devices installed on federal-aid roadways after the sunset dates to be evaluated according to MASH 2016. Due to the updated safety performance criteria for roadside hardware and the desire for user agencies to continue to use the thrie-beam bullnose as part of their safety inventory, an effort to full-scale crash test the thrie-beam bullnose system to the MASH TL-3 safety requirements was undertaken at the Midwest Roadside Safety Facility (MwRSF). The results and findings of that effort are provided herein.

2 Thrie-Beam Bullnose System Details

Thrie-beam guardrail, which was mounted at a height of 803 mm and supported by twenty-nine posts, was used to construct the bullnose barrier, as shown in Fig. 2. The total bullnose installation was 4,502-mm wide by 20.2-m long. The width of the system was believed to be a worst-case scenario because the rail element is forced to bend through an angle of 180° as the vehicle progresses into the system. Thus, a narrow system would subject the rail element to higher stresses and as a result cause higher vehicle deceleration. A one-half barrier system was utilized for the testing program in order to reduce costs and construction time. Note that a dual-cable, end anchorage system was used on each free end of the bullnose system to provide tensile and compressive resistance at the end of the one-half barrier system, thus mimicking the behavior of a complete installation.

The bullnose system was constructed with twenty-nine posts, which included fifteen posts on Side A and fourteen posts positioned on Side B. Side A of the system was slightly longer to span a below grade obstruction located at the test site that prevented installation of the BCT foundation tubes. Side A of the system contained two BCT posts, six UBSP posts, five W6 × 8.5 standard guardrail posts, and two BCT anchorage posts, respectively, from the nose of the system. Side B of the system contained two BCT posts, six UBSP posts, four W6 × 8.5 standard guardrail posts, and two BCT anchorage posts, respectively, from the nose of the system. The lower portion of the UBSP consisted of a foundation tube with the lower base plate. The upper portion of the UBSP consisted of a post with the upper base plate. The upper and lower halves of the UBSP were connected with a series of four bolts that were designed to allow the upper portion of the post to disengage at specific force levels when loaded in the lateral and longitudinal directions.

All guardrail used in the bullnose barrier consisted of 12-gauge steel thrie beam. Eleven 3,810-mm long sections of thrie beam were spliced together with a standard lap splice on each interior end. The first three rail sections were cut with slots in the valleys, as shown in Fig. 2. Slots were included in order to allow the guardrail to spread apart and more effectively capture the impacting vehicle as well as weakening the guardrail to reduce the formation of large kinks that could impact the side of a vehicle during a crash. The nose section of the rail consisted of a 3,810-mm long section bent into a 1,580-mm radius. The second rail section on each side was bent to form a 10,400-mm radius curve. The bullnose system was constructed with the rail lap splices oriented based on the traffic flow on each side of the system, such that the upstream rail segment overlapped the downstream rail segment. All of the full-scale crash testing for the thrie-beam bullnose system detailed herein was conducted on a splice configuration for median installations. Thus, for a median installation, the thrie-beam bullnose lap splicing proceeded in a continuous manner all the way around the system in the direction of oncoming traffic. A cable anchor system was used between post no. 1 and the thrie beam on each side of the system in order to develop the tensile strength of the thrie-beam guardrail downstream of the post no. 2.

In the course of the development of the bullnose barrier system, the researchers found that it was necessary to add a set of steel cable retention devices to contain impacting vehicles in the event of rail fracture, as shown in Fig. 2. A 4.42-m long by 16-mm diameter, 6 × 19 XIPS IWRC cable was added behind the top and middle humps of the nose section of thrie-beam rail. A 6 × 19 XIPS IWRC cable was chosen such that either one of the two cables could contain the impacting vehicle. Cables were only placed behind the first rail section because that was the only section that had failed in previous testing. It was believed that the rail sections after the nose section would be active in containing the vehicle, and therefore, the use of longer cable lengths was deemed unnecessary.

Note that the downstream ends of the thrie-beam bullnose installation were configured with a bi-directional, trailing-end anchorage system. The guardrail anchorage system was utilized to simulate the anchorage tensile and compressive anchorage provided by continuous guardrail or attachment to an approach guardrail transition. The dual anchorage would not be required in a typical installation unless the bullnose system was configured with a free end similar to the test setup. The anchorage system consisted of timber posts, foundation tubes, anchor cables, bearing plates, rail brackets, and channel struts, which were components from the hardware used in a crashworthy, downstream, trailing-end terminal (Mongiardini et al. 2013a, Mongiardini et al. 2013b, Stolle et al. 2015).

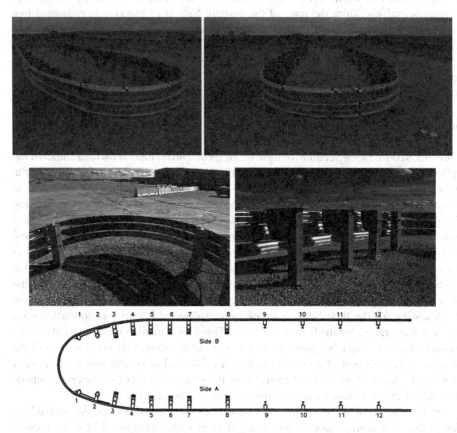

Fig. 2. Thrie-beam bullnose system details

3 Mash Evaluation Requirements

Thrie-beam bullnose systems must satisfy impact safety standards. For new hardware, these safety standards consist of the guidelines and procedures published in MASH 2016. The thrie-beam bullnose system is classified as a non-gating, redirective, crash cushion for the purposes of evaluation. In MASH 2016, as many as ten full-scale crash tests are potentially required to evaluate this type of hardware, as shown in Table 1.

Table 1. MASH TL-3 Test Matrix for the Thrie-Beam Bullnose

Test no.	Vehicle	Speed (km/h)	Angle (deg)	Impact point	Other notes
3–30	1100C	100	0	Center of nose @ ¼ offset	–
3–31	2270P	100	0	Center of nose	–
3–32	1100C	100	5/15	Center of nose	–
3–33	2270P	100	5/15	Center of nose	–
3–34	1100C	100	15	CIP for capture/redirection	CIP selected at post no. 2 of the system
3–35	2270P	100	25	CIP for capture/redirection	CIP selected at post no. 3 of the system
3–36	2270P	100	25	CIP @ transition to rigid structure	Deemed non-critical if using MASH TL-3 compliant AGT
3-37a	2270P	100	25	CIP for reverse direction	Deemed non-critical
3-37b	1100C	100	25	CIP for reverse direction	CIP selected 2,876 mm upstream from the end post in the system
3–38	1500A	100	0	Center of nose @ ¼ offset	Deemed non-critical dependent on 1500A estimation procedure

Out of the ten required crash tests, three tests were deemed non-critical. Test designation no. 3–36, on the transition to a rigid structure, was not required as it was assumed that the bullnose system will use MASH 2016 TL-3 approved thrie-beam approach guardrail transitions for attachment to any rigid structures. Test designation no. 3–38 is intended to evaluate the performance of mid-sized sedan vehicles with terminals and crash cushions. However, MASH 2016 uses an analytical estimation of 1500A vehicle decelerations based on the results of test designation no. 3–31 to determine whether or not this test is required. Thus, test designation no. 3–38 was deemed non-critical until the results from an analytical estimation of 1500A vehicle decelerations were known.

The final non-critical test was test designation no. 3–37 for reverse-direction impacts, which includes test designation 3–37a with the 2270P vehicle and test designation 3–37b with the 1100C vehicle. For systems utilizing a breakaway cable system like end terminals and the thrie-beam bullnose system, MASH 2016 recommends test designation no. 3–37b as the critical test to evaluate snag of the 1100C vehicle on the cable anchorage system. Thus, test no. 3–37a was deemed non-critical. Previous research regarding small car impacts on trailing-end anchorages for the Midwest Guardrail System (MGS) determined a CIP using LS-DYNA computer simulations for wedging the 1100C vehicle underneath the rail and snagging on the downstream cable anchor system for the MGS (Mongiardini et al. 2013a, Mongiardini et al. 2013b, Stolle et al. 2015). This CIP was determined to be 2,876 mm upstream from the end post in the system. Based on the similarity of the cable anchorage used in the MGS trailing-end anchorage system and the thrie-beam bullnose anchorage, the researchers elected to use the same CIP for test no. 3−37b to maximize the probability of vehicle and wheel snag on the cable anchor system.

Test designation nos. 3–34 and 3–35 are required to be conducted at a CIP where the behavior of the crash cushion transitions between capture and redirection of the impacting vehicle. In order to identify CIPs for these impacts, MwRSF reviewed previous testing of the thrie-beam bullnose system and end terminals and compared this data with the MASH 2016 impact conditions for test designation nos. 3–34 and 3–35. When testing the steel and wood post bullnose according to NCHRP Report No. 350 criteria, MwRSF conducted a test similar to MASH 2016 test designation no. 3–35 with the 2000P vehicle impacting the system at post no. 2 at an angle of 20°, and it proved to be a very difficult test to pass. For MASH 2016 test designation no. 3–35, the impact angle increased to 25°, and the mass of the pickup truck increased. Thus, there was reason to expect that the bullnose system would capture the vehicle rather than redirect it when impacted at post no. 2. It was also expected that the potential was greater for the system to redirect the vehicle when nearing post no. 5, which was when the rail became parallel to the road, and the impact was 3.8 m from the anchor. This behavior would correlate with post no. 3 for a standard guardrail end terminal, which is typically used for the beginning of length of need (LON) in terminal impacts. Thus, it was determined that the CIP for MASH test designation no. 3–35 be located at post no. 3, which was halfway between the cable anchor at post no. 1 and the assumed beginning of LON/redirection point at post no. 5.

A small car test of the CIP of the transition from capture to redirection was not conducted on the thrie-beam bullnose under NCHRP Report No. 350 as it was not required. Due to the lack of previous thrie-beam bullnose data related to this test, the researchers reviewed MASH end terminal testing for the MSKT (Griffith 2016) and the SoftStop (Griffith 2012). Note that these systems have posts at 1,905-mm post spacing, while the bullnose uses 953-mm post spacing. Test designation no. 3–34 on the SoftStop impacted at post no. 1 in the system (at the impact head). Test designation no. 3–35 on the SoftStop was conducted at the beginning of LON, and the impact point was at post no. 3. For the MSKT, test designation no. 3–34 was conducted with the impact point at post no. 2, and test designation no. 3–35 was conducted with the impact point at post no. 3. Thus, it appeared that test designation no. 3–34 has been typically conducted upstream from the system's LON point based on the test having a reduced impact angle of 15° and the lighter vehicle mass than test designation no. 3–35. Because the thrie-beam bullnose uses as similar cable anchorage to existing end terminal designs near the nose, it was believed that test designation no. 3–34 should be conducted upstream from the LON for end terminals due to the lighter vehicle mass and reduced impact angle, which would correspond to impacting upstream of the third post on a typical end terminal and the fifth post on the thrie-beam bullnose. Additionally, it was noted previously that test designation no. 3–35 should be conducted on the thrie-beam bullnose with the CIP located at post no. 3. Placement of the CIP at post no. 1 would eliminate the potential for vehicle redirection as the cable anchorage is connected to that post. Thus, post no. 2 on the thrie-beam bullnose was selected as the

CIP for evaluation of the transition of the system's behavior from capture to redirection for test designation no. 3–34.

4 Full-Scale Crash Testing

The results from the MASH TL-3 full-scale crash testing on the thrie-beam bullnose system are summarized below. Note that the bullnose system was modified following a failed test result in test no. MSPBN-4 by the addition of a third retention cable in the nose of the system. Thus, test nos. MSPBN-1 through MSPBN-4 were conducted on Version 1 of the thrie-beam bullnose with two retention cables in the nose of the system, and test nos. MSPBN-5 through MSPBN-8 were conducted on Version 2 of the thrie-beam bullnose with two retention cables in the nose of the system. The modification of the bullnose design during the testing will be discussed further during the individual test descriptions.

4.1 Test No. MSPBN-1

Test no. MSPBN-1 was conducted on the thrie-beam bullnose system according to MASH 2016 test designation no. 3–35 with a 2270P vehicle at an impact speed of 101.2 km/h and an angle of 25.1°. Test no. MSPBN-1 was conducted to examine the CIP of the bullnose system where its behavior transitioned from capture to redirection. After initial impact, the vehicle was contained and safely redirected by the bullnose system, as shown in Fig. 3. As the vehicle redirected, UBSP nos. A5 through A8 disengaged due to fracture of the base plate bolts. This post disengagement created limited vehicle pocketing and snag at post nos. A9 and A10, which were the first two standard W6 × 8.5 line posts in the system. However, this behavior did not compromise vehicle capture or stability and did not negatively affect the occupant risk values. The vehicle came to rest 20.1 m downstream from and 300 mm behind the initial impact point after the brakes were applied. Test no. MSPBN-1 met all the safety requirements for MASH TL-3.

Fig. 3. Test No. MSPBN-1

4.2 Test No. MSPBN-2

Test no. MSPBN-2 was conducted on the thrie-beam bullnose system according to MASH 2016 test designation no. 3–34 with an 1100C vehicle at an impact speed of 99.9 km/h and an angle of 14.7°. Test no. MSPBN-2 was conducted to evaluate the impact performance of the bullnose at the point where the device transitions from capture to redirection. After initial impact, the vehicle was captured and safely redirected by the bullnose system, as shown in Fig. 4. As the vehicle redirected, BCT post no. 2 and UBSP post nos. 3 through 6 were deflected laterally, but none of the posts disengaged. The cable anchorage at post no. 1 remained engaged as well. The vehicle exited and came to rest 30.3 m downstream from and 7.7 m in front of the initial impact point after brakes were applied. Note that some of the vehicle damage observed in the test was due to vehicle impact on an earth berm used to shield a portion of the test site after exiting the system. Test no. MSPBN-2 met all the safety requirements for MASH 2016 TL-3.

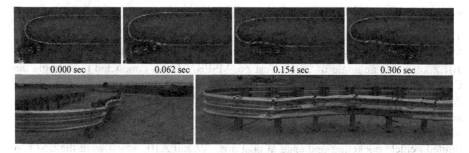

Fig. 4. Test No. MSPBN-2

4.3 Test No. MSPBN-3

Test no. MSPBN-3 was conducted on the thrie-beam bullnose system according to MASH 2016 test designation no. 3–32 with an 1100C vehicle at an impact speed of 100.9 km/h and an angle of 15.1°. Test no. MPSNB-3 was conducted to evaluate the bullnose behavior during oblique impacts on the end or nose of the system. Following the initial impact, the nose of the bullnose system wrapped around the front of the small car, as shown in Fig. 5. The lower peak of the thrie beam was pushed below the bumper and fractured, while the top two peaks of the thrie beam engaged the vehicle above the bumper and captured the vehicle. As the vehicle continued into the system, the thrie-beam rail was deformed and pulled downstream, and the breakaway posts in the system disengaged. These two actions dissipated the kinetic energy of the small car and decelerated it. The small car impacted the back side of post nos. B3 through B5, which further decelerated the small car. The vehicle was brought to a controlled stop approximately 0.800 s after impact. The vehicle came to rest 7 m from the initial impact location in the middle of the bullnose. Test no. MSPBN-3 met all the safety requirements for MASH 2016 TL-3.

Fig. 5. Test No. MSPBN-3

4.4 Test No. MSPBN-4

Test no. MSPBN-4 was conducted on the thrie-beam bullnose system according to MASH 2016 test designation no. 3–30 with an 1100C vehicle at an impact speed of 99.9 km/h and an angle of 1.3° with a 1/4 offset towards Side A of the system. During the test, the vehicle was initially captured by the thrie-beam nose of the system. However, as the vehicle proceeded into the system, the slot between the upper two corrugations of thrie beam and the lowest corrugation opened. This behavior allowed the top two thrie-beam corrugations to push upward and the lower thrie-beam corrugation to move downward and rupture. Eventually, the rail fracture and separation compromised the capture of the vehicle front end and allowed the vehicle to penetrate the system, as shown in Fig. 6. Thus, the system was not acceptable under the MASH 2016 TL-3 criteria.

Fig. 6. Test No. MSPBN-4

Following test no. MSPBN-4, the researchers reviewed the results from the test to determine the factors that influenced the failure of the system as well as to suggest potential modifications to the thrie-beam bullnose system to improve its performance. The performance of the system was also compared with previous NCHRP Report No. 350 test designation no. 3–30 testing on the same thrie-beam bullnose system in test no. USPBN-3 (Schmidt et al. 2010b). The mass of the small car and corresponding kinetic energy of MASH test no. 3–30 increased 25% as compared to previous NCHRP Report 350 testing. The increased kinetic energy imparted into the system may have contributed to the fracture of the lowest corrugation of the thrie beam in test no. MSPBN-4 and increased the potential for vehicle underride of the guardrail. The 1100C front-end geometry and structure also appeared to be different than that of the 820C vehicle tested previously in test no. USPBN-3, as shown in Fig. 7. These geometric differences may have altered the vehicle engagement and interlock with the nose rail and changed the nose rail slot tab tearing and corrugation separation. The 1100C vehicle had a more vertical profile, which may have contributed to the more rapid slot tab tearing and rail opening as well as the vertical motion of the upper two corrugations of the thrie beam up and over the front of the vehicle. The previous two factors led to more rapid rupture of the thrie-beam slot tabs in the nose section, opening of the thrie beam between the middle and lower rail corrugations, upward movement of the upper two rail corrugations, and rupture of the lower rail corrugation, which allowed the vehicle to penetrate the system.

Fig. 7. Vehicle profiles – 820C versus 1100C

Based on this analysis, it was determined that the best option for improving the system performance would be to add a third nose cable behind the lowest thrie-beam corrugation, as shown in Fig. 8. Adding a third cable to the lowest corrugation would improve the capture behavior with respect to the 1100C vehicle now that it had also been observed to fracture the rail in the nose section. The addition of a third nose cable was not expected to affect the performance of the thrie-beam bullnose system relative to the three previously successful MASH 2016 crash tests (test nos. MSPBN-1 through MSPBN-3), and it was determined that these three tests would not need to be rerun if the proposed modification was successful. All subsequent tests were conducted on a modified version of the bullnose system with a third nose cable.

Fig. 8. Thrie-beam bullnose with three retention cables in nose

4.5 Test No. MSPBN-5

Test no. MSPBN-5 was conducted on the modified thrie-beam bullnose system according to MASH 2016 test designation no. 3–30 with an 1100C vehicle at an impact speed of 100.9 km/h and at an angle of 0.3° with a 1/4 offset towards Side A of the system. During the test, the vehicle was captured by the thrie-beam nose of the system, as shown in Fig. 9. As the vehicle proceeded into the system, the thrie-beam rail and nose cables remained wrapped around the front of the vehicle. The deformation of the thrie-beam panels and the disengagement of several breakaway posts on both sides of the system decelerated the vehicle to a safe and controlled stop. The vehicle came to rest in the interior of the system 9.2 m downstream from the initial impact. Test no. MSPBN-5 met all the safety requirements for MASH 2016 TL-3.

Fig. 9. Test No. MSPBN-5

4.6 Test No. MSPBN-6

Test no. MSPBN-6 was conducted on the modified thrie-beam bullnose system according to MASH 2016 test designation no. 3–31 with an 2270P vehicle at an impact speed of 102.1 km/h and at an angle of 0.3°. During the test, the vehicle was captured by the thrie-beam nose of the system as the thrie-beam rail and nose cables wrapped around the front of the vehicle, as shown in Fig. 10. The deformation of the thrie-beam panels and the disengagement of several breakaway posts on both sides of the system decelerated the vehicle to a safe and controlled stop. The vehicle came to rest in the interior of the system 17.0 m downstream from initial impact. Test no. MSPBN-6 met all the safety requirements for MASH 2016 TL-3.

Fig. 10. Test No. MSPBN-6

4.7 Test No. MSPBN-7

Test no. MSPBN-7 was conducted on the modified thrie-beam bullnose system according to MASH 2016 test designation no. 3–33 with an 2270P vehicle at an impact speed of 101.6 km/h and at an angle of 15.1°. During the test, the vehicle was captured by the system, and the thrie-beam rail and nose cables wrapped around the front of the vehicle. The deformation of the thrie-beam panels and the disengagement of breakaway posts on both sides of the system decelerated the vehicle to a safe and controlled stop, as shown in Fig. 11. It was noted that late in the impact event, approximately 0.540 s after impact, the vehicle engaged some of the buildup of breakaway post debris and the first non-breakaway post on the left side of the system, which caused the vehicle to climb up the posts and roll to the right slightly as it was brought to a stop. This behavior did not cause issues with vehicle capture or the overall stability of the vehicle, nor did the override of the post debris cause any contact or tearing of the floorboard. The vehicle came to rest 15.83 m downstream of initial impact. Test no. MSPBN-7 met all the safety requirements for MASH 2016 TL-3.

4.8 Test No. MSPBN-8

Test no. MSPBN-8 was conducted on the modified thrie-beam bullnose system according to MASH 2016 test designation no. 3-37b with an 1100C vehicle at an impact speed of 101.8 km/h and at an angle of 25.0° at a CIP determined to maximize the potential for vehicle snag on the cable anchorage when impacted in the reverse

Fig. 11. Test No. MSPBN-7

direction. During the test, the vehicle was captured and redirected by the thrie beam on the side of the bullnose system, vehicle interaction with the cable anchorage was not observed, and the anchorage remained intact, as shown in Fig. 12. Test no. MSPBN-8 met all the safety requirements for MASH 2016 TL-3.

Fig. 12. Test No. MSPBN-8

4.9 1500A Analysis Procedure

Following the full-scale crash testing, an analysis procedure was performed using the accelerometer data from test no. MSPBN-6 to determine if test designation no. 3–38 was required for evaluating of the thrie-beam bullnose system. MASH 2016 allows the use of an analytical estimation procedure for the 1500A vehicle based on the results from test designation no. 3–31 to determine whether or not this test is required. The procedure consists of estimating the occupant risk values for the 1500A test based on the acceleration trace obtained from test designation no. 3–31. The estimated OIV and

ORA values for the 1500A vehicle impact with the thrie-beam bullnose system were 7.43 m/s and 13.75 g's, respectively. These values were well below the MASH limits. As such, test designation no. 3–38 was deemed non-critical for the thrie-beam bullnose system and was not conducted.

5 Summary and Conclusions

The thrie-beam bullnose barrier system was evaluated to determine its compliance with MASH TL-3 evaluation criteria. The thrie-beam bullnose consisted of a guardrail envelope comprised of thrie-beam panels mounted at a height of 803 mm and supported by UBSPs. The nose rail section and the first two adjacent rail sections were radiused and slotted, and the fourth and fifth rail sections were slotted. Retention cables were placed behind the nose rail section to aid in vehicle capture. The combination of the radiused and slotted rail segments and the breakaway posts allowed the system to safely capture and decelerate vehicles impacting near the end or nose of the system, while redirecting vehicles impacting along the side.

A total of seven tests were successfully conducted to complete the MASH TL-3 test matrix for evaluation of the thrie-beam bullnose system. These tests were test designation nos. 3–30, 3–31, 3–32, 3–33, 3–34, 3–35, and 3–37b. A failed full-scale crash test in test no. MSPBN-4 (test designation no. 3–30) led the researchers to modify the system with an additional retention cable in the nose of the system. This modification allowed the remaining crash tests to be successfully conducted. Based on the results of the successful full-scale crash tests conducted and the results of the estimation procedure for test designation no. 3–38, the modified thrie-beam bullnose system with a third nose cable met all of the safety requirements for MASH TL-3.

The MASH 2016 TL-3 thrie-beam bullnose system detailed herein was evaluated using a basic test configuration on level terrain. Real-world installations involve other design considerations, including soil grading around the barrier, transitioning to other barrier systems, and the location of hazards relative to the barrier among others. Additionally, there are three foreseeable field applications for the bullnose barrier system: (1) the protection of the gap between twin bridges; (2) gore area protection; and (3) protection of narrow median hazards, such as bridge piers and overhead sign support structures. For each of the three bullnose applications, there are installation and design factors that should be addressed before the system can properly be used in these situations. The researchers provided recommendations regarding implementation of the thrie-beam bullnose system in the research reports detailing the full-scale crash testing (Bielenberg et al. 2020a, Bielenberg et al. 2020b).

Acknowledgements. The authors wish to acknowledge the Midwest Pooled Fund Program for sponsoring and supporting this research effort and MwRSF personnel for constructing the barriers and conducting and analyzing the crash tests.

References

American Association of State Highway and Transportation Officials (AASHTO). Manual for Assessing Safety Hardware (MASH), Second Edition, Washington, D.C (2016)

American Association of State Highway and Transportation Officials (AASHTO). Manual for Assessing Safety Hardware (MASH), Washington, D.C (2009)

Arens, S.W., et al.: Investigating the Use of a New Universal Breakaway Steel Post, Report No. TRP-03-218-09, Midwest Roadside Safety Facility, University of Nebraska-Lincoln, Lincoln, Nebraska (2009)

Bielenberg, R.W., Faller, R.K., Reid, J.D., Rohde, J.R., Sicking, D.L., Keller, E.A.: Concept Development of a Bullnose Guardrail System for Median Applications, Report No. TRP-03-73-98, Project No. SPR-3(017)-Year 7, Midwest Roadside Safety Facility, University of Nebraska-Lincoln, Lincoln, Nebraska (1998a)

Bielenberg, R.W., et al.: Phase II Development of a Bullnose Guardrail System for Median Applications, Report No. TRP-03-78-98, Project No. SPR-3(017)-Years 7 and 8, Midwest Roadside Safety Facility, University of Nebraska-Lincoln, Lincoln, Nebraska (1998b)

Bielenberg, R.W., et al.: Phase III Development of a Bullnose Guardrail System for Median Applications, Report No. TRP-03-95-00, Project No. SPR-3(017)-Years 7 and 8, Midwest Roadside Safety Facility, University of Nebraska-Lincoln, Lincoln, Nebraska (2000)

Bielenberg, R.W., Reid, J.D., Faller, R.K.: NCHRP Report No. 350 compliance testing of a bullnose median barrier system, Paper No. 01–0204. Transp. Res. Rec. **1743**, 60–70 (2001)

Bielenberg, R.W., Faller, R.K., Ammon, T.J., Holloway, J.C., Lechtenberg, K.A.: MASH testing of bullnose with break away steel posts (Test Nos. MSPBN-1-3), Final Report, Report No. 03-389-20, Midwest Roadside Safety Facility, University of Nebraska-Lincoln, Lincoln, Nebraska (2020a)

Bielenberg, R.W., Ahlers, T.J., Faller, R.K., Holloway, J.C.: MASH Testing of Bullnose with Breakaway Steel Posts (Test Nos. MSPBN-4 through MSPBN-8), Final Report to the Midwest Pooled Fund Program, Transportation Research Report No. TRP-03-418-20, Project No.: TPF-5(193) Supplement #123, Midwest Roadside Safety Facility, University of Nebraska-Lincoln, Lincoln, Nebraska (2020b)

Griffith, M.S.: Federal Highway Administration. Eligibility Letter No. HSST/CC-115A for Trinity SOFT-STOP Terminal for line posts with 8" wood offset blocks, to Brian Smith, Trinity Highway Products, LLC (2012)

Griffith, M.S.: Federal Highway Administration. Eligibility Letter No. HSST-1/CC-126 for MASH Sequentially Kinking Terminal (MSKT), to Kaddo Kothmann, Road Systems, Inc. (2016)

Mongiardini, M., Faller, R.K., Reid, J.D., Sicking, D.L., Stolle, C.S., Lechtenberg, K.A.: Downstream Anchoring Requirements for the Midwest Guardrail System, Report No. TRP-03-279-13, Midwest Roadside Safety Facility, University of Nebraska-Lincoln, Lincoln, Nebraska (2013)

Mongiardini, M., Faller, R.K., Reid, J.D., Sicking, D.L.: Dynamic evaluation and implementation guidelines for a non-proprietary w-beam guardrail trailing-end terminal, Paper No 13-5277. Transp. Res. Rec. J. Transp. Res. Board **2377**, 61–73 (2013b)

Ross, H.E., Sicking, D.L., Zimmer, R.A., Michie, J.D.: Recommended procedures for the safety performance evaluation of highway features, national cooperative highway research program (NCHRP) Report 350. Transp. Res. Board (1993)

Schmidt, J.D., Sicking, D.L, Faller, R.K., Reid, J.D., Bielenberg, R.W., Lechtenberg, K.A.: Investigating the Use of a New Universal Breakaway Steel Post – Phase 2, Report No. TRP-03-230-10, Midwest Roadside Safety Facility, University of Nebraska-Lincoln, Lincoln, Nebraska (2010a)

Schmidt, J.D., Sicking, D.L, Faller, R.K., Reid, J.D., Bielenberg, R.W., Lechtenberg, K.A.: Investigating the Use of a New Universal Breakaway Steel Post – Phase 3, Report No. TRP-03-244-10, Midwest Roadside Safety Facility, University of Nebraska-Lincoln, Lincoln, Nebraska (2010b)

Stolle, C.S., Reid, J.D., Faller, R.K., Mongiardini, M.: Dynamic strength of a modified W-Beam BCT trailing-end termination, Paper No. IJCR 886R1, Manuscript ID 1009308. Int. J. Crashworthiness **20**(3), 301–315 (2015)

Safety Performance Evaluation of Modified Thrie-Beam Guardrail

Robert Bielenberg[✉], Ron Faller, and Karla Lechtenberg

Midwest Roadside Safety Facility, University of Nebraska, Lincoln, NE, USA
rbielenberg2@unl.edu

Abstract. Modified thrie-beam guardrail is a non-proprietary, longitudinal barrier system originally developed and evaluated under the National Cooperative Highway Research Program (NCHRP) Report 350 Test Level 3 (TL-3) and Test Level 4 (TL-4) safety performance criteria. Modified thrie-beam can be configured as both a single-sided roadside barrier and a dual-sided median barrier. The single-sided roadside configuration of the modified thrie-beam guardrail consists of 12-gauge thrie-beam panels mounted at a height of 864 mm and supported by 2,057-mm long W6 × 8.5 posts and W14 × 22 blockouts with an angled cutout in the web. The dual-sided modified thrie-beam guardrail is largely identical to the single-sided configuration, except that the blockouts and thrie-beam guardrail panels are mirrored on the backside of the system. Modified thrie-beam guardrail provides for increased vehicle capture and reduced deflection as compared to typical W-beam guardrails.

In 2009, the American Association of State Highway and Transportation Officials (AASHTO) implemented an updated standard for the evaluation of roadside hardware. The new standard, the Manual for Assessing Safety Hardware (MASH), improved the criteria for evaluating roadside hardware beyond the previous NCHRP Report No. 350 standard through updates to test vehicles, test matrices, and impact conditions. A research effort was undertaken to evaluate the both the single-sided and dual-sided median barrier configurations of modified thrie-beam guardrail to MASH Test Level 3 through full-scale crash testing. This paper details the performance evaluation of the thrie-beam guardrail through MASH TL-3 full-scale crash tests and recommendations for implementation of the system.

Keywords: Highway safety · Crash test · Compliance test · MASH 2016 · TL-3 · Roadside appurtenances · Thrie beam

1 Introduction

In 2016, the American Association of State Highway and Transportation Officials (AASHTO) implemented an updated standard for the evaluation of roadside hardware. The standard, called the Manual for Assessing Safety Hardware 2016 (MASH 2016) (AASHTO 2016), improved the criteria for evaluating roadside hardware beyond the previous National Cooperative Highway Research Program (NCHRP) Report No. 350 (Ross et al. 1993) standard through updates to the test vehicles, test matrices, and impact conditions. In an effort to encourage state departments of transportation and hardware developers to advance their hardware designs, the Federal Highway Administration (FHWA) and AASHTO have collaborated to develop a MASH implementation policy that includes sunset dates for various categories of roadside hardware. The new policy required that devices installed on federal aid roadways after the sunset dates must have been evaluated to MASH 2016.

The New Jersey Department of Transportation (NJDOT) and the California Department of Transportation (Caltrans) currently use the modified thrie-beam guardrail, which was previously evaluated to NCHRP Report No. 350 Test Levels 3 (TL-3) and 4 (TL-4). Additionally, these states desire to use a dual-sided version of the system for median applications that has yet to be evaluated to MASH or NCHRP Report No. 350. It was previously determined to be acceptable under NCHRP Report No. 350 by the FHWA based on crash testing of the single-sided system.

The original testing and evaluation of the modified thrie-beam guardrail was performed by the Texas A&M Transportation Institute (TTI) (Ivey et al. 1982). The original development of the modified thrie-beam guardrail stemmed from a desire to develop a barrier capable of safely redirecting bus-type vehicles, while still providing safe performance for passenger car impacts. Testing of standard thrie-beam guardrail during early research found that the performance of the standard thrie beam was marginal, as it captured and redirected the bus but allowed the vehicle to roll over. Thus, a modified thrie-beam guardrail was developed that utilized 356-mm deep M14 × 17.2 blockouts with an angled cutout and increased the top rail height to 864 mm. A thrie-beam backup plate was included between the thrie beam and the blockout at posts at non-splice locations to reduce the potential for stress concentrations that could arise as the thrie beam wrapped around the edge of the blockout during the impact. The modified thrie beam was evaluated by impacting the barrier with a 9,090-kg International school bus at 89.8 km/h and an angle of 15.0°. The modified thrie beam safely redirected the bus with a dynamic deflection of 875 mm. A subsequent test was conducted to evaluate the performance of small car on the system in terms of vehicle snag and capture. A 1,032-kg Honda Civic was used to impact the barrier at 100.6 km/h and an angle of 15.0°. The small car was safely redirected with a dynamic deflection of 244 mm. No vehicle snag on the system posts was noted. A second test with a Honda Civic involved a vehicle impacting at 99.5 km/h and 18.1° on the repaired barrier from the first test demonstrated very similar performance.

Several previous research efforts have evaluated modified thrie-beam guardrail under NCHRP Report No. 350. In 1995, TTI performed test designation no. 3–11 on a modified thrie-beam guardrail similar to the system detailed above, except the blockout section was changed to a W14 × 22 section (Mak et al. 1999). The modified thrie-beam guardrail system successfully contained and redirected the vehicle and met all evaluation criteria set forth in NCHRP Report No. 350 for TL-3. The maximum dynamic deflection of the guardrail was 1,036 mm. The relatively large dynamic deflection sustained by the guardrail system and snagging of the left wheel assembly on post no. 17 was somewhat unexpected given the stiffness of the thrie-beam rail element and the 356-mm deep blockout. Review of the high-speed film showed that post nos. 16 through 18 were severely twisted from the vehicle impact as the thrie-beam rail element deflected. The added moment arm from the deep blockout aggravated the torsional moment acting on the posts. As the posts twisted, the resistance to rail motion provided by the posts decreased, which increased the dynamic deflection of the guardrail. The torsional collapse of the posts allowed the left-front wheel assembly of the vehicle to come into direct contact with post no. 17.

Finally, two tests have been conducted on modified thrie beam under NCHRP Report No. 350 TL-4 impact criteria. TTI tested the modified thrie beam with W14 × 22 blockouts with an impact of a 8,000-kg single-unit truck at a speed of 78.9 km/h and an angle of 15.7° (Buth et al. 1999). The 8000S single-unit truck was safely and stably redirected with a maximum dynamic deflection of 710 mm. A subsequent test of the modified thrie beam was conducted according to NCHRP Report No. 350 TL-4 for Trinity Industries that used a slightly modified blockout with a different shape for the angled cutout (Griffith 2014). In this test, a 7.883-kg single-unit truck impacted the barrier at a speed of 80.8 km/h and an angle of 14.9°. The test resulted in a successful redirection of the 8000S vehicle with a dynamic deflection of 664 mm.

Review of previous testing of the modified thrie-beam system suggested the barrier may potentially meet MASH TL-3 criteria. However, the increased mass and kinetic energy of the MASH 2270P test vehicle has been shown to increase impact loading and dynamic deflection of guardrail systems. Additionally, no testing has been conducted on a dual-sided modified thrie-beam system. Thus, a need existed to evaluate the modified thrie-beam system under MASH 2016 criteria to determine its dynamic deflection, working width, and crashworthiness under MASH TL-3. If the modified thrie-beam system proved successful under TL-3 impact conditions, further study may be warranted regarding its performance under TL-4 impacts with the 10000S vehicle.

The objective of this research performed at the Midwest Roadside Safety Facility (MwRSF) was to conduct full-scale crash testing and evaluation of the modified thrie-beam guardrail system according to TL-3 of the MASH 2016 impact safety standards. The effort sought to evaluate both the single-sided and dual-sided median versions of the barrier. Two full-scale crash tests were conducted on the modified thrie-beam guardrail according to MASH 2016 test designation nos. 3–10 and 3–11. The researchers proposed conducting MASH test designation nos. 3–10 and 3–11 on the critical configuration of the barrier such that only two tests were required. Test designation no. 3–10 was conducted on the dual-sided, median version of the modified thrie beam and test designation no. 3–11 was conducted on the single-sided configuration. The test results were evaluated, and conclusions and recommendations were made pertaining to the safety performance of the system. Specific recommendations were made regarding implementation of the system, including transitioning of the modified thrie beam to crashworthy thrie-beam approach guardrail transitions (AGTs) and transitioning the modified thrie-beam transition from its 864-mm height to the 787-mm height of the Midwest Guardrail System (MGS).

2 Mash Evaluation Requirements

Longitudinal barriers, such as the modified thrie-beam guardrail, must satisfy impact safety standards in order to be declared eligible for federal reimbursement by the FHWA for use on the National Highway System (NHS). For new hardware, these safety standards consist of the guidelines and procedures published in MASH 2016. According to TL-3 of MASH 2016, longitudinal barrier systems should be subjected to two full-scale vehicle crash tests, as summarized in Table 1.

Table 1. MASH TL-3 test matrix for the thrie-beam bullnose

Test no.	Barrier configuration	Vehicle	Speed (mph)	Angle (deg)	Impact point
3–10	Dual-sided median	1100C	62	25	CIP- Based on MASH recommendation
3–11	Single-sided roadside	2270P	62	25	CIP- Based on MASH recommendation

Because NJDOT and Caltrans desired to evaluate both the single-sided roadside and dual-sided median versions of modified thrie-beam guardrail, MwRSF proposed to run test designation nos. 3–10 and 3–11 on the critical configuration of the barrier such that only two tests were required for evaluation. Test designation no. 3–10 (test no. MTB-2) was conducted on the dual-sided, median version of the modified thrie beam as this system configuration would tend to increase loading and occupant risk values for the small car vehicle and increase the propensity for vehicle snag on the post due to the higher stiffness and reduced dynamic deflection of the dual-sided system. Conversely, test designation no. 3–11 (test no. MTB-1) was conducted on the single-sided configuration because the 2270P vehicle will impart increased barrier loading on the components of a single-sided system. Additionally, the potential for the torsional buckling of the system posts that could lead to increased barrier deflection and post snag, as was observed in the original, NCHRP Report 350 test designation no. 3–11 testing of the modified thrie-beam guardrail, would be more prevalent in the single-sided configuration. Finally, evaluation of the single-sided, modified thrie-beam configuration with the 2270P vehicle would also produce the maximum dynamic deflection and working width values for the barrier system. Previous evaluation of the T-39 thrie-beam barrier for both roadside and median versions followed a similar methodology (Griffith 2006). Critical impact points (CIPs) for each full-scale crash test were developed using longitudinal CIP guidance provided in MASH based on the post dynamic yield force per unit length of the barrier and the effective plastic rail moment.

3 Modified Thrie-Beam System Details

Both side-sided roadside and dual-sided median configurations of the modified thrie-beam guardrail were constructed for full-scale testing and evaluation. The single-sided roadside configuration of the modified thrie-beam guardrail was 53.66 m long and consisted of 12-gauge thrie-beam panels supported by W6 × 8.5 posts and W14 × 22 blockouts with an angled cutout in the web. For the single-side roadside version of the modified thrie-beam guardrail evaluated in test no. MTB-1, post nos. 3 through 27 were 2,057-mm long W6 × 8.5 A36 steel posts spaced 1,905 mm apart with W14 × 22 blockouts and an embedment depth of 1,168 mm, as shown in Fig. 1. The blockouts were attached with two diagonally-opposed, 16-mm diameter, A307 Grade A bolts, and the thrie-beam rail elements were attached to the blockout with one 16-mm diameter, A307 Grade A button head bolt. Post nos. 3 through 27 featured 12-gauge, thrie-beam rails with additional post bolt slots at half-post spacing intervals. The mounting height was 864 mm to the top of the thrie-beam rail. Rail splices were located at posts. The lap splice connections between the rail sections were configured to reduce vehicle snag potential at the splice. The modified thrie-beam guardrail utilized 305-mm long, 12-gauge thrie-beam backup plates at each post location without a rail splice.

The upstream and downstream ends of the single-sided roadside configuration of the modified thrie-beam guardrail installation were configured with a non-proprietary end anchorage system (Mongiardini et al. 2013a, Mongiardini et al. 2013b, Stolle et al. 2015). The guardrail anchorage system had a comparable anchorage strength and capacity to other crashworthy end terminals. The anchorage system consisted of breakaway cable terminal (BCT) timber posts, foundation tubes, anchor cables, bearing plates, rail brackets, and channel struts. Due to the 864-mm height of the modified thrie-beam guardrail, a 10-gauge, symmetric W-beam to thrie-beam transition section was used to transition down to a 12-gauge, W-beam rail segment with a top mounting height of 765 mm at each end of the system. This selection allowed for anchorage of the system using typical trailing-end anchorage hardware. The only required modification was altering the hole location for the post bolt in the BCT posts to adjust for the 22-mm height difference.

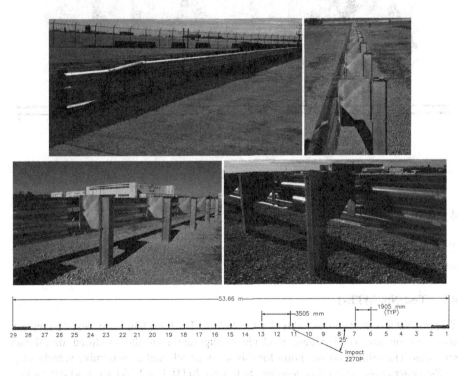

Fig. 1. Single-sided, roadside modified thrie-beam system, test no. MTB-1

The dual-sided, modified thrie-beam guardrail system evaluated in test no. MTB-2 was identical to the single-sided modified thrie-beam system with the exception of a second set of blockouts and thrie-beam rails installed on the backside of the line posts, as shown in Fig. 2. In addition, the upstream and downstream ends of the guardrail installation were configured with dual, non-proprietary, end anchorage systems on each end of the barrier to develop rail tension on both sides of the system.

Fig. 2. Dual-sided, median modified thrie-beam system, test no. MTB-2

4 Full-Scale Crash Testing

The results from the MASH TL-3 full-scale crash tests on the modified thrie-beam guardrail system are summarized below.

4.1 Test No. MTB-1

Test no. MTB-1 was conducted under the MASH TL-3 guidelines for test designation no. 3–11 on the single-sided, roadside configuration of the modified thrie-beam guardrail. The critical impact point for this test was selected to maximize vehicle snag on the system posts and splice loading. In test no. MTB-1, a 2,269-kg quad-cab pickup truck impacted the MTB guardrail at a speed of 101.2 km/h, an angle of 25.4°, and at an impact point 3,505-mm upstream from post no. 13, as shown in Fig. 3. During the test, the pickup truck was captured and redirected by the thrie beam. During the redirection of the vehicle, torsional collapse of some of the W-section blockouts was observed similar to that seen in the original NCHRP Report No. 350 testing of the system. The torsional collapse of the blockouts did not compromise the overall test result. However, it may have led to increased wheel snag on the posts and disengagement of the right-front wheel. Additionally, the collapse of the blockouts appeared to allow the lower portion of the thrie-beam guardrail to contact the flange and web of the blockout and the post flanges at post nos. 12 and 13. The contact at post no. 13 was sufficient to cause a small tear just downstream from the thrie-beam splice at that post. However, this tear did not adversely affect the barrier system's performance. The stability and trajectory of the vehicle were acceptable. Prior to coming to a stop, the test vehicle impacted portable barriers that were used to shield other areas of the test facility downstream from the barrier. This contact was well after vehicle exit and resulted in minor damage to the front of the test vehicle. The vehicle came to rest 86.0-m downstream from the impact point and 4.4-m laterally in front of the barrier after brakes were applied.

Fig. 3. Full-scale crash test no. MTB-1

Barrier damage was moderate, as shown in Fig. 4, and consisted of contact marks, deformation, disengaged rail elements, and bending, kinking, rotation, and twisting of the steel posts. The total length of vehicle contact along the barrier was approximately 11,544 mm, which spanned from 50 mm upstream from post no. 10 to 64 mm downstream from post no. 16. Various kinks and dents were observed on the rail between post no. 9 and post no. 17. As noted previously, a 102-mm long tear was found at the bottom edge of the rail 203 mm downstream from post no. 13. Damage to the system posts occurred mainly on post nos. 10 through 15 as these posts were deflected laterally in the soil and displayed permanent deformation. Post nos.

11 through 14 also displayed significant torsional deformation. Contact marks from tire/wheel snag were visible on post nos. 12 through 14. No damage was observed on post nos. 1 and 2, 17 through 22, and 24 through 29.

Fig. 4. Barrier damage, test no. MTB-1

Damage to the vehicle was moderate, as shown in Fig. 5. MASH 2016 defines intrusion or deformation as the occupant compartment being deformed and reduced in size with no observed penetration. Penetration of the occupant compartment is not allowed. There were no penetrations into the occupant compartment, and none of the established MASH 2016 intrusion limits were violated. Most of the damage was concentrated on the right-front corner and right side of the vehicle where impact had occurred. The front bumper cover was crushed inward and partially torn away from the vehicle, and the front bumper mounts were bent backward. The grille and both headlights were disengaged from the vehicle. The right-front wheel assembly was torn from the vehicle. The right-front fender was crushed inward. The right side of vehicle was deformed or scratched along its entirety. The right taillight was crushed. The right-side shocks were bent backward. The right-side, sway-bar, end link was disconnected from the lower control arm. The right-side steering knuckle disengaged from the vehicle. The right-side lower control arm broke, and the upper control arm bent backward. The steering gear box broke apart, and the right-side tie rod was bent.

Fig. 5. Vehicle damage, test no. MTB-1

The analysis of the test results for test no. MTB-1 showed that the system adequately contained and redirected the 2270P vehicle with controlled lateral displacements of the barrier. Detached elements, fragments, or other debris from the test article did not penetrate or show potential for penetrating the occupant compartment, or present an undue hazard to other traffic, pedestrians, or work-zone personnel. Deformations of, or intrusions into, the occupant compartment that could have caused serious injury did not occur. The test vehicle did not penetrate nor ride over the barrier and remained upright during and after the collision. Vehicle roll, pitch, and yaw angular displacements were deemed acceptable because they did not adversely influence occupant risk nor cause rollover. The calculated occupant impact velocities (OIVs) and maximum 0.010-s average occupant ridedown accelerations (ORAs) in both the longitudinal and lateral directions, as determined from the accelerometer data, were within the MASH 2016 suggested limits. After impact, the vehicle exited the barrier at an angle of 15.0°, and its trajectory did not violate the bounds of the exit box. Therefore, test no. MTB-1 was determined to be acceptable according to the MASH 2016 safety performance criteria for test designation no. 3–11.

4.2 Test No. MTB-2

Test no. MTB-2 was conducted under the MASH TL-3 guidelines for test designation no. 3–10 on the dual-sided median configuration of the modified thrie-beam guardrail. The critical impact point for this test was selected to maximize vehicle snag on the system posts and splice loading. In test no. MTB-2, a 1,095-kg small car impacted the barrier system at a speed of 101.5 km/h, an angle of 24.9°, and at an impact point 2,256-mm upstream from post no. 13, as shown in Fig. 6. The actual point of impact was 41 mm upstream from the target location. During the test, the vehicle was captured and redirected by the modified thrie-beam guardrail. As the vehicle was redirected, the right-front rim and tire of the vehicle snagged on post no. 13 in the system. However, the wheel snag did not adversely affect vehicle stability or the occupant risk values.

Fig. 6. Full-scale crash test no. MTB-2

After exiting the system, the vehicle came to rest 57.2 m downstream from the impact point and 15.8 m laterally in front of the barrier after brakes were applied.

Barrier damage was moderate, as shown in Figs. 7, and consisted primarily of bending, kinking, denting, and contact marks on the front face of the rail and deformation and displacement of the posts. The length of vehicle contact along the barrier was approximately 5,528 mm, which spanned from 559 mm upstream from post no. 12 to 965 mm downstream from post no. 14. Kinking, denting, and deformation of the thrie-beam rail was observed on the impact face thrie-beam rail between post no. 11 and post no. 15. Only minimal deformation was noted on the backside rail. Post damage was limited primarily to post nos. 12 through 14, which displayed minor bending of the post section near ground line. Contact marks were noted on the flange of post no. 13 due to wheel and tire contact. Soil gaps and displacement of the posts in the soil were found at post nos. 11 through 16. No damage was observed on the remainder of the posts.

Fig. 7. Barrier damage, test no. MTB-2

Damage to the vehicle was moderate, as shown in Fig. 5. MASH 2016 defines intrusion or deformation as the occupant compartment being deformed and reduced in size with no observed penetration. Penetration of the occupant compartment is not allowed. There were no penetrations into the occupant compartment, and none of the established MASH 2016 intrusion limits were violated. The majority of the damage was concentrated on the right-front corner and the right side where impact had occurred. The hood kinked on the right side. The front bumper detached, and the right-front quarter panel was deformed inward and scraped. The right-front door was deformed inward along its length and dented near the handle. The right-rear door was dented and scraped along its length. The right-rear quarter panel was crushed inward and scraped along its length. The right-rear wheel well was crushed inward, and the right-front wheel was dented due to contact with post no. 13. The right taillight was broken, and the cover was disengaged. The windshield was cracked and buckled outward slightly. The rest of the window glass and roof were undamaged. The right-side spring perch was bent. The right lower control arm was bent backward. The front cross member of the vehicle was bent upward near the midpoint (Fig. 8).

Fig. 8. Vehicle damage, test no. MTB-2

The analysis of the test results for test no. MTB-2 showed that the system adequately contained and redirected the 1100C vehicle with controlled lateral displacements of the barrier. Detached elements, fragments, or other debris from the test article did not penetrate or show potential for penetrating the occupant compartment, or present an undue hazard to other traffic, pedestrians, or work-zone personnel. Deformations of, or intrusions into, the occupant compartment that could have caused serious injury did not occur. The test vehicle did not penetrate nor ride over the barrier and remained upright during and after the collision. Vehicle roll, pitch, and yaw angular displacements were deemed acceptable because they did not adversely influence occupant risk nor cause rollover. The calculated occupant impact velocities (OIVs) and maximum 0.010-s average occupant ridedown accelerations (ORAs) in both the longitudinal and lateral directions, as determined from the accelerometer data, were within the MASH 2016 suggested limits. After impact, the vehicle exited the barrier at an angle of 13.4°, and its trajectory did not violate the bounds of the exit box. Therefore, test no. MTB-2 was determined to be acceptable according to the MASH 2016 safety performance criteria for test designation no. 3–10.

5 Summary and Conclusions

The modified thrie-beam guardrail system was full-scale crash tested and evaluated to determine its compliance with MASH 2016 TL-3 evaluation criteria in both a single-sided, roadside configuration and a dual-sided, median configuration. The single-sided, roadside configuration of the modified thrie-beam guardrail consisted of a 12-gauge thrie-beam panels mounted at a height of 864 mm, supported by 2,057-mm long W6 × 8.5 posts spaced at 1,905 mm, and incorporating W14 × 22 blockouts with an angled cutout in the web. The dual-sided, median configuration guardrail was identical to the single-sided, roadside configuration, except that the blockouts and thrie-beam guardrail panels are mirrored on the backside of the system.

Review of the system configurations and test requirements led the researchers to determine that test designation no. 3–11 was critical for evaluation of the single-sided, roadside configuration in order to maximize structural loading of the barrier system, evaluate the potential for collapse of the wide flange of the blockouts, and determine the maximum dynamic deflection and working width. Test designation no. 3–10 was selected to evaluate the dual-sided, median configuration as it would tend to produce increased loading and occupant risk values for the small car and increase the propensity for vehicle snag on the post due to the higher stiffness and reduced dynamic deflection of the dual-sided configuration. Based on the results of the two successful full-scale crash tests conducted, the modified thrie-beam guardrail system meets all safety requirements for MASH 2016 TL-3 for both single-sided roadside and dual-sided median configurations.

As noted previously, the modified thrie-beam guardrail system was previously successfully tested to NCHRP Report No. 350 TL-4. Based on its previous use as a TL-4 system, users may desire to use the modified thrie-beam guardrail as a TL-4 barrier under MASH as well. While the design of the modified thrie-beam guardrail system may have increased capacity as compared to standard W-beam guardrails due to its mounting height and use of thrie-beam rail elements, there are concerns with its ability to meet the MASH TL-4 safety criteria. Test designation no. 4–12 for MASH TL-4 consists of a 10,000-kg single-unit truck (SUT) impacting the barrier at 90 km/h and an angle of 15°. This test differs significantly from test designation no. 4–12 in NCHRP Report No. 350, which consisted of a 8,000-kg SUT vehicle impacting the barrier at 80 km/h and an angle of 15°. The increased mass and speed for MASH test designation no. 4–12 has led to increased barrier loads during crash testing of TL-4 barriers. Additionally, rigid barrier heights required to meet MASH TL-4 have increased to 914 mm in order to capture and contain the SUT vehicle. Based on the increased MASH TL-4 requirements, it is unknown if the modified thrie-beam guardrail can effectively meet MASH TL-4 without further research and/or full-scale crash testing.

6 Recommendations for Transitioning to Approach Guardrail Transitions

The modified thrie-beam guardrail systems detailed herein was evaluated using a basic test configuration on level terrain in both roadside and median configurations. Real-world installations will have other considerations for the application of the design that should be addressed, including working width guidance, alternative blockouts, installation adjacent to curbs, grading requirements, flaring of the barrier system, transitioning to W-beam guardrail systems, and attachment of end terminals and trailing-end anchors. The researchers provided recommendations regarding implementation of the modified thrie-beam guardrail system in the research report detailing the full-scale crash testing (Bielenberg et al. 2020). In addition to these issues, transitioning of the modified thrie-beam guardrail to approach guardrail transitions (AGTs) is commonly required. Further guidance on this issue is outlined in this section.

With regards to attachment of the modified thrie-beam guardrail to a thrie-beam AGT, it is recommended that the modified thrie-beam guardrail be attached to a MASH-compliant thrie-beam AGT that is crashworthy at both the upstream stiffness transition and the attachment to the bridge rail or parapet. MwRSF has previously developed an upstream stiffness transition for use when transitioning between the Midwest Guardrail System (MGS) and thrie-beam AGTs (Rosenbaugh et al. 2010,

Lechtenberg et al. 2012). This upstream stiffness transition should be applicable to the modified thrie beam as well, because the barrier system would have similar or greater stiffness than the MGS system. Details on attachment of the upstream stiffness transition from the MGS to a variety of crashworthy thrie-beam AGTs were described in the original research reports.

A schematic outlining the basic parts of a thrie-beam AGT and upstream stiffness transition to the MGS is shown in Fig. 9. Application of the MGS upstream stiffness transition to the connection of the modified thrie beam to a MASH compliant thrie-beam AGT should not require transitioning of the rail element as the modified thrie beam and the AGT both use thrie-beam rail elements. However, in order to apply the previously-developed, upstream stiffness transition to a crashworthy thrie-beam AGT, several minor adjustments to the basic schematic in Fig. 9 are needed.

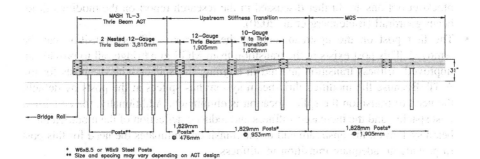

Fig. 9. Schematic of upstream stiffness transition from MGS to MASH TL-3 thrie-beam AGT

- The MASH TL-3 thrie-beam AGT region on the downstream end of the transition can use the post spacing and rail configuration of any MASH TL-3 compliant AGT. It should be noted that the selected MASH TL-3 compliant AGT should be compatible with the bridge rail/end buttress being used.
- In the upstream stiffness transition region, the 1,905-mm long, 10-gauge W-beam to thrie-beam transition section and the 1,905-mm long, 12-gauge thrie beam are replaced by a single 3,810-mm long thrie-beam section.
- In the upstream stiffness transition region, it is recommended to use the same W6 × 8.5 or W6 × 9 posts at the same spacing used in the original MASH-tested design. Note that end users could elect to use up to 2,057-mm long posts in that region as well if it was desired to limit the number of different post types in the system. For example, many thrie-beam AGTs use 1,981-mm long posts at reduced post spacing and the modified thrie beam uses 2,057-mm long posts. As such, it may be desired to use one of these post alternatives to limit the number of post types in inventory. It is believed that this increase in the post depth would not negatively

affect the upstream stiffness transition region as the modified thrie beam is already using 2,057-mm long posts.
- In the upstream stiffness transition region, it is recommended to use 152-mm × 305-mm × 483-mm timber blockouts. These blockouts are required in the upstream stiffness transition to reduce vehicle snag on the posts in that region. During MASH TL-3 testing of the upstream stiffness transition, researchers observed significant wheel snag with the small car and pickup truck on the posts in the upstream stiffness transition area where the vehicle engaged in the ½ post spacing adjacent to the W-beam to thrie-beam transition section. As such, there is concern with reducing blockout depth in that region. Additionally, it is not recommended to use the W14x22 blockouts from the modified thrie beam in that region due to their tendency to collapse in the web and potentially reduce their effective depth, which may similarly increase the snag concern. Potential alternative blockout options are further discussed in the research report on the modified thrie-beam guardrail (Bielenberg et al. 2020).
- The first post on the upstream end of the upstream stiffness transition can be removed. This post exists in the transition from MGS to a bridge rail to provide an improved stiffness transition and aid in aligning the splices with the posts for the AGT. Because the modified thrie-beam system has splices at the posts by default, the need to transition the splice location is eliminated. Additionally, the consistent post spacing and the increased stiffness and reduced deflection of the modified thrie-beam system on the upstream end of the transition eliminates the need for this post to provide an adequate transition in stiffness.
- Following the single 3,810-mm long, thrie-beam section, the modified thrie beam can be attached. The modified thrie beam will start at the 787 mm mounting height of the AGT and then transition to the standard modified thrie beam height of 864 mm over a minimum distance of 7.62 m to 15.24 m.
- Following the height transition for the modified thrie beam, standard modified thrie beam, as evaluated in this research study, is applied.
- Note that the use of curbs within the transition region would follow guidance published previously relative to AGTs and curbs (Winkelbauer et al. 2014).

As an example, a conversion from an existing AGT from MGS to a bridge rail has been completed in Fig. 10. The existing AGT design consisted of the MGS guardrail, the MASH TL-3 tested upstream stiffness transition, and a MASH TL-3 compliant thrie-beam transition to bridge rail, commonly called the "Iowa Transition," that utilizes 1,981-mm long, W6 × 8.5 or W6 × 9 posts at ¼ post spacing (Polivka et al. 2006). The example shown in Fig. 10 implements the transition guidance above to an existing AGT design.

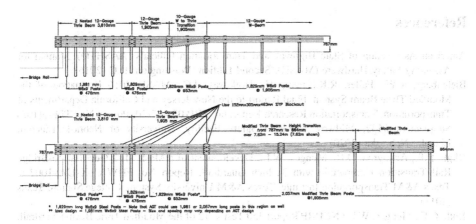

Fig. 10. Modified thrie-beam transition to thrie-beam AGT example

Alternatively, MwRSF has evaluated an 864-mm tall AGT that uses the standardized end buttress, which was developed through the Midwest Pooled Fund Program (Rosenbaugh et al. 2019). If desired, end users could apply this AGT configuration to attach to the modified thrie beam without a height transition. The basic configuration of the transition would be the same as the 787-mm tall AGT transition detailed previously, except that there would be no height transition and the 864-mm tall modified thrie beam would be attached directly following the single 3,810-mm long, thrie-beam section. In order to use this alternative, the AGT would have to be attached to the standardized end buttress designed for 864-mm tall AGTs.

End users may also be interested in the attachment of the modified thrie beam to MASH TL-2 compliant thrie-beam AGTs. Currently, only one thrie-beam AGT has been evaluated to MASH TL-2. The thrie-beam AGT shown in Fig. 8 was evaluated to MASH TL-2 through three full-scale crash tests at TTI (Bligh et al. 2011). This TL-2 thrie-beam AGT was identical to the previous MASH TL-3 upstream stiffness transition for thrie-beam AGTs developed at MwRSF when upstream from the downstream end of the W-to-thrie transition section. As such, the basic guidance provided previously for transitioning from modified thrie beam to MASH TL-3 AGTs would also apply to transitioning to the MASH TL-2 AGT system. Additional details and guidance for transitioning from modified thrie-beam guardrail to the MASH TL-2 AGT is provided in the research report (Bielenberg et al. 2020).

It should be noted that the proposed transition designs recommended herein are based on the best currently available transition research and engineering judgment. Further analysis and full-scale crash testing would be required to fully verify the performance of the transitions.

Acknowledgements. The authors wish to acknowledge the New Jersey Department of Transportation and the California Department of Transportation for sponsoring and supporting this research effort and MwRSF personnel for constructing the barriers and conducting and analyzing the crash tests.

References

American Association of State Highway and Transportation Officials (AASHTO). Manual for Assessing Safety Hardware (MASH), Second Edition, Washington, D.C (2016)

Bielenberg, R.W., Faller, R.K., Jiang, H., Holloway, J.C.: MASH 2016 Evaluation of the Modified Thrie Beam System, Final Report to the New Jersey and California Departments of Transportation, Transportation Research Report no. TRP-03-417-20, Project No.: TPF-5(193) Supplement #132, Midwest Roadside Safety Facility, University of Nebraska-Lincoln, Lincoln, Nebraska (2020)

Bligh, R.P., Arrington, D.R., Menges, W.L.: Development of a Mash TL-2 Guardrail-to-Bridge Rail Transition Compatible with 31-Inch Guardrail, Report No. FHWA/TX-12/9-1002-8, Texas A&M Transportation Institute, Texas A&M University System, College Station, Texas (2011)

Buth, C.E., Menges, W.L.: NCHRP Report 350 Test 4–12 of the Modified Thrie Beam Guardrail, Report No. FHWA-RD-99-065, Texas Transportation Institute, Texas A&M University System, College Station, Texas (1999)

Ivey, D.L., Robertson, R., Buth, C.E., McDevitt, C.F.: Test and Evaluation of W-beam and Thrie-Beam Guardrails, FHWA Contract DOT-FH-11-9485, Texas Transportation Institute, Texas A&M University System, College Station, Texas (1982)

Griffith, M.S.: FHWA Eligibility Letter B-248, Federal Highway Administration (2014)

Griffith, M.S.: Eligibility Letter HSA-10/B-148 for: T-39 Thrie-Beam Guardrail, Federal Highway Administration (FHWA) (2006)

Lechtenberg, K.A., Mongiardini, M., Rosenbaugh, S.K., Faller, R.K., Bielenberg, R.W., Albuquerque, F.D.B.: Development and Implementation of the Simplified MGS Stiffness Transition. Transp. Res. Rec. J. Transp. Res. Board **2309**, 1–11 (2012)

Mak, K.K., Bligh, R.P., Menges, W.L.: Testing of State Roadside Safety Systems: Volume I: Technical Report, Report No. FHWA-RD-98-036, Texas Transportation Institute, Texas A&M University System, College Station, Texas (1999)

Mongiardini, M., Faller, R.K., Reid, J.D., Sicking, D.L., Stolle, C.S., Lechtenberg, K.A.: Downstream Anchoring Requirements for the Midwest Guardrail System, Report No. TRP-03-279-13, Midwest Roadside Safety Facility, University of Nebraska-Lincoln, Lincoln, Nebraska (2013a)

Mongiardini, M., Faller, R.K., Reid, J.D., Sicking, D.L.: Dynamic evaluation and implementation guidelines for a non-proprietary W-beam guardrail trailing-end terminal, Paper No. 13-5277. Transp. Res. Rec. J. Transp. Res. Board **2377**, 61–73 (2013b)

Polivka, K.A., et al.: Performance Evaluation of the Guardrail to Concrete Barrier Transition – Update to NCHRP 350 Test No. 3–21 with 28 in. C.G. Height (2214T-1), Report No. TRP-03–175–06, Midwest Roadside Safety Facility, University of Nebraska-Lincoln, Lincoln (2006)

Rosenbaugh, S.K., Lechtenberg, K.A., Faller, R.K., Sicking, D.L., Bielenberg, R.W., Reid, J.D.: Development of the MGS Approach Guardrail Transition Using Standardized Steel Posts, Report No. TRP-03–210–10, Midwest Roadside Safety Facility, University of Nebraska-Lincoln, Lincoln (2010)

Rosenbaugh, S.K., Faller, R.K., Schmidt, J.D., and Bielenberg, R.W.: Development of a 34-in. Tall thrie-beam guardrail transition to accommodate future roadway overlays, Paper No. 19–03446. J. Transp. Res. Board Transp. Res. Rec. **2673**(2), 489–501 (2019). https://doi.org/10.1177/0361198118825464

Ross, H.E., Sicking, D.L., Zimmer, R.A., Michie, J.D.: Recommended Procedures for the Safety Performance Evaluation of Highway Features, National Cooperative Highway Research Program (NCHRP) Report 350, Transportation Research Board, Washington, D.C (1993)

Stolle, C.S., Reid, J.D., Faller, R.K., Mongiardini, M.: Dynamic Strength of a modified W-Beam BCT trailing-end termination, Paper No. IJCR 886R1, Manuscript ID 1009308. Int. J. Crashworthiness 20(3), 301–315 (2015)

Winkelbauer, B.J., et al.: Dynamic Evaluation of MGS Stiffness Transition with Curb, Report No. TRP 03-291-14, Midwest Roadside Safety Facility, University of Nebraska Lincoln, Lincoln, Nebraska (2014)

In-Service Performance Evaluation of Iowa's Sloped End Treatments

Cody Stolle[1(✉)], Jessica Lingenfelter[2], Khyle Clute[3], and Robert Bielenberg[1]

[1] Midwest Roadside Safety Facility, The University of Nebraska-Lincoln, Lincoln, NE, USA
cstolle2@unl.edu
[2] Kawasaki Motors Manufacturing Corp., Lincoln, USA
[3] Iowa Department of Transportation, Lincoln, USA

Abstract. Sloped end treatments (SETs) were historically developed as low-cost, low-maintenance end treatments for rigid features like concrete barriers and bridge rails. Crash testing indicated that sloped end treatments are associated with significant instability for impacting vehicles, but their historical crash performance was not known.

An in-service performance evaluation (ISPE) was performed to evaluate vehicle crashes with sloped end treatments in Iowa between 2008 and 2017. Researchers generated a geographic inventory of 658 sloped end treatment locations, reviewed crash narratives and scene diagrams for crashes near these sloped end treatments, and calculated an estimated crash rate and crash cost for these sloped end treatments. A total of 30 SET-related crashes were identified, resulting in one fatal crash and one severe injury crash. Eighteen of these crashes climbed the SET feature and seven crashes were associated with vehicle rollover, with eight unknown crash outcomes.

The estimated sloped end treatment average crash cost was approximately $178,260. For comparison, crash costs for other fixed objects in the same vicinity as sloped end treatments, typically associated with roads with higher speed limits, averaged $67,449 per crash. Benefit cost was calculated for replacing select groups of sloped end treatments with various crash cushions. Nineteen of the crashes occurred on a total of seven bridges, indicating most impacts were in "black spot" locations. Researchers recommended prioritization of removal and replacement of some sloped end treatments based on crash history.

Keywords: In-service performance evaluation · Concrete sloped end treatment · Concrete barriers · MASH · Injury analysis · Benefit-to-cost ratio

1 Introduction

Concrete barrier sloped end treatments (SETs) are used in many states such as Iowa for terminating ends of concrete barriers. Historically, before crashworthy end treatment options were developed, SETs offered a safety benefit compared to terminating concrete barriers with blunt ends. SETs are also generally inexpensive to install and require no routine maintenance and minimal repair. SETs can be cast in place and doweled into an existing concrete barrier and can be installed in conjunction with a curb. Examples of SETs from the state of Iowa are shown in Fig. 1.

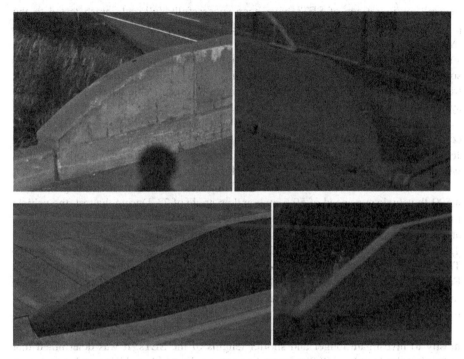

Fig. 1. Examples of concrete SETs in Iowa

Since the adoption of the National Cooperative Highway Research Program (NCHRP) Report No. 350 (Ross et al. 1993) and the American Association of State Highway and Transportation Officials (AASHTO) *Manual for Assessing Safety Hardware* (2016), many SETs were removed and replaced with newer, crashworthy end treatment options. However, SETs are still preferred in some locations with:

- low ADT and limited crash history;
- limited space due to intersections, driveways, or other fixed obstacles;
- curbs and gutters which could adversely affect crashworthiness of other features; or
- exceptional difficulty for end treatment repair or maintenance.

Some of the SETs have been successfully full-scale crash tested, but typical test conditions consist of level, flat terrain and test vehicles typically experience significant roll angle displacements during the tests. The safety performance of SETs may be different when installed in conjunction with bridge ends or adjacent to slopes. It is uncertain what risk, if any, is posed to occupants of vehicles during crashes with real distributions of impact conditions and roadside geometries. An in-service performance evaluation (ISPE) of these features has not been conducted prior to this study. Therefore, the Iowa Department of Transportation (IaDOT) funded research to determine the ISPE of existing concrete SETs and recommend warrants for replacing SETs, based on factors such as cost-effectiveness, site limitations, or crash history.

2 Background

Several configurations of SETs have been successfully full-scale crash tested to the prevailing criteria in the U.S. during the 1970s through 2000s, such as NCHRP 230 (Michie 1981), NCHRP 350 (Ross et al 1993), and MASH (2016) criteria. During some of these tests, vehicles experienced high roll angles, instability, or rollover, and some vehicles came to rest on the non-traffic side of the SET. Although SETs are not traditionally defined as gating terminals, vehicle traversal to the non-traffic side of the system was nonetheless deemed acceptable.

A summary of the full-scale testing involving SETs is provided below. Testing was conducted at the Southwest Research Institute (SwRI) in the 1970s to evaluate the New Jersey SET (NJSET) (Bronstad et al. 1976), shown in Fig. 2(a). Two full-scale tests were performed, evaluating the NJSET performance at low impact angles.

Test no. CMB-17A featured the NJSET impacted 30 ft from the leading end by a 4,500 lb vehicle at a speed of 59.6 mph and an angle of 7°. During the test, the vehicle impacted the NJSET, slid along the top of the barrier until the barrier installation ended, and regained contact with the ground. Test no. CMB-17B was performed with a 4,500 lb vehicle impacting the NJSET 26 ft from the leading end at 64.1 mph and 10°. The vehicle rode over the NJSET and landed behind the barrier, and remained upright. For both test nos. CMB-17A and CMB-17B, the test vehicles experienced significant roll displacement and instability, but test results were considered successful because rollovers did not occur.

Subsequent computer simulations of alternative barrier geometries were conducted, mostly to investigate longer and shorter lengths of the SETs. It was determined that increasing the length of SETs resulted in marginally stable vehicles and was not an improvement compared to shorter SETs. Therefore, it was recommended that taper length be shortened to reduce cost.

In NCHRP Report No. 358 (Ross et al. 1994), a series of work zone and temporary barrier applications were evaluated. Full-scale tests and simulations were conducted on two types of concrete barrier SETs: conventional SET (CSET), shown in Fig. 2(b), and New York SET (NYSET), shown in Fig. 2(c). Full-scale tests were performed with small cars, weighing approximately 1,970 lb (894 kg). Three tests were performed with the NYSET and three were performed with the CSET. Impact speeds ranged between 30 and 45 mph and impact angles were either 0 or 30°. Two of the six tests impacted the SET on the leading end and four impacted 2 ft from the leading end. Four of the six tests resulted in vehicle rollover. The remaining two tests, nos. 7110-5 and 7110-8, both of which impacted the SET end-on, resulted in marginally stable vehicles. After reviewing these tests, it was found that the guide plate attached to the right-front wheel contacted the pavement before the wheel, which resisted vehicle rollover. Simulations were utilized to determine the validity of this finding: simulations with the guide plate predicted no rollover, and those without predicted rollover. Researchers concluded that an end-on impact at 45 mph with a SET would result in vehicle rollover.

Researchers conducted computer simulation on additional impact conditions and SET designs using the CSET model, because it was simpler than the NYSET model but had similar test outcomes. An 1,800 lb test vehicle was simulated impacting CSETs of varying taper lengths at varying impact angles, locations, and speeds, for a total of 84 simulations. From simulation results, it was recommended that SETs be at least 20 ft long and be used on roadways with speed limits less than or equal to 45 mph.

Texas Transportation Institute (TTI) developed a low-profile concrete barrier and associated SET for the Texas Department of Transportation (TXDOT) (Beason et al. 1992), and evaluated it with NCHRP 230 test vehicles and impact angles, at work zone speeds of 45 mph. The barrier was 20 in. tall and utilized a rectangular profile, shown in Fig. 2(d). Three full-scale tests were performed on the low-profile SET (LPSET). Test no. 1949A-1 impacted the LPSET 6.5 ft from the end at 16.3° and 44.7 mph, and the LPSET redirected the vehicle. Test no. 1949A-2 impacted the LPSET end-on at 45.1 mph, with the centerline of the right wheels aligned with the centerline of the LPSET. The right side of the vehicle rode along the top of the barrier, then the vehicle lost contact with the barrier and exited the system. Test no. 1949A-3 impacted the LPSET end-on at 46.5 mph, with the centerline of the vehicle aligned with the centerline of the LPSET. The vehicle rode atop the barrier before coming to rest. Thus, the LPSET was determined to be successful according to NCHRP 230 test criteria.

Fig. 2. Full-scale crash tested concrete SETs, (a) NJSET, (b) CSET, (c) NYSET, and (d) LPSET

TTI re-evaluated the LPSET according to NCHRP 350 TL-2 criteria in 1998 (Beason et al. 1998). Test no. 414038-1 was performed with a 1990 Ford Festiva impacting the LPSET 3 ft from the end at of 44.1 mph and 15.8°. During the test, the right rear tire rolled on top of the barrier and the vehicle came to rest on the traffic side of the barrier. Test no. 414038-2 featured a 1990 Ford Festiva impacting the nose of the LPSET at 15.1° and 42.8 mph. The vehicle traveled up the LPSET and came to rest on the non-traffic side of the concrete barrier. Thus, the LPSET was determined to be successful according to NCHRP 350 TL-2 test criteria.

In 2013, TTI re-tested a modified, non-pinned version of the LPSET according to MASH TL-2 criteria (Beason et al.). Test no. 490023-5 was performed with the car impacting the LPSET 33 in. from the end at 43.9 mph and 15.2°. During this test, the vehicle rode up the LPSET and came to rest on the non-traffic side of the barrier. Test no. 490023-7 was performed with a 2270P pickup truck impacting the LPSET at 45.0 mph and 25.3°. The impact location was 78.0 in upstream of the splice location, coinciding with where the LPSET reached a height of 18 in. The vehicle was successfully redirected and came to rest on the traffic side of the barrier. Thus, the LPSET was determined to be successful according to MASH impact conditions.

3 Data Collection and Review Methodology

The first step to executing the ISPE involved identifying crashes involving SETs. An inventory of SET locations in the state of Iowa was not available, and creating a statewide, all-inclusive inventory was not within the scope of this project. Researchers reviewed every road in three counties in Iowa to every SET, as well as identify the features that SETs were attached to. Then, researchers extrapolated those findings to generate a statewide estimate of SET locations using Iowa DOT's bridge inventories.

The virtual tour utilized Google Earth and Google Street View (Google) to virtually tour every road in Johnson, Polk, and Linn counties to identify locations of SETs. Researchers annotated the locations, types of road characteristics, and features connected to the SETs (i.e., concrete barriers). For these three counties, it was found that 93% of SETs were located on bridges or overpasses, 5% were located on entrance or exit ramps, and 2% were located on other roadways. Because SETs were primarily connected to bridges, which are maintained and indexed by Iowa DOT, researchers believed that most SETs would be located adjacent to bridges and a thorough review of every road in Iowa was unnecessary to develop a representative, statewide sample of SET locations and crash data.

Iowa DOT provided researchers access to a bridge inventory database, which included bridge number, latitude, longitude, and features and structures in conjunction with each bridge. The bridge inventory was tabulated as a combination of maintenance, asset management, and inspection databases. Each bridge in the inventory was inspected using Google Street View (Google) to determine which bridges were attached to concrete SETs. A total of 658 unique SETs were identified in conjunction with 183 bridges. Additionally, a small number of entrance or exit ramps transitioning between low-ADT, lower-speed roads and higher-speed roads such as freeways utilized

SETs at intersections. The geo-terrestrial mapping software ArcGIS was used to tabulate the locations and unique indices of each SET.

Next, Iowa DOT provided crash data for a 10-year span extending from 2008 to 2017, filtered based on crash location for crashes which occurred within 305 m of the location of any SET. The large radius was used in the event that the crash location noted on the crash report indicated one of: (a) final position of the vehicle; (b) initial loss of control or initial roadside departure; (c) pre-crash contributing circumstances; and (d) representative location of impact close to the SET. Per Iowa DOT, it was assumed that no new SETs were constructed or repaired during the 10-year window, and thus no data was excluded from review due to road construction or closure throughout the 10-year (3,653-day) span. Careful review of crash reports, scene diagrams, crash narratives, and crash locations led to the identification of 30 impacts with SETs and 2,346 non-SET, fixed-object crashes. Crash results are summarized in Table 1.

Table 1. Summary of SET Crashes

Year	Total number of crashes in Iowa	Number of SET crashes in Iowa	Percentage of crashes that involved SETs
2008	59,918	3	0.005%
2009	55,494	0	0.000%
2010	54,396	6	0.011%
2011	48,793	5	0.010%
2012	47,882	4	0.008%
2013	50,009	0	0.000%
2014	52,102	3	0.006%
2015	54,624	2	0.004%
2016	55,848	2	0.004%
2017	55,180	5	0.009%
Total	534,246	30	0.006%
Average	53,425	3	0.006%

For the 10 years of crash data, SET crashes accounted for 0.006% of all crashes, or an average of 3 SET crashes per year. In comparison, an average of 53,425 reported crashes occurred annually in Iowa.

4 Exposure Calculation

Researchers investigated the crash risk associated with SETs, which was calculated using exposure, or number of opportunities for vehicles to engage a SET. Researchers established a series of assumptions regarding the SET exposure:

- Many bridges were associated with more than one SET, and the most common configuration was four SETs attached to one bridge. It was believed that in a given crash, only one SET would be struck, and vehicles passing by each SET would only be counted one time.
- Only crashes in which a vehicle impacted the upstream SET were considered. Reverse-direction impacts were either likely to impact the attached feature (e.g., bridge rail) upstream from the SET, or traverse the trailing SET without damage. Two-directional traffic flow was assumed to be equally distributed, with half traffic passing by the feature in one travel direction, and half in the opposite direction. Thus, for two-directional traffic flow, the exposure for each end of the bridge would be ½ the total ADT.
- For two-directional traffic, it was assumed that 60% of the crash risk was associated with features on the right side of the road, and 40% with features on the left side. For one-directional traffic, it was assumed that left- and right-side departures would be equally weighted (50% each) (Mak et al. 2010).

The exposure for each SET was calculated and used to find the average SET exposure as well as the total exposure. The equation utilized to calculate exposure is shown in Eq. 1.

$$\text{Exposure}_i = (\text{AADT})_R * (\text{Traffic Factor})_R * (\text{Side Factor})_i * \text{Time} \quad (1)$$

Where:
Exposure_i = number of opportunities to crash into the ith SET
AADT_R = annualized average daily traffic at SET (vehicles/day)
Traffic Factor_R = for road adjacent to i^{th} SET:
two-way traffic: 0.5
one-way traffic: 1.0 for upstream end, 0.0 for downstream end
Side Factor_i = Run-off-road risk per SET:
two-way traffic: 0.6 for right-side, 0.4 for left-side
one-way traffic: 0.5 (left and right)
Time = years of traffic data (days)

A total of 98 out of the 183 bridges had four SETs, located on each approach and departure, with two-way traffic. Six bridges featured one-way traffic but still utilized four SETs, two on the upstream side of the bridge and two on the downstream side. Eight bridges utilized three SETs each, with the remaining end consisting of an approach guardrail transition. Of bridges with two SETs, 17 bridges utilized SETs adjacent to only one lane; 2 bridges utilized SETs only at catty corners; 10 bridges utilized SETs only on one end of the bridge on left- and right-sides with two-way traffic; and 5 bridges utilized two SETs on one bridge end, with one-way traffic. A total of 5 bridges featured a single SET. The remaining SETs were located at ramps leading to or from high-speed roadways. An example of a bridge with four SETs and the associated numbering scheme is shown in Fig. 3. Additional documentation for the locations of the SETs used in conjunction with Iowa bridges is provided by Lingenfelter et al. (2020).

AADT data was available for 632 of the 658 SETs (96%), and the total exposure for the 10-year span was 4,802,002,443 vehicles. An estimated total exposure of 4,999,553,176 vehicles was calculated by scaling the total exposure by 1.04 (658/632). The estimated average exposure per SET was 7,598,105 vehicles and the estimated average exposure per SET per year was 759,811 vehicles. Lastly, the estimated exposure-based crash rate of SETs was calculated by comparing the estimated total exposure to the total number of SET crashes. Therefore, the exposure rate of SETs was 4,999,553,176 to 30, or 166,651,773 to 1.

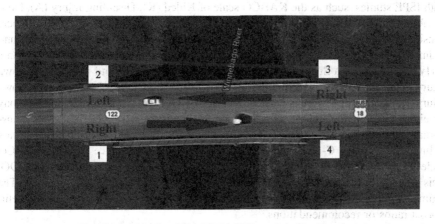

Fig. 3. Bridge No. 1710.2S122 SETs and traffic direction

5 Crash Outcomes and Crash Costs

The database of crashes involving SETs was limited in scope, so researchers compared the SET results to the baseline dataset of crashes involving other fixed objects in the vicinity of sloped end treatments. Results of the crash results involving SET and baseline, non-SET fixed object crashes are shown in Table 2. The SET and non-SET, baseline, fixed-object impact databases were compared to determine if there were any unique crash characteristics related to SET impacts; however, no statistically-significant differences were identified. Nonetheless, SET crashes were generally associated with lower-speed roads and not in snowy, icy, or slushy conditions.

It was noted that each of the SETs which were involved in a crash were located in one of three locations: (1) at the upstream, right-side of the bridge; (2) at the upstream end the left-side of a one-direction roadway; or (3) at the upstream end of a concrete barrier adjacent to a ramp connected to a rigid, concrete bridge rail. No impacts were identified in which a vehicle crossed from the right-side lanes past the left-side lanes and impacted the SET, nor were "reverse-direction" impacts in which the downstream, trailing SET contributed to the crash outcome. Unrelated impacts, including traversing an SET after a separate impact with a bridge rail, were not considered.

The maximum injury level sustained by any occupant of a crash was recorded in Iowa crash report forms as one of: Fatal, Suspected Serious/Incapacitating Injury, Suspected Minor/Non-Incapacitating Injury, Possible/Unknown Injury, and Property Damage Only (PDO). These values were associated with estimated cumulative societal cost per crash category which were used to investigate the benefit-to-cost associated with treating the sloped ends or replacing it with an alternative, crashworthy feature to MASH requirements. This injury scale is slightly different from values commonly used with ISPE studies, such as the KABCO scale of Killed (K), Disabling Injury (A), Non-Disabling Injury (B), Minor Injury (C), and Property Damage Only (O). Researchers considered benefit-to-cost and crash cost estimates using both Iowa's injury scale and estimated crash costs, as well as using the Federal Highway Administration's (FHWA's) Value of a Statistical Life (VSL) estimates for 2016 and mapping the Iowa injury data to a KABCO distribution, as shown in Tables 2 and 3. Most of the Iowa injury distribution data were consistent with KABCO, except that "minor" injuries may be either C-level (i.e., minor, non-transportable injury) or B-level (moderate, transportable but non-incapacitating). Thus, researchers estimated that 80% of Iowa's "Minor" injury were C-level and 20% were B-level, in accordance with the KABCO scale. In addition, the "Possible/Unknown" injury identified was assumed to be PDO. This was the most economically-favorable condition for retaining existing SETs. The estimated injury distribution did not have a significant outcome on the resulting benefit-to-cost ratios or recommendations.

The estimated total crash cost and average cost per crash were estimated using the Iowa injury distribution classification and Iowa's estimated costs per crash, as shown in Table 3. For SET crashes, the average cost per crash was approximately 2.6 times the average cost per crash of other fixed objects in the non-SET, fixed object control group, which were located within 305 m of the SETs. Despite this observation, the SET impacts were less common in conjunction with roads with elevated speed limits, which suggests that the average impact speed for SETs was lower than the average speed at impact with other fixed objects. However, this correlation is assumed, as speed limit may not be a good predictor of actual impact speed.

Table 2. Summary of SET and Non-SET crash datasets

Crash characteristic		SET crashes		Non-SET crashes (Baseline)	
		Crashes	Percent	Crashes	Percent
Total crashes		30	–	2,346	–
Speed limit	40 kmph	6	20.0%	416	17.7%
	48 kmph	6	20.0%	220	9.4%
	56 kmph	13	43.3%	481	20.5%
	64 kmph	1	3.3%	40	1.7%
	72 kmph	2	6.7%	229	9.8%
	80 kmph	0	0.0%	23	1.0%
	89 kmph	2	6.7%	582	24.8%
	97 kmph	0	0.0%	355	15.1%
Weather	Clear	15	50.0%	1,025	43.7%
	Cloudy	8	26.7%	666	28.4%
	Rain	7	23.3%	242	10.3%
	Snow	0	0.0%	321	13.7%
	Other	0	0.0%	92	3.9%
Road conditions	Dry	21	70.0%	1,286	54.8%
	Wet	8	26.7%	403	17.2%
	Slush, Snow, & Ice	1	3.3%	620	26.4%
	Other	0	0.0%	37	1.5%
Vehicles	Passenger Car	15	50.0%	1,313	56.0%
	SUV (light truck)	7	23.3%	444	18.9%
	Pickup (light truck)	4	13.3%	299	12.7%
	Van (light truck)	3	10.0%	125	5.3%
	Large Vehicle	1	3.3%	96	4.1%
	Other	0	0.0%	69	2.9%
Injuries (Iowa Scale)	PDO	22	73.3%	1,527	65.1%
	Possible/Unknown	1	3.3%	440	18.8%
	Minor	5	16.7%	289	12.3%
	Major	1	3.3%	70	3.0%
	Fatal	1	3.3%	20	0.9%
Injuries (KABCO Scale)	O	23	76.7%	1,578	67.3%
	C	4	13.3%	606	25.8%
	B	1	3.3%	72	3.1%
	A	1	3.3%	70	3.0%
	K	1	3.3%	20	0.9%

Table 3. Crash costs for baseline, Fixed-Object, Non-SET and SET crashes

Injury level	Cost per crash outcome	SET crashes		Non-SET Fixed-Object crashes (Reference)	
		No. crashes	Total cost	No. crashes	Total cost
PDO	$7,400	22	$162,800	1,527	$11,299,800
Possible/Unk	$35,000	1	$35,000	440	$15,400,000
Minor	$65,000	5	$325,000	289	$18,785,000
Major	$325,000	1	$325,000	70	$22,750,000
Fatal	$4,500,000	1	$4,500,000	20	$90,000,000
Total	–	30	$5,347,800	2,346	$158,234,800
Average cost per crash		$178,260		$67,449	

Each of the 30 SET crashes were reviewed to evaluate how the vehicle trajectory and reaction after contacting the SET were related to injury outcome, as shown in Table 4. For every crash, researchers attempted to determine if the impacting vehicle remained stable or rolled over during or after impact with the SET, as well as if the vehicle was redirected (i.e., impacted along the side of the SET and remained on the traffic side of the feature) or climbed the SET. For crashes in which the vehicle climbed the SET, researchers subsequently evaluated if the vehicle came to rest on the traffic side, on top of the barrier, or on the non-traffic side of the barrier. Scene diagrams, crash narratives, and crash report data were not conclusive to determine if the vehicle was redirected by or climbed up the SET in 8 of the 30 crashes. None of the crashes with unknown outcomes were associated with severe or fatal injuries. Of the 22 crashes with known vehicle response outcomes, four crashes (18%) were associated with redirection and 18 crashes (82%) were associated with climbing behaviors. This distribution is understandable due to the extremely short length of longitudinal SET exposure which would not be associated with climbing, compared to the much larger likelihood of engaging the sloped face between the vehicle's impacting front two wheels.

Table 4. Injury vs. outcome for SET crashes

Vehicle action			Injury					Total
Crash outcome	Vehicle stability	Location of final rest	O	C	B	A	K	
Redirect	Non-Rollover	Traffic side	–	–	–	–	–	–
	Rollover	Traffic side	4	–	–	–	–	4
Total Redirected by SET			4	–	–	–	–	4
Climb	Non-Rollover	Traffic side	1	–	–	–	–	1
		Non-Traffic side	3	1	–	1	–	5
		Top of barrier	3	–	–	–	–	3
	Rollover	Traffic side	2	2	1	–	–	5
		Non-Traffic side	3	–	–	–	1	4
		Top of barrier	–	–	–	–	–	–
Total Climbed SET			12	3	1	1	1	
Unknown			7	1	–	–	–	8
Total			23	4	1	1	1	30

Two severe crashes were identified which involved the sloped end treatments, and both of the severe crashes involved a vehicle impacting the SET at the upstream end and climbing the end treatment. The first crash was a non-rollover event in which a vehicle rode up and vaulted off of the SET, over the edge of the overpass at the bridge deck, and laned on the travel lanes below the overpass. Based on preliminary kinematics analysis, it appeared that the impacting vehicle was traveling faster than the posted 56 km/h posted speed limit at the point of impact. The second crash involved a vehicle impacting the SET while performing a steering maneuver, causing the vehicle to ride up the barrier and trip over the edge of the bridge deck onto the travel lanes below. The crash outcome resulted in a fatality as the impacting vehicle laned on the occupant compartment.

One additional crash was worth noting in detail, which involved a non-rollover, non-injury climbing event. The vehicle came to rest with the left-side wheels straddling the barrier and the right-side wheels located on a pedestrian walkway sidewalk, which was located to the side of a bridge rail. This impact was noted as "non-traffic side" trajectory because the right-side wheels were supported by the sidewalk after the crash. It was noted that were it not for the sidewalk, the vehicle would have likely rolled over and off the overpass.

Of the four crashes that were noted to redirect the impacting vehicle with the SET, all four resulted in vehicle rollover. Note that a rollover was defined as the vehicle completing a minimum of a 90-degree turn along the longitudinal axis. In each of the redirection-related rollovers, the vehicle "tipped" over, resulting in minimal injuries. However, rollover crashes tend to be disproportionately severe compared to non-rollover crashes. It is likely that the impact speeds in each of these four crashes were very low. Seven of the 22 crashes with known vehicle responses (32%) resulted in rollover and the vehicle remained upright during 15 crashes (68%).

6 Discussion

Results of the Iowa SET ISPE indicated that when crashes involved SETs, crash results were more severe on average than for other fixed objects; nonetheless, crashes involving the SETs were very uncommon. The majority of the 658 SETs were not impacted in the ten-year evaluation window. Moreover, all 30 of the SET impacts occurred on a total of 18 unique bridges or connecting ramps, and nineteen of the crashes occurred on a total of seven bridges. Results suggested that a good predictor of the likelihood of a future SET crash was existing crash history, and therefore treatment of the SETs which are involved in crashes may be a cost-effective method for preventing future severe crashes with SETs.

Identifying crashes involving SETs was very challenging; reviewing and conclusively evaluating post-crash vehicle behaviors proved to be equally difficult. Iowa DOT's crash reports between 2008 and 2017 were generally consistent with crash reporting recommendations found in the Model Minimum Uniform Crash Criteria or MMUCC (e.g., 2017), but there are no hazard classification fields specific to the SETs and first responders who complete the majority of the crash reports are not trained to identify specific roadside hardware. An extensive process was required first to locate

SETs in the state, to collect and review all crash report candidates within a radius of the SET and determine which crashes involved the end treatment, and classify results. The extensive time commitment and effort required to perform the identification and filtering process is likely to be difficult for DOTs to perform. These challenges have been noted previously in comparable ISPE studies (Albuquerque et al. 2013, Stolle et al. 2012). In order for ISPE studies to be expedited in the future, improved training, feature recognition, and/or alternative data collection such as photographs may be required to add confidence during the identification and evaluation phases. Nonetheless, adding additional steps, effort, or burden on first responders by lengthening crash report form completion may result in fewer reported crashes, longer documentation efforts, poor data collection, and increased risk of injury or harm from law enforcement which are primarily unshielded while performing crash report documentation. Current data collection may require an hour per crash report on the scene plus additional follow-up, documentation, and summary at departmental levels. Severe crashes generally require significantly more effort, as do crashes in which at least one party is suspected of violating local, state, or national ordinances. A careful balancing act is required between the burden placed on first responders responsible for collecting crash data, and improving data quality, usability, and clarity.

SETs were only used in Iowa in locations for which significant space constraints existed and where few, if any, MASH-crashworthy features could be installed to shield the bridge ends. SETs were noted to reduce the loading to the ends of bridge rails and may reduce the severity of impacts which could have otherwise impacted the blunt end of an unshielded feature. It may be that the low number of observed crashes and large exposure of vehicles per recorded crash also reflects that, without a definite impact including ramping up a bridge rail, rollover, or launching event, drivers involved in low-speed impacts with the SETs choose not to report impacts and instead drove away with minimal damage. As well, the short length of the SETs means that there is a very narrow region of vulnerability to a frontal impact before the impact outcome would transition to redirective behavior along the length of the attached bridge rail.

A benefit-to-cost (B/C) analysis was performed to determine if it was cost-economical to replace any or all of the SET treatments with MASH-crashworthy hardware, when site conditions permitted. The results of the benefit-to-cost analysis indicated that if the locations of the SET crashes were known in advance, replacing the SETs could result in a positive benefit-cost outcome greater than 4.0. However, for every SET treatment scheme that attempted to predict future crash locations based on attributes of the crashes that had occurred, the benefit-to-cost analysis indicated low benefit, less than 2.0 and often less than 1.0. Results of the benefit-to-cost analysis are discussed in greater detail by Lingenfelter (2020). The low B/C outcome was related to the low number of reported crashes and few approved end treatments for low-volume and low-speed roadways, and sparse information about the safety performance and installation costs of lower service level hardware which was therefore not conducive for B/C replacement analyses. All end treatments considered in this study were approved to either MASH or NCHRP Report No. 350, had been previously studied to estimate the distribution of injury outcomes, and were associated with state DOT historical data for installation, repair, and maintenance costs (Savolainen et al. 2017). Crash cost estimation factors were evaluated using both Iowa DOT crash cost data and estimates of the Value of a Statistical Life (VSL) available by the Federal Highway Administration (USDOT 2016, USDOT 2015). Although the benefit-to-cost analysis for removing and replacing SETs with crashworthy end treatment hardware was generally very low, researchers recommended that alternative end treatments be considered for locations in which vehicles ramping off of the end treatment may result in severe outcomes (e.g., overpasses over highways) or in locations where one or more crashes was observed at the bridge ends. Of the 30 total SET crashes in 10 years, all crashes occurred at a total of 18 bridges, and 19 crashes occurred on just seven bridges, indicating that crash history may be the best predictor of where future crashes will occur.

Currently, little guidance exists which could guide the implementation of lower service-level hardware for use on higher-speed roadways and in close proximity to the roadway, such as MASH TL-1 crashworthy end treatments. For many of the locations in which SETs were installed in Iowa, insufficient space, grading, or other considerations (e.g., pedestrian obstructions or sight line concerns) affected the selection of treatments. Iowa had observed that although SETs are non-energy-absorbing, they may be preferrable to allowing an errant vehicle to impact into a blunt end of a bridge rail, and thus were considered to be a "best-available" strategy in the absence of alternative guidance. It is recommended that further research regarding installation of safety treatments in constrained locations be performed to determine what is the optimal selection criteria or considerations for treatments.

As noted, because trailing-end SETs were not noted in the crash reports to contribute to occupant injury or vehicle instability, trailing-end SETs were determined to be a safe downstream end treatment and were not considered in this evaluation. As well, no cross-centerline crashes involving errant vehicles impacted an SET located adjacent to the opposite-direction crashes. All SET-related crashes identified in Iowa were located adjacent to upstream ends of bridge features or ramps by adjacent, same-direction traffic lanes. The only SETs impacted on the left-side of the road were associated with one-way traffic flow. Although crashes with SETs had a greater potential to be severe, the prioritization of the replacement or treatment of some SETs may be low.

7 Conclusions

The in-service performance of Iowa's sloped end treatments was investigated for 658 SETs attached to bridge rails or ramps connected to bridges. For the ten-year span between 2008 and 2017, a total of 30 SET crashes were identified, associated with 19 unique bridges. Although insufficient detail was provided to classify the vehicle reactions for eight of the crashes, a total of 22 crash outcomes were identified. Of these known crashes, four (18%) resulted in vehicle redirection whereas 18 crashes (82%) resulted in the vehicle climbing the barrier; as well, seven crashes (32%) resulted in rollover and the vehicle remained upright during 15 crashes (68%). Resulted indicated that when SET crashes occurred, undesirable outcomes were common; however, despite a large exposure and close proximity to the roadway, the crash rate was very low for SETs, and thus the cost-effectiveness of treating or replacing SETs with other treatments was very low.

Iowa DOT had implemented SETs only in locations with generally low posted speed limits and limited availability for alternative treatments. In addition, many were located adjacent to highway ramps and near stoplight conditions, which may have further reduced the average speed at impact with the SET as well as the total number of impacts. Application in situations with different conditions, particularly including higher impact speeds, could produce different and more severe outcomes.

Identifying which crashes involved SETs and their contribution to the vehicle trajectory, stability, injury outcome, and prevailing circumstances leading up to the crash proved to be very difficult. Of the nearly 2,500 candidate fixed-object crashes within a 305-m radius of an SET, only 30 SET crashes were identified and each required careful inspection of the crash narrative, scene diagram, vehicle damage, and injury information. Improvements to crash reporting documentation may be possible to expedite this effort and improve the results, but should balance the concerns, costs, and requirements of first responders when implementing new solutions.

References

Albuquerque, F.D.B., Sicking, D.L.: In-Service safety performance evaluation of roadside concrete barriers. J. Transp. Saf. Secur. 5(2) (2013). https://doi.org/10.1080/19439962.2012.715618

Beason, W.L.: Development of an End Treatment for a Low-Profile Concrete Barrier. Report No. TX-92-1949-2, Texas Transportation Institute, Texas A&M University, College Station, Texas (November 1992)

Beason, W.L., Brackin, M.S., Bligh, R.P., Menges, W.L.: Development and Testing of a Non-Pinned Low-Profile End Treatment, Report No. TX-13-9-1002-12-7, Texas Transportation Institute, Texas A&M University, College Station, Texas (October 2013)

Beason, W.L., Menges, W.L., Ivey, D.L.: Compliance Testing of an End Treatment for the Low-Profile Concrete Barrier. Report No. TX-98-1303-S, Texas Transportation Institute, Texas A&M University, College Station, Texas (April 1998)

Bronstad, M.E., Calcote, L.R., Kimball Jr., C.E.: Concrete Median Barrier Research Volume 2–Research Report. Report No. FHWA-RD-77-4, Southwest Research Institute, San Antonio, Texas (March 1976)

Google Maps: Google. https://maps.google.com

Lingenfelter, J.A.: Master's Thesis in Mechanical Engineering, University of Nebraska-Lincoln (2020)

Lingenfelter, J.A., Stolle, C.S., Bielenberg, R.W., Klute, K.: In-Service Performance Evaluation of Concrete Sloped End Treatments in Iowa. Midwest Roadside Safety Facility, University of Nebraska-Lincoln, Lincoln, Nebraska (2020)

Mak, K.K., Sicking, D.L., Benicio de Albuquerque, F.D., Coon, B.A.: Identification of Vehicular Impact Conditions Associated with Serious Ran-Off-Road Crashes. NCHRP Report No. 665, National Cooperative Highway Research Program, Washington, D.C. (2010)

Manual for Assessing Safety Hardware (MASH), Second Edition: American Association of State Highway and Transportation Officials (AASHTO), Washington, D.C. (2016)

Michie, J.D.: Recommended Procedures for the Safety Performance Evaluation of Highway Appurtenances, NCHRP Report No. 230, National Cooperative Highway Research Program, Washington, D.C. (March 1981)

Model Minimum Uniform Crash Criteria (MMUCC) Guidelines, Fifth Edition: National Highway Traffic Safety Administration (NHTSA), Washington, D.C. (2017)

Ray, M.H., Weir, J.: Unreported collisions with post-and-beam guardrails in Connecticut Iowa, and North Carolina. J. Transp. Res. Board **1743**, 111–119 (2000). https://doi.org/10.3141/1743-15

Revised Departmental Guidance 2016: Treatment of the Value of Preventing Fatalities and Injuries in Preparing Economic Analyses. U.S. Department of Transportation, Washington, D.C. (August 2016). https://www.transportation.gov/sites/dot.gov/files/docs/2016%20Revised%20Value%20of%20a%20Statistical%20Life%20Guidance.pdf

Ross, H.E., Krammes, R.A., Sicking D.L., Tyer, K.D., Perera, H.S.: Recommended Practices for Use of Traffic Barrier and Control Treatments for Restricted Work Zones. National Cooperative Highway Research Program Report No. 358, Washington, D.C. (1994)

Ross, H.E., Sicking, D.L., Zimmer, R.A.: Recommended Procedures for the Safety Performance Evaluation of Highway Appurtenances. NCHRP Report No. 350, National Cooperative Highway Research Program, Washington, D.C. (1993)

Savolainen, P.T., Barnwal, A., Kirsch, T.J.: Crash Cushion Selection Criteria. Final (September 2017)

Report to the Iowa Department of Transportation, Iowa State University Center for Transportation Research and Education, Ames, Iowa

Stolle, C.S., Sicking, D.L.: Cable Median Barrier Failure Analysis and Prevention, Report No. TRP-03-275-12, Midwest Roadside Safety Facility, University of Nebraska-Lincoln, Lincoln, Nebraska (December 2012)

TIGER Benefit-Cost Analysis (BCA) Resource Guide: U.S. Department of Transportation, Washington, D.C. (March 2015). https://www.transportation.gov/sites/dot.gov/files/docs/Tiger_Benefit-Cost_Analysis_%28BCA%29_Resource_Guide_1.pdf

A Synthesis of 787-mm Tall, Non-proprietary, Strong-Post, W-beam Guardrail Systems

Scott Rosenbaugh[✉], Robert Bielenberg, and Ronald Faller

Midwest Roadside Safety Facility, The University of Nebraska,
Lincoln, NE, USA
srosenbaugh2@unl.edu

Abstract. Since its development in the early 2000's, 787-mm tall W-beam guardrail, commonly referred to as the Midwest Guardrail System (MGS), has proven to be one of the most versatile and robust roadside barrier systems in use today. The MGS has been designed and successfully crash tested in a wide range of configurations including both roadside and median systems, steel and wood post systems, blocked out and non-blocked systems, and variable post spacing systems. The MGS has also been successfully crash tested adjacent to steep roadside slopes, with an omitted post, in combination with curbs, and in special applications such as culvert mounted installations. Since the MGS is a non-proprietary barrier system, these various configurations can be used by roadway agencies all around the world.

This paper contains details and drawings encompassing a wide range of 787-mm tall W-beam guardrail configurations that have been developed and evaluated to the safety standards of the American Association of State Highway and Transportation Officials (AASHTO) Manual for Assessing Safety Hardware (MASH). The various configurations are discussed in terms of key components and performance characteristics, such as test level and working width. Finally, implementation guidance is provided for the proper selection, layout, and installation of the various MGS configurations.

Keywords: Guardrail · W-beam · Midwest Guardrail System · MASH · Barrier

1 Introduction

W-beam guardrail systems are some of the most popular and frequently used vehicle barrier systems. For over 50 years, these barrier systems have been redirecting errant vehicles and preventing impacts into roadside hazards. However, by the 1990's the increased size of the United States vehicle fleet was beginning to push the containment limits of the existing W-beam guardrail configurations. Multiple full-scale vehicle crash tests with heavy, high center-of-gravity passenger vehicles, such as pickup trucks and vans, resulted in rail ruptures and vehicle rollovers. Thus, W-beam guardrails needed to be updated to handle the larger vehicle fleet.

In the early 2000's, the Midwest Guardrail System (MGS) was developed through modifications to the existing G4(1S) W-beam guardrail system. These modifications included raising the height to the top of the rail to 787 mm, a reduction in the post embedment depth to 1,016 mm, and moving the rail splices from post locations to mid-spans (Polivka et al. 2004) and (Sicking et al. 2002). Additionally, the depth of the blockouts was increased to 305 mm to aid in the containment of taller vehicles and to reduce vehicle snag on system posts. Details for the MGS are shown in Fig. 1.

It should be noted that the MGS was developed in the United States using US Customary units. Thus, all of the components, post spacing, and offsets were originally designed and evaluated in US Customary units. For this paper, efforts were made to convert all distances over to SI units with a hard conversion. Figure 1 contains dimensions in both unit conventions to illustrate this conversion. However, if a component has a standard shape that is referred to in US Customary units, that component name was not changed. For example, standard steel guardrail posts are referred to as W6 × 8.5 posts with the understanding that similar SI shapes exist.

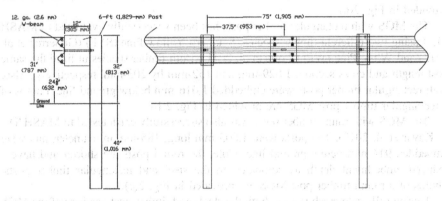

Fig. 1. Characteristics of standard strong-post MGS, 787-mm tall W-beam guardrail

The MGS was originally developed and successfully crash tested to the Test Level 3 (TL-3) safety performance standards of NCHRP Report No. 350 (Ross et al. 1993). Since its inception, the MGS has proven to be a very robust barrier system as multiple variations and special applications of the MGS have been developed and successfully crash tested. Even with the adoption of a new crash testing standard, the Manual for Assessing Safety Hardware (MASH) (AASHTO 2016), which utilizes larger vehicles and higher impact severities than NCHRP Report No. 350, the MGS and its variations have continued to provide crashworthy results. Although many agencies across the world still utilize NCHRP 350 hardware, this paper is focused on only MASH evaluated systems.

The purpose of this paper is to provide a resource documenting the numerous non-proprietary, strong-post, 787 mm tall W-beam guardrail configurations and special applications. Details required for the proper installation of each system configuration are provided herein, including drawings, component details, performance levels, working widths, and installation guidance. Note, working width is defined as the lateral distance from the front face of the guardrail to the furthest extent of the deflected system or the vehicle position during redirection, whichever is greater.

2 Post Variations

Strong-post W-beam guardrail systems, which include both steel and timber posts, have been the most widely used guardrail systems for decades. Accordingly, both steel- and wood-post MGS systems have been successfully tested and evaluated to MASH TL-3. Regardless of the post type, the MGS utilizes the same rail height and post spacing as shown in Fig. 1. The only differences would be the shape of the blockout, which is cut to fit flush against the post, and the length of the guardrail attachment bolt.

The steel post version of the MGS utilizes 1,829-mm long W6 × 9 or W6 × 8.5 posts, which are embedded 1,016 mm below ground line. These two post sections have nearly identical flange widths and section depths and provide similar bending strengths. Thus, both post sections provide similar performance when used as a guardrail post. The steel-post MGS was successfully crash-tested to MASH test 3–10 (Polivka et al. 2006a) and MASH test 3–11 (Polivka et al. 2006b). Photos of a steel-post MGS are provided in Fig. 2(a).

The MGS with rectangular timber posts has been successfully evaluated to MASH TL-3 using two different timber species, Southern Yellow Pine (SYP) (Gutierrez et al. 2013) and White Pine (WP) (Stolle et al. 2011). Both timber species utilized the same post length and cross section, 1,829-mm and 152-mm by 203-mm, respectively. Also, both rectangular timber posts were embedded 1,016 mm below ground line. Photos of a rectangular timber post MGS are provided in Fig. 2(b).

The MGS with round timber posts was also successfully crash tested to MASH TL-3 (Kovar et al. 2019). The posts were 1,803 mm long, 185-mm in diameter, and were embedded 914 mm below ground line. Note, the round posts are shorter and have a reduced embedment depth as compared to the steel and rectangular timber posts. Photos of a round timber post MGS are provided in Fig. 2(c).

During full-scale crash testing, both the steel- and timber-post versions of the MGS performed similarly in terms of vehicle behavior, occupant ridedown accelerations (ORA), and occupant impact velocities (OIV), which were all well within the MASH limits. However, there are a few notable differences between the systems. Wood posts tend to fracture under extreme loading conditions while steel posts plastically deform and continue to apply a resistance force to the vehicle. Consequently, the working width for timber-post systems is slightly higher than that of the steel-post system, as shown in Table 1. Thus, the required clear distance behind the barrier, which should be free of all hazards, depends on the post type.

The timber species utilized for the posts also factor into the system performance. Rectangular WP posts have roughly 37% less strength than rectangular SYP posts. This reduction in strength resulted in increases in the number of fractured posts, the deformed length of the guardrail system, and the system working width, as shown in Table 1. Therefore, it is recommended that rectangular timber posts only be made from timbers with strength greater than or equal to that of Select Structural WP unless further evaluation is performed.

Fig. 2. MGS installations with (a) steel posts, (b) rectangular timber posts, and (c) round timber posts.

As noted previously, the round timber posts have a reduced length and embedment depth. MASH crash testing with 178-mm diameter, round timber posts with a standard 1,016-mm embedment depth resulted in too many posts fracturing and the vehicle ultimately overrode the guardrail (Bligh 2019). The minimum diameter was increased and the embedment depth was decreased to reduce post fracture. Thus, the 185-mm diameter and 914-mm embedment depth are critical to the performance of the round timber post MGS.

Table 1. Comparison of Steel- and Timber-Post MGS

System	Steel-Post MGS	Rectangular timber-post MGS	Rectangular timber-post MGS	Round timber-post MGS
Posts	W6 × 8.5	152-mm by 203-mm SYP	152-mm by 203-mm WP	185-mm diameter
Reference	(Polivka et al. 2006a)	(Gutierrez et al. 2013)	(Stolle et al. 2011)	(Kovar et al. 2019)
Performance level	MASH TL-3	MASH TL-3	MASH TL-3	MASH TL-3
MASH 3–11 results				
Working width	1,234 mm	1,367 mm	1,483 mm	1,580 mm
Contact length	10.3 m	10.5 m	11.1 m	13.3 m
No. deflected posts	6	6	7	10
No. fractured posts	–	4	5	7

3 Blockout Variations

The MGS was originally designed with 305-mm deep blockouts for two reasons: 1) to reduce the likelihood and severity of vehicle snag on guardrail posts and 2) to maintain rail height as the post deflects backward, making the system more likely to capture large vehicles. However, 787-mm tall W-beam guardrail has been successfully crash tested to MASH TL-3 standards while utilizing 203-mm blockouts and in non-blocked configurations. All three blockout variations are shown in Fig. 3.

The 787-mm W-beam guardrail system with 203-mm blockouts differs from the standard MGS only in the depth of the blockout and length of the attachment bolt. This configuration was crash tested to MASH test designation 3–10 to evaluate any potential for the 1100C small car to snag on the guardrail posts. The test passed and resulted in vehicle and system behavior similar to that of the MGS with 305-mm blockouts (Bligh et al. 2011). Thus, 203-mm blockouts can be utilized with steel- or timber-post versions of the MGS.

The non-blocked MGS eliminated the timber blockout, but a 305-mm long segment of W-beam was placed between the rail and the post to prevent contact between the post flanges and the rail, which could initiate tearing of the rail. The non-blocked system was evaluated to MASH TL-3 and satisfied all safety criteria for both full-scale crash tests (Schrum et al. 2013) and (Reid et al. 2013). The working width for the non-blocked system was measured to be 1,097 mm.

During evaluation of the non-blocked system, the small car contacted the posts earlier in the impact event and caused the vehicle to exit the system in a non-tracking manner (i.e., the vehicle was yawing toward the guardrail as it exited the system). Additionally, snag on the posts resulted in a longitudinal OIV value of −9.53 m/s, which was more than double that of the MGS with 305-mm blockouts but still within the MASH limit of 12.2 m/s. These differences illustrate the benefits of utilizing blockouts within a guardrail system. As such, the use of blockouts is recommended for guardrail installations whenever the roadside can accommodate the increased width.

The non-blocked system has not been evaluated with timber posts, which may further increase vehicle snag, vehicle decelerations, and the risk for vehicle instabilities. Therefore, it is not recommended to utilize a non-blocked system with timber posts until further analysis is performed.

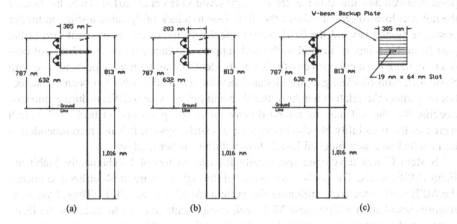

Fig. 3. MGS details for (a) 305-mm blockouts, (b) 203-mm blockouts, and (c) non-blocked configurations.

4 Placement Adjacent to Slopes

The strength and stiffness of W-beam guardrail is heavily dependent on post-soil resistance forces. Placing the system on or adjacent to a fill slope reduces the amount of soil behind the post, lowers the post-soil resistance, and can negatively affect the performance of the system. Thus, it is generally recommended for guardrail posts to be installed with at least 610 mm of level terrain behind the system to ensure the guardrail performs as designed and evaluated. However, there are instances where placing guardrail adjacent to slopes is necessary due to limited roadside widths.

Four different W-beam guardrail configurations have been successfully developed and crash tested to MASH TL-3. Although all four systems utilized 787-mm tall W-beam rail, they had varying post lengths, blockout depths, allowable slopes, and placements relative to the slope break point (SBP), as shown and detailed in Table 2.

The guardrail system denoted as System A in Table 2 was the first guardrail system developed for use adjacent to slopes and tested to MASH TL-3. It utilized 305 mm blockouts and 2,743-mm long steel posts centered on the SBP of a 2H:1V fill slope (Polivka et al. 2008) and (Wiebelhaus et al. 2010). Dynamic testing of various posts located at the SBP of a 2H:1V slope illustrated that 2,286-mm long rectangular timber posts would provide similar strength and stiffness to the full-scale crash tested system (McGee et al. 2010). Full-scale testing of System B and additional dynamic post testing on slopes has since justified the use of 2,438-mm steel posts as well. Thus, both are listed as alternative post options for System A. Additionally, System C illustrated the crashworthiness of a non-blocked guardrail system on slopes, so both 203-mm blockouts and non-blocked configurations of System a are considered alternative options.

System B is the only guardrail system to be successfully crash tested with the posts installed beyond the SBP. It utilized 203-mm blockouts and 2,438-mm long W6 × 8.5 posts centered 381 mm down a 2H:1V slope (Abu-Odeh et al. 2013). Thus, the face of the rail was located directly above the SBP. Due to a lack of dynamic testing on timber posts positioned beyond the SBP, which effectively increases the moment arm in the post from the center of the rail to the soil support and increases the possibility of post fracture, there have not been any timber posts identified as alternative posts for use with System B. Non-blocked guardrail located beyond the SBP has not yet been evaluated, but may affect the relative height of the W-beam rail as system deflects during impacts. Specifically, the rail may be pulled downward as the posts deflect backward, which increases the possibility of vehicle override. As such, System B is not recommended to be installed as a non-blocked installation without further analysis.

System C was developed specifically for use on top of Mechanically Stabilized Earth (MSE) walls. The posts were extended through the wire mesh utilized to anchor the MSE wall, effectively stiffening the system (McGee et al. 2011). Thus, System C requires installation within these MSE wall components in order to maintain the listed system performance, specifically the reduced working width. Placement on a standard fill slope would likely result in system deflections and working widths similar to System D.

System D represents a standard MGS with 1,829-mm long posts installed at the SBP of a 2H:1V slope (Haase et al. 2016) and (Lechtenberg et al. 2011). The system was tested with steel guardrail posts, but rectangular timber posts installed at the SBP should provide similar performance. Thus, the 1,829-mm SYP timber post is listed as an alternative post for System D. Additionally, based on the performance of System C, both 203-mm blockouts and non-blocked were considered alternative blockout options for System D. Due to the shorter post length of System D, a significant increase in working width was observed. It is important to note that MASH requires guardrail systems to be tested within strong soils to evaluate critical loading to barrier components. If the system were installed in a weaker soil, the deflections and working width would increase even further and may eventually become excessive and lead to vehicle instability, loss of vehicle containment, or rail rupture. Thus, System D is only recommended to be utilized in strong soil conditions similar to the soil specified by MASH. Installations sites with weaker or sandy soils are encouraged to utilize Systems A or B.

During the full-scale testing of System A, a MASH 3–11 test was conducted on a system with a 705-mm top rail height. The 2270P pickup overrode this lower-height guardrail, thus failing the test. Subsequently, it is recommended for all W-beam guardrails adjacent to slopes to be installed with a minimum rail height of 787 mm until further analysis is conducted. Additionally, guardrail systems have only been evaluated on slopes as steep as 2H:1V. Thus, these guardrail systems should be limited to slopes of 2H:1V or flatter until further evaluation is performed.

Finally, the recommendations listed herein for W-beam guardrail placed on or adjacent to steep slopes are applicable only to guardrail length of need installations.

Special guardrail applications such as omitted posts, roadway curbs, and guardrail stiffness transitions have not yet been designed or evaluated for use on slopes. As such, it is not recommended to install these specialized guardrail applications on or adjacent to steep slopes without further analysis. Similarly, guardrail end terminals require specific grading to function properly. It is recommended that guidance from the individual end terminal manufacturer be sought after and followed concerning placement near or on slopes.

Table 2. Details for 787-mm W-beam guardrail adjacent to slope

System	A	B	C	D
Layout				
Reference	(Polivka et al. 2008) and (Wiebelhaus et al. 2010)	(Abu-Odeh et al. 2013)	(McGee et al. 2011)	(Haase et al. 2016) and (Lechtenberg et al. 2011)
Performance Level	MASH TL-3	MASH TL-3	MASH TL-3	MASH TL-3
Full-Scale Tests	MASH 3-11	MASH 3-10 MASH 3-11	MASH 3-10 MASH 3-11	MASH 3-11
Post	2,743-mm W6x8.5	2,438-mm W6x8.5	1,829-mm W6x8.5	1,829-mm W6x8.5
Post Spacing	1,905 mm	1,905 mm	1,905 mm	1,905 mm
Blockout	305-mm Blockout	203-mm Blockout	Non-Blocked	305-mm Blockout
Slope	2H:1V	2H:1V	3H:1V	2H:1V
Post Locations	Centered on SBP	Centered 381 mm Down Slope	Centered on SBP	Centered on SBP
Working Width	1,585 mm	1,402 mm	1,148 mm	1,966 mm
Alternative Posts	2,438-mm W6x8.5 or 2,286-mm Timber*	-	-	1,829-mm Timber*
Alternative Blockouts	Non-Blocked or 203-mm Blockout	305-mm Blockout	-	Non-Blocked or 203-mm Blockout
Allowable Slopes	2H:1V or Flatter	2H:1V or Flatter	-	2H:1V or Flatter

*Timber Posts should be 152 mm × 203 mm and have strength equal to or greater than SYP grade 1

5 Median Configuration

A 787-mm W-beam median barrier was developed and successfully crash tested to MASH TL-3 safety criteria. The median system utilized standard 2.66-mm thick W-beam guardrail, spices at mid-span locations, and 203 mm deep blockouts mounted on both sides of standard steel guardrail posts, as shown in Fig. 4. Although the system deflections were similar to those of the standard roadside systems, the increased width of the median system resulted in a 1,397 mm working width (Abu-Odeh et al. 2014).

Testing and evaluation of the median system was conducted with 203-mm blockouts, but 305-mm. blockouts should provide similar system performance. During the small car test on the median system, a tear extended through two-thirds of the rail. Eliminating the blockouts could result in small increases to the rail loads and cause complete rupture of the rail. Additionally, vehicle snag on posts directly attached to two rail elements may result in excessive decelerations to the small car. Therefore, it is not recommended to install the median system in a non-blocked configuration without further analysis.

The median barrier system was evaluated on level terrain, and its performance on sloped terrain remains unknown. Thus, the median barrier system should only be installed in median with a slope of 10:1 or flatter until further evaluation is performed. Finally, the ends of the system need to be properly treated with a crashworthy median guardrail terminal capable of anchoring a dual-sided, W-beam system.

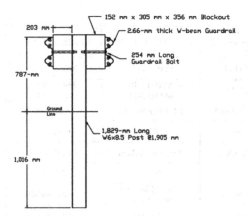

Fig. 4. Cross section of 787-mm tall, W-beam, median barrier configuration

6 MGS with an Omitted Post

Occasionally within a guardrail installation, obstructions within the ground prevent the proper installation of a post. At these locations, it is often desired to omit the guardrail post leaving a 3,810-mm span between the posts adjacent to the obstruction. Subsequently, a steel post MGS installation with a single missing post was subjected to MASH 3–11 test to evaluate critical rail loadings and possible vehicle instabilities. The system contained and redirected the vehicle with a working width of 1,273 mm, and the system satisfied MASH TL-3 safety criteria (Lingenfelter et al. 2016).

The omission of a post effectively weakens the guardrail system and results in increased system deflections, rail loads, and vehicle pocketing. The omission of multiple posts within the contacted region of a guardrail installation system may lead to excessive displacements, loads, and/or pocketing that may ultimately lead to system failure. Therefore, until further evaluation is completed on multiple missing posts within a system, it is recommended that at least eight posts be installed between omitted posts to ensure proper system performance.

Since the performance of timber and steel posts are so similar, the same guidelines should be utilized for the omission of a post within a guardrail system for either post type. Additionally, utilization of either 203-mm or 305-mm blockouts are acceptable adjacent to the omitted post location. However, the increased deflections associated with an omitted post may be problematic for a non-blocked system due to the increased risk for rail rupture and exposure of the vehicle floor pan to contact with posts. Thus, it is not recommended to omit posts within a non-blocked system until further evaluation is conducted.

Finally, these guidelines on omitted post are only intended for standard, length-of-need installations. Specialized W-beam applications such as guardrail adjacent to slopes, guardrail stiffness transitions, or guardrail end terminals are sensitive systems that may be negatively affected by a post omission. Thus, it is not recommended to omit posts within these specialized guardrail regions without further analysis.

7 Mash TL-2 Guardrail

A TL-2 version of standard strong-post guardrail has been developed by doubling the post spacing to 3,810-mm. The system was comprised of 787-mm tall W-beam, standard 1,829-mm W6 × 8.5 posts, and 203-mm deep blockouts. The system was successfully tested and evaluated to MASH TL-2 and had a 1,125-mm working width (Sheikh and Menges 2014). Both steel and rectangular timber posts are compatible with the TL-2 system. However, it is not recommended to install the TL-2 system on slopes, with omitted posts, or in a non-blocked configuration without further analysis.

8 Placement Adjacent to Curbs

The MGS has been successfully crash tested in combination with roadside curbs. A steel-post MGS configuration was evaluated with the face of the guardrail located 152 mm behind a 152-mm tall, AASHTO type B curb. The system utilized 305-mm blockouts and the W-beam rail was mounted at a height of 787 mm above the roadway. Soil back fill was installed behind the curb, which increased the embedment depth of the posts to 1,168 mm. This configuration, as shown in Fig. 5, was successfully crash tested to MASH TL-3 performance criteria and had a working width of 1,232 mm (Ronspies et al. 2020).

Placement of the MGS closer to the face of the curb has typically been considered to enhance system performance. As the guardrail is moved closer to the curb, the vehicle interacts sooner with the guardrail and the effects of the vehicle wheels overriding the curb are reduced. Greater rail-to-curb offsets would lead to increased vehicle vaulting after it overrode the curb. In fact, a previous MASH TL-3 crash test with the MGS offset 2,438 mm from the curb resulted in the 2270 pickup truck rolling over the barrier (Thiele et al. 2009). Therefore, the MGS should be considered crashworthy with the face of the rail offset between 0 and 152 mm from the face of the curb.

Reduced curb heights would have less of an effect on a vehicle's trajectory, so the MGS paced within 152 mm of a curb is considered crashworthy with all curbs heights at or below 152 mm. Since the MGS has performed similarly with both steel and timber posts, standard rectangular timber posts should also be considered an option for use with the MGS placed adjacent to curb. Blockouts used within W-beam guardrail systems function to reduce vehicle snag and maintain rail height as the system deflects backward during an impact. With the increased embedment depth for guardrail posts placed behind curbs, these blockout functions may become critical. Thus, it is not recommended to install non-blocked MGS installations behind roadway curbs until further evaluation is conducted. Finally, it is not recommended to install curbs and special applications of the MGS, like stiffness transitions and terminals, unless the combination of these roadside structures has been properly evaluated.

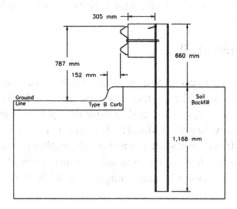

Fig. 5. MGS installed in combination with roadway curb

The MGS placed adjacent to curb has also been evaluated with an omitted post. The installation was identical to that shown in Fig. 5 except that a single post was omitted creating a 3,810 mm elongated span. During MASH test 3–10 on the installation, the W-beam rail ruptured at the splice location within the elongated span and the 1100C small car penetrated the guardrail (Rosenbaugh et al. 2019). To prevent rail tearing in the weakened area from the omitted post, a second W-beam rail was added to the system such that 11.43 m of nested W-beam was located at the location of the omitted post. The nested system was successfully crash tested to MASH TL-3 performance criteria with a working width of 1,062 mm (Rosenbaugh et al. 2019). Thus, for omitted post locations within MGS adjacent to curb, the W-beam rail should be nested adjacent to the omitted post location. The nested rail section should be at least 11.43 m in length and extend past a minimum of two posts on each side of the elongated span.

9 Culvert Mounted Configurations

Two 787-mm tall, strong-post W-beam guardrail systems have been developed for attachment to low-fill, concrete box culverts. The first system consisted of W6 × 9 posts spaced at 1,905 on-center and 203-mm deep blockouts. Each post was welded to a 22-mm thick baseplate and mounted to the top surface of the culvert using four threaded anchor rods epoxied into the top slab of the culvert. The guardrail system was evaluated with 229 mm of soil fill and the posts were offset 457 mm from the inside face of the culvert headwall, as shown in Fig. 6(a). The system was successfully crash tested to MASH TL-3 and had a 1,250 mm working width (Williams and Menges 2011). This guardrail system was previously crash tested at a reduced height of 686 mm and resulted in complete rupture of the W-beam rail (Williams et al. 2008). Thus, the system should not be installed with a top rail height below 787 mm until further evaluation is conducted. Finally, this system can be directly connected to standard MGS installations as long as adequate soil grading is provided behind the adjacent system's posts.

The second culvert-mounted guardrail system also consisted of W6 × 9 posts welded to a baseplate and had a soil fill thickness of 229 mm. However, the baseplate was only 13 mm thick and was designed to deform and allow the post to rotate back under lateral impact loads. The posts were spaced at 953 mm on-center, commonly referred to as half-post spacing, and offset 305 mm from the inside face of the culvert headwall. The posts were anchored to the top slab of the culvert utilizing four 25-mm through bolts. The second guardrail system was evaluated with 305-mm blockouts. This system was also successfully crash tested to MASH TL-3 performance criteria and had a working width of 1,290 mm (Asadollahi Pajouh et al. 2020). Note, the system was evaluated with five standard 1,016-mm embedded guardrail posts at half-post spacing placed both upstream and downstream of the culvert mounted system to transition into MGS at standard 1,905-mm spacing. This transition region is recommended for use until further evaluation can be conducted on the crashworthiness of the system without these transition posts.

Both of the culvert mounted guardrail systems were evaluated with 229 mm of soil fill on the culvert. Installing the posts with shallower embedments shortens the moment arm of the post and stiffens the response of each post. This, in turn, can lead to increased rail loads and pocketing which may degrade the performance of the system. Additionally, shorter embedment depths would increase the propensity for wheel snag on the posts as the lower section of the post cannot displace as much. Thus, both systems require a minimum soil fill depth of 229 mm. Soil fill deeper than 229 mm should not adversely affect the performance of the system.

Both guardrail systems are considered crashworthy with either 203-mm or 305-mm blockouts. However, it is not recommended to install these systems in a non-blocked configuration due to vehicle snag concerns on the stiffened posts. Both systems can be

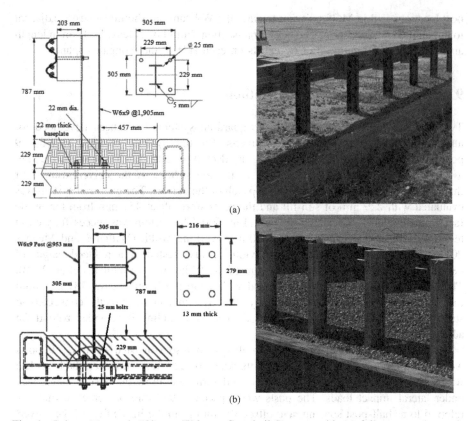

Fig. 6. Culvert Mounted, 787-mm W-beam Guardrail Systems with (a) full-post spacing and (b) half-post spacing

attached to the top slab of a culvert using either through bolts or epoxy anchors, as long as the top slab thickness and epoxy strength are capable of developing the full strength of the threaded anchor rods. Finally, both systems should be installed with the posts offset a minimum of 305 mm from the inside face of the culvert headwall. Reduced offsets can limit the posts from rotating backward as they would impact the headwall, which would cause sudden increases in stiffness to the system. Previous crash testing of a similar guardrail system offset only 25 mm from the headwall resulted in limited post displacement, excessive vehicle snag on the posts, and vehicle instability ultimately resulting in vehicle roll over (Polivka et al. 2002).

10 Placement of Luminaire Poles Near the MGS

A study was recently conducted to explore the crashworthiness of breakaway luminaire poles placed within the working width of an MGS installation. Computer simulation of MASH impacts lead to the selection of a minimum offset of 1,041 mm between the face of the W-beam guardrail and the face of the luminaire pole. This critical offset was crash tested to MASH TL-3 with the luminaire poles located in longitudinal potions to maximize interaction with guardrail posts and maximize vehicle snag potential. Both crash tests satisfied MASH TL-3 safety performance criteria (Asadollahi Pajouh et al. 2017). The study was conducted with a 15.2 m tall aluminum pole supported by a breakaway transformer base. Poles of a similar or smaller height and weight and with a similar breakaway base would likely behave similarly. Thus, the same 1,041 mm minimum offset could be applied to these alternative luminaire configurations. Further, since the offset is measured from the face of the guardrail, the same minimum offset can be used for MGS installations with 305-mm blockouts, 203-mm blockouts, or for non-blocked installations. This study was conducted with a steel post MGS configuration. Timber posts provide similar strength and stiffness, so the same 1,041 mm offset would apply to timber posts MGS configurations as well.

11 Additional Systems

Other MASH crashworthy, W-beam guardrail systems exist beyond the systems presented herein. Some are proprietary, some utilize different guardrail heights, and some were left out simply due to a lack of space in the paper (e.g., guardrail in mow strips, MGS long-span, trailing-end terminals, shallow foundations for guardrail posts, and guardrail stiffness transitions). In addition, there are numerous 787-mm W-beam systems that were initially developed and evaluated under the prior safety standards of NCHRP Report 350 that have yet to be evaluated to MASH including MGS with reduced post spacing, MGS on approach slopes, and flared MGS.

The details and installation guidance provided for the various guardrail configurations herein incorporate the knowledge currently available on the testing and evaluation of 787-mm guardrail systems. However, it is recognized that the future will bring new developments, new configurations, and new testing and evaluations on the performance of W-beam guardrail. Future testing may include evaluations at higher test levels and/or tests conducted under different evaluation criteria, such as the European standard, EN 1317. The guidance provided herein should be used in combination with any future knowledge to optimize the performance of W-beam guardrail systems.

Acknowledgements. The authors wish to acknowledge several sources that made a contribution to this project: (1) the MwRSF and TTI for conducting and documenting all of the full-scale crash tests referenced herein, and (2) FHWA, NCHRP, and numerous State Departments of Transportation for sponsoring all of the research related to W-beam guardrail referenced herein.

References

Abu-Odeh, A.Y., Ha, K., Liu, I., Menges, W.L.: MASH TL-3 testing and evaluation of the W-beam guardrail on slope, Report No. 405160-20. Texas A&M Transportation Institute, College Station, Texas (2013)

Abu-Odeh, A.Y., Bligh, R.P., Mason, M.L., Menges, W.L.: Development and evaluation of a MASH TL-3 31-in. W-beam median barrier, Report No. 9–1002-12-8. Texas A&M Transportation Institute, College Station, Texas (2014)

American Association of State Highway Transportation Officials (AASHTO). Manual for assessing safety hardware (MASH), 2nd Edition, Washington, D.C (2016)

Asadollahi Pajouh, M., Bielenberg, R.W., Schmidt, J.D., Lingenfelter, J., Faller, R.K., Reid, J.D.: Placement of breakaway light poles located directly behind Midwest Guardrail System, Report no. TRP-03-361-17. Midwest Roadside Safety Facility, University of Nebraska-Lincoln, Lincoln, Nebraska (2017)

Asadollahi Pajouh, M., Bielenberg, R.W., Rasmussen, J.D., Bai, F., Faller, R.K., Holloway, J.C.: Dynamic testing and evaluation of culvert-mounted, strong-post MGS to TL-3 guidelines of MASH 2016. Report no. TRP-03-383-20. Midwest Roadside Safety Facility, University of Nebraska-Lincoln, Lincoln, Nebraska (2020)

Bligh, R.P., Abu-Odeh, A.Y., Menges, W.L.: MASH test 3–10 on 31-inch W-beam guardrail with standard offset blocks. Report no. 9-1002-4, Texas A&M Transportation Institute, College Station, Texas (2011)

Bligh, R.P., Menges, W.L., Griffith, B.L., Schroeder, G.E., Kuhn, D.L.: MASH evaluation of TXDOT roadside safety features – phase II. Report no. 0-6946-R2. Texas A&M Transportation Institute, College Station, Texas (2019)

Gutierrez, D.A, Lechtenberg, K.A., Bielenberg, R.W., Faller, R.W., Reid, J.D., Sicking, D.L.: Midwest Guardrail System (MGS) with Southern Yellow Pine posts. Report no. TRP-03-272-13. Midwest Roadside Safety Facility, University of Nebraska-Lincoln, Lincoln, Nebraska (2013)

Haase, A.J., Kohtz, J.E., Lechtenberg, K.A., Bielenberg, R.W., Reid, J.D., Faller, R.K.: Midwest Guardrail System (MGS) with 6-ft posts placed adjacent to a 1V:2H fill sSlope, Report no. TRP-03-320-16. Midwest Roadside Safety Facility, University of Nebraska-Lincoln, Lincoln, Nebraska (2016)

Kovar, J.C., Bligh, R.P., Griffith, B.L., Kuhn, D.L., Schroeder, G.E.: MASH test 3–11 evaluation of modified TXDOT round wood post guardrail system. Report no. 0-6968-R4. Texas A&M Transportation Institute, College Station, Texas (2019)

Lechtenberg, K.A., Faller, R.K., Rohde, J.R., Reid, J.D.: Nonblocked Midwest Guardrail System for wire-faced walls of mechanically stabilized earth. Transp. Res. Rec. J. Transp. Res. Board **2262**, 94–106 (2011)

Lingenfelter, J.L., Rosenbaugh, S.K., Bielenberg, R.W., Lechtenberg, K.A., Faller, R.K., Reid, J. D.: Midwest Guardrail System (MGS) with an omitted post, Report no. TRP-03-326-16. Midwest Roadside Safety Facility, University of Nebraska-Lincoln, Lincoln, Nebraska (2016)

Mcgee, M.D., Lechtenberg, K.A., Bielenberg, R.W., Faller, R.K., Sicking, D.L., Reid, J.D.: Dynamic impact testing of wood posts for the MGS placed adjacent to a 2H:1V fill slope, Report no. TRP-03-234-10. Midwest Roadside Safety Facility, University of Nebraska-Lincoln, Lincoln, Nebraska (2010)

Mcgee, M.D., Faller, R.K., Rohde, J.R., Lechtenberg, K.A., Sicking, D.L., Reid, J.D.: Development of an Economical Guardrail System for use on wire-faced, MSE walls, Report no. TRP-03-235-11. Midwest Roadside Safety Facility, University of Nebraska-Lincoln, Lincoln, Nebraska (2011)

Polivka, K.A., Faller, R.K., Sicking, D.L., Rohde, J.R., Reid, J.D., Holloway, J.C.: NCHRP 350 development and testing of a guardrail connection to low-fill culverts, Report no. TPR-03-114-02. Midwest Roadside Safety Facility, University of Nebraska-Lincoln, Lincoln, Nebraska (2002)

Polivka, K.A., et al.: Development of the Midwest Guardrail System (MGS) for standard and reduced post spacing and in combination with curbs, Report no. TRP-03-139-04. Midwest Roadside Safety Facility, University of Nebraska-Lincoln, Lincoln, Nebraska (2004)

Polivka, K.A., Faller, R.K., Sicking, D.L., Rohde, J.R., Bielenberg, R.W., Reid, J.D.: Performance evaluation of the Midwest Guardrail System – update to NCHRP 350 test no.

3–11 with 28" c.g. height (2214MG-2). Report no. TRP-03-171-06. Midwest Roadside Safety Facility, University of Nebraska-Lincoln, Lincoln, Nebraska (2006a)

Polivka, K.A., Faller, R.K., Sicking, D.L., Rohde, J.R., Bielenberg, R.W., Reid, J.D.: Performance evaluation of the Midwest Guardrail System – update to NCHRP 350 test no. 3–10 (2214MG-3). Report no. TRP-03-172-06. Midwest Roadside Safety Facility, University of Nebraska-Lincoln, Lincoln, Nebraska (2006b)

Polivka, K.A., Sicking, D.L., Faller, R.K., Bielenberg, R.W.: Midwest Guardrail System adjacent to a 2:1 slope. Transp. Res. Rec. J. Transp. Res. Board **2060**, 74–83 (2008)

Reid, J.D., Bielenberg, R.W., Faller, R.K., Lechtenberg, K.A.: Midwest guardrail system with blockouts. Transp. Res. Rec. J. Transp. Res. Board **2377**, 1–13 (2013)

Ronspies, K.B., Rosenbaugh, S.K., Bielenberg, R.W., Faller, R.K., Stolle, C.S.: Evaluation of the MGS placed 6 in. behind a 6-in. tall AASHTO Type B curb to MASH TL-3, Report no. TRP-03-390-20. Midwest Roadside Safety Facility, University of Nebraska-Lincoln, Lincoln, Nebraska (2020)

Rosenbaugh, S.K., Stolle, C.S., Ronspies, K.B.: MGS with curb and omitted post: evaluation to MASH 2016 test designation no. 3-10, Report no. TRP-03-393-19. Midwest Roadside Safety Facility, University of Nebraska-Lincoln, Lincoln, Nebraska (2019)

Ross, H.E., Sicking, D.L., Zimmer, R.A.: NCHRP Report 350: recommended procedures for the safety Performance evaluation of highway features. National Cooperative Highway Research Program, Transportation Research Board, Washington, D.C (1993)

Schrum, K.D., Lechtenberg, K.A., Bielenberg, R.W., Rosenbaugh, S.K., Faller, R.K., Reid, J.D., Sicking, D.L.: Safety performance evaluation of the non-blocked Midwest Guardrail System (MGS). Report no. TRP-03-262-13. Midwest Roadside Safety Facility, University of Nebraska-Lincoln, Lincoln, Nebraska (2013)

Sheikh, N.M., Menges, W.L.: MASH test 2–11 of the 31-inch W-beam guardrail with 1.5-ft post spacing, Report no. 602921-1. Texas A&M Transportation Institute, College Station, Texas (2014)

Sicking, D.L., Reid, J.D., Rohde, J.R.: Development of the midwest guardrail system. Transp. Res. Rec. J. Transp. Res. Board **1797**, 44–52 (2002)

Stolle, C.J., Lechtenberg, K.A., Faller, R.K., Rosenbaugh, S.K., Sicking, D.L., Reid, J.D.: Evaluation of the midwest guardrail system (MGS) with white pine wood posts. Report no. TRP-03-241-11. Midwest Roadside Safety Facility, University of Nebraska-Lincoln, Lincoln, Nebraska (2011)

Thiele, J.C., Lechtenberg, K.A., Reid, J.D., Faller, R.K., Sicking, D.L., Bielenberg, R.W.: Performance limits for 6-in. high curbs placed in advance of the MGS using MASH vehicles Part II: full-scale crash testing, Report no. TRP-03-221-09. Midwest Roadside Safety Facility, University of Nebraska-Lincoln, Lincoln, Nebraska (2009)

Wiebelhaus, M.J., et al.: Development and evaluation of the Midwest Guardrail System (MGS) placed adjacent to a 2:1 fill slope, Report no. TRP-03-185-10. Midwest Roadside Safety Facility, University of Nebraska-Lincoln, Lincoln, Nebraska (2010)

Williams, W.F., Bligh, R.P., Bullard, D.L., Menges, W.L.: Crash testing and evaluation of W-beam guardrail on box culvert. Report no. 405160-5-1. Texas A&M Transportation Institute, College Station, Texas (2008)

Williams, W.F., Menges, W.L.: MASH test 3–11 of the W-beam guardrail on low-fill box culvert. Report no. 405160-23-2. Texas A&M Transportation Institute, College Station, Texas (2011)

A Synthesis of MASH Crashworthy, Non-proprietary, Weak-Post, W-beam Guardrail Systems

Scott Rosenbaugh[✉], Robert Bielenberg, and Ronald Faller

Midwest Roadside Safety Facility, The University of Nebraska,
Lincoln, NE, USA
srosenbaugh2@unl.edu

Abstract. Since its development in the early 2000's, 787-mm tall W-beam guardrail, commonly referred to as the Midwest Guardrail System (MGS), has proven to be one of the most versatile and robust roadside barrier systems in use today. The MGS is typically configured as a strong-post system utilizing either W6 × 8.5 steel posts or 152 mm × 203 mm wood posts. However, multiple weak-post MGS configurations have been developed using S3 × 5.7 steel posts. These weak-post systems absorb energy through post bending instead of post rotation through soil, so they are favored in various special applications such as bridge rails, culvert mounted systems, and placement within pavements or mow strips. These weak-post guardrail systems are non-proprietary and can be used by roadway agencies all around the world.

This paper contains details and drawings encompassing various weak-post MGS configurations that have been developed and evaluated to the American Association of State Highway and Transportation Officials (AASHTO) Manual for Assessing Safety Hardware (MASH) safety standards. The systems are discussed in terms of key components and performance characteristics, such as test level and working width. Finally, implementation guidance is provided for the proper layout and installation of the weak-post MGS configurations.

Keywords: W-beam · Guardrail · Weak-Post · Midwest Guardrail system · MASH · Barrier

1 Introduction

W-beam guardrail systems are some of the most popular and frequently used vehicle barrier systems. For over 50 years, these barrier systems have been redirecting errant vehicles and preventing impacts into roadside hazards. However, by the 1990's the increased size of the vehicle fleet was beginning to push the containment limits of the existing W-beam guardrail configurations. Multiple full-scale vehicle crash tests with heavy, high center-of-gravity passenger vehicles, such as pickup trucks and vans, resulted in rail ruptures and vehicle rollovers. Thus, in the early 2000's, the Midwest Guardrail System (MGS) was developed through modifications to the existing G4(1S) W-beam guardrail system to handle the larger vehicle fleet. These modifications included raising the height to the top of the rail to 787 mm, a reduction in the post embedment depth to 1,016 mm, and moving the rail splices from post locations to mid-spans

(Polivka et al. 2004) and (Sicking et al. 2002). Since its inception, the MGS has proven to be one of the most versatile and robust roadside barrier systems in use today.

The MGS is typically configured as a strong-post system utilizing either W6 × 8.5 steel posts or 152 mm × 203 mm wood posts, and the barrier system absorbs impact energy through rotation of the posts through the soil in which the posts are embedded. However, over the past decade, multiple weak-post MGS configurations have been developed using weak posts comprised of S3 × 5.7 steel sections. These weak-post systems absorb energy through post bending instead of rotation through soil, Additionally, due to the reduced width of the post flange, weak-post systems use 8-mm diameter attachment bolts, half the diameter of the standard 16-mm bolts in strong-post MGS systems. Square washer plates measuring 44-mm by 3-mm thick are used to prevent the smaller bolt head from prematurely pulling through the slot in the rail. Weak-post systems do not utilize blockouts to offset the rail from the face of the post, but they do typically include backup plates placed between the rail and the posts to prevent rail tearing. Photos showing these standard characteristics are shown in Fig. 1. These features have proven advantageous in a number of guardrail special applications.

Fig. 1. Photos of typical weak-post guardrail components and post-to-rail attachment

All of the weak-post guardrail systems detailed herein have been evaluated to the safety performance criteria of the American Association of Highway Transportation Officials (AASHTO) Manual for Assessing Safety Hardware (MASH) (AASHTO 2016). Further, these weak-post guardrail systems are non-proprietary and can be used by roadway agencies all around the world.

It should be noted that the original MGS and its weak-post variations were developed in the United States using US Customary units. Thus, all of the components and post spacings were originally designed and evaluated in US Customary units. For this paper, efforts were made to convert all distances over to SI units with a hard conversion. However, if a component has a standard shape that is referred to in US Customary units, that component name was not changed. For example, the standard MGS weak-posts are referred to as S3 × 5.7 posts with the understanding that similar SI shapes exist.

The purpose of this paper is to provide a resource documenting the numerous non-proprietary, weak-post, W-beam guardrail configurations and their special applications. Details required for the proper installation of each system configuration are provided herein, including drawings, component details, performance levels, working widths, and installation guidance. Note, working width is defined as the lateral distance from the front face of the guardrail to the furthest extent of the deflected system or the vehicle position during redirection, whichever is greater.

2 Modified G2 Weak Post Guardrail

The first weak-post guardrail system successfully evaluated to MASH safety performance criteria was a modified version of the original G2 weak post guardrail system. This modified G2 weak-post guardrail system consisted of W-beam guardrail mounted to S3 × 5.7 posts at a height of 813 mm. Back up plates consisting of 305-mm long sections of W-beam rail were placed between the rail and each post. In addition to the attachment bolts, the W-beam rail was also supported by 13-mm diameter support bolts located at the bottom of the rail. The posts had 610 mm tall by 203 mm wide by 6 mm thick soil plates welded to the back flange to increase soil resistance and promote post bending. The posts were spaced at 3,810 mm on-center, commonly referred to as double post spacing as it is twice the typical post spacing of 1,905 mm. The modified G2 guardrail system is shown in Fig. 2.

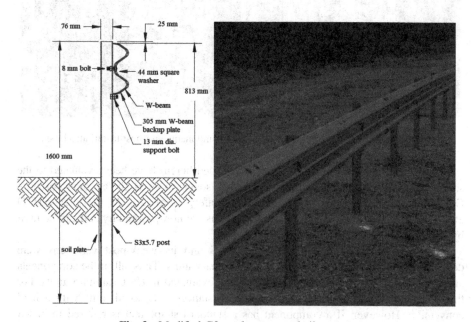

Fig. 2. Modified G2 weak-post guardrail system

The modified G2 weak-post guardrail system was successfully crash tested to both MASH test 3–10 (Bullard et al. 2017) and MASH test 3–11 (Bullard et al. 2010). Thus, the system has pass requirements for a MASH Test Level 3 (TL-3) barrier.

Due to the system having weaker posts at double post spacing, the Modified G2 weak-post system was significantly softer than standard MGS. Subsequently, the weak-post system had higher lateral deflection as compared to standard strong-post MGS. The modified G2 weak-post system had a working width of 2,743 mm, while the strong-post MGS had a working width of 1,234 mm (Polivka et al. 2006). Thus, the modified G2 weak-post guardrail system should only be used along roadsides with enough clear space behind the rail to avoid additional impact events. Additionally, the vehicle contact length, or the length of the damaged section of the guardrail, during MASH test 3–11 was approximately 42 m for the G2 system, while the contact length for the strong-post MGS was only 10.3 m. Thus, vehicle impacts into the modified G2 weak-post system will result in a greater amount of damaged components that will need to be replaced.

The modified G2 weak-post guardrail system does have a number of advantages over standard, strong-post MGS. The smaller posts spaced at double post spacing represents significant material cost savings. Additionally, the softer weak-post system imparts lower impact forces to the errant vehicle, thus reducing the chance of injury to the vehicle occupants.

Unfortunately, there have not been any guardrail end terminals or anchorages designed for weak-post, W-beam systems. Thus, the weak-post system would need to be transitioned to a strong-post configuration in order to anchor/terminate the installation. There is a 25 mm height difference between this weak-post system and the nominal 787 mm height of standard, strong-post MGS. If the two systems were to be connected, the height of the W-beam rail should be transitioned over a length of 3.8 m. However, due to the difference in stiffness and expected deflections between the modified G2 weak-post guardrail system and standard strong-post MGS, a stiffness transition would be required to gradually change the stiffness between systems and prevent excessive vehicle pocketing, which can lead to rail rupture, vehicle override, and/or vehicle instability. At this time, a stiffness transition has not yet been designed for the connection of these two guardrail systems.

3 Bridge Rails

Weak-post guardrail systems possess a few characteristics that make them attractive for use as bridge railings. First, they are much less costly to install than concrete railings, steel tube railings, and other guardrail railings that utilize larger section posts. Second, the weak-posts limit the amount of shear and bending moment that can be transferred to the bridge deck at each post location. Thus, the risk of damaging the deck during an impact event can be greatly reduced. Third, because weak-post systems do not include blockouts, they have minimal footprints help maximize the traversable width of the roadway. Finally, if the W-beam bridge railing has similar stiffness to that of the adjacent MGS, the two systems can be directly attached to each other without the need for a guardrail transition.

Three different bridge rails have been developed to MASH standards using 787-mm tall W-beam guardrail and weak, S3 × 5.7 posts. All three utilize a combination of post bending and tensile loads in the deformed guardrail, also referred to as membrane action, to redirect errant vehicles. The main differences between the three railings are in the location of the post on the deck, the post connections to the bridge deck, and the shape of the backup plate.

The first weak-post bridge rail system was the MGS bridge rail. The weak-post MGS bridge rail utilized steel tube sockets mounted to the side of the bridge deck to support the S3 × 5.7 posts spaced at 953 mm on-center (Thiele et al. 2010) and (Thiele et al. 2011). An 11-mm thick top mounting plate was welded to the front of the HSS4 × 4 × 3/8 sockets, and the assembly was attached to the deck with a vertical, 25-mm diameter bolt, as shown in Fig. 3. The bottom of the socket was bolted to an L7 × 4 × 3/8 angle to provide support for reverse bending and longitudinal loads. Two 6-mm thick shim plates were welded on the upstream and downstream sides of the post so that the post fit snuggly inside the socket. The weak-post MGS bridge rail utilized the same 305-mm long W-beam backup plates as the modified G2 weak-post guardrail system discussed earlier.

This system was successfully crash tested to MASH TL-3 and had a working width of 1,351 mm. The system was specifically designed with the 953-mm post spacing, or half post spacing, to provide similar system stiffness to the standard strong-post MGS, which had a working width of 1,234 mm. Because the system deflections and working widths were similar, no transition was necessary and the bridge rail could be directly connected to the adjacent MGS off the bridge.

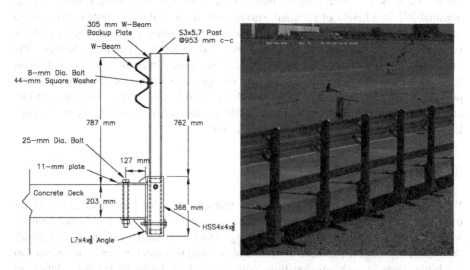

Fig. 3. Details and photo of the Weak-Post MGS Bridge Rail

The T631 bridge rail was a top-mounted system that utilized a 16-mm thick baseplate, a 6-mm thick washer plate, and four 16-mm diameter bolts to mount the S3 × 5.7 posts to the top of the bridge deck, as shown in Fig. 4. A 13-mm diameter shelf bolt helped support the rail vertically. The T631 bridge rail can be configured with two different post spacings. A 1,905-mm post spacing was successfully tested and evaluated to MASH TL-2 with a 762-mm working width (Williams et al. 2014a). A MASH 3–11 test was conducted on the 1,905 mm spacing, but the rail ruptured and the test failed (Williams et al. 2015). Subsequently, the post spacing was reduced to 953 mm and resulted in the T631 being successfully tested and evaluated to MASH TL-3 with a 1,466-mm working width (Williams et al. 2014b).

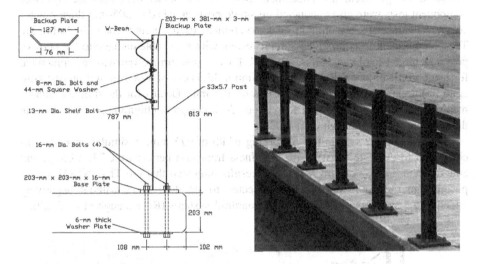

Fig. 4. Details and photo of the T631 bridge rail

The third weak-post W-beam bridge rail was developed for the Nebraska Department of Transportation (NDOT) to be a MASH TL-2 bridge rail targeted for use on low volume roads. The railing was to be completely side mounted with no components or hardware on the deck surface. NDOT regularly used C-channels along the edges of their low-volume bridge decks as a cast-in formwork that provides confinement and strength to the edges of the deck. The NDOT did not desire to have hardware extruding outward from the edge channels, so any attachment hardware would have to extend through the edge channel.

The system was adapted from the Weak-Post MGS Bridge Railing, so it also used HSS4 × 4 × 3/8 sockets, 6-mm thick shims on the upstream and downstream sides of the S3 × 5.7 posts, and 305-mm long W-beam backup plates. However, the post spacing for the TL-2 bridge rail was increased to 1,905 mm from the 953-mm spacing used in the TL-3 MGS bridge rail. The NDOT bridge railing sockets were welded to a side mounting plate and bolted directly to the side of the bridge deck. The side bolts were threaded into coupling nuts, which were cast directly into the side of the bridge deck on the inside face of the edge channel, and threaded anchor rods extended from the coupling nuts toward the center of the deck. Using this attachment design, the tensile loads in the attachment bolts during an impact event are directly transferred through the coupling nuts and into the threaded anchor rods. Thus, the edge of the deck is only subjected to compression loads at the bottom of the socket mounting plate as the post is laterally loaded. Without tensile or shear loads in the deck edge, the risk of deck damage during impacts was minimized. The NDOT bridge rail and this unique attachment are shown in Fig. 5.

The NDOT weak-post bridge rail was successfully crash tested to MASH TL-2 evaluation criteria while mounted to a critically weak 178-mm thick cast-in-place concrete deck (Rosenbaugh et al. 2020). There was no observable damage to the simulated bridge deck, the attachment bolts, or the attachment sockets. Dynamic component tests were also conducted on single posts attached to 305-mm thick precast concrete beam slabs, a common deck component used on rural roads in Nebraska. These component tests provided similar results with no visible damage to the deck or post attachment hardware. Thus, the NDOT weak-post bridge rail was recommended for use on cast-in-place deck with a minimum thickness of 178 mm and precast concrete beam slabs with a minim thickness of 305 mm. Details for the different attachment components and hardware for various deck design configurations is provided in the test report (Rosenbaugh et al. 2020).

The tested TL-2 system had a working width of 975 mm. A simulation study was conducted to evaluate the need for a stiffness transition between the TL-2 bridge rail and adjacent TL-3, standard MGS. The results indicated that the TL-2 NDOT weak-post bridge rail could be directly connected to standard MGS without negatively affecting the safety performance of the guardrail systems (Rosenbaugh et al. 2020).

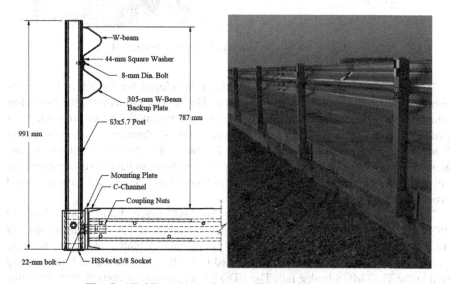

Fig. 5. NDOT weak-post bridge railing details and photo.

The three weak-post bridge rails presented herein were all designed with similar components and performed similarly by relying on the posts bending near ground line and the rail providing redirective force through tensile, membrane action. With the successful crash testing conducted on the various post spacings, all three designs should be considered MASH TL-2 crashworthy with a post spacing of 1,905 mm and MASH TL-3 crashworthy with a post spacing of 953 mm. Further, simulation studies conducted on the connections of both TL-3 and TL-2 bridge rail configurations to adjacent MGS showed no need for a guardrail stiffness transition. Thus, all three of the railings, in either of the two post spacings for TL-2 or TL-3 configurations, can be directly connected to MGS with a 1,905 mm post spacing between the outer most S3 × 5.7 weak post and the adjacent W6 × 8.5 strong post.

It should be noted that since these weak-post, W-beam bridge rails rely on tensile loads and membrane action to redirect vehicles, proper guardrail anchorage must be incorporated into the MGS installations adjacent to both ends of the bridge railing in order for the system to perform as intended. Bridge railing designers should also note that traditional methods used to design steel post and beam bridge railings do not account for high railing deformations and the contributions of membrane action to the redirective forces of the railing. Thus, traditional methods like the inelastic beam procedure provided in the AASHTO LRFD Bridge Design Specifications (AASHTO 2017) do not apply to deformable railings and should not be used to design these types of W-beam bridge rails.

4 Culvert Mounted Guardrail Systems

Multiple MASH TL-3 weak-post, 787 mm tall W-beam guardrail systems have been developed for attachment to low-fill, concrete box culverts. These culvert mounted guardrail systems were adapted from the weak-post MGS bridge rail design. As such, they utilized the same 787-mm tall W-beam guardrail, 1,118-mm long S3 × 5.7 posts, 953-mm post spacing, 305-mm long W-beam backup plates, and HSS4 × 4 × 3/8 sockets as the weak-post MGS bridge rail. However, the socket attachment hardware was modified to allow the sockets to be placed at various locations on the culvert.

The first set of guardrail systems were designed to mount the socket to the outside face of culvert headwalls (Schneider et al. 2014) and (Rosenbaugh et al. 2014). Multiple attachment configurations were developed including a top, single bolt attachment and a side-mounted attachment, as shown in Fig. 6. The top, single bolt attachment was designed to be nearly identical to the weak-post MGS bridge rail, as it utilizes the same top mounting plate and gusset as the bridge rail system. However, instead of a through bolt, the culvert mounted system used an epoxy-anchored, 29-mm diameter, threaded rod embedded 254 mm into the headwall. Additionally, the bottom of the socket was anchored to the outer face of the headwall with two epoxy-anchored, 13-mm diameter threaded rods, embedded to a depth of 114 mm. Open vertical slots were cut into the bottom mounting plate such that the socket assembly could be lowered into position after the epoxied anchors were installed.

The side mounted attachment to culvert headwalls was developed to remove all hardware from the top surface of the culvert. The upper and lower mounting plates are identical in size, but the upper mounting plate had to be reinforced with triangular gusset plates to reinforce the welds and prevent bending in the upper plate. The upper mounting plate was epoxy anchored into the headwall with two 19-mm diameter threaded rods at a depth of 229 mm, while the lower attachment bolts remained the same from the top- single bolt attachment design. Note, vertical slots were not necessary for the lower mounting plate in the side-mounted attachment design as the socket assembly is installed laterally over the epoxy anchors.

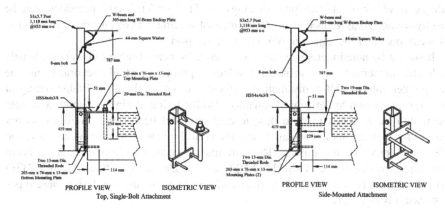

Fig. 6. Top, single-bolt attachment and side-mounted attachment for weak-post MGS mounted to culvert headwall

Two attachment designs have been developed for mounting weak-post guardrail to the top slab of a culvert. The first top-mounted system utilized a cylindrical concrete foundation that was originally developed for use as a socketed foundation for cable barrier posts (Rosenbaugh et al. 2015a). The cylindrical foundations were 305 mm in diameter and the HSS4 × 4 × 3/8 socket was placed in the center of the foundation. The concrete foundations were reinforced with both no. 4 hoops and vertical no. rebar. The vertical rebar extended out the bottom the foundation and were embedded 178 mm into the top culvert slab with epoxy. The height of the concrete cylinder was variable between 267 mm and 914 mm as it was intended to match the depth of the soil fill above the culvert. Note, the length of the steel socket remained 356 mm (359 mm when including the 3-mm cap welded to the base of the socket) regardless of the height of the cylinder in order to support the 1,118-mm long post and ensure proper rail height. Details for the cylindrical concrete foundations and installation photos are shown in Fig. 7 (Rosenbaugh et al. 2019).

The second top-mounted attachment system utilized a steel socket assembly fabricated from an HSS4 × 4 × 3/8 tube, 6-mm thick reinforcing plates, a 13-mm thick base plate, and 10-mm thick gusset plates, as shown in Fig. 8 (Rosenbaugh et al. 2019). The steel socket and the reinforcing plates have variable lengths dependent upon the soil fill depth on the culvert. The socket extends from the base plate to 51 mm above the ground line, while the reinforcing plates are placed on the front and backside of the socket and extend from the baseplate to 152 mm below the ground line. A 16-mm diameter bolt is used to support the post in the socket and ensure it is inserted to the proper embedment depth and the W-beam rail is at a nominal height of 787 mm. The base plate is attached to the culvert using four 19-mm threaded anchor rods, which are epoxied into the top-slab with an embedment depth of 178 mm.

Dynamic component testing was conducted on all of the culvert mounted weak-post guardrail systems described herein. Post and socket assemblies from each system were installed on a simulated culvert and impacted in both the lateral and longitudinal directions to evaluate the strength of the socket attachments. In each test, the post bent plastically near the top of the socket while the socket and the culvert remained undamaged. Thus, the socket attachment designs showed adequate strength to support the weak posts, and all three guardrail systems should perform identically to the weak-post MGS bridge railing. Accordingly, each of the culvert mounted, socketed, weak-post guardrail systems were considered to be crashworthy to MASH TL-3 (Schneider et al. 2014 and Rosenbaugh et al. 2019). Since the these culvert mounted systems use the same posts and post spacing as the weak-post MGS guardrail system, each of these culvert mounted systems can be directly connected to adjacent strong-post MGS installations located on both sides of the culvert.

The dynamic tests were conducted on simulated culverts with 229-mm tall by 305-mm wide headwalls on top of 229-mm thick top slabs. Both the headwall and culvert top slab contained minimal reinforcement established via a survey of multiple state DOTs to identify a critical configuration (Schneider et al. 2014). Culverts at actual installation sites should provide similar characteristics or increased strengths to properly support these side- and top-mounted guardrail systems. Further details on culvert strength was provided in Schneider et al. (2014).

All of the socket attachment configurations utilized epoxy anchorage into the culvert. As such, these weak-post guardrail systems can be attached to both new and existing culvert structures. If desired, the anchors could be cast into place during fabrication of new culverts. Additionally, each of these culvert mounted systems utilized the exact same post assembly as the weak-post MGS bridge rail (i.e., a 1,118-mm long S3 × 5.7 post with 6-mm shims welded to the upstream and downstream sides of the post). Thus, agencies only need to stock one type of post to install/repair the weak-post MGS bridge rail and all of the culvert mounted, weak-post guardrail systems detailed herein.

When utilizing the side-mounted designs or the top-mounted designs in close proximity to the culvert headwall, the headwall should not extend more than 50 mm above the ground. Headwalls extending further than 50 mm may act as vertical curbs and could pose a stability hazard as the vehicle wheels traverse over them. Weak-post guardrail systems have not yet been evaluated in combination with curbed roadways.

Soil fill on top of concrete box culverts often contains fill slopes. As such, testing of the top-mounted attachment designs was conducted with the socket assemblies located at the slope-break-point of 2H:1V soil fill slopes. Since these tests demonstrated adequate strength, the top-mounted sockets may be use adjacent to soil fill slopes of 2H:1V or flatter. If steeper fill slopes are desired, the slope break point should be located a minimum of 600 mm laterally behind the sockets.

To date, all of the socketed, weak-post MGS variations have been evaluated with level terrain in front of the barrier. The introduction of an approach slope may negatively affect the performance of these systems in terms of vehicle capture and stability. Thus, it is recommended that approach slopes of 10H:1V or flatter be placed in front of the top-mounted, socketed, weak-post MGS on culverts.

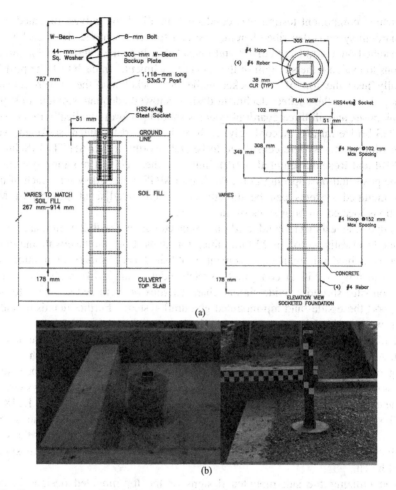

Fig. 7. Top-mounted, cylindrical concrete foundations for weak-post guardrail on culverts (a) details and (b) installation photos

5 Guardrail Systems for Use in Mow Strips

Often times, guardrail systems are installed with a thin pavement surrounding the guardrail posts as a means of vegetation control or erosion prevention, commonly referred to as mow strips. Strong-post guardrail systems require leave-outs within the mow strip at post locations to allow the post to rotate back through the soil and absorb impact energy. However, weak-post guardrail systems absorbed energy through plastic bending of the posts at ground line, so the need for large leave-outs would not be necessarily.

To date, a few different configurations of weak-post guardrail installed within mow strips have been evaluated. All of these systems consisted of 787-mm tall W-beam, S3 × 5.7 posts spaced at 953 mm on-center, and the same 8-mm bolt and 44-mm square

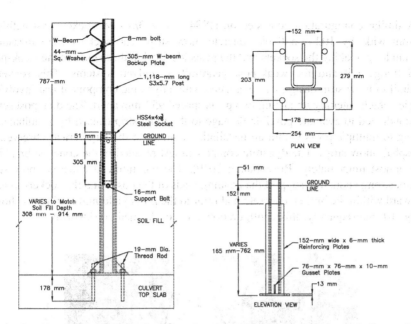

Fig. 8. Details of the top-mounted, steel socket assemblies for weak-post guardrail on culverts

washer to attach the rail to the post. The first configuration consisted of HSS4 × 4 × 3/8 steel sockets installed down the center of a 1,219-mm wide by 152-mm thick asphalt mow strip. The sockets were 762-mm deep and included 254-mm wide by 229-mm tall soil plates to distribute load. This configuration was full-scale crash tested in accordance with MASH test 3–11. The test was successful in that it satisfied all MASH safety performance criteria, but a large crack opened in the asphalt down the center of the mow strip, as shown in Fig. 9 (Rosenbaugh et al. 2015b). The crack had a maximum opening width of 38 mm and a length of 18.3 m. Many of the sockets within the crack were found to be loose, so repairs to the asphalt mow strip would be necessary in addition to repairing the guardrail installation. As such, the tested configuration was considered to be MASH TL-3 crashworthy, but extensive repairs would be necessary after an impact event.

Fig. 9. Post-test damage to socketed weak-post guardrail in 152-mm asphalt mow strip.

A similar configuration consisted on HSS4 × 4 × 3/8 steel sockets cast within a 914-mm wide by 102-mm thick concrete mow strip. The sockets were measured 356 mm long, making the sockets and the posts identical to those used in the weak-post MGS bridge rail and the weak-post guardrail on culvert systems. This socketed guardrail in mow strip design was evaluated with a dynamic component test involving a bogie vehicle laterally impacting two posts spaced 953 mm apart. The dual post setup was conducted to evaluate possible damage to the mow strip caused by simultaneous loading to multiple posts within an installation (similar to the large crack observed in the asphalt mow strip). The dynamic component test resulted in the concrete breaking apart almost immediately (Rosenbaugh 2019). The unreinforced concrete mow strip was not strong enough to support the bending loads of the posts, and the sockets rotated backward within the broken concrete pad prior to the posts yielding/bending, as shown in Fig. 10. Subsequently, this configuration was not deemed crashworthy.

Fig. 10. Post-test damage to dual socketed posts in 102-mm thick concrete mow strip

Finally, a dual post test was conducted on two S3 × 5.7 posts inserted into holes within a 1,219-mm wide by 102-mm thick unreinforced concrete mow strip and driven to a depth of 1,016 mm. Note, this configuration did not utilize a steel socket or sleeve and the post did not have a soil plate. The dual-post impact resulted in the post bending backward with only minor spalling of the concrete adjacent to the back of the posts. The combination of the post embedment depth and the 102-mm thick concrete pad proved strong enough to support the full plastic load of the weak posts. Thus, a weak-post MGS with a 953-mm post spacing and a 1,016-mm embedment depth should be MASH TL-3 crashworthy when installed within concrete mow strip with a thickness of at least 102 mm (Fig. 11).

Fig. 11. Post-test damage to dual posts with 1,016 mm embedment depth in 102-mm thick concrete mow strip.

6 Future Developements

Although multiple weak-post guardrail systems have been developed and proven crashworthy to MASH criteria, there has yet to be a terminal or anchorage system developed for weak-post guardrail systems. Currently, all of the guardrail systems described herein must be transitioned to strong-post guardrail prior to terminating or anchoring the installation. The development of a crashworthy weak-post guardrail terminal would eliminate the need for strong posts in these installations.

Similarly, all of the approach guardrail transitions (AGTs), which attach W-beam guardrail to rigid parapets or bridge railings, utilize strong-post guardrail. Although many weak-post guardrail systems can be directly attached to strong-post MGS without the need for a transition, this connection has never been evaluated adjacent to an AGT. The transition between weak-post and strong-post may negatively affect the performance of an AGT if this post transition is adjacent to the upstream end of the AGT. Further evaluation of the connection between weak-post guardrail and AGTs is needed.

W-beam guardrail is often placed on curbed roadways. Curbs have been shown to affect vehicle trajectories and the performance of roadside barriers located behind curbs. Strong-post MGS systems have been evaluated adjacent to roadside curbs, but the performance of weak-post guardrail systems in combination with roadside curbs has not been evaluated. Further, the effect of curbs on guardrail performance may be critical for weak-post systems as they do not utilize blockouts, which help maintain rail height during deflection. Thus, research and testing is needed on weak-post guardrail systems in combination with roadside curbs.

All of the weak-post, W-beam guardrail systems described herein were designed and evaluated to MASH safety performance criteria. However, many countries utilize alternative barrier standards, such as the European standard, EN 1317. For weak-post, W-beam guardrail systems to be used throughout the world, testing and evaluations should be conducted in accordance with the other international barrier standards.

Acknowledgements. The authors wish to acknowledge several sources that made a contribution to this project: (1) the Midwest Roadside Safety Facility (MwRSF) and Texas A&M Transportation Institute (TTI) for conducting and documenting all of the full-scale crash tests referenced herein, and (2) FHWA, NCHRP, and numerous State Departments of Transportation for sponsoring all of the research related to W-beam guardrail referenced herein.

References

American Association of State Highway Transportation Officials (AASHTO): Manual for assessing safety hardware (MASH), 2nd Edition, Washington, D.C. (2016)

American Association of State Highway Transportation Officials (AASHTO): AASHTO LRFD Bridge Design Specifications, 8th Edition, Washington, D.C. (2017)

Bullard, D.L., Bligh, R.P., Menges, W.L., Haug, R.R.: NCHRP Web-Only Document 157: Volume I: evaluation of existing roadside safety hardware using updated criteria–Technical report, National Cooperative Highway Research Program, Transportation Research Board, Washington D.C. (2010)

Bullard, D.L., Menges, W.L., Kuhn, D.L.: MASH test 3–10 of PennDOT G2 weak post W-beam guardrail, Report no. 608221-1, Texas A&M Transportation Institute, College Station, Texas (2017)

Polivka, K.A., et al.: Development of the Midwest Guardrail System (MGS) for standard and reduced post spacing and in combination with curbs, Report no. TRP-03-139-04, Midwest Roadside Safety Facility, University of Nebraska-Lincoln, Lincoln, Nebraska (2004)

Polivka, K.A., Faller, R.K., Sicking, D.L., Rohde, J.R., Bielenberg, R.W., Reid, J.D.: Performance evaluation of the Midwest Guardrail System – update to NCHRP 350 test no. 3-11 with 28 c.g. height (2214MG-2). Report no. TRP-03-171-06, Midwest Roadside Safety Facility, University of Nebraska-Lincoln, Lincoln, Nebraska (2006)

Rosenbaugh, S.K., Lechtenberg, K.A., Faller, R.K., Bielenberg, R.W.: Weak-post W-beam guardrail attachment to culvert headwalls. Transp. Res. Rec. J. Transp. Res. Board **2437**. Washington D.C (2014)

Rosenbaugh, S.K., Schmidt, T.L., Faller, R.K., and Reid, J.D.: Development of socketed foundations for S3 × 5.7 posts, Report no. TRP-03-293-15, Midwest Roadside Safety Facility, University of Nebraska-Lincoln, Lincoln, Nebraska (2015a)

Rosenbaugh, S.K., Faller, R.K., Lechtenberg, K.A., Holloway, J.C.: Development and evaluation of weak-post W-beam guardrail in mow strips, Report no. TRP-03-322-15, Midwest Roadside Safety Facility, University of Nebraska-Lincoln, Lincoln, Nebraska (2015b)

Rosenbaugh, S.K., Asadollahi Pajouh, M., Faller, R.K.: Top-mounted sockets for weak-post MGS on culverts, Report no. TRP-03-368-19, Midwest Roadside Safety Facility, University of Nebraska-Lincoln, Lincoln, Nebraska (2019)

Rosenbaugh, S.K., DeLone, J.A., Faller, R.K., Bielenberg, R.W.: Development and testing of a bridge rail for low-volume roads, Report no. TRP-03-407-20, Midwest Roadside Safety Facility, University of Nebraska-Lincoln, Lincoln, Nebraska (2020)

Schneider, A.J., Rosenbaugh, S.K., Faller, R.K., Sicking, D.L., Lechtenberg, K.A., Reid, J.D.: Safety performance evaluation of weak-post, W-beam guardrail attached to culvert, Report no. TRP-03-277-14, Midwest Roadside Safety Facility, University of Nebraska-Lincoln, Lincoln, Nebraska (2014)

Sicking, D.L., Reid, J.D., Rohde, J.R.: Development of the Midwest guardrail system. Transp. Res. Rec. J. Transp. Res. Board **1797**. Washington D.C (2002)

Thiele, J.C., et al.: Development of a low-cost, energy-absorbing bridge rail, Report no. TRP-03-226-10, Midwest Roadside Safety Facility, University of Nebraska-Lincoln, Lincoln, Nebraska (2010)

Thiele, J.C., et al.: Development of a low-cost, energy-absorbing bridge rail. Transp. Res. Rec. J. Transp. Res. Board **2262**. Washington D.C (2011)

Williams, W.F., Bligh, R.P., Menges, W.L., Kuhn, D.L.: Crash test and evaluation of the TxDOT T631 bridge rail, Report no. 9-1002-12-10, Texas A&M Transportation Institute, College Station, Texas (2014a)

Williams, W.F., Bligh, R.P., Menges, W.L., Kuhn, D.L.: MASH TL-3 crash testing and evaluation of the TxDOT T631 bridge rail, Report no. 9-1002-12-12, Texas A&M Transportation Institute, College Station, Texas (2014b)

Williams, W.F., Bligh, R.P., Odell, W., Smith, A., Holt, J.: Design and full-scale testing of low-cost Texas Department of Transportation Type T631 Bridge Rail for MASH Test , Level 2 and 3 applications, Transportation Research Record: Journal of the Transportation Research Board, No. 2521, Washington D.C (2015)

The Safety Highway Geometry Based on Unbalanced Centripetal Acceleration

Creso de Franco Peixoto[✉] and Maria Teresa Françoso

The State University of Campinas UNICAMP, Campinas, Sao Paulo State, Brazil
cresopeixoto@gmail.com

Abstract. Road accidents take a significant part in the human lives losses and financial resources in worldwide sense. In order to devise new strategies to combat this problem there is a continuous search for new approaches from researchers, dealers and government agencies. Recent changes in the Brazilian traffic code, such as the increase of the score limit on the driver qualification and the restrictions on speed enforcement, imposed a new operational scenario. Speeding should increase and, consequently, reflecting more accidents and severity. A proposed model in this work has several equations, correlating road geometry parameters and accident record indexes. This model can be applied as an objective policy tool of physical improvements and speed reduction focusing on accident severity and amount. It was taken the unbalanced centripetal acceleration on horizontal curves as the main derived term, generated mainly by excess of speed in relation to the allowed one, under the Physics and human restrictions criteria. There were selected some Brazilian road segments to apply this model, conceiving specific equations and evaluating confidence under determination coefficient, that resulted a good level to the selected place.

Keywords: Road geometry · Road accidents · Unbalanced centripetal acceleration

1 Introduction

The mode of road transportation is particularly unique among others, in relation of its great social spectrum and due to the easiness of vehicle purchase: low cost and financing facility. A rout that a pavement is not always offered and required. It is almost celibate in door-to-door paths.

However, the annual global statistics on road accidents indicate values in the order of 1.2 million deaths and cost government approximately 3% of the Gross Domestic Product (GDP) of each country, highlighting near 5% for low- to the middle-income ones (Global Road Safety Partnership 2016). The road is the label of the 10th cause of the world mortality statistics in 2000, rising to the 8th place on the podium of human tragedy in 2016 (World Health Organization 2018). Therefore, road accidents, from severe to just material damage ones, carry heavy human and economic burdens in global sense.

Researches and programs developed since the moment that car gained a home place are trying to minimize the negative features of this way of transport, side-by-side of the continuous growing of accident number and severity. It is mandatory a continuous development of new methods and procedures. This statement can be exemplified taking the first traffic code, conceived by William Phelps Eno to the New York City in 1909 (ENO Center of Transportation 2019), the car crash dummies from John Paul Stapp in 1949 (The New York Times Magazine 2012) and the evolution of horizontal curve design (Fitzpatrick and Khal 1992).

Heuristics is also an important tool to find the adequate questions and approaches for road accident mitigation or minimizing, as: Why does the concentration of accident kind changes in time and place? How the vehicle design can reduce severity of the injured ones? What is the influence of the road to the accidents? Then, drivers, roads and vehicles need to be adequately analyzed and evaluated as generators factors to reach these requirements.

The road design needs to be conceived under a certain level of risk. According to Report 480 of the National Highway Cooperation Research Program (Transportation Research Board 2002) the performance-based motorway design process considers environment and feasible solutions as fundamentals, not giving to safety the main position. According to Hauer (2000) there are common design procedures without any relationship to the road risk rate. The minimum visibility distance in vertical curves is based on conjectures about the height of the object and model of calculation.

In Brazil, the speed allowed is not effectively respected. Control equipment cannot be hidden and users have easy information about inspection sites. This fact generates incompatibility between design premises and minimum safety level. Thus, a model that associate speed excess to the risk level becomes excellent piece on the adequate road offer gear. The approval of the Brazilian Law 14.071 (BRAZIL 2020) imposed changes to the Brazilian Traffic Code, as the increase in the limit of points on the qualification of the Brazilian driver and the restrictions on speed enforcement. It is a new operational scenario, where the speeding should increase and, consequently, the severity and death amount of road accidents.

2 Objectives

In this work there were analyzed geometric parameters, in an isolated or joined way, associated to the effective speed and car crash indexes. The relationship of these features and figures is based on Physics and math correlation. It is focused mainly on unbalanced centripetal acceleration (UCA), that is generated by gravity acceleration times side friction coefficient, between tires and pavement surface. The length and steepness of a previous tangent from a horizontal curve induces excessive speed. A horizontal curve in sequence imposes high centripetal acceleration.

It was applied a conceived model in this work to some Brazilian highway segments, allowing to evaluate results and to propose a math base of costs and benefits, in relation to possible future accidents and road improvements.

The main focus of this research is to conceive a model between unbalanced centripetal acceleration to accident data indexes, considering some relevant geometry features and allowing to be adjusted by each studied road segment. Taking in consideration that it is common violent crashes in horizontal curves, when driver oversteers trying to avoid the imminent impact, the excessive centripetal acceleration turns to be the main *actor* and the *cast* composed by the opposite factors to minimize the vehicle loss of control, as superelevation and friction factor, to avoid the fatal skid. There is also a particular Brazilian feature: drivers have, in general, the information on speed enforcement location, in parallel to the discredit in Applied Physics of the vehicle stability.

Then, the objectives are:
General Objective focus on:

- To analyze and correlate highway geometry parameters and accident indexes of medium to high severity under distinguished features aiming to conceive a relationship model.

The specific objectives include:

- To develop a generic model that associates selected geometric features, accident indexes and speed.
- To evaluate conceived equations of the proposed model, with accident indexes and selected features, having the effective speed of any studied road segment as a centripetal acceleration generator, called here as Road Risk Index (RRI) or Road Risk Index Scenario (RIS).
- To test the general proposed model on some segment of road or highway, allowing to predict accidents and to compare accident costs with any proposed improvement.

3 References Revision

There were selected some references on this matter, considered as main ones, because there is a huge number of researches on this issue over decades. The main focus of this research is to evaluate the influence of unbalanced centripetal acceleration on the accident rate of each selected road, indirect or direct ways.

It is easy to think that drivers just want to get there faster. But drivers tend to keep the vehicle at constant speed on highways, something as "cruising speed". So, the magnitude of the desired speed is strongly influenced by sensory issues or bad premises. The vehicle in high speed generates a lot of fast images before driver's eyes. This fact, tend to generate some restraint regarding the speeding, but it is an insufficient warning to guarantee safety. As the alignment of the path varies in front of the driver eyes, it does not characterize enough information to guarantee safe movement. In the case of geometry of new roads, conceived under high design speed, the horizontal radii are of high magnitude, offering a larger field of vision. So, it tends to give some safety, when the speed is controlled mainly by viewer field and body sensation. In the case of a road part, from a long tangent to a bended horizontal curve, it is easy to conclude that to control just by the view scenario it is a great mistake. In former roads, there are places of effective pitfalls to induce drivers to a movement of high risk, when they not respect traffic signs. Drivers evaluating the speed limit just by field of vision, sensation of centripetal force in their bodies and skidding noise. They are likely within accident scenery.

The horizontal radius is of outstanding importance in this field of knowledge. It is common to find researches evolving it. Hauer (2000) established relationship between accident indexes and radii values, where radii greater than 500 m indicate reduced influence on the accident generation.

On long tangent road part followed by reduced horizontal radius, there is a scenario of increased operational risk. Matthews and Barnes (1988) correlated the horizontal curve radius to the extension of the preceding tangent and the number of accidents per million vehicle-kilometers.

Lamm and Choueri (1987) developed several analyzes of the correlation between the accident rate and the horizontal radius of curvature. In this reference, a limit of accident rate (AR) is adopted, to 2 accidents per 10^6 vehicles per kilometer. This limit of AR is associated to safety level of 99.9998% that one accident does not occur, estimated by: $1-(2 \times 10^{-6}) = 0.999998$. For the safety level taken as reference, a 350 m radius is critical, as Fig. 1 shows. There is evidence of risk stability to radius greater than 2000 m.

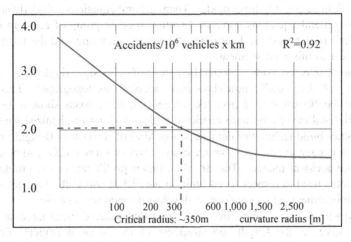

Fig. 1. Accident Rate (AR) as Function of the radius of Horizontal Curve Adapted from: Lamm and Choueiri (1987)

Lammand Choueri (1987) also reported the correlation between the accident rate and the slope of the studied segment. For the safety level taken as a reference, 2 acc/10^6 vehicle x km, there is a critical slope of 6.5% and little influence of the slope between 2 to 4% regarding risk. Slopes greater than 6.5% generate considerable increase in operational risk.

Vayalamkuzhi and Amirthalingan (International Conference on Transportation Planning and Implementation Methodologies for Developing Countries 2014) developed equations correlating the speed of operation to the following quantities: slope, superelevation, access points, average daily volumes (ADV) and access points through beds, based on accident prediction models, valid for double lanes. In Eq. 1, a Poisson regression is shown, associated to the operating speed on the curve, the operating speed on the tangent, superelevation and longitudinal slope. This research associates the magnitudes to accidents figures, not specifically to those who have victims or not.

$$Vop_{MC} = 15.138 - 0.684 \times i - 0.285 \times e + 0.718 \times Vop_T; R^2 = 0.64 \quad (1)$$

Vop_{MC}: operation speed at the middle length of the curve; I: longitudinal slope; e: superelevation;

Vop_T: operation speed at the tangent; R^2: math correlation coefficient.

4 Research Method

Developing a model that integrates geometry features, speed and accident indexes it is relevant to envisage an archetype where each feature is a particular piece studied in integrated or isolated ways, aiming to match accident indexes. It is a conceptual approach. Some other questions, as human behavior, math and physics principles are also included, in direct or indirect modes. Then, general equations will derivate from it. When these general equations correlate to some specific group of accident data and places of their occurrences, it determines the specific constants and the math correlation, valid just to this road segment.

The kind of selected roads is, preferably, some of old design, single lane, two ways, high density of daily traffic nowadays and mountainous topography. This is very common in the Brazilian road network. In general, these roads show a lot of very bended horizontal curves, because restricted budget and not mechanized earth movement and other build techniques demanded this kind of road offer. Bringing its use to nowadays, it is easy to conclude that excessive level of service (LOS) imposes traffic jam and high accident indexes. The Brazilian driver profile adds extra problems: low credulity on the need to respect speed limit, in parallel of soft traffic law and low level of police enforcement, in comparison to developed countries scenario.

The adopted geometric quantities were those that induced speed increase before a horizontal curve, as the length and steepness of the previous tangent. Two other parameters are those that require centripetal acceleration to keep vehicle on the road, curvature and superelevation from the horizontal curve in sequence to the selected tangent. The operational greatnesses are the design and effective speeds and crash indexes.

These selected geometric quantities are those that induce speed increasing and, consequently, generate a favorable environment for skidding and, in turn, for accidents. Longitudinal acceleration is a factor induced by decreasing slope and length. Thus, excessive centripetal acceleration is easily imposed by the road when an in sequence horizontal curve is reached. The centripetal acceleration, due to transverse friction at the tire-pavement contact, acts as the main character. This acceleration, called unbalanced one, is the one added to that generated by the transverse inclination of the floor, named superelevation, resulting the full centripetal acceleration.

The physical features must be obtained from topographic survey, allowing highest precision. Otherwise, when the road design is available, that effectively turned to be "as built", it is the ideal tool to obtain this information. *Google Maps* and *Google Street* values are of low-precision products, however they can be applied to evaluate road network in a fast way, because it is as a cheaper and easier tool to detect road parts that need to be surveyed accurately way, as a selection filter. So, *Google Maps* allows to find horizontal curve radius, length and slope of each tangent. *Google Street*, the superelevation. The horizontal error is lower than vertical one because length is, in general, higher than altitude differences. The worst value to be found this way, the superelevation, occurs because image of Google Street comes from perspective projection. The Google Vertical Model has low accuracy, because a 3D Model from NASA is applied. It is important to highlight that the major part of the Brazilian road network do not have available design, mainly those former ones. But, if a math relationship between geometry data from Google and crash indexes show high value of math correlation coefficient, it is a good to optimum information to conceive a road risk software prediction with low cost and easy to apply to any road. This can help a lot, to drivers during travel, with an on line information and to highway planners, for maintenance basic design.

Design speed is the limit of the selected road segment and the effective speed is the real one. The design speed is that from signalization. The effective speed has two values: the average and the highest value of each sample of values. Speed measurements must be done in free LOS; in other words, when the observed vehicle does not have its movement influenced by another nearby. Cars are preferable in this study, because the car, among other vehicles, appear in high participation level, in relation of Brazilian roads. The speed measurements occur with the personnel in the selected place, during daylight and in an ambience of high visibility and dry surface of pavement. The speed is the value of the length between two marks divided by its lapse of each selected vehicle. The lapse comes from a cell phone, avoiding any equipment that could draw attention, as radar or chronograph. There were avoided for speed measurements any road part with speed checkers. It highlights that the effective speed is an information of the at present moment and crash indexes came from the past. Thus, it is accepted here that the speeds currently measured are practically the same of the period that the referred accidents occurred.

The accident number and severity should be analyzed under indexes. These values are obtained from Government data, from at least 10 years of surveys, preferably, avoiding random influence. The main kind of accident in this study is that where there are victims. The preferable accident road data are those that shows exact place of each sinister. Brazilian Federal database shows just the kilometer of each event, imposing the acceptance of it in this work. The data of each road segment, composed of several road parts, characterized by the group of features related before, are correlated in math basis in regard of a relationship equation and defining the determination coefficient. Different types of regressions can be applied, from linear to exponential ones.

5 Proposed Models

The proposed model, correlating geometry features, speed and accident indexes, has a physical composition of a tangent before a very bended horizontal curve. The kinematics of the full movement encompass constant acceleration on the straight line and decelerating on the circular, as shown at Fig. 2. The "Risk Trajectory" presents A and B as the extremes of the straight line and C as the slip dot point. This is *The Umbrella Handle Model*.

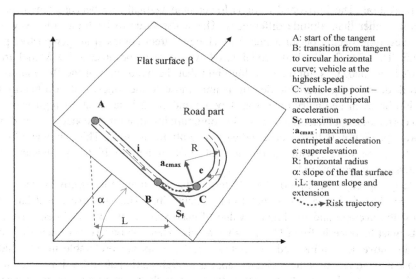

Fig. 2. The umbrella handle model Source: Authors.

The accident rate (**AR**) can be estimated by Eq. 2, a classic format of association of the number of events, time interval, extension and traffic volume, according to Silva & Ferreira (INTERNATIONAL CONFERENCE ON ROAD AND RAIL INFRASTRUCTURE 2018). The structure of Eq. 2 indicates the **AR** which can be changed to the Injured rate (**IR**) replacing the number of accidents for injured ones or Death rate (**DR**), substituting the numerator for the number of dead.

$$AR = \frac{(accident.number)}{(average.annual.daily.traffic) \times 365 \times (years) \times (studied.extension)} \quad (2)$$

AR: accident rate

The proposed road risk relationships are associated with not compensated centripetal acceleration, in direct or indirect ways. Proposals for studying correlations in an integrated manner, are presented in Eqs. 3; 4 and 5. Equation 3 shows the Free Road Risk Index (RRI_F) with a free approach and for the average effective speed for light vehicles. This is an indirect way of unbalanced centripetal acceleration (UCA). In this approach, the insertion of factors and their correlations are under conjecture. The longer and steeper the downward slope (i) on the straight line (TE), bigger is the final speed (S_E). As the circular radius (R) and superelevation (e) are getting smaller, greater is the UCA. A normalization factor (F) avoids very high or very low values.

$$RRI_F = F \times S_E \times \frac{TE}{R} \times \left(1 - \frac{i}{100}\right) \times \left(1 - \frac{e}{100}\right) \quad (3)$$

RRI$_F$: road risk index under free relationship; F: normalization factor; S$_E$: average effective speed of light vehicles (m/s); TE: tangent extension (m); R: horizontal radius of the in sequence curve (m); i: longitudinal tangent slope (%); e: superelevation (%)

The average and maximum velocities will be obtained by taking field values, according to samples taken as representative ones, in a place upstream of the point between the preceding tangent and the succeed horizontal curve. Equation 4 shows the Scientific Road Risk Index (RRI$_S$), proportional to the side friction coefficient (ft$_{max}$), which generates UCA. The total centripetal acceleration, calculated by the square of the speed divided by the radius, is equal to the sum of the centripetal accelerations for superelevation and side friction. The maximum side friction coefficient (ft$_{MAX}$) varies under distinguished factors, as tire pressure and wear level, regularity and roughness of the pavement among other factors. So, it is almost impossible to define an exact value of it. This approach conceives an environment of imminent side skidding, with loss of control and the occurrence of an accident. In this case, the Road Risk Index has a scientific approach, restricted to the laws of Physics.

$$RRI_S = k \times ft_{MAX} = k \times \left(\frac{S_E^2}{g} \times \frac{1}{R} - \frac{e}{100}\right) \quad (4)$$

RRI$_S$: road risk index under scientific relationship; ft$_{MAX}$; maximum side friction coefficient; S$_E$: average effective speed for light vehicles at the tangent end (m/s); g: gravity acceleration (m/s^2); R: horizontal radius of the downstream curve (m); e: superelevation (%); k: numerical coefficient

Equation 5 is based on the Eq. 3, without vehicle speed. It is the determination of the Free Road Risk Index Scenario (RIS$_F$). In this case, when **AR** and **RIS$_F$** composes a regression, just road physical properties are evaluated, having speed and UCA as supporting characters. In other words, this way of analysis concentrates efforts just on the road factors influence.

$$RIS_F = F \times \frac{TA}{R} \times \left(1 - \frac{i}{100}\right) \times \left(1 - \frac{e}{100}\right) \quad (5)$$

RIS$_F$: F: normalizing factor; road risk index Scenario, under free math relationship; TE: tangent extension (m); R: horizontal radius of the in sequence curve (m); i: longitudinal tangent slope (%); e: superelevation (%)

Equation 6 is the Scientific Road Risk Index Scenario (RIS$_S$), based on the Eq. 4, but without speed, as the same approach of Eq. 5. In this model, the extension of the previous tangent and its slope are not considered. It starts from the premise that the movement that is conducive to the generation of accidents is already conceived, under AR and RIS$_S$ figures compared.

$$RIS_S = k \times \left(\frac{1}{g \times R} - \frac{e}{100}\right) \qquad (6)$$

RIS_S: road risk index under scientific relationship; gravity acceleration (m/s^2); R: horizontal radius of the downstream curve (m); e: superelevation (%); k: numerical coefficient

It highlights that these equations are the result of attempts made over the past few years on this model, to seek correlation between the quantities, in addition to the determination based on applied Physics. The surveys were carried out on the basis of an adequate representation of the Brazilian environment, where old highways, whose capacity offered at the time of starting operation were less to much less than the present demand. At the same time, drivers who have restricted training or are little credulous in the problem of speeding, drinking alcohol or other drugs before driving, as well as the great contemporary problem: the improper usage of cell phones complement the scenario.

The regressions were made of isolated or integrated parameters to accident indexes, considering or not the Physics approach. In this work, lots of references were consulted to offer theory basement.

6 Results

The *Umbrella Handle Model* was applied to two distinguished Brazilian road segments. A group of students developed a Completion of Course Work on a segment of the SP 98 Paulo Rolim Loureiro, an old road crossing Serra do Mar Mountain Range. Caio Almeida de Oliveira was the student coordinator of this group. Another student, Rafael Fogarolli developed a Completion of Course Work on a segment of the SP 105 Rubens Pupo Pimentel, a road on an undulating terrain. All students were in the final semester of Civil Engineering Degree Course. Both works were supervised by one of the authors of this work.

The research on the SP 105 resulted in several correlation studies between RIS_F, IR and UCA. A comparison between RIS_F from the "as built" and Google Maps was made. The average annual daily traffic was taken from the website of a Road Department of Sao Paulo State (Departamento De Estradas De Rodagem Do Estado De Sao Paulo 2020) And Accidents Reports from The Website Infosiga (Sao Paulo 2020). Time was taken for speed estimation in situ, from some selected horizontal curves of this road. Geometry Features were taken under consultation with the regional highway department of the city of Campinas. Figure 3 shows a relationship between RIS_F and UCA on a sample of 10 Horizontal Curves of SP 105 road. These values of UCA were calculated since the average speed measurements. The lapse was taken at the "entrance" of each curve, in other words, just before the point between the previous tangent and the in sequence horizontal curve. Horizontal radius and superelevation of each curve; steepness and length of each tangent were taken from an "as built" of the road segment. The best relationship achieved was under exponential regression. The math correlation coefficient near **0.6**. The samples must be larger in later stages of this research, aiming to increase the coefficient as well as confirm the effectiveness of the correlations according to the model under study.

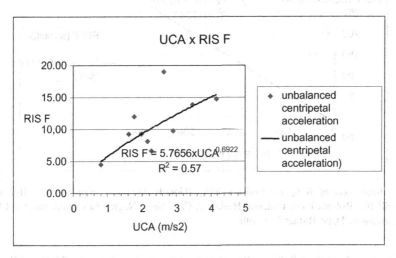

Fig. 3. Math regression between RIS_F and UCA – sample of 10 Horizontal Curves SP 105 Rubens Pupo Pimentel Highway, City: Serra Negra; Sao Paulo State; BRAZIL RIS_F: Road Risk Index Scenario; UCA: unbalanced centripetal acceleration (m/s^2) Source: Authors; speed and geometry measurements: Rafael Fogarolli

The SP 105 physical road properties were taken from *Google Maps, Earth and Street* to be compared to those obtained from the "as built" values. A sample of 31 pairs of values were applied in Fig. 4. Its best relationship, a linear model, shows that there is a constant error between these values. The math correlation coefficient, about **0.74**; shows that it is possible to obtain better regressions between RIS_F; RIS_S; RRI_F and RRI_s associated to IR values. The samples must be larger in later stages of this research, aiming to increase the coefficient as well as confirm the effectiveness of the correlations according to the model under study.

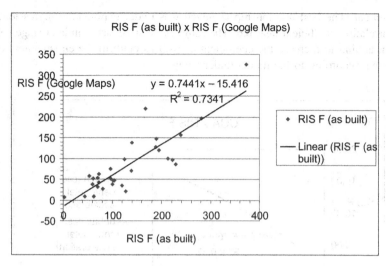

Fig. 4. Regression of RIS_F (as built) x RIS_F (Google Maps) – sample of 31 Horizontal Curves SP 105 Rubens Pupo Pimentel Highway, City: Serra Negra; Sao Paulo State; BRAZIL Source: Authors. Data: Rafael Fogarolli

The research on the SP 98 resulted in several regressions between RISF and IR. The average annual daily traffic was taken from the website of a Road Department of Sao Paulo State (Departamento De Estradas De Rodagem Do Estado De Sao Paulo 2020), accidents reports from the website Infosiga (Sao Paulo 2020) and a complete 10 year-accident file was received from a Sao Paulo road department member. Nowadays, INFOSIGA does not show more than 4 years of data. The time for estimating the speed was obtained in situ, from some selected horizontal curves of this road. Geometry Features were taken under consultation with the regional highway department of Sao Paulo. All kinds of accidents were considered. Figure 5 shows a regression of RISF and IR for 9 pairs of values.

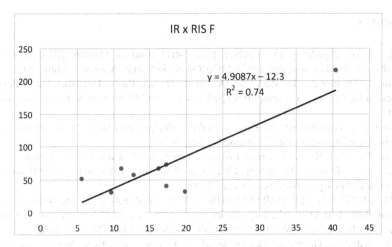

Fig. 5. Regression of RISF (to allowed speed to cars) x IR – sample from 9 Horizontal Curves SP 98 Paulo Rolim Loureiro, Mogi das Cruzes too Bertioga; Sao Paulo State; BRAZIL Source: Caio Almeida de Oliveira et Alii.

7 Conclusions

This research emphasizes the importance to apply the Umbrella Handle Model to evaluate road operation safety according to specific equations between IR and DR associated to the UCA, allowing conceive a tool to reduce severity and amount of accidents.

The Umbrella Handle Model allows to take in consideration several features of each selected road segment aiming to correlate to correspondent IR or DR. This kind of evaluation has the UCA inserted in the analyses in indirect way, allowing to quantify the impact of any speed excess and helping to develop approaches of improvements that could minimize accident indexes. This also can offer arguments to introduce new speed enforcement campaigns about the negative impact of Brazilian excessive freedom on vehicular driving.

It is relevant to highlight that just establish a speed design for each segment of the track development, usual for decades, does not guarantee safety over the life of the enterprise, because speed control and driving pattern tend to change.

The next step of this research is to conceive new specific equations under this proposed model, evaluating the importance of each feature. Another approach is to get a large amount of data from distinguished roads and segments, aiming to develop a general equation that could help highway designers.

Acknowledgements. We would like to thank Cassio Eduardo Lima de Paiva, from The School of Civil Engineering, Architecture and Urban Planning of State University of Campinas, for the suggestion of the *Umbrella Handle* nomenclature regarding the studied geometry as well as other important observations.

We are also grateful for the financial support of the Coordination for the Improvement of Higher Education Personnel (CAPES), Ministry of Education Foundation (MEC) of the Brazilian Federal Government, through the Graduate Support Program (PROAP), whose function is to finance activities of postgraduate courses, providing better conditions for the training of human resources.

References

BRAZIL (2020): Presidency of Republic. Law 14071 from 14 October 2020. http://www.planalto.gov.br/ccivil_03/_ato2019-2022/2020/lei/l14071.htm. Accessed 5 June 2021

Departamento De Estradas De Rodagem Do Estado De Sao Paulo: Estatística De Tráfego Da Secretaria De Logística E Transportes (2020). http://200.144.30.103/vdm/. Accessed 10 Nov 2020

Eno Center of Transportation: Our History: William Phelps Eno (2019). https://www.enotrans.org/about-eno/mission-history/. Accessed 5 July 2019

Fitspatrick, K., Khal, K.: A Historical Literature review of Horizontal Curve Design. College Station: Texas Transport Institution (1992). https://d2dtl5nnlpfr0r.cloudfront.net/tti.tamu.edu/documents/1949-1.pdf. Accessed 21 July 2020

GLOBAL ROAD SAFETY PARTNERSHIP (2016): Road Safety: In the Western Pacific Region. WPR/2016/DHN/022. https://www.who.int/violence_injury_prevention/road_safety_status/2015/Road_Safety_WPRO_English.pdf. Accessed 20 July 2020. Geneva

Hauer, E.: Safety in Geometric Design Standards I: Three Anecdotes. Mainz: International Symposium on Highway Geometric Design (2000). https://www.researchgate.net/publication/228903642_Safety_in_geometric_design_standards. Accessed 2 Nov 2018

Hauer, E.: Safety of Horizontal Curves–Review of Literature for the Interactive Highway Safety Design Model, Disponível em (2000). http://www.roadsafetyresearch.com. Accessed 4 Nov 2018

International Conference on Road and Rail Infrastructure: Accident Prediction Models Considering Pavement Quality. Zadar (Croatia): International Conference on Road and Rail Infrastructure (2018)

International Conference on Transportation Planning and Implementation Methodologies for Developing Countries, Mumbai: Development of Comprehensive Crash Models for Four-lane Divided Highways in Heterogeneous Traffic Condition. Elsevier Press Ltd., New York (2016). https://www.sciencedirect.com/science/article/pii/S2352146516307323. Accessed 12 July 2019

Lamm, R., Choueiri, E.M.: Comparative Analysis of Operation Speeds and Accident Rates on Two-Lane rural Highways. Dissertation Submitted to the Department of Civil Engineering for the Degree of Doctor of Philosophy. Clarkson University. Potsdam, New York (1987)

Matthews, L.R., Barnes, J.W.: Relation between Road Environment and Curve Accidents. In: Proceedings, 14th ARRB Conference, Part 4, pp. 105–120 (1988)

SAO PAULO: INFOSIGA SP Sao Paulo–State Government (2020). http://painelderesultados.infosiga.sp.gov.br/dados.web/ViewPage.do?name=acidentes_naofatais&contextId=8a80809939587c0901395881fc2b000. Accessed 10 Nov 2020

The New York Times Magazine: Who Made That Crash Test Dummy? New York: The New York Times Company (2012). https://www.nytimes.com/2012/05/20/magazine/who-made-that-crash-test-dummy.html. Accessed 20 July 2020

Transportation Research Board: NCHRP REPORT 480: A Guide to Best Practices for Achieving Context Sensitive Solutions. Washington (2002). http://onlinepubs.trb.org/onlinepubs/nchrp/nchrp_rpt_480.pdf. Accessed 6 May 2019

World Health Organization: The top 10 causes of death (2018). http://www.who.int/en/news-room/fact-sheets/detail/the-top-10-causes-of-death. Accessed 24 May 2018

Stiffening Guidance for Temporary Concrete Barrier Systems in Work Zone and Construction Situations

Karla Lechtenberg[✉], Chen Fang, and Ronald Faller

Midwest Roadside Safety Facility, University of Nebraska-Lincoln,
Lincoln, NE, USA
kpolivka2@unl.edu

Abstract. Portable concrete barrier (PCB) systems are utilized in many situations, including placement adjacent to vertical drop-offs, work zones, and construction areas. Free-standing PCB systems are known to have relatively large deflections when impacted, which may be undesirable when dealing with limited space behind the barrier, such as on a bridge deck or with limited lane width in front of the barrier system. The risk of injury to workers in the work zone or construction area from displaced PCB systems is great. In order to allow PCB systems to be used in space-restricted locations, a variety of PCB stiffening options have been used, including beam stiffening and bolting or pinning the barriers to the pavement. This research evaluated the safety performance of the New Jersey Department of Transportation (NJDOT) PCB, Type 4 in various stiffening configurations according to the Test Level 3 (TL-3) criteria set forth in MASH 2016. All tested configurations successfully met all the requirements of MASH 2016 test designation no. 3–11 which was a 2,270-kg pickup truck impacting at a speed of 100 km/hr and 25°. This research further reviewed and correlated stiffening techniques to dynamic deflection and working widths, as well as barrier segment damage. Implementation guidance is provided for the various PCB stiffening options to help protect the workers in work zone and construction areas.

Keywords: Crash test · Temporary concrete barrier · Portable concrete barrier · MASH · Stiffening · Bolting · Pinning · TL-3

1 Introduction

Portable concrete barrier (PCB) systems are utilized in many situations, including placement adjacent to vertical drop-offs, work zones, and construction areas. Free-standing PCB systems are known to have relatively large deflections when impacted, which may be undesirable when dealing with limited space behind the barrier, such as on a bridge deck or with limited lane width in front of the barrier system. Whenever a traffic control plan is developed that utilizes a PCB system, it is important to define acceptable barrier deflection criteria to minimize risk of injury to workers in the work zone or construction area. The deflection criteria should be selected based on risk of injury for vulnerable road users and work zone workers, risk of the barriers or vehicles falling off an unshielded bridge deck edge, and to optimize expense and intrusion in accordance with posted speed limits. In order to allow PCB systems to be used in space-restricted locations, a variety of PCB stiffening options have been used, including beam stiffening and bolting or pinning the barriers to the pavement.

The New Jersey Department of Transportation (NJDOT) currently uses a New Jersey shape, Precast Concrete Curb, Concrete Barrier, which will be referred to as PCB. Vertical, I-beam connection pins are inserted into keyways in consecutive barriers to provide continuity between barrier joints. Additional deflection control features are utilized depending on the limits of allowable deflection. The *2015 NJDOT Roadway Design Manual* (NJDOT 2015) provides guidance on allowable deflections for various connection types, as shown in Table 1. The PCB, connection key, and joint class configurations are discussed in greater detail in the following sections.

Table 1. 2015 NJDOT roadway design manual PCB guidance.

Connection type	Use	Joint treatment*
A	Allowable movement over 1041 mm	Connection Key and barrier end sections fully pinned
B	Maximum allowable deflection of 711 mm (Cannot be used with traffic on both sides of the barrier.)	Connection Key, 152 mm by 152 mm box beam, and barrier end sections fully pinned
C	Maximum allowable movement of 280 mm	Connection Key, construction (back) side of all sections pinned, and barrier end sections fully pinned

*Barrier end sections fully pinned – first and last PCB segments of the entire run regardless of connection type have pins in every anchor recess on both sides.

The guidance provided in the *2015 Roadway Design Manual* was based on test data obtained from previous testing standards, which needed to be updated to be consistent with current crash testing standards and a changing vehicle fleet (NJDOT 2015). Crash testing of other PCB systems under the Test Level 3 (TL-3) criteria of the *Manual for Assessing Safety Hardware, Second Edition* (AASHTO 2016) has indicated that dynamic barrier deflections can increase significantly when compared to dynamic deflections based on older crash test data. Thus, a need existed to investigate the performance of the NJDOT PCB system in various configurations in order to provide updated design guidance. Therefore, the objective of this research effort was to evaluate the safety performance of the NJDOT PCB, Type 4 in various configurations according to the TL-3 criteria set forth in MASH 2016.

2 Test Requirements and Evaluation Criteria

Longitudinal barriers, such as PCBs, must satisfy impact safety standards in order to be declared eligible for federal reimbursement by the Federal Highway Administration (FHWA) for use on the National Highway System (NHS). For new hardware, these safety standards currently consist of the guidelines and procedures published in MASH 2016. Note that there is no difference between MASH 2009 and MASH 2016 for most longitudinal barriers, such as the PCB systems tested in this project, except that additional occupant compartment deformation measurements are required by MASH 2016. According to TL-3 of MASH 2016, longitudinal barrier systems must be subjected to two full-scale vehicle crash tests, as summarized in Table 2.

Table 2. MASH 2016 TL-3 crash test conditions for longitudinal barriers.

Test article	Test designation no.	Test vehicle	Vehicle weight (kg)	Impact conditions Speed (km/h)	Angle (deg.)
Longitudinal barrier	3–10	1100C	1100	100	25
	3–11	2270P	2270	100	25

In test no. 7069-3, a rigid, F-shape, concrete bridge rail was successfully impacted with a small car weighing 816 kg and traveling at 96.7 km/h and 21.4° according to the American Association of State Highway and Transportation Officials (AASHTO) *Guide Specifications for Bridge Railings* (Buth et al. 1990). In the same manner, test nos. CMB-5 through CMB-10, CMB-13, and 4798-1 showed that rigid, New Jersey, concrete safety shape barriers struck by small cars have been shown to meet safety performance standards (Bronstad et al. 1976 and Buth et al. 1986). In addition, in test no. 2214NJ 1, a rigid, New Jersey, ½-section, concrete safety shape barrier was impacted by a passenger car weighing 1,170 kg at 97.8 km/h and 26.1° according to the TL-3 standards set forth in MASH 2009 (Polivka et al. 2006). Furthermore, temporary, New Jersey safety shape, concrete median barriers have experienced only slight barrier deflections when impacted by small cars and behave similarly to rigid barriers as seen in test no. 47 (Fortuniewicz et al. 1982). As such, the 1100C passenger car test was deemed not critical for testing and evaluating this PCB system.

For test designation no. 3–11, the test vehicle impacts into the test article at a critical impact point (CIP), which is a location on the test article expected to maximize the risk of the test failing to pass MASH safety evaluation criteria. This could mean maximizing the risk of vehicle rollover or instability, penetration behind the test article by the vehicle, exceeding occupant risk value tolerances, or some combination thereof. Initial vehicle impact was to occur 1.3 m upstream from the centerline of the joint between barrier nos. 4 and 5 which was selected using Table 2 of MASH 2016.

3 Design Details

The NJDOT Construction Barrier Curb, Type 4 (Alternative B) is representative of the typical PCB system used by NJDOT to create work zones and construction areas, as shown in Fig. 1. Each PCB segment measured 813-mm tall and 6.1 m long and utilized an I-beam connection key for the barrier-to-barrier connection. The concrete mix for the barrier sections required a minimum 28-day compressive strength of 25.5 MPa.

A minimum concrete cover of 38 mm was used along all rebar in the barrier. All of the steel reinforcement in the barrier was ASTM A615 Grade 60 rebar and consisted of four No. 6 longitudinal bars, eight No. 4 bars for the vertical stirrups, four No. 6 lateral bars, and nine No. 4 bars for the anchor hole reinforcement loops. The I-beam connection keys consisted of 13-mm thick, ASTM A36 steel plates welded together to form the key shape. A connection socket consisting of three ASTM A36 steel plates welded on the sides of an ASTM A500 Grade B or C steel tube was configured at each end of the PCB section. The I-beam connection key was inserted into the steel tubes of two adjoining PCBs to form the connection.

NJDOT has used grout wedges placed at the toes between adjacent PCB segments to limit the rotation at the connection, as shown in Fig. 1. The grout wedges between the PCB segments were intended to allow the entire barrier system to act as a continuous element, such that the load disperses throughout all barrier segments rather than being concentrated on those in the impact zone. The non-shrink grout wedges consisted of a non-shrink grout mix with a minimum 1-day compressive strength of 7 MPa.

Fig. 1. Barrier system.

3.1 Pin and Bolted Anchorages

Each barrier section of NJDOT PCBs consists of five pin anchor recesses on traffic side and four pin anchor recesses on the back side, as shown in Fig. 2. The pin anchors are 25-mm diameter by 381-mm long, ASTM A36 steel pins that were inserted into 32-mm diameter drilled holes in the road surface. During installation when pin anchors were used, the PCB segments were connected with the I-beam connection key and then pulled in a direction parallel to the longitudinal axis, removing slack in the joints. After removal of the joint slack, 32-mm diameter holes were drilled into the road surface using the pin anchor recesses as guides. Finally, the steel pins were embedded to a depth of 127 mm.

Each barrier section of NJDOT PCBs also consists of five bolt anchor pocket recesses on the traffic side and five bolt anchor pocket recesses on the back side, as shown in Fig. 2. The bolt anchors are 25-mm diameter ASTM F1554 Grade 36 threaded rods that were epoxied into 29-mm diameter drilled holes in road surface. During installation when bolted anchors were used, the PCB segments were connected with the I-beam connection key and then pulled in a direction parallel to the longitudinal axis, removing slack in the joints. After removal of the joint slack, 29-mm diameter holes were drilled into the road surface using bolt anchor recesses as guides. Finally, the anchor rods were embedded to a depth of 178 mm and epoxied into the road surface. The bond strength of the epoxy used to anchor the bolt anchor rods to the road surface was 10 MPa. A plate washer and nut were attached to each threaded rod to secure the PCBs in place.

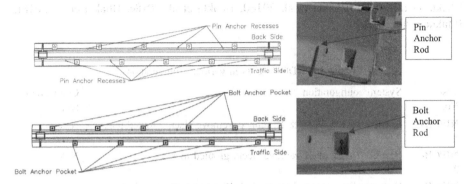

Fig. 2. Pin anchors and bolt anchors on barriers.

3.2 Box-Beam Rail as Stiffener

Box beam rail stiffeners were mounted on the back face of the system at each joint, as shown in Fig. 3, to reduce lateral deflections and prevent PCB segment separation when deflected and suspended over the edge of a bridge deck. Due to the high tensile capacity of the steel, PCB segments and the box beam act as a composite bending member, with the concrete in compression on the traffic face of the barrier and the steel in tension. Each box beam stiffener was a 3.7-m long, 152-mm × 152-mm × 5-mm ASTM A500 Grade C box beam. The box beam rails were mounted on the PCB segments with 19-mm diameter by 432-mm long ASTM A307 Grade A bolts without square necks and 19-mm diameter ASTM A563A nuts. A 19-mm diameter ASTM F844 fender washer was placed between the PCB segment and the bolt head on the traffic side and a 203-mm × 203-mm × 13-mm ASTM A36 steel plate was placed between the nut and the box-beam rail stiffener on the back side. Additional mounting details are provided in the test report (Bhakta 2018e).

Fig. 3. Box beam stiffeners.

4 New Jersey PCB Tests

For the seven full-scale crash tests, various NJDOT PCB system configurations consisting of ten PCB segments was constructed, as listed in Table 3. Two anchoring techniques and two stiffening techniques were evaluated. Anchoring techniques include use of pins and bolts. Stiffening techniques include use of box beam rails and non-shrink grout wedges placed between the PCB segments. Further details on each test can be found in the published research reports (Bhakta et al. 2018a, Bhakta et al. 2018b, Bhakta et al. 2018c, Bhakta et al. 2018d, Bhakta et al. 2018e, Bhakta et al. 2018f, Bhakta et al. 2018g).

Table 3. NJDOT PCBs in various configurations.

Test No.	System configuration	Connection type
NJPCB-1	Barriers 1, 3, 5, 7, 9, & 10 pinned, remove slack, grouted toes	N/A
NJPCB-2	Barriers 1–10 bolted, remove slack, grouted toes	N/A
NJPCB-3	Free-standing system, barriers 1 & 10 pinned, remove slack, no grouted toes	A
NJPCB-4	Free-standing system, barriers 1 & 10 pinned, remove slack, grouted toes	N/A
NJPCB-5	Free-standing system, barriers 1 & 10 pinned, box-beam stiffened all joints (8 joints), remove slack, grouted toes	B
NJPCB-6	Barriers 1 & 10 pinned, barriers 2–9 pinned back side only, remove slack, grouted toes	C
NJPCB-7	Barriers 1 & 10 pinned, Barriers 2–9 pinned front side only, remove slack, grouted toes	N/A

N/A = Not Applicable

4.1 Test No. NJPCB-1

During test no. NJPCB-1, the 2,274 kg pickup truck impacted the NJDOT PCB system configuration at a speed of 100.7 km/h and at an angle of 24.7°, resulting in an impact severity of 155.8 kJ. The vehicle was successfully contained and smoothly redirected with moderate damage to both the barrier and vehicle, as shown in Figs. 4 and 5. Barrier damage consisted of contact marks on the front face of the PCB segments, concrete spalling, and concrete cracking on barrier nos. 4, 5, and 6. The maximum dynamic barrier deflection and working width were 343 mm and 953 mm, respectively. All occupant risk measures were within the recommended limits, and the occupant compartment deformations were also deemed acceptable. Therefore, NJDOT barriers, Type 4 (Alternative B), consisting of NJDOT PCB barriers joined with a connection key, joint slack removed, grouted toes, and every other barrier pinned on both the traffic side and back side, successfully met all the safety performance criteria of MASH test designation no. 3–11.

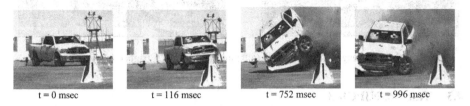

Fig. 4. Sequential view of vehicle behavior, test no. NJPCB-1.

Fig. 5. Barrier damage – barrier no. 4 (left) and barrier no. 5 (right) – test no. NJPCB-1.

4.2 Test No. NJPCB-2

During test no. NJPCB-2, the 2,264 kg pickup truck impacted the NJDOT PCB system configuration at a speed of 100.7 km/h and at an angle of 24.5°, resulting in an impact severity of 152.6 kJ. The vehicle was successfully contained and smoothly redirected with moderate damage to both the barrier and the vehicle, as shown in Figs. 6 and 7. Barrier nos. 4 and 5 were fractured and experienced concrete spalling and cracks. The maximum dynamic barrier deflection and working width were 125 mm and 610 mm, respectively. All occupant risk values were found to be within limits, and the occupant compartment deformations were also deemed acceptable. Therefore, NJDOT barriers, Type 4 (Alternative B), consisting of NJDOT PCB barriers joined with a connection key, joint slack removed, grouted toes, and every barrier bolted on both the traffic and back side, successfully met all the safety performance criteria of MASH test designation no. 3–11.

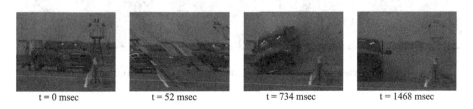

Fig. 6. Sequential view of vehicle behavior, test no. NJPCB-2.

Fig. 7. Barrier damage – barrier no. 4 (left) and barrier no. 5 (right), test no. NJPCB-2.

4.3 Test No. NJPCB-3

During test no. NJPCB-3, the 2,268 kg pickup truck impacted the NJDOT PCB system configuration at a speed of 100.2 km/h and at an angle of 25.8°, resulting in an impact severity of 166.6 kJ. The vehicle was successfully contained and smoothly redirected with moderate damage to both the barrier and the vehicle, as shown in Figs. 8 and 9. Barrier nos. 3, 4, 5, and 6 experienced concrete spalling and cracking, with most of the damage concentrated on the downstream end of barrier no. 4 and upstream end of barrier no. 5. The maximum dynamic barrier deflection and working width were 968 mm and 1,577 mm, respectively. All occupant risk values were found to be within limits, and the occupant compartment deformations were also deemed acceptable. Therefore, NJDOT barriers, Type 4 (Alternative B), consisting of free-standing NJDOT PCB barriers joined with a connection key, joint slack removed, and only the end barriers pinned on both the traffic and back sides, successfully met all the safety performance criteria of MASH test designation no. 3–11.

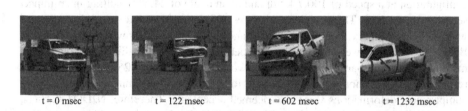

Fig. 8. Sequential view of vehicle behavior, test no. NJPCB-3.

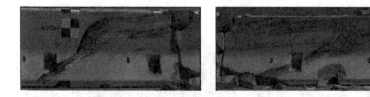

Fig. 9. Barrier damage – barrier no. 4 (left) and barrier no. 5 (right), test no. NJPCB-3.

4.4 Test No. NJPCB-4

During test no. NJPCB-4, the 2,268 kg pickup truck impacted the NJDOT PCB system configuration at a speed of 101.0 km/h and at an angle of 24.5°, resulting in an impact severity of 153.7 kJ. The vehicle was successfully contained and smoothly redirected with moderate damage to both the barrier and the vehicle, as shown in Figs. 10 and 11. Barrier nos. 4 and 5 experienced concrete spalling and cracking. The maximum dynamic barrier deflection and working width were 1,034 mm and 1,643 mm, respectively. All occupant risk values were found to be within limits, and the occupant compartment deformations were also deemed acceptable. Therefore, NJDOT barriers, Type 4 (Alternative B), consisting of free-standing NJDOT PCB barriers joined with a connection key, joint slack removed, grouted toes, and only the end barriers pinned on both the traffic and back sides, successfully met all the safety performance criteria of MASH test designation no. 3–11.

Fig. 10. Sequential view of vehicle behavior, test no. NJPCB-4.

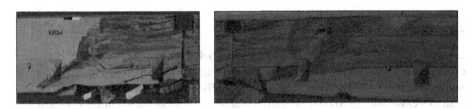

Fig. 11. Barrier damage – barrier no. 4 (left) and barrier no. 5 (right), test no. NJPCB-4.

4.5 Test No. NJPCB-5

During test no. NJPCB-5, the 2,268-kg pickup truck impacted the NJDOT PCB system configuration at a speed of 100.8 km/h and at an angle of 24.9°, resulting in an impact severity of 157.7 kJ. The vehicle was successfully contained and smoothly redirected with moderate damage to both the barrier and the vehicle, as shown in Figs. 12 and 13. Barrier nos. 4 and 5 experienced concrete spalling and cracking. The maximum dynamic barrier deflection and working width were 838 mm and 1,448 mm, respectively. All occupant risk values were found to be within limits, and the occupant compartment deformations were also deemed acceptable. Therefore, NJDOT barriers, Type 4 (Alternative B), consisting of box-beam stiffened, free-standing NJDOT PCB barriers joined with a connection key, joint slack removed, grouted toes, and only the end barriers pinned on both the traffic and back sides, successfully met all the safety performance criteria of MASH test designation no. 3–11.

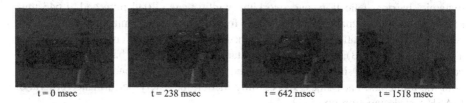

Fig. 12. Sequential view of vehicle behavior, test no. NJPCB-5.

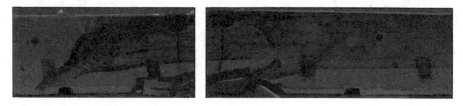

Fig. 13. Barrier damage – barrier no. 4 (left) and barrier no. 5 (right), test no. NJPCB-5.

4.6 Test No. NJPCB-6

During test no. NJPCB-6, the 2,268-kg pickup truck impacted the NJDOT PCB system configuration at a speed of 101.3 km/h and at an angle of 25.1°, resulting in an impact severity of 161.3 kJ. The vehicle was successfully contained and smoothly redirected with moderate damage to both the barrier and the vehicle, as shown in Figs. 14 and 15. Barrier nos. 3, 4, 5, and 6 experienced spalling and cracking. The maximum dynamic barrier deflection and working width were 386 mm and 1,041 mm, respectively. All occupant risk values were found to be within limits, and the occupant compartment deformations were also deemed acceptable. Therefore, NJDOT barriers, Type 4 (Alternative B), consisting of NJDOT PCB barriers joined with a connection key, joint slack removed, grouted toes, the end barriers pinned on both the traffic and back sides, and the interior barriers pinned only on the back side, successfully met all the safety performance criteria of MASH test designation no. 3–11.

Fig. 14. Sequential view of vehicle behavior, test no. NJPCB-6.

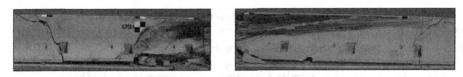

Fig. 15. Barrier damage – barrier no. 4 (left) and barrier no. 5 (right), test no. NJPCB-6.

4.7 Test No. NJPCB-7

During test no. NJPCB-7, the 2,268-kg pickup truck impacted the NJDOT PCB system configuration at a speed of 101.0 km/h and at an angle of 25.2°, resulting in an impact severity of 162.0 kJ. The vehicle was successfully contained and redirected with moderate damage to both the barrier and the vehicle, as shown in Figs. 16 and 17. Barrier nos. 4 and 5 experienced spalling, cracking, and fracture. The maximum dynamic barrier deflection and working width were 290 mm and 899 mm, respectively. All occupant risk values were found to be within limits, and the occupant compartment deformations were also deemed acceptable. Therefore, NJDOT barriers, Type 4 (Alternative B), consisting of NJDOT PCB barriers joined with a connection key, joint slack removed, grouted toes, the end barriers pinned on both the traffic and back sides, and the interior barriers pinned only on the traffic side, successfully met all the safety performance criteria of MASH test designation no. 3–11.

Fig. 16. Sequential view of vehicle behavior, test no. NJPCB-7.

Fig. 17. Barrier damage – barrier no. 4 (left) and barrier no. 5 (right), test no. NJPCB-7.

5 Discussion and Analysis

A summary of full-scale crash testing on seven configurations of the NJDOT PCB system is shown in Tables 4 and 5. The tests were separated into groups for comparison: 1) free-standing barriers without anchorage (test nos. NJPCB-3 and NJPCB-4); 2) free-standing, box-beam stiffened barriers (test no. NJPCB-5); 3) barriers using pinned anchorage (test nos. NJPCB-1, NJPCB-6 and NJPCB-7), and 4) barriers using bolted anchorage (test no. NJPCB-2). These tests were compared to similar New York PCB systems previously evaluated and included: a) test no. NYTCB-1, which was a box-beam stiffened system; b) test no. NYTCB-2, which was a free-standing system without removal of joint slack or grouted toes; c) test no. NYTCB-5, which was a system with only the back-side pinned and without removal of joint slack or grouted toes (Stolle et al. 2008 and Lechtenberg et al. 2010). Results from these tests included the impact conditions and impact severity as well as dynamic barrier deflection, permanent set barrier deflection, working width (as measured from the original front face of the barrier), and the clear space behind the barrier. The working width is defined as the distance from the original front of the barrier to the furthest point of either the backside of the barrier including deflection or vehicle intrusion beyond the backside of the barrier. The clear space behind the barrier is used by NJDOT to define the maximum deflection of the back of the barrier from its original position. The schematic diagrams shown in Fig. 18 indicate how the dynamic deflection, permanent set deflection, and working width for each crash test was defined.

Table 4. Full-scale crash test results – barrier deflections.

Test No.	Connection type	Anchored barriers	No. of Anchor Pins/Bolts	Dynamic deflection (mm)	Permanent set (mm)	Working width (mm)	Clear space behind barrier (mm)
Free-Standing Systems							
NYTCB-2	A	1 & 10	18	1024	1003	1633	1024
NJPCB-3	A	1 & 10	18	968	930	1577	968
NJPCB-4	N/A	1 & 10	18	1034	965	1643	1034
Free-Standing, Box-Beam Stiffened Systems							
NYTCB-1	N/A	1 & 10	18	701	660	1311	701
NJPCB-5	B	1 & 10	18	838	826	1448	838
Pinned Anchorage Systems							
NYTCB-5	N/A	1–10 (Back)	40	521	229	889	279
NJPCB-1	C	1, 3, 5, 7, 9 & 10	54	343	159	953	343
NJPCB-6	C	1 & 10, 2–9 (Back)	50	386	95	1041	386
NJPCB-7	C	1 & 10, 2–9 (Front)	58	290	159	899	290
Bolted Anchorage System							
NJPCB-2	D	1–10	100	125	-13	610	0

Table 5. Full-scale crash test results – vehicle behavior and impact severities.

Test No.	Roll (degrees)	Pitch (degrees)	Vehicle mass (kg)	Impact speed (km/h)	Impact angle (degrees)	Impact severity (kJ)
Free-Standing Systems						
NYTCB-2	-12.4	-10.6	2,279	98.5	25.8	161.6
NJPCB-3	-17.2	-9.0	2,268	100.2	25.8	166.6
NJPCB-4	-16.2	-14.2	2,268	101.3	24.5	153.7
Free-Standing, Box-Beam Stiffened Systems						
NYTCB-1	-10.5	-11.4	2,275	99.5	24.6	151.0
NJPCB-5	-7.9	-12.5	2,268	100.8	24.9	157.7
Pinned Anchorage Systems						
NYTCB-5	41.8	-21.2	2247	103.5	26.2	180.9
NJPCB-1	-39.9	-12.8	2,274	100.7	24.7	155.8
NJPCB-6	28.9	-12.2	2,268	101.3	25.1	161.3
NJPCB-7	-29.2	-18.6	2,268	101.0	25.2	162.0
Bolted Anchorage System						
NJPCB-2	20.7	-12.0	2,264	100.7	24.5	152.6

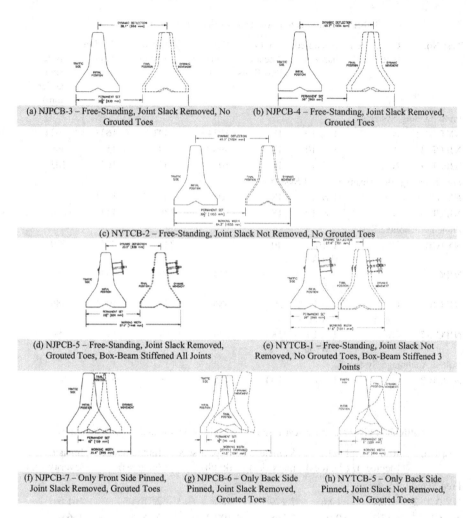

Fig. 18. Barrier Deflection Comparison, Test Nos. (a) NJPCB-3, (b) NJPCB-4, (c) NYTCB-2, (d) NJPCB-5, (e) NYTCB-1, (f) NJPCB-7, (g) NJPCB-6, and (h) NYTCB-5

5.1 Free-standing Systems

A review of the data from test nos. NJPCB-3, NJPCB-4, and NYTCB-2 found that there was little to no benefit in terms of barrier deflection and clear space requirements for a free-standing PCB system due to the removal of joint slack and/or the use of grouted barrier toes. This can be seen in the fact that dynamic deflections and the clear space behind barrier for all three tests were very similar. The primary cause of the lack of observed benefit for the modified PCB joints was the absence of barrier reinforcement in the toes of both the New York and New Jersey PCB segments. The lack of reinforcement led to spalling and disengagement of the barrier toes when they were loaded by adjacent barrier segments which caused increased rotation and motion of the barrier joints. This toe disengagement increased the deflection by altering the effective contact surface and barrier moment for both the New Jersey and New York PCB crash tests.

5.2 Free-standing, Box-Beam Stiffened Systems

A review of the data from test nos. NJPCB-5 and NYTCB-1 found that there was little to no benefit in terms of barrier deflection and clear space requirements for a free-standing, box-beam stiffened PCB system due to the removal of joint slack and/or the use of grouted barrier toes. This can be seen in the fact that dynamic deflections and the clear space behind barrier for both tests were very similar. The primary cause of the lack of observed benefit for the modified PCB joints was the absence of barrier reinforcement in the toes of both the New York and New Jersey PCB segments. The lack of reinforcement led to spalling and disengagement of the barrier toes when they were loaded by adjacent barrier segments, which caused increased rotation and motion of the barrier joints. This toe disengagement increased the deflection by altering the effective contact surface and barrier moment for both the New Jersey and New York PCB crash tests.

5.3 Pinned Anchorage Systems

A review of the data from test nos. NJPCB-6, NJPCB-7, and NYTCB-5 suggested that pinning the barriers on the traffic side of the PCB segments provided two benefits as compared to pinning on only the back side. First, pinning the traffic side of the PCB produced lower deflections for test no. NJPCB-7 as compared to test no. NJPCB-6. Second, in both tests of the back-side pinned barriers, the impacting vehicle climbed the barrier face significantly and rolled away from the barrier face as it was redirected. This was due to the back-side pins constraining the back of the PCB segments and causing increased vertical barrier rotation, which promotes vehicle climb and instability. Test no. NJPCB-7 with the barrier pinned on the traffic face of the barrier showed improved vehicle stability with less roll and vehicle climb while the vehicle was in contact with the barrier. Previous research by CALTRANS and MwRSF has noted that anchoring of PCB segments on the traffic side of the barrier improved stability as well. Thus, it would be recommended that pinning of the NJ PCB segments be done on the traffic side of the PCB system whenever possible to promote improved vehicle stability with similar or reduced deflection.

5.4 Bolted Anchorage System

Even though no previous bolted-down New Jersey PCB crash tests exist, it is anticipated that little to no benefit would be observed in reduced barrier deflections and clear space requirements due to joint slack removal and/or use of grouted toes in conjunction with bolted-down barriers. The primary cause of anticipated lack of observed benefit was the absence of barrier reinforcement in the toes of the New Jersey PCB segments. The lack of steel reinforcement led to concrete fracture near the barrier toes when they were loaded by adjacent barrier segments, which caused increased rotation of the barrier joints. This concrete toe disengagement did not reduce the expected benefit by altering the effective contact surface and barrier moment for the PCB crash test.

5.5 General System Configurations

The PCB segments used in these tests have a relatively small gap between adjacent barrier segments. Thus, improvement of the joint response through removal of joint slack and use of grouted toes provided less benefit than would be expected for other PCB systems which utilize joint spacings up to 102 mm. Finally, barrier system behavior and associated barrier deflections can vary from test to test due to the natural variability of a wide variety of factors involved in full-scale crash testing. These factors would include slight differences in impact conditions, differing test vehicle model years, slight variations in steel and concrete strengths, and variation of the cracking and damage observed on the barrier segments, among others. Thus, some variability would be expected in the barrier performance even for basically identical systems. Because the lack of improvement in PCB deflections and clear space behind the barrier due to removal of joint slack and use of grouted toes is primarily due to the fracture and disengagement of the barrier toes, redesign or modification of the barrier to further reinforce the barrier toes may improve the performance of the barrier when removal of joint slack and use of grouted toes is used in the barrier assembly.

6 Conclusions and Recommendations

Anchoring and stiffening techniques for PCBs were evaluated according to MASH using full-scale crash tests. Anchoring techniques included the use of pins and bolts, and stiffening techniques included the use of box beam rails and non-shrink grout wedges placed at the toes between barriers. These techniques were intended to limit barrier deflections and to implement and update NJDOT's PCB installation guidance if necessary. According to TL-3 evaluation criteria in MASH 2016, two tests are required for evaluation of longitudinal barrier systems: (1) test designation no. 3–10 – an 1100C small car and (2) test designation no. 3–11 – a 2270P pickup truck. However, only the 2270P crash test was deemed necessary as other prior small car tests were used to support a decision to deem the 1100C crash test not critical. Seven full-scale crash tests were conducted with the 2270P vehicle.

All tested system configurations had ten 6.1-m long PCBs with connection keys and all end sections fully pinned. All systems successfully met all the requirements of MASH 2016 test designation no. 3–11. Further, in the NJDOT Roadway Design Manual, the allowable deflection is determined by the clear space behind the barrier, which is defined as the maximum deflection of the back of the barrier from its original position. For connection type A, which is the free-standing configuration, the NJDOT allowable movement guidance is 1044 mm. For the test nos. NJPCB-3 and NJPCB-4, the clear space behind the barrier was 968 mm and 1034 mm, respectively. For connection type B, which is the free-standing, box-beam stiffened configuration, the NJDOT allowable movement guidance is 711 mm. For the test no. NJPCB-5, the clear space behind the barrier was 838 mm. For connection type C, which is the construction side of all PCB segments pinned, the NJDOT allowable movement guidance is 279 mm. For the test nos. NJPCB-1 and NJPCB-6, the clear space behind the barrier was 343 mm and 386 mm, respectively. For bridge parapets, which is the bolted configuration, the NJDOT allowable movement guidance is 0 mm. For test no. NJPCB-2, the clear space behind the barrier was 0 mm. Limited reductions in PCB deflections and clear space behind the barrier were observed with joint slack removal and use of grouted toes. Again, this finding is primarily due to the fracture and disengagement of the barrier toes. If larger reductions in PCB deflections and clear space are desired, PCB redesign or modification would be required, including reinforcement of the barrier toes, which may improve the effectiveness of joint slack removal and the use of grouted toes.

Acknowledgements. The authors wish to acknowledge several sources that made a contribution to this project: (1) New Jersey Department of Transportation for sponsoring this project and (2) MwRSF personnel for constructing the barriers and conducting the crash tests.

References

American Association of State Highway and Transportation Officials: Manual for Assessing Safety Hardware, Washington, D.C (2009)

American Association of State Highway and Transportation Officials: Manual for Assessing Safety Hardware, Second Edition., Washington, D.C (2016)

Bhakta, S.K., Lechtenberg, K.A., Faller, R.K., Reid, J.D., Bielenberg, R.W., Urbank, E.L.: Performance evaluation of New Jersey's portable concrete barrier with a pinned configuration and grouted toes – Test no. NJPCB-1, TRP-03-338-18, Midwest Roadside Safety Facility, University of Nebraska-Lincoln, Lincoln, Nebraska (2018a)

Bhakta, S.K., Lechtenberg, K.A., Faller, R.K., Reid, J.D., Bielenberg, R.W., Urbank, E.L.: Performance evaluation of New Jersey's portable concrete barrier with a bolted configuration and grouted toes – Test no. NJPCB-2, TRP-03-340-18, Midwest Roadside Safety Facility, University of Nebraska-Lincoln, Lincoln, Nebraska (2018b)

Bhakta, S.K., Lechtenberg, K.A., Faller, R.K., Reid, J.D., Bielenberg, R.W., Urbank, E.L.: Performance evaluation of New Jersey's portable concrete barrier with a free-standing configuration – Test no. NJPCB-3, TRP-03-355-18, Midwest Roadside Safety Facility, University of Nebraska-Lincoln, Lincoln, Nebraska (2018c)

Bhakta, S.K., Lechtenberg, K.A., Faller, R.K., Reid, J.D., Bielenberg, R.W., Urbank, E.L.: Performance evaluation of New Jersey's portable concrete barrier with a free-standing configuration and grouted toes – Test no. NJPCB-4, TRP-03-371-18, Midwest Roadside Safety Facility, University of Nebraska-Lincoln, Lincoln, Nebraska (2018d)

Bhakta, S.K., et al.: Performance evaluation of New Jersey's portable concrete barrier with a box-beam stiffened configuration and grouted toes – Test no. NJPCB-5, TRP-03-372-18, Midwest Roadside Safety Facility, University of Nebraska-Lincoln, Lincoln, Nebraska (2018e)

Bhakta, S.K., Lechtenberg, K.A., Faller, R.K., Reid, J.D., Bielenberg, R.W., Urbank, E.L.: Performance evaluation of New Jersey's portable concrete barrier with a back side pinned configuration and grouted toes – Test no. NJPCB-6, TRP-03-373-18, Midwest Roadside Safety Facility, University of Nebraska-Lincoln, Lincoln, Nebraska (2018f)

Bhakta, S.K., Lechtenberg, K.A., Faller, R.K., Reid, J.D., Bielenberg, R.W., Urbank, E.L.: Performance evaluation of New Jersey's portable concrete barrier with a traffic-side pinned configuration and grouted toes – Test no. NJPCB-7, TRP-03-374-18, Midwest Roadside Safety Facility, University of Nebraska-Lincoln, Lincoln, Nebraska (2018g)

Bronstad, M., Calcote, L., Kimball Jr, C.: Concrete median barrier research. Vol. 2: FHWA-RD-77-4, Office of Research and Development, Federal Highway Administration, Southwest Research Institute, San Antonio, Texas (1976)

Buth, C., Campise, W.L., Griffin, L., Love, M., Sicking, D.: Performance limits of longitudinal barrier systems, volume I: summary report, Report No. FHWA/RD-86/153, Federal Highway Administration, Office of Safety and Traffic Operations R&D, Texas Transportation Institute, Texas A&M University, College Station, Texas (1986)

Buth, C., Hirsch, J.: Performance level 2 bridge railings. Transportation Research Record No. 1258, Transportation Research Board, National Research Council, Washington, D.C. (1990)

Fortuniewicz, J.S., Bryden, J.E., Phillips, R.G.: Crash tests of portable concrete median barrier for maintenance zones. Report No. FHWA/NY/RR-82/102, Office of Research, Development, and Technology, Federal Highway Administration, Performed by the Engineering Research and Development Bureau, New York State Department of Transportation, New York (1982)

Lechtenberg, K.A., Faller, R.K., Reid, J.D., Sicking, D.L.: Dynamic Evaluation of a Pinned Anchoring System for New York State's Temporary Concrete Barriers – Phase II, TRP-03-224-10. Midwest Roadside Safety Facility, University of Nebraska-Lincoln, Lincoln, Nebraska (2010)

New Jersey Depatment of Transportation: Roadway Design Manual, Trenton, NJ (2015)

Polivka, K.A., et al.: Performance Evaluation of the Permanent New Jersey Safety Shape Barrier – Update to NCHRP 350 Test No. 3-10 (2214NJ-1), TRP-03-177-06, Midwest Roadside Safety Facility, University of Nebraska-Lincoln, Lincoln, Nebraska (2006)

Stolle, C.J., et al.: Evaluation of Box Beam Stiffening of Unanchored Temporary Concrete Barriers, TRP-03-202-08, Midwest Roadside Safety Facility, University of Nebraska-Lincoln, Lincoln, Nebraska (2008)

Safety Performance Evaluation of a Non-proprietary Type III Barricade for Use in Work Zones

Karla Lechtenberg[1(✉)], Ronald Faller[1(✉)], Jennifer Rasmussen[2], and Mojdeh Asadollahi Pajouh[1(✉)]

[1] Midwest Roadside Safety Facility, University of Nebraska-Lincoln, Lincoln, NE, USA
kpolivka2@unl.edu
[2] Safe Roads Engineering Inc., Whitchurch-Stouffville, Canada

Abstract. Work-zone traffic control devices, such as Type III barricades, must satisfy impact safety standards to improve safety and minimize risk for the motoring public traveling within work zones and on our highways and roadways. More specifically, this study focused on developing improved methods and products for addressing safety and mobility in work zones by evaluating new technologies and methods, thereby enhancing the safety and efficiency of traffic operations and highway workers. The non-proprietary, Type III barricade had three horizontal High-Density Polyethylene (HDPE) panels with a 1,219-mm × 762-mm × 2-mm aluminum sign attached to the top two barricade panels. The barricade panels were attached to two Perforated Square Steel Tubing (PSST) uprights, which were inserted into two PSST vertical stubs welded to two PSST legs. All PSST was galvanized ASTM 1011 Grade 55 steel. A 22.7-kg sandbag was placed on top of the end of each leg. A Type A/C warning light was attached to the front of the top barricade panel and upright at both upright locations. All of the impacts on the non-proprietary Type III barricade systems resulted in acceptable safety performance according to MASH 2016. Since the system is non-proprietary, any manufacturer could provide the components as long as they had similar dimensions and material grade as the as-tested system. It is anticipated that the Type III barricade without an attached aluminum sign panel would perform equivalent to or better than the system tested with a sign panel.

Keywords: Work zone · Construction area · Type III barricade · MASH · Non-proprietary

1 Introduction

A wide variety of traffic control devices are used in work zones. These devices are used to enhance the safety of the work zones by controlling the traffic through these hazardous areas. Unfortunately, the devices themselves may be potentially hazardous to occupants of errant vehicles. Thus, the Federal Highway Administration (FHWA) and the Manual on Uniform Traffic Control Devices (MUTCD) require that work-zone traffic control devices must demonstrate acceptable crashworthiness in order to be used on the National Highway System (NHS). The National Cooperative Highway Research

Program (NCHRP) Report No. 350, Recommended Procedures for the Safety Performance Evaluation of Highway Features set forth the first guidelines for the safety performance of work-zone traffic control devices. This document recommended that work-zone traffic control devices should be subjected to two full-scale crash tests with a small passenger car. If a device showed a propensity to penetrate into the occupant compartment, NCHRP Report No. 350 recommended that an additional crash test should be conducted with a pickup truck. Even though penetration into the occupant compartment was the primary concern for virtually all temporary sign support systems, the FHWA had not required crash testing with a pickup truck for any work-zone traffic control device under NCHRP Report No. 350.

In 2009, the American Association of State Highway and Transportation Officials (AASHTO) published the Manual for Assessing Safety Hardware (MASH), which replaced NCHRP Report No. 350 as the new safety performance guidelines used for crash testing and evaluating roadside safety devices. MASH requires that work-zone traffic control devices be crash tested with both a small car and a full-size pickup truck. Historically, work-zone traffic control devices have been specifically developed to meet NCHRP Report No. 350 safety evaluation guidelines for the 820-kg small car impact condition. Hence, certain parameters of current crashworthy traffic control devices may cause these devices to have an unacceptable safety performance when impacted with larger vehicles. In addition, many of the traffic control devices approved under NCHRP Report No. 350 were designed to bridge the windshield and strike the roof for taller systems or to breakaway and pass over the top of the car without contacting the windshield or remain intact and travel out in front of an impacting vehicle. However, this behavior was dependent upon the front-end profile of the 820- kg small car. Vehicles with longer or taller front-end profiles could allow the traffic control devices to contact the windshield and produce undesirable behavior. Therefore, the devices found in work zones along the NHS may not be crashworthy with many vehicles larger than the 820- kg small car.

Through a project funded jointly through Dicke Safety Products, the Mid-America Transportation Center, and the Smart Work Zone Deployment Initiative (SWZDI) from 2008 to 2010, several NCHRP Report 350 crashworthy work-zone traffic control devices were evaluated (Schmidt et al. 2009). When these devices were subjected to the new MASH crash testing and safety performance criteria, several of the work-zone sign stands produced undesirable results, including windshield and floorboard penetration as well as excessive windshield and roof deformation. This testing program indicated that devices tested under previous NCHRP Report 350 safety performance standards may not perform acceptably with the new MASH safety performance standards.

In an effort to encourage state departments of transportation (DOTs) and hardware developers to advance hardware designs, the FHWA and AASHTO collaborated to develop a MASH implementation policy that included sunset dates for various roadside hardware categories. Further, work-zone traffic control devices must satisfy current MASH impact safety standards to improve safety and minimize risk for the motoring public traveling within work zones and on our highways and roadways.

NCHRP Project 03–119 is currently being conducted to evaluate the in-service safety performance of breakaway sign supports, luminaires, and work-zone devices, and evaluate these devices to MASH 2016 (Marzougui 2015). This NCHRP Project began with identifying devices, an agency survey, and contact with practitioners to identify current practices related to the use of work-zone traffic control devices. Researchers identified a list of non-proprietary safety work-zone devices in common use as well as insights on safety performance issues associated with each of them. However, not many work-zone traffic control devices will be evaluated under the NCHRP Project. Therefore, a need exists to crash test and evaluate the MASH 2016 crashworthiness performance of non-proprietary work-zone traffic control devices. Evaluating new or existing technologies and methods would provide products for addressing safety and mobility in work zones, thereby enhancing safety and efficiency of traffic operations and highway workers.

2 System Selection and Design Details

Based on the background research in NCHRP Project No. 03–119, state DOT survey, and additional review of NCHRP Report 350 crash tests, two general categories of work-zone traffic control devices were determined to be needs for the state DOTs: (1) Type III barricades and (2) portable signs. Almost all non-proprietary portable signs have historically been made of Perforated Square Steel Tubing (PSST). However, one system, the Wisconsin temporary gore sign, is made of wood. Several non-proprietary Type III barricades have been previously tested to NCHRP Report 350 and include steel angle or PSST legs and uprights. Since some non-proprietary work-zone sign stands have already been tested to MASH 2016, the highest need was determined to be a non-proprietary Type III barricade, none of which had been evaluated to MASH 2016 crash test standards.

Following a review of state DOT standards, several different variations of the Type III barricade systems exist. The final system selected for evaluation was designed to be the most representative and useful to the state DOTs. The test installation consisted of two Type III barricades, as shown in Fig. 1. Each Type III barricade consisted of three horizontal High Density Polyethylene (HDPE) panels, measuring 2,428 mm in length, with an 1,219-mm × 762-mm × 2-mm aluminum sign attached to the top two barricade panels. The barricade panel was targeted to have nominal cross-sectional dimensions of 203 mm tall × 25 mm thick. However, the dimensions of the barricade panel vary between manufacturers, and the supplied barricade panel was 210 mm 19 mm. The barricade panels were attached to two 44-mm × 1.9-mm thick PSST uprights, which were inserted into two 51-mm × 1.9-mm thick × 152-mm long PSST vertical stubs that were each welded to one of the two legs. The legs were 51-mm × 1.9-mm thick × 1,524-mm long PSST. All PSST used was galvanized ASTM 1011 Grade 55 steel with a minimum yield strength of 414 MPa. A 23-kg sandbag was placed on top of both ends of each leg. A Type A/C warning light was attached to the front of the top barricade panel and to the upright at both upright locations.

Fig. 1. Type III barricade details.

3 Test Requirements and Evaluation Criteria

Category 2 work-zone traffic control devices, such as Type III barricades, must satisfy impact safety standards in order to be declared eligible for federal reimbursement by the FHWA for use on the NHS. For new hardware, these safety standards consist of the guidelines and procedures published in MASH 2016. Note that there is no difference between MASH 2009 and MASH 2016 for work-zone traffic control devices, such as Type III barricades tested in this project. According to TL-3 of MASH 2016, work-zone traffic control devices must be subjected to three full-scale vehicle crash tests, as summarized in Table 1.

Table 1. MASH 2016 TL-3 crash test conditions for work-zone traffic control.

Test article	Test designation no.	Test vehicle	Vehicle weight (kg)	Impact conditions Speed (km/h)	Angle (degrees)
Work-Zone Traffic Control Device	3–70	1100C	1,100	30	CIA
	3–71	1100C	1,100	100	CIA
	3–72	2270P	2,270	100	CIA

CIA - Critical Impact Angle is the worst case impact condition in which the traffic control device will be deployed along the roadway.

The low-speed test, test designation no. 3–70, is intended to evaluate the breakaway, fracture, or yielding mechanism of the device. The high-speed tests, test designation no. 3–71 and 3–72, are intended to evaluate vehicular stability, test article trajectory, and occupant risk factors. Since most work-zone traffic control devices have a relatively small mass, less than 100 kg, the high-speed crash tests are more critical due to the propensity of the test article to penetrate into the occupant compartment. Therefore, since the Type III barricade weighed less than 100 kg, test designation no. 3–70 was not required. MASH 2016 recommends test designation nos. 3–71 and 3–72 be conducted both perpendicular to the device (0°) and parallel to the device (90°), as both orientations may occur along roadsides. A procedure for testing multiple work-zone traffic control devices in one test run has been developed. Therefore, the barricade was evaluated at two impact angles, 90° (System A) and 0° (System B), in one full-scale crash test. The devices were spaced 18.3 m apart and each device impacted at the quarter points on the front bumper. Thus, two MASH 2016 test designation nos. 3–71 and 3–72 crash tests were conducted at two CIAs.

4 Type III Barricade Full-Scale Crash Tests

4.1 Test No. WZNP-1

Test no. WZNP-1 was conducted on a Type III barricade in accordance with MASH 2016 test designation no. 3–71. Two Type III barricades were placed 18.3 m apart on level terrain with one sandbag on the end of each leg. The 1,100-kg small car impacted System A, oriented at 90° or perpendicular to the vehicle, at a speed of 104.2 km/h and System B, oriented at 0° or head-on to the vehicle, at a speed of 98.6 km/h. The impact speed of System A was above the velocity range for impact speed according to MASH 2016. However, the higher velocity represents a worst-case scenario and was targeted to be high to insure that the second impact was also within the velocity range.

During the test, the 1100C small car impacted and disengaged both barricades from the legs, as shown in Fig. 2 and Fig. 3. The systems readily activated in a predicable manner and allowed the 1100C vehicle to continue traveling without any major obstruction of the windshield. System damage, as shown in Fig. 4, consisted of bends and buckles in the vertical uprights and panels, tears in the sandbags, and pull out of bolts. Vehicle damage was minimal, as shown in Fig. 5. The windshield was cracked across its entirety with localized crush on the lower right side and on the left side adjacent to the A-pillar. Deformations of, or intrusions into, the occupant compartment that could have caused serious injury did not occur. The test vehicle remained upright during and after the collisions. Therefore, test no. WZNP-1 was determined to be acceptable according to the MASH 2016 safety performance criteria for test designation no. 3–71.

Safety Performance Evaluation of a Non-proprietary Type III Barricade 873

Fig. 2. Sequential photographs, test no. WZNP-1, System A.

Fig. 3. Sequential photographs, test no. WZNP-1, System B.

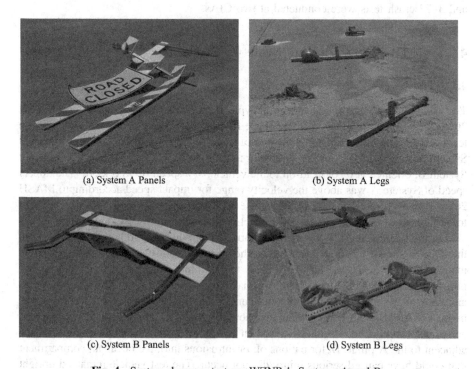

Fig. 4. System damage, test no. WZNP-1, Systems A and B.

Fig. 5. Vehicle damage, test no. WZNP-1.

4.2 Test No. WZNP-2

Test no. WZNP-2 was conducted on a Type III barricade in accordance with MASH 2016 test designation no. 3–72. Two Type III barricades were placed 18.3 m apart on level terrain with one sandbag on the end of each leg. The 2,268-kg pickup truck impacted System A, oriented at 90° or perpendicular to the vehicle, at a speed of 104.2 km/h and System B, oriented at 0° or head-on to the vehicle, at a speed of 100.8 km/h. The impact speed of System A was above the velocity range for impact speed according to MASH 2016. However, the higher velocity represents a worst-case scenario and was targeted to be high to insure that the second impact was also within the velocity range.

During the test, the 2270P pickup truck impacted and disengaged both barricades from the legs, as shown in Fig. 6 and Fig. 7. The systems readily activated in a predicable manner and allowed the 1100C vehicle to continue traveling without any major obstruction of the windshield. System damage, as shown in Fig. 8, consisted of punctured sandbags, bent vertical uprights and legs, bent and torn barricade panels, and bolt pull out in the barricade panels. Vehicle damage was minimal, as shown in Fig. 9, and was concentrated on the right-front corner where the vehicle impacted System A. Deformations of, or intrusions into, the occupant compartment that could have caused serious injury did not occur. The test vehicle remained upright during and after the collisions. Therefore, test no. WZNP-2 was determined to be acceptable according to the MASH 2016 safety performance criteria for test designation no. 3–72.

Safety Performance Evaluation of a Non-proprietary Type III Barricade   875

Fig. 6. Sequential photographs, test no. WZNP-2, System A.

Fig. 7. Sequential photographs, test no. WZNP-2, System B.

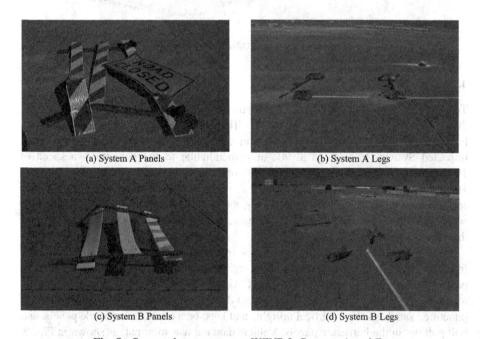

(a) System A Panels   (b) System A Legs

(c) System B Panels   (d) System B Legs

Fig. 8. System damage, test no. WZNP-2, Systems A and B.

Fig. 9. Vehicle damage, test no. WZNP-2.

5 Conclusions

The objective of this research effort was to evaluate the performance of a non-proprietary work-zone safety device, such as a work-zone sign support or barricade. Recommendations on the performance and usage of work-zone devices was based on the background research in NCHRP Project No. 03–119, a state DOT survey, and additional review of NCHRP Report 350 crash tests. A non-proprietary, Type III barricade that is commonly used by state DOTs was selected for the full-scale crash testing program. The Type III barricade was evaluated according to the MASH TL-3 test designation nos. 3–71 and 3–72 safety criteria through four full-scale crash tests at 0-degree and 90-degree impact angles.

Test no. WZNP-1 was conducted on a Type III barricade in accordance with MASH 2016 test designation no. 3–71. Two identical Type III barricades were placed 18.3 m apart on level terrain with one sandbag on the end of each leg. During the test, the 1100C small car impacted and disengaged both barricades. The systems readily activated in a predicable manner and allowed the 1100C vehicle to continue traveling without any major obstruction of the windshield. There were no detached elements nor fragments which showed potential for penetrating the occupant compartment nor presented undue hazard to other traffic. Deformations of, or intrusions into, the occupant compartment that could have caused serious injury did not occur. The test vehicle remained upright during and after the collisions. Vehicle roll, pitch, and yaw angular displacements were deemed acceptable, because they did not adversely influence occupant risk nor cause rollover. After impact, the vehicle's trajectory did not violate the bounds of the exit box. Therefore, test no. WZNP-1 was determined to be acceptable according to the MASH 2016 safety performance criteria for test designation no. 3–71.

Test no. WZNP-2 was conducted on a Type III barricade in accordance with MASH 2016 test designation no. 3–72. Two Type III barricades were placed 18.3 m apart on level terrain with one sandbag on the end of each leg. During the test, the 2270P pickup truck impacted and disengaged both barricades. The systems readily activated in a predicable manner and allowed the 2270P vehicle to continue traveling without any major obstruction of the windshield. There were no detached elements nor fragments which showed potential for penetrating the occupant compartment nor presented undue hazard to other traffic. Deformations of, or intrusions into, the occupant compartment that could have caused serious injury did not occur. The test vehicle remained upright during and after the collisions. Vehicle roll, pitch, and yaw angular displacements were deemed acceptable, because they did not adversely influence occupant risk nor cause rollover. After impact, the vehicle's trajectory did not violate the bounds of the exit box. Therefore, test no. WZNP-2 was determined to be acceptable according to the MASH 2016 safety performance criteria for test designation no. 3–72. Thus, this non-proprietary Type III barricade satisfied all of the requirements for the crash tests in the TL-3 test matrix and, therefore, is a MASH TL-3 crashworthy device.

6 Recommendations

The Type III barricade evaluated was non-proprietary. Thus, the components could be provided by any manufacturer. When assembling this Type III barricade, hardware parts and materials that are similar to those used in the as-tested system should be utilized. For example, the Type III barricade panels consisted of three horizontal HDPE panels, measuring 2,428 mm in length. The barricade panel was targeted to have nominal cross-sectional dimensions of 203 mm tall × 25 mm thick. However, the dimensions vary between manufacturers, and the supplied barricade panel was 210 mm × 19 mm. Therefore, HDPE panels that are similar to those in the as-tested installation or with the nominal dimensions could also be used for this Type III barricade.

Sandbags, weighing approximately 23 kg, should be placed on the ends of each leg. One Type A/C warning light was attached to the top and front-side of the HDPE panels at each PSST upright on each barricade to evaluate a worst-case configuration with attachments. Thus, two warning lights were attached to each barricade. Utilizing one or no warning lights would also be acceptable. The warning lights were attached to the front side of the top barricade panel but could also be attached to the backside of the top barricade panel, as that would be a less critical configuration.

An aluminum sign panel can be attached to the Type III barricade, with a maximum sign size and location similar to the as-tested installation. Smaller aluminum sign panels attached with a top height that is even with the top barricade panel or lower, or omitting the aluminum sign panel would also be acceptable configurations.

Acknowledgements. The authors wish to acknowledge several sources that made a contribution to this project: (1) the Smart Work Zone Deployment Initiative (SWZDI), the FHWA, the Iowa DOT and the other pooled fund state partners, and the NCHRP for sponsoring this project; (2) George Mason University, the lead agency for NCHRP project; (3) the technical advisory committee of Matt Neemann – Nebraska DOT, Brian Smith – Iowa DOT, Michael Seifert – Wisconsin DOT, and Matt Rauch – Wisconsin DOT for their guidance through the project; (4) TrafFix Devices Inc. of San Clemente, California for donating the barricade panels; and (5) MwRSF personnel for constructing the systems and conducting the crash tests.

References

AASHTO: Manual for Assessing Safety Hardware, American Association of State Highway and Transportation Officials, Washington, DC (2009)

AASHTO: Manual for Assessing Safety Hardware, Second Edition, American Association of State Highway and Transportation Officials, Washington, DC (2016)

Marzougui, D.: NCHRP Project 03-119: Application of MASH Test Criteria to Breakaway Sign and Luminaire Supports and Crashworthy Work-Zone Traffic Control Device, National Cooperative Highway Research Program, Washington, D.C. (2015). http://apps.trb.org/cmsfeed/TRBNetProjectDisplay.asp?ProjectID=3857

Schmidt, J.D., Flores, J., Asadollahi Pajouh, M.A., Faller, R.K., Lechtenberg, K.A.: MASH 2016 Evaluation of a Non-Proprietary Type III Barricade: MASH Test Designation No. 3-72, Report No. TRP-03-416-20, Midwest Roadside Safety Facility, University of Nebraska-Lincoln, Lincoln, Nebraska (2020)

Schmidt, J.D., Langel, T.J., Asselin, N., Pajouh, M.A., Faller, R.K.: MASH 2016 Evaluation of a Non-Proprietary Type III Barricade, Report No. TRP-03-394-18, Midwest Roadside Safety Facility, University of Nebraska-Lincoln, Lincoln, Nebraska (2018)

Schmidt, J.D., Sicking, D.L., Lechtenberg, K. A., Faller, R.K., Holloway, J.C.: Analysis of Existing Work-Zone Devices with MASH Safety Performance Criteria, Mid-America Transportation Center, University of Nebraska-Lincoln, Lincoln, Nebraska (2009)

Ross, H.E., Sicking, D.L., Zimmer, R.A., Michie, J.D.: NCHRP Report 350: Recommended Procedures for the Safety Performance Evaluation of Highway Features. National Cooperative Highway Research Program, Washington, DC (1993)

ITS Design and Implementation Strategies

Unified ITS Environment in the Republic of Tatarstan

Rifkat Minnikhanov[1(✉)], Maria Dagaeva[1,2], Sofya Kildeeva[1], and Alisa Makhmutova[1,2]

[1] "Road Safety" State Company, Kazan, Russia
its.center.kzn@gmail.com
[2] Kazan National Research Technical University named after A.N. Tupolev–KAI, Kazan, Russia

Abstract. The Republic of Tatarstan has always been a pioneer among other regions of Russia in the development of innovative and digital technologies. Tatarstan is changing and adapting to be relevant in healthcare, economy, and transport areas. Challenges of modern transport systems cannot be solved with strategies and tools from the past. Intelligent transport systems (ITS) are being actively introduced to overcome the challenges of modern life. The article describes the development of ITS in the Republic of Tatarstan. There are considered issues related to existing technologies, priority areas of development, and the implementation of a unified ITS environment. Much attention is paid to the ITS subsystems, which are being developed in order to ensure the safety and comfort of the residents of the Republic.

Keywords: Road safety · Technologies · Systems design and development · Intelligent transport systems

1 Mobility in the Republic of Tatarstan

The Republic of Tatarstan plays a key role in transport links between the Eastern and European parts of Russia, as well as in communication with other countries. For many years, the Republic of Tatarstan has maintained the position of an innovative region, a testing ground for the development of digital technologies, and it has experience in the field of innovations. The national interests of the republic for the long term are the implementation of accessible transport infrastructure, improving the quality of transport services and the environmental situation in the region, development of safe environment, sustainable mobility, and development additional conditions for a comfortable life for residents. The leadership of the republic understands that the complexities of the modern transportation system cannot be overcome using the strategies and tools of the past generation. To overcome barriers in the new reality, information technologies and intelligent transport systems (ITS) are being actively introduced in the Tatarstan.

Socio-economic transformations in society place serious demands on the transport system. Mobility in transportation is one of the key demand of modern society. Due to this transport system must be adapted to social needs, providing transportation in ways that are safe and economically efficient. Moreover, whole system must satisfy demands of environmental rationality and the new logic of sustainability paradigms.

From this viewpoint, assessment of an efficiency and flexibility of transport system in the region can be based on intelligent and sustainable patterns of mobility. In this paper we evaluate effect of various national projects implemented in transport environment in Republic of Tatarstan and try to make conclusion about their impact on the growth of mobility in the region. The accessibility, environmental friendliness of movement, its safety and comfort characterize a high level of mobility. We will take these four metrics as evaluation criteria.

2 Road Safety

2.1 Traffic Enforcement

There is the state registration of the main indicators of the state of road safety on the territory of the Russian Federation. Such indicators are: the number of road traffic accidents, victims of citizens, vehicles, vehicle drivers; violators of traffic rules, administrative offenses and criminal offenses in the field of road traffic, as well as other indicators reflecting the state of road safety and the results of activities to ensure it.

The system of state accounting ensures the organization and implementation by federal executive bodies, executive bodies of the constituent entities of the Russian Federation and local self-government bodies of work on the formation and implementation of state policy in the field of road safety.

To achieve the safety indicator, it is necessary to reduce social risk and use preventive measures. Monitoring compliance with traffic rules disciplines drivers, stabilizes the speed limit in free areas, improves adherence to traffic lane, forces pedestrians to take advantage of the crosswalk, increases the accuracy of crossing intersections at a permissive traffic signal. All of these factors lead to a significant reduction of the number of serious accidents, saving drivers, passengers and pedestrians from road traffic accidents with serious injury or death. It is necessary to automate this process to cover a large area and create 24/7 control.

In modern settlements the use of traffic enforcement cameras is absolutely necessary. They are becoming an integral part of the city's intelligent transport system.

The Republic of Tatarstan is the first region in the Russian Federation to implement a system for automatic detection of traffic violations in 2008.

Currently, the traffic enforcement is actively using in general 1150 stationary, portable and mobile complexes.

With the initial introduction in 2008 of traffic enforcement, improvements could be observed in key road safety indicators (Fig. 1). We can observe a gradual decline in key indicators from 2008 to 2014, however, it was very slow. This can be attributed to the fact that potential violators are gradually adapting to the traffic rules control. This leads to a decrease in the social efficiency of local automatic control of compliance with the established speed limit. If there had not been an increase in the number of devices, the efficiency would have been very low.

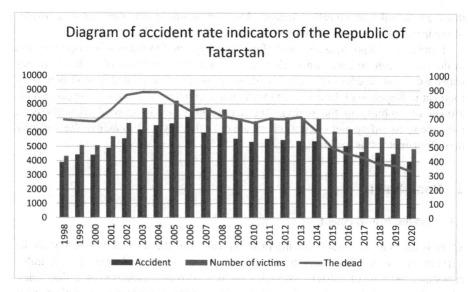

Fig. 1. Statistics on indicators of road safety in the Republic of Tatarstan. The axis on the left represents the number of accidents, the axis on the right shows the metric for the number of fatalities.

Therefore, the system must constantly change and modernize. This project "Total" control has been implemented in the Republic of Tatarstan in 2018, on three federal highways. It's goal to provide permanent control on federal and regional highways of the Republic of Tatarstan. In 2018–2019, 100 traffic cameras operated in automatic mode: the devices monitored 290 road sections with a total control length of 924 km. In 2020, the project was further developed on four regional highways of the republic. In addition to the existing sensors, 73 devices have been installed, which monitor an additional 256 road sections with a total length of 402 km. The devices are interconnected and record both the instantaneous speed and the average speed between the sensors, forming continuous control zones along the entire route. These zones encourage law-abiding behavior not only in a specific place where the speed control device is installed, but throughout the entire road section. Average speed control tends to wean from the habit of "imitation of careful driving", when drivers sharply reduce their speed in front of the camera, and after passing it, they immediately accelerate, thus increasing the risk of accidents. From Fig. 1 we can see how effective this measure was. Following the results of 2020, 86.3% of the total number of all decisions issued by the State Traffic Inspectorate are recorded by these automated systems.

2.2 Video Surveillance Systems

Traffic enforcement is supported also by city's video surveillance systems. In 2014, the government of the Russian Federation approved the concept of construction and development of the "Safe City" hardware and software complex to improve the overall level of public safety, also including the roads safety. Activities within the "Safe City" concept are being implemented simultaneously with the "Smart City" project for digitalization of the urban environment. As at December 2020, there are 47,037 u. of CCTV. They conduct continuous video surveillance in crowded places, for example, in residential buildings, educational and cultural institutions, bus and railway stations and airports. Also, the face recognition system, which has been launched in test mode in Almetyevsk city. The entire video stream from cameras is stored on the assets of telecom operators. Only law enforcement agencies have access the broadcast.

However, without smart automated analysis all stored massive amount of information can be useless. One of the main drawbacks of traditional video surveillance systems is poor performance and responce when scaling the system to the city level. In the case of tens or more incoming video streams, the operator is no longer able to monitor the current situation in real time.

Currently, the republic is working to develop a video analysis module, work of which represented in Fig. 2. The module is designed to recognize road users (pedestrians and vehicles), draw up their trajectories, determine the categories of vehicles, count the number of passing vehicles, identify incidents (traffic congestion and traffic accidents). Today the module processes video streams from 11 cameras of the "Safe City" hardware and software complex, the module is being adapted to run on 144 CCTV cameras at the at the crossroads.

Fig. 2. a) Marking the trajectories of objects at the intersection in the video analytics block b) Recognition of vehicle types at an intersection

2.3 Weight-In-Motion Systems

Next important question that directly affects the safety of the trip is the quality of the roads.

The active construction and repair of roads leads to an increase in traffic intensity. However, the increased traffic of overloaded freight vehicles on the roadway is one of the main reasons for its early destruction. The introduction of Weight-In-Motion systems (WIM) in the Republic of Tatarstan has become a new, dynamically developing direction in the field of monitoring compliance with the permitted maximum mass and axle loads of vehicles. Automatic points of weight and dimension control began their tests in the fall of 2019. The procedure is performed automatically, without requiring the vehicle to stop. If the measurement data contains a violation of the current standards for dimensions, total weight or axle load, the system sends the vehicle's state registration number, its photograph and metrological print to a unified information center. There the vehicle is identified, the presence of a permit is checked, and in case of its absence, a fine is issued. The system has high efficiency in real time and does not require the presence of traffic police officers and technical personnel on the road. During the test period, more than 32 thousand vehicle passes with excess weight and dimensions were identified. A significant part of these passages are with special permits. However, some are violators, in 2020 there were 349 of them. The system monitors the entire flow of vehicles.

Also, in the Republic, WIM systems integrated with video enforcement cameras and electronic toll collection systems are used. Integration of mentioned systems provides enforcement and a control of transport flow in accordance to following parameters: weight and size characteristics; speeding; contraflow driving; lack of onboard unit or toll ticket.

2.4 Environmental Monitoring on the Roads

The climate of the Republic of Tatarstan is temperate continental, characterized by hot summers and moderately cold winters. Winter in Tatarstan lasts from late November to late March. In general, winter is characterized by moderately cold weather, but frequent invasions of cold from the northeast, which are accompanied by frosty (up to -30 °) and little cloudy weather. Often, drivers must select a speed that would ensure road safety in difficult road and meteorological conditions.

There is a system that monitors meteorological information in order to improve road safety in adverse weather conditions. The system is designed to implement the collection, processing, storage and transmission of data on actual and predicted meteorological parameters. The system is a complex of a meteorological observation system, a complex of video equipment and data transmission facilities. The system covers public federal and regional roads passing through the territory of the Republic of Tatarstan. The collected data is necessary to ensure the functioning of other subsystems (services) of the ITS, the maintenance of roads and the provision of the necessary weather data to consumers. Accurate information allows to prepare in advance for hazardous weather phenomena (ice, precipitation, strong wind), which helps to reduce the number of road accidents. In the event of unfavourable weather conditions, an automatic e-mail and SMS distribution notifies the enterprises of the road transport complex. In the event of a forecast of difficult meteorological conditions, the traffic police recommend drivers to refrain from traveling outside of settlements unless absolutely necessary, reminds of strict observance of the traffic rules and urges to be extremely attentive and cautious.

2.5 Navigation Systems Based on GLONASS+112

Dispatching contributes to the safety of vehicles, passengers and drivers, since it can help reduce the number of failures in work, prevent the development of emergency situations. In addition, in the event of transport problems and a threat to the health of traffic participants, the dispatcher receives an alarm signal and promptly responds to events.

The complex of the regional navigation and information transport monitoring system is implemented in the Republic through the regional navigation and information transport monitoring system based on GLONASS/GPS. It works within the Unified State Information System "GLONASS+112" (USIS "GLONASS+112"), which also include:

1. Unified communication system of emergency operational services "112" (System 112);
2. Regional Navigation and Information System;
3. Geoportal and geoinformation systems of ministries and departments of the Republic of Tatarstan.

USIS "GLONASS+112" is operating in the Republic of Tatarstan since 2009. This is a comprehensive solution that allows a single interdepartmental information environment, to monitor forces and means, automated processing of telephone calls from victims and routing of such calls between divisions of various operational services, the establishment and processing of a single incident record card for all operational services participating in the work on incident.

This approach has proven its effectiveness due to the one-time creation and mass replication of the product where it is needed. System-112 of the Republic of Tatarstan on the basis of USIS "GLONASS+112" was the first in Russia to pass successful state tests in 2013. In the period from 2010 to 2019, specialists in the reception and processing of emergency calls to the System-112 in the republic received and processed more than 13.4 m calls, sent more than 5.9 m incident cards to emergency operational services.

The regional navigation and information system is responsible for transport dispatching. All vehicles carrying out passenger transportation on intermunicipal routes (662 vehicles) and transportation of children: school buses (734 vehicles), buses of children's sports and youth schools (140 vehicles), as well as additional educational institutions (129 vehicles), are equipped with navigation equipment, a hands-free kit and a panic button, and are also connected to the "GLONASS+112" system in order to increase the safety level of passenger transportation. Satellite navigation equipment, which is installed on a vehicle, determines the geographic coordinates and parameters of the vehicle's movement. In addition, 416 vehicles of the Ministry of Health of the Republic of Tatarstan (ambulance) are equipped with GLONASS equipment. A vehicle that is not available in the system is not allowed to enter the route. The tasks of regional navigation and information system in the field of road safety are:

1. Prompt response in case of accidents;
2. 24/7 vehicle monitoring;
3. Parameter control: speed, schedule, route;
4. Organization of interdeparrtment cooperation;
5. Formation of a register of stopping points, routes and timetables.

The following results were obtained from the implementation of USIS "GLONASS +112" in the field of transport dispatching:

1. Departures from the route were reduced by 81%;
2. Speed violations decreased by 91%;
3. Deviation from the schedule decreased by 55%;
4. Fuels consumption decreased by 20%.

2.6 Conclusion

To summarize all road safety activities, we can rely on statistics. In the period from 2008 to 2020 the number of fatalities in road accidents decreased from 713 people to 331, that is, almost 2.5 times. It can be concluded that the measures taken by ITS are working and are providing positive changes in the field of road safety.

3 Reliability and Availability

3.1 Adaptive Traffic Control System and Priority System for Public Transport and Emergency Services

The availability of transportation largely depends on the capacity of the roads and organisation of the traffic flows. The first adaptive traffic control system (ATCS) in the Russian Federation was put into operation in capital of the Tatarstan, Kazan, in 2011. From 2011 to 2020 the system expanded from 15 to 162 traffic lights. Nowadays, the system includes 700 traffic detectors (this is, in average, 5 detectors per traffic light), which update traffic information in real time.

The general principle of ATCS operation is to develop an optimal strategy for managing individual traffic lights and groups of intersections within a given traffic control system. Traffic management strategy is developed every 4 s, depending on the data on the movement of vehicles in the detection zones.

The system can make decisions about the control strategy locally, at the level of the traffic light object. Many intersections are equipped with all-round cameras, which allows remotely assess the efficiency of the system and make corrections if necessary. The information display board, which is part of the ATCS, informs the drivers of vehicles about possible traffic jams and congestions at certain sections of the road network. Taking into account the information received, the driver himself chooses the ways to bypass the problem area. In Kazan, 21 displays of variable information have been installed.

By means of controlled road signs, the ATCS system redirects traffic flows to exits and transport hubs with less congestion or reduces the flow rate to prevent congestion at the exit. In the event of an accident, the system can prohibit passage to this section, thereby preventing the formation of a dead traffic jam, in which road users would have to be until the moment of elimination of the consequences of an accident.

As part of an adaptive traffic control system, it assumes the operation of a priority system for public transport. The system allows, in the long term, to reduce the time of movement of passenger transport along the established route by reducing the waiting time at regulated intersections. The flexibility of the system is the ability to assess the need to provide priority for travel for each route unit, depending on the data provided by the dispatch system.

The project of the priority system for the movement of special transport vehicles proposed for implementation is also based on the adaptive traffic control system in Kazan. At the pilot stage, the system will include 12 vehicles of the Ministry of Emergency Situations from two fire departments and 18 traffic lights with 2 fixed routes.

The principle of operation of the priority system for emergency service transport is an evolutionary development of the concept of priority for selected transport groups. The work of the system based on specialized software, tracking the location of the vehicle, transmits this information to the system, requesting the priority of travel. The key difference between this software module and the public transport priority module is that it uses software for monitoring the direction of vehicle movement and an approximation that eliminates geolocation errors.

An analysis of the functioning of the transport system during the implementation of ATCS in cities shows the following efficiency of traffic management:

- reduction of transport delays at intersections by 30–50%, due to the optimization of traffic light modes;
- increase of the average speed of the vehicle by 10–15% between intersections by reducing the length of queues waiting for a permitting traffic signal;
- reduction of travel time on the road network by 10–20%;
- increase of road network efficiency by 15–25%;
- improvement in the healthiness of the city's air by 20–25% as a result of the reduction of pollution caused by exhausted gases of engines (by reducing vehicle stops, increasing the average speed).

The result of implementation is an average reduction in travel time for vehicles of the fire service up to 37%. Taking into account the strict requirements for the response time of emergency services to a call signal, this optimization gives services 3–4 min of actual time to implement rescue and other routine actions.

3.2 Geoinformation for Monitoring and Managing the Condition of Roads and Infrastructure

Within the functioning of the Geoinformation system of the road industry of the Ministry of Transport and Road Facilities of the Republic of Tatarstan, the work of a number of subsystems has been implemented.

Geographic information system functions:

- control of road equipment operation using GLONASS satellite navigation equipment;
- maintaining a database of regulatory and reference information, including the visual formation of road passports;
- determination of the type of work, time and place of work;
- obtaining in real time pictures from cameras installed on road vehicles operating at sites;

- control over the implementation of work plans, instructions for eliminating deficiencies in the maintenance of highways.

The geographic information subsystem acts as a supporting subsystem and is used to form an image of an electronic map, consisting of various information layers. The collection of the database is carried out by a mobile laboratory. While driving a vehicle on the road, with installed sensors, GLONASS satellite navigation equipment, video cameras, digital video material is recorded. Based on different data, a technical passport of the road is created. When diagnosing highways, specialists receive complete, objective and reliable information about the condition of the roads. Defects are determined, the presence and condition of road objects are assessed. Further, the information is processed, digitized and displayed on a topographic basis.

In addition to the collected information, there is also information about road infrastructure objects (stops, crossings, and other objects).

3.3 Automated System for Dispatching Passenger Transport in Kazan

Since 2006, the administration of Kazan has been purposefully dealing with the issue of using modern informatics, navigation and mobile communications in passenger city transport. New technologies have made it possible to approach the tasks of traffic control and management in a different way. The use of equipment for satellite navigation systems and geoinformation technologies allows to transmit information about the location of the vehicle and its compliance with the schedule.

All urban passenger transport in Kazan operates under the control of an automated dispatch control system in the GLONASS satellite system in an agreed uniform timetable. Buses, trams and trolleybuses in the city of Kazan is equipped with on-board navigation kits, namely:

- 58 bus routes with 774 units,
- 7 tram routes for 82 units,
- 11 trolleybus routes for 140 units.

The length of the routes is: buses - more than 1.2 thousand km, trams −159 km, trolleybuses −273 km, metro −18.3 km.

The automated navigation system consists of several software modules.

To improve the quality of service at almost all major passenger-forming stops electronic information boards have been installed (245 u.), providing data on the exact actual time of arrival of the rolling stock.

The navigation data obtained by information system is relayed to the regional navigation information center of the Republic of Tatarstan in a unified protocol of interaction between telematic platforms of monitoring and control systems of mobile objects. Also, information about the movement of public transport is transmitted to the IS "Open Kazan", where any developer of mobile products can receive information for their transport application.

Passengers, using a cell phone, will be able to receive information on timetables, travel schedules and the approach of buses to any stop on request. They will receive operational information directly from the database.

4 Environmental Monitoring

In order to ensure environmental safety, prevent the negative impact of economic and other activities on the environment, ensure the realization of the rights of citizens to a favorable environment, the Ministry of Ecology and Natural Resources of the Republic of Tatarstan has been working since 2007 to form and ensure the functioning of the territorial monitoring system for atmospheric air. The environmental monitoring system was created on the basis of the "Environmental Map" program of the Republic of Tatarstan, which allows to form a single picture of the violations identified and control over the implementation of the issued orders.

Environmental monitoring in the republic is carried out with the help of 23 observation points for atmospheric air pollution of the Department for Hydrometeorology and Environmental Monitoring of the Republic of Tatarstan and 16 automatic air pollution control stations of the Ministry of Ecology and Natural Resources of the Republic of Tatarstan. It is planned to automate the monitoring of surface and groundwater pollution, soil pollution, precipitation and radiation control. Within the framework of the observation and laboratory control system, 186 laboratories, posts, control stations function in the Republic of Tatarstan.

The measurements of the Ministry are carried out in those territories where there is a direct influence of large industrial facilities for a timely response and prevention of damage to the environment. Actual data on the concentrations of pollutants in the air are recorded by automatic stations every 20 min in automatic mode and displayed on the GIS "Ecological Map of the Republic of Tatarstan".

Over 5 years (from 2015 to 2019), the number of excess pollutants in the air of the Republic of Tatarstan, recorded by automatic air pollution control stations of the Ministry, decreased in 2.5 times (from 4 thousand excess in 2015 to 1, 5 thousand exceedances for 9 months of 2019). The decrease in the amount of excess pollutants in the air was due to round-the-clock monitoring, detection and suppression of violations of environmental legislation in the field of air protection.

5 Transport System's User Comport

5.1 Mobility as a Service

One of the important directions of ITS development is the creation of various mobile applications, both for drivers and for all road users to implement the concept of Mobility as a Service (MaaS). This contributes to satisfaction with the convenience of new electronic services, the ticketing system, the increasing level of comfort on the way and during transfers, as well as the proliferation of new modes of shared transport.

Since 2012, the republic has implemented the possibility of receiving resolutions issued based on the results of the operation of complexes for the automated recording of administrative offenses in the field of road traffic by e-mail. More than 116 thousand vehicles of legal entities and 153 thousand citizens are connected to this service. In 2019, this technology continued to develop: to find out about the fine for violation of traffic rules on the day of its issuance and pay it without leaving your home, you can contact the Portal of State and Municipal Services of the Republic of Tatarstan in the section "Traffic police services" using the standard USIA account. The popularity of electronic information is growing: more than 85 thousand people have connected to the Portal.

The whole fleet of all urban public transport operates under the control of an automated dispatch control system in the GLONASS satellite system in an agreed uniform timetable. You can track the movement of public transport in real time in various applications and on the city's transport website. All public transport in the city has an automated fare collection system. A transport card is a single means of payment that provides travel by suburban railway transport, trams, trolleybuses, buses, and metro. Online bank cards are also accepted for payment.

Since 2015, a unified city parking space has been operating with differentiated payment for use through a mobile application and SMS and automatic control using technical means of photo and video recording. Since 2019, a car-sharing service for 400 vehicles has been operating in Kazan. In addition, this year in Kazan, a scooter-sharing system (stationless rental of electric scooters) for 1600 u. was launched, which operates thriough the Whoosh, Urent and Greenbee mobile applications. To date, there are about 400 parking spots for scooters and bicycles. The length of the equipped cycle paths is more than 12 km.

5.2 Driverless Technologies

Driverless technology is a kind of pinnacle in the development of intelligent transport systems and related infrastructure components. Companies that develop these technologies are predicting that such cars will become a reality on city streets over the next three to four years.

The Republic of Tatarstan supports the global trend of working on driverless transport systems. In August 2018, Yandex, on the basis of an agreement with the Government of the Republic of Tatarstan, launched self-driving taxi in the city of Innopolis. Innopolis became the first city where you can use a driverless car for everyday travel.

KAMAZ presented its first unmanned bus. The presentation of the vehicle called "Shuttle" took place at the Moscow International Motor Show in 2016. The test drive of the self-driving bus for 12 passengers took place at the FIFA World Cup in Kazan. During the entire match, this vehicle shuttled between the stadium and the fan zone. The first test drives of the self-driving vehicle KAMAZ-4308 have begun on the KAMAZ factory premises. A truck without a driver will master the logistics of delivering cabs from a press-frame plant to an automobile plant. The project for the transportation of components using robotic vehicles was named "Odyssey". So far, we are talking about logistics operations on the roads within the perimeter of the industrial site of KAMAZ. In general, such machines can be used in any industry where shuttle transportation is required along predetermined routes.

6 Unified Transport System

The increase in the number of applications, the fragmentation of systems, as well as the emerging need on the part of the consumer for safety, "green" environment and smart mobility makes one think about the integration of all elements into a single digital space. Within the framework of the national project "Safe and High-Quality Roads", it is envisaged that the ITS management system at the regional level is based on a single management platform that will serve as the main integration platform.

As part of the development and implementation of a unified platform for transport management system, it is planned to obtain information from various information transport systems of the Republic of Tatarstan, mentioned in this paper. The data will be processed in a unified transport system management platform; a modular architecture is planned.

First of all, modules were implemented that collected data from the existing information systems of the Kazan city, provided visualization of target indicators of ITS efficiency: harmful emissions, traffic intensity, public transport performance, etc. The congestion of roads and road events was visualized on the map, and the automatic generation of analytical reports on the development of ITS was provided. The perimeter maintains a register of incidents and work requests with the ability to control the execution of orders.

At the second stage, the work will focus on creating tools for dispatching ITS for emergency situations, which implies the performance of functions for displaying events and messages. It is supposed to calculate and display indicators by the number and types of incidents.

In the transport forecasting and modeling module, forecast values for passenger and freight traffic will be determined. Also, the developed models of the Situation Center related to the road transport complex, built on the basis of data from the Unified Transport System, will also be displayed. Also, a module will be created that is responsible for integration with information systems. It is planned to connect the information systems of the other big city of Republic - Naberezhnye Chelny.

The modules of the third stage can be referred to the task of transport planning automation. Development is planned:

1) a coordinated traffic control module for automating traffic control of traffic flows
2) a configuration module for scenario traffic control plans, and
3) a module of an electronic integrated traffic management scheme.

The modules of the fourth stage are aimed at ensuring the future of digital mobility - the movement of unmanned vehicles and vehicles with a high level of automation. An analysis of toll billing and services is also planned. Additionally, it is planned to pay special attention to the information security.

The unified system planned for completion in 2023 will have to be a digital platform for organizing the interconnected functioning of all ITS subsystems and services of the Republic of Tatarstan road network as a whole.

The trend continues to structure and unite all elements and systems in a unified space.

7 Conclusions

Thus, it should be noted that the development of intelligent transport systems plays an important role in ensuring road safety in the Republic of Tatarstan. To create comfortable, stable and safe conditions on public roads for the population, today it is important to develop and introduce a single digital space.

It is also relevant to combine all disparate elements of the road traffic infrastructure into a single intelligent transport system. To ensure this, work is underway in various directions: technical, organizational and scientific.

References

Atteih, A.S., AlQahtani, S.A., Nazmy, A.: Emergency Management Information System: Case Study, GM, Unicom for Communication Technologies (2010)

Brodsky, G., Aivazov, A.: Automated traffic control in an urban environment. World Roads **26**, 2–3 (2007)

Dagaeva, M., Garaeva, A., Anikin, I., Makhmutova, A., Minnikhanov, R.: Big spatio-temporal data mining for emergency management information systems. IET Intell. Transp. Syst. **13**(11), 1649–1657 (2019)

Decision No. 585/2014/EU of the European Parliament and of the Council of 15 May 2014 on the deployment of the interoperable EU-wide eCall service Text with EEA relevance (2014)

Gómez-García, J.A., Moro-Velázquez, L., Godino-Llorente, J.I.: On the design of automatic voice condition analysis systems. Part II: Review of speaker recognition techniques and study on the effects of different variability factors. Biomed. Sign. Process. Control **48**, 128–143 (2019)

Kamath, U., Liu, J., Whitaker, J.: Deep Learning for NLP and Speech Recognition, pp. 1–621. Springer, Cham (2019).https://doi.org/10.1007/978-3-030-14596-5

Minnikhanov, R., Dagaeva, M., Farrakhov, I.: Piezoelectric and tensometric sensors in WIM control in the Republic of Tatarstan, "Vestnik", no. 4, pp. 127–134 (2017)

Panova, I., Ivankin, I., Tsiulin, S.: Gross weight control of vehicles in motion and its legislation appliance in Russian Federation. Logistics Managing Logistic Chains **73**, 65–84 (2016)

National State Standard 54620. Global Navigation Satellite System. Accident emergency response system. Automotive emergency call system. General technical requirements (2011)

Riveiro, M., Lebram, M., Elmer, M.: Anomaly detection for road traffic: a visual analytics framework. IEEE Trans. Intell. Transp. Syst. **18**(8), 2260–2270 (2017)

Schapire, R.E., Singer, Y.: Boostexter: A boosting-based system for text categorization. Mach. Learn. **39**(2–3), 135–168 (2000)

Road Pricing and Tolling

Assessment of the Potential Implementation of High-Occupancy Toll Lanes on the Major Freeways in the United Arab Emirates

Ahmed Shabib[1], Mahmoud Khalil[2], and Muamer Abuzwidah[2(✉)]

[1] Sharjah Research Institute of Sciences and Engineering, Sharjah, UAE
[2] University of Sharjah, Sharjah, UAE
mabuzwidah@sharjah.ac.ae

Abstract. The High Occupancy Toll (HOT) lanes would provide a great opportunity for enhancing the transportation sector through potential reduction of roadway congestions, reducing the overall sector costs, and mitigating the carbon footprint. Based on the literature review, HOT lanes system is expected to provide a great opportunity to enhance the transportation sector through potential reduction of roadway congestions, reducing the overall sector costs, and mitigating the carbon footprint resulting from this sector. In this study, a micro-simulation was conducted using VISSIM software on a selected segment of a major freeway. The results indicated that the HOT lanes could reduce the delay time and overall travel time by around 19.02 and 23.44%, respectively, while the average travel speed could increase by up to 24.98%. Additionally, it was found that HOT lanes can significantly reduce the fuel consumption by around 12.7% and that resulted in air emissions reduction by 19.9 and 15.9% for the CO and the NOx, respectively.

Keywords: HOT lanes · Micro-Simulation · PTV VISSIM · Roadway performance · Carbon footprint

1 Introduction

The world population reached around 7.5 b in 2017 and is projected to increase by another one billion before 2030 (United Nations 2017). This population growth, associated with an increase in living standards and urbanization levels, impose multiple implications on the transportation sector in the shape of increased traffic volumes, congestions, and accidents. Overall, around 1.35 m fatalities and 50 m injuries are recorded annually as a result of traffic accidents, in which about 93% of those fatalities occur in the low- and middle- income countries despite having only around 60% of vehicles worldwide (World Health Organization (WHO)). Additionally, road traffic injuries are the leading cause of death for children and young adults aged between 5 and 29 years old and are the eighth leading cause of death for all age groups (surpassing HIV and tuberculosis). In addition to the fatalities, road accidents cause significant financial losses to individuals and their families, as well the countries as whole. For instance, Individuals may require some rehabilitation time following an injury caused by a traffic accident, while countries suffer financially from accidents due the required costs for repairment of affected roads. According to the World Health Organization, road accidents cost around 520 b USD globally, which represents around 1–3% of the global gross domestic product.

Multiple factors contributing to road accident have been reported throughout the literature over the years. Human factors have been identified as the main causes of road accidents, followed by vehicle and equipment as well as environmental factors (World Health Organization, 2004). Despite the fact that they are complex, vehicle accidents broadly depend on the characteristics of drivers (Rolison et al. 2018) and (Shabib et al. 2020). In the case of young drivers, it was found that the skill level, inexperience, and risk-taking behaviors (excessive speed, reckless driving, drugs, and alcohol consumption, as well as traffic violations) were the main factors being road collisions compared to other group ages. Additionally, loss of control and failure to detect other vehicles were reported as primary causes of road collisions for young drivers (Elbaz et al. 2020); (Braitman et al. 2008). However, it should be noted that those factors vary among genders within the same group, for instance, young male drivers were reported to be more likely involved in road collisions due to risk-taking behaviors compared to young female drivers (Curry et al. 2012); (Elawady et al. 2020). On the contrary, compared to young drivers, road collisions involving older drivers typically involve human errors at significant road segments, i.e., intersections and sharp turns (Langford and Koppel 2006); (Alhmoudi et al. 2017). It was also found that the failure to yield right of way, failure to site objects properly, failure to comply with signs and signals efficiently, as well as improper turns and lane changes were commonly reported in road accidents involving older drivers (McGwin, Jr and Brown 1999); (Najah et al. 2020). Those critical failures and errors while driving are mainly due to age-related features, such as decline in visual, mobility functioning, and perceptive abilities, in addition to other medical conditions such as heart diseases and strokes (Rolison et al. 2018).

As a result of the significant implications resulting from road accidents, in addition to the difficulty of managing the complex human factors leading to the majority of those collisions, highway safety has been prioritized in the agendas of most governments worldwide. For instance, the 2030 Agenda for Sustainable Development have set a target to half the number of fatalities and injuries resulting from road collisions by 2020 (World Health Organization). Despite the unlikeliness of meeting the target, significant efforts have been applied which led to sustaining the rate of fatalities over recent years, which is considered as a significant success compared to the no-action (conventional) situation. Additionally, significant enhancement has been achieved in the highway safety sectors of multiple countries over recent years. In 2018, around 25,100 fatalities from road accidents were reported from the 28 European Union (EU) members, which represents around 21% decrease in fatalities compared to 2010 (European Commision 2019). Moreover, in the United States, it was reported that the number of road fatalities in 2017 was around 130.1 per each million vehicles, which is a significant decrease compared to the 177.1 fatalities per 1 m vehicles reported in 2005 (OCED 2017).

Those enhancements in reducing the fatalities and injuries from road accidents, as well as reducing the number of accidents itself, were due to various factors including better education, stricter enforcements, and mainly the significant development of the transportation network components and the infrastructure itself. For instance, in addition to improvements to mobility and productivity, intelligent transportation systems (ITS) improve the overall transportation safety as a result of integrating advanced communication technologies within the transportation infrastructure and vehicles (Federal Highway Administration,). Advanced vehicle technologies include mainly automated vehicles in which critical aspects of safety-control functions occur without the need for driver input and connected vehicles where connected safety applications will provide drivers with complete awareness to hazards that are commonly unforeseen. In addition to vehicular technologies, advanced technologies in the infrastructure aid the safety enhancement of the transportation sector, such as the implementation of high- occupancy toll (HOT) lanes. HOT lanes are a type of high-occupancy vehicle (HOV) lanes (lanes where only vehicles with specific criteria related to their occupancy are allowed) but with the allowance for single-occupant vehicles (SOV) and all other vehicles that do not meet the HOV criteria to utilize those lanes with a pre-specified toll (fee) [13]. This toll is based on variable pricing in order to maintain a reliable performance of those HOT lanes at all times, i.e., maintaining the travel speed on the HOT lanes as a minimum of the posted speed limit. HOT lanes have been proven to be more efficient than traditional HOV lanes as they encourage carpooling and other transit alternatives while offering vehicles that do not meet the HOT occupancy requirements another option by means of toll payment (Federal Highway Administration). Additionally, the adjacent general-purpose lanes also benefit from the implementation of HOT lanes as a result of the reallocation of the vehicles in the corridor. However, the safety of the HOT lanes requires adequate design in order to ensure that the implementation of those lanes would reduce road collisions and not lead to deteriorated conditions. Most importantly, the implementation of those HOT lanes provides a significant opportunity for meeting the sustainable development plans set in the transportation sectors for reducing greenhouse gas (GHG) emissions. The transportation sector contributes with nearly one third of the global anthropogenic GHG emissions (Barth et al. 2015). It was also projected that the global carbon emissions due to transportation would increase by 112% between 2000 and 2030 (Desertot et al. 2013).

Due to the multiple promising safety, economical, and environmental benefits of HOT lanes, multiple governments have initiated various projects constituting the implementation of those lanes. Additionally, the implementation of such renovation in the transportation infrastructure could benefit various countries undergoing rapid increase in living standards and urbanization, such as the United Arab Emirates (UAE). In 2018, around 2,500 total road accidents were reported in on the UAE emirates, Dubai, with around 1,800 injuries and 140 fatalities (Government of Dubai 2018). Due to the number of road collisions, the UAE has initiated a road safety management plan that aims at decreasing road fatalities by around 3% per 100,000 inhabitants by 2021. In addition to roadway safety, according to the Roads and Transport Authority (RTA) in Dubai, the total carbon emissions from the transportation sector were 831,983 tCO2e in 2016, which is estimated to reach 957,222 tCO2e by 2036 (Dubai Roads and Transport Authority,2017). Therefore, HOT lanes would provide a great opportunity for the UAE for enhancing the transportation sector, acquiring additional revenues and reducing the overall sector costs, as well as mitigating the carbon footprint resulting from the transportation sector.

However, the main challenge confronting the implementation of HOT lanes in the UAE is public perception. As previously mentioned, in order for HOT lanes to maintain reliable performance, the paid toll must be variable and allowed to reach big amounts. However, the increase of the required toll might decrease the public perception and utilization of the HOT lanes, leading to continuous and maybe increased congestions. Therefore, the purpose of the present paper is to obtain a preliminary visualization of the potential outcome of implementing HOT Lanes in one of the UAE's main Freeways-Sheikh Mohammed Bin Zayed Road (E311). A well-derived literature review of previously conducted studies evaluating the performance and public perception of HOT lanes is provided. Following, the results of a well-engineered field survey constructed and distributed to a wide spectrum of drivers in the UAE were analyzed in order to obtain a deeper insight regarding public perception to HOT lanes in the UAE. Following, a sample network including HOT lanes is simulated for the E311 road using the PTV VISSIM. Finally, the paper concludes with critical remarks for decision- makers in the UAE.

2 Literature Review

HOT lanes provide drivers the possibility to gain access to alternative road facilities with superior services but with a charged fee (Burris and Stockton 2004). Those lanes are separated lanes within a highway, hence can ease the congestion on the adjacent general-purpose lanes (Konishi and Mun 2010). Additionally, the revenues from implementing HOT lanes charged fees can be later used in the maintenance of those lanes or other road segments. However, despite the benefits of HOT lanes suggested through the literature, the public perception to such facilities remains a main challenge.

2.1 Simulation and Analysis

Few studies have been conducted throughout the literature with a purpose of evaluating current HOT lanes, as well as simulating and analyzing potential implementations of HOT lanes. Those studies are key for obtaining an insight regarding the most suitable configuration for HOT lanes components in order to enhance their performance from multiple perspectives, such as highway safety, environmental benefits, and financial Feasibility. A recent study conducted safety, operational, and design analyses for managed toll (HOT) lanes based on a 9-mile corridor of HOT and general-purpose lanes in Miami-Dade County, Florida (Saad et al. 2019). PTV VISSIM 11 was utilized to replicate the network in a microsimulation environment taking into consideration traffic data, geometric design, and driving behavior. Additionally, several safety measures of effectiveness (i.e., speed standard deviation, time-to-collision, and conflict rate) and operational performance measures (i.e., level of service, average speed, average delay) were analyzed using statistical models. After comparing multiple safety measures, it was found that HOT lanes were safer than the general-purpose lanes since they had higher time-to-collision, higher post-encroachment-time, and lower maximum deceleration. Also, analysis of conflicts proposed that one accessibility level is the safest option in a 9-mile corridor, while analysis of safety measures proposed that a weaving length between 1,000 feet and 1,400 feet per lane change should be considered. Similarly, (Yuan et al. 2019) investigated the safety effects of weaving length, traffic conditions, and driver characteristics on the mandatory lane-change behavior of commuters in managed lanes, including HOT lanes. The investigation constituted three weaving length per lane change values, two traffic volumes (peak and off-peak), as well as speed harmonization and non-speed harmonization. 54 drivers were selected for running a driving simulator test in order to obtain more accurate results. The results showed that a weaving length of 1,000 feet for the entrance segment is recommended if space is limited, while 1,400 feet is the optimum. As for the exit segment, 1,000 feet weaving length was recommended, while a 1,400 length was found to be more dangerous compared to 600- and 1,000-feet lengths. In another study, (Toledo and Sharif 2019) evaluated the importance of information provided to drivers through variable message signs on their choices regarding the utilization of HOT or general- purpose lanes. A mixed-effects regression model was developed in order to predict the choice of drivers whether to utilize the HOT lanes or not based on expected travel times. It was found that the efficiency of information provided to drivers significantly affects their expected travel times, hence affecting their choice of travel lanes. More precisely, in the presence of information, drivers use the toll rate as an indication for expected travel time, however, those toll rates are used to keep the lanes flowing regardless of the traffic conditions on the general-purpose lanes.

Moreover, (Ma et al. 2018) investigated the presence of time-varying interdependencies between HOT and general-purpose lanes volumes as well as the dynamic toll rate. The analysis was conducted based on empirical data obtained from loop detectors and toll logs on the Washington State Route 167 and then using a time- varying parameter vector autoregressive model. Analysis results confirmed the time-varying interdependencies within the investigated parameters. For morning period traffic, the response of toll rates to the HOT lane demand is positive and reaches its peak between 6 and 7 AM. Additionally, it was found that an increased usage of general-purpose lanes will lead to same patterns for the HOT lanes, however, the opposite is not necessarily true. (Sajjadi and Kondyli 2017) conducted an operational analysis of two HOT lane segments located in South Florida. Macroscopic capacity analysis and microscopic calibration of two sites, namely one-lane and two-lane segments separated by flexible pylons (FPs), were conducted using VISSIM microsimulation. Results indicated that the percent drop in capacity for the case of one-lane FP site is 7.6% while the flow did not substantially change after the breakdown in the case of two-lane FP site. The Wiedemann car-following parameters (CC0 = 3.9 ft, CC1 = 1.9 s, CC2 = 26.25 ft, CC4 = −0.35, and CC5 = 0.35) provided the best fit for the one-lane FP site, while the combination (CC0 = 4.92 ft, CC1 = 1.9 s, CC2 = 39.37 ft, CC4 = −0.7, and CC5 = 0.7) parameters is recommended for the two-lane FP site. Furthermore, (Schultz et al. 2016) conducted an analysis aiming at identifying various improvements to average speeds in the HOT lanes in Utah, United States to 55 mph. The study included a thorough investigation of the current utilization of both HOT and general-purpose lanes by user (vehicle) type. Additionally, violation data were examined to determine violator reduction rates methods that will increase the flow speed in the HOT lanes. Multiple scenarios including HOT lanes education campaigns, increase enforcements, and increased JOT lanes peak toll rates. It was found that an optimum combination of all three scenarios would cause a significant increase of 9.6 mph in the average HOT lanes flow speed.

In addition to conventional evaluation frameworks for HOT lanes, (Boyles et al. 2015) evaluated an improved framework in which departure time choice is incorporated alongside stochasticity in demand. Two driver classes were taken into consideration: strategic drivers that time their departures to minimize generalized costs, and captive drivers that have random departure times, thus opting to choose the general-purpose lanes. A traffic flow model was developed containing two bottlenecks and comparing four toll algorithms. Following adequate analysis, it was found that incorporating departure time choices would enhance the choice of any tolled lane configurations compared to the un-tolled setup. (Toledo et al. 2015) conducted a real-time simulation-based control framework that dynamically determines the varying toll rates. The purpose of the study was to optimize the objectives of operators under various constraints, such as smooth toll rate changes and maintain adequate levels of service on HOT lanes. A macroscopic simulation model was utilized to predict flow conditions and predict travel time of using HOT lanes. Additionally, multiple models were incorporated in order to predict the vehicle arrival process upstream of HOT lanes and the choice of drivers whether to use those lanes or not. The whole process was then embedded within an optimization algorithm that determines the optimized toll rate based on a given function. A case study implementation of the developed algorithm showed promising potential for further useful HOT lanes settings. Similarly, (Jang et al. 2014) proposed a methodology where toll prices for HOT lanes are dynamically determined based on changes in traffic conditions due to multiple parameters. Those parameters included expected delays, available capacity for toll-paying commuters and commuters' value of time. The proposed methodology included an objective function of reducing delays and maximizing revenues. Moreover, the methodology was tested on a 14-mile freeways segment in the San Francisco Bay Area, however, due to differences in pricing strategies, it was not feasible to validate the performance of the proposed methodology.

From an environmental perspective, the Virginia Department of Transportation, in cooperation with the Federal Highway Administration, studied the environmental consequences of improving the I-95 by constructing HOT lanes. The construction location extended approximately 46 miles and included converting two existing high occupancy vehicle (HOV) lanes to two HOT lanes. For purposes of the environmental analyses, computations for construction "footprint" impacts have been prepared assuming the entire median as the impact area, even though the entire median will not be impacted. The study found that implementing HOT lanes would be environmentally beneficial by means of increasing the average travel speed, hence reducing air emissions. Additionally, an air quality analysis showed that implementing such project would not result in any violations of the National Ambient Air Quality Standards for Ozone (O3), carbon monoxide (CO), or fine particulate matter (PM2.5). (Zhang et al. 2009) developed an HOT lane module using microscopic traffic simulation software, VISSIM 4.30, which was used for emulating and investigating the Washington State Route 167 HOT Lane Pilot Project and compared with HOV lanes implementation. It was found that average travel time was improved by 8–40% under different scenarios for HOT lanes, as well as improvements in the corridor flow speed's.

2.2 Public Perception

The optimum method used to obtain an insight regarding public perception to any project, policy, or development is conducting an extensive field survey involving a wide spectrum of respondents. In the current case of HOT lanes, and managed lanes in general, it is essential that the respondents of the survey vary in socio- economic and demographic characteristics, such as age, gender, nationality, and average income. A study was conducted in Atlanta, United States, in order to investigate the preferences and choices of commuters regarding HOT lanes and resulting carpooling benefits (Guensler et al. 2019). The investigation was related to the current I-85 HOT corridors in Atlanta and was based on a questionnaire-based survey distributed to around 12,000 households/commuters. From the distributed surveys, a total of 642 surveys were retrieved and around 300 were utilized for the study following data screening. The obtained data was then comprehensively analyzed in order to develop classification trees and logistic regression models that would explain HOT lanes and carpooling choices of the commuters. The developed models indicated that the preference of commuters to utilize HOT lanes with a toll or carpool was affected by various factors such as age, gender, average income, education level, and commute distance. More importantly, the study showed that implying HOT lanes would not always boost the choice of carpool formation, thus not reaching their maximum potential benefits. Another study was conducted in a Middle Eastern country in order to examine the feasibility of implementing HOT lanes and the behavior of drivers towards such implementation (Abulibdeh and Zaidan 2018) The main focus of the study was to develop a relationship between drivers potential WTP for HOT lanes to avoid congestions on major highways with various socio-economic and driver characteristics, as well as trip conditions. A questionnaire was developed and distributed to drivers with distinct characteristics, and around 6,000 responds were used in further data analysis. Three trip conditions, namely urgency, speed, and distance, were mainly utilized to determine the WTP of drivers for HOT lanes through the survey, as well as the influence of multiple characteristics. The statistical analysis showed that the WTP for HOT lanes typically decreased when trip conditions improve, with urgency having the greatest impact, followed by travel speed and distance. Also, it was found that the income was the most statistically correlated driver characteristic to the WTP for HOT lanes, followed by age, gender, vehicle ownership, and trip frequency. Moreover, an investigation of the public desirability for HOT lanes was conducted in the Greater Toronto Area through data collection via well-established surveys (Finkleman et al. 2011). A total of 4,000 surveys were distributed in five selected sample areas where respondents were asked about their WTP for HOT lanes in order to escape traffic congestions under eight distinct trip conditions. Preliminary analysis of the results indicated that there was a considerable public support for HOT lanes in the area. Additionally, in six of the eight investigated trip conditions, most of the respondents firmly opted for utilizing HOT lanes rather than enduring congestions. However, further analysis using the one-way analysis of variance (ANOVA) test found that the mean WTP for HOT lanes of respondents varied by trip condition. Furthermore, a study was conducted in California, United States, in order to assess the level of support from residents towards a range of charged transportation facilities, including HOT lanes (Dill and Weinstein 2007). The results of the study provided some degree of optimism for the implementation of HOT lanes based on the positive support of the respondents. Noticeably, the majority of the respondents reacted positively towards the idea of HOT lanes when their environmental benefits were stated clearly. Another similar survey was conducted in Texas, where the public perception of HOT lanes was investigated (Podgorski and Kockelman 2006). Due to the confronted outfalls in the transportation funding in Texas, as well as the needs for system improvements, the Texas Department of Transportation visualized HOT lanes as solution for the funding gap. However, investigating the public perception to such implementation was essential prior to employment in order to avoid any potential losses. A telephone-based survey was conducted with around 2,100 Texas residents statewide in order to ensure the variability of the sample investigated. The public opinions regarding the implementation of HOT lanes varied by region, where Austinites residents showed positive support to the idea, while residents of the Lower Rio Grande Valley were less supportive. Moreover, binomial, and multinomial logit models were estimated in order to assess the impacts of various demographic and travel characteristics on the opinion of respondents. Male older drivers that typically live more than 50 miles away from their workplace were the most supportive respondents to the implementation of HOT lanes.

3 Micro Simulation

In order to evaluate the potential of implementing HOT lanes in the UAE, micro-simulation via PTV VISSIM was conducted on a selected segment of the E311 freeway. The micro-simulation approach provides an acceptable prediction of the performance of HOT lanes once applied. The simulation is conducted for the base conditions, i.e., without HOT lanes, and with the implementation of HOT lanes in order to compare both freeway behavior and environmental performance.

3.1 Study Area

Figure 1 shows the location of the E311 freeway selected for implementation. The E311 spans around 140 km, connecting multiple emirates in the UAE. However, for simulation purposes, an 8-km segment span of the E311 has been selected in which two emirates are connected: Sharjah and Dubai. The selected segment was selected as it is considered a representative segment of the most highly congested freeways in the UAE.

3.2 Simulation Scenarios

As previously mentioned, micro-simulation runs have been conducted for both the base scenario and HOT lanes-scenario. The following characteristics and assumptions have been taken into consideration for the simulations:
- Weaving length before and after interchange entrances and exits of 1000 ft (300 m) have been selected based on literature studies (Saad et al. 2019).

- The selected segment constitutes multiple road characteristics, such as the number of lanes. The number of lanes vary within the segment at different locations based on interchange exists. However, most of the segment consists of 5-lanes per direction.
- Traffic volume of 3000-vehicle per lane have been assumed for the simulations, with an operational speed of 110 km/h and lane width of 3.6 m.

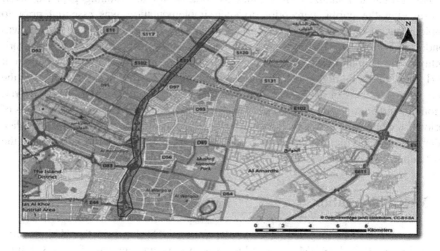

Fig. 1. Location of the E311 freeway in the UAE

- Minor roads consisting of two lanes with an operational speed of 100 km/h and 3.6 m lane width.
- The vehicle composition was assumed to consist of 15 and 85% of light and heavy vehicles, respectively.
- Each scenario, i.e., base and HOT lanes scenarios, have been simulated 10 times and average results have acquired in order to eliminate the randomness of the simulation and obtain more accurate results.

The computation of the dynamic toll rate during the simulation of HOT lanes via PTV VISSIM was based on the default function shown as following:

$$P_{(Toll)} = 1 - \frac{1}{1 + e^{1 - U_{(Toll)}}} \qquad (2)$$

Where P(Toll) is the dynamic toll rate, is a logit coefficient (assumed by default as 0.05), and U(Toll) is the managed lane utility, which can be calculated as following:

$$U_{Toll} = Costcoefficient \times TollRate + TimeCoefficient * TimeGain + BaseUtility \qquad (3)$$

Moreover, Fig. 2 shows the difference in geometry between the base and HOT lanes scenarios at a certain location of the selected freeway segment. The first scenario shows the base conditions of the segment where no HOT lanes are implemented, and all lanes are for general purposes. The second scenario where HOT lanes implemented shows the HOT lanes (highlighted in red) on the two left lanes of each direction that start following a 300m waving length following an interchange exit. It should be noted that the other locations highlighted in green, and red represent of the conflict points of the segment, such as the merging of minor roads with the freeway. The green areas represent the priority to the drivers on the main freeway before the merging points.

The roadway performance parameters include delay time, level of service (LOS), travel speed, and travel time. Table 1 shows the roadway performance results of 10 runs for each scenario using PTV VISSIM micro-simulation. It was found that the average delay time for both base and HOT scenarios was around 95.64 and 77.45 s, respectively. This translates to an approximate 19.02% reduction in delay time when HOT- lanes are implemented, leading to an improvement in the LOS of the freeway segment from LOS F to LOS E. Moreover, the travel speed in the base scenario was around 58.62 km/h on average, while the travel speed was found to be around 73.27 km/h when HOT lanes are implemented, representing a 24.98% increase in travel speed. Additionally, the average total travel time decreased by around 23.44% when HOT lanes are implemented, where the travel time in the base and HOT scenarios were 2124.34 and 1626.33 h, respectively.

3.3 Roadway Performance

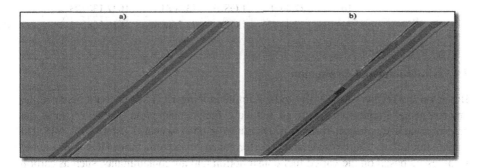

Fig. 2. Comparison in geometry between the base and HOT scenarios as implemented in PTV VISSIM

Those results show that the implementation of HOT lanes in the UAE could positively enhance the performance of freeways by decreasing congestions and increasing the average flow speed. Such enhancement would lead to increased roadway safety by decreasing significantly decreasing certain types of accidents, particularly vehicle rear-end collisions.

Table 1. Roadway performance results of base and HOT-Lanes scenarios roadway performance parameters

Simulation scenario	Run	Delay time	Level of service	Travel speed (km/h)	Travel time (h)
Base scenario	1	95.608	LOS F	58.611	2124.89859
	2	95.6073	LOS F	58.5488	2123.252195
	3	95.7439	LOS F	58.6935	2122.900345
	4	95.7346	LOS F	58.6122	2124.886657
	5	95.7412	LOS F	58.7082	2123.497466
	6	95.6262	LOS F	58.5648	2122.685907
	7	95.6057	LOS F	58.6074	2126.331135
	8	95.6082	LOS F	58.6007	2127.065585
	9	95.5055	LOS F	58.5971	2123.192821
	10	95.5982	LOS F	58.6767	2124.738776
	Average	95.63788	LOS F	58.62204	2124.344948
HOT lanes scenario	1	77.4345	LOS E	73.26375	1626.538955
	2	77.4085	LOS E	73.3008	1626.318793
	3	77.3936	LOS E	73.2294	1626.002882
	4	77.4854	LOS E	73.1772	1625.503781
	5	77.5168	LOS E	73.2698	1626.155774
	6	77.4085	LOS E	73.1782	1625.929836
	7	77.5026	LOS E	73.3679	1622.205273
	8	77.4687	LOS E	73.2983	1630.412331
	9	77.4258	LOS E	73.2732	1627.084099
	10	77.4426	LOS E	73.3081	1627.130793
	Average	77.4487	LOS E	73.266665	1626.328252

3.4 Environmental Evaluation

Based on the results of multiple studies, implementing HOT lanes could enhance the environmental performance through reducing fuel consumption. Table 2 shows the environmental evaluation results of 10 runs for each scenario using PTV VISSIM micro-simulation. The environmental performance of implementing HOT lanes could be assessed based on the varied fuel consumption and air contaminants, such as carbon oxide (CO) and nitrogen oxides (NOx). Based on the simulation results, it was found that implementing HOT lanes would reduce fuel consumption by around 12.7% with around 5455.03 US liquid gallons of fuel consumed compared to 4761.71 US liquid gallons consumed in the base scenario. This reduction in fuel consumption is mainly attributed to the reduced delay and travel times as a result of increased average flow speed. Moreover, the CO and NOx emissions from the base scenario were around 381.22 and 74.19 kg respectively. In contrast, the CO and NOx emissions from implementing HOT lanes were around 305.33 and 62.35 kg, respectively, which can be translated to around 19.9 and 15.9% reductions compared to the base scenario. Furthermore, using the global warming potentials of air emissions provided by the

Intergovernmental Panel on Climate Change (IPCC), those reductions can be translated to around 452.32 kg CO2 eq. it should be noted that this reduction is a result of implementing HOT lanes on only an 8-km segment of one of the freeways in the UAE, thus, future implementing of HOT lanes on all freeways could significantly enhance the environmental performance of the transportation sector in the country.

Table 2. Environmental evaluation results of Base and HOT-lanes scenarios environmental evaluation parameters

Simulation scenario	Run	Fuel consumption (US liquid gallon)	CO (g)	NO_x (g)
Base scenario	1	5454.4174	381263.7983	74180.0812
	2	5455.322523	381187.4057	74152.24996
	3	5452.585359	381456.7043	74170.04047
	4	5451.536032	381084.9621	74155.633
	5	5458.002037	380813.3117	74132.87098
	6	5460.216318	381295.0431	74162.60581
	7	5452.588167	380817.8027	74152.30166
	8	5456.827978	381443.8491	74356.73075
	9	5453.804694	381313.194	74204.94307
	10	5454.989456	381493.9408	74207.07264
	Average	**5455.028996**	**381217.0012**	**74187.45295**
HOT lanes scenario	1	4760.402153	305277.1437	62365.63091
	2	4760.36726	304953.2216	62317.30932
	3	4767.166542	305706.9736	62306.98256
	4	4766.70678	305283.2717	62365.28068
	5	4767.03397	305783.2324	62324.508
	6	4761.30843	305036.5589	62300.68882
	7	4760.285958	305258.3847	62407.67602
	8	4755.296248	305204.5379	62315.56671
	9	4759.913132	305619.1564	62360.94039
	10	4758.643256	305163.7458	62399.28931
	Average	**4761.712373**	**305328.6227**	**62346.38727**

4 Future Recommendations

Based on the literature, the High Occupancy Toll (HOT) lanes could offer a great opportunity for the UAE in order to mitigate their environmental problems as well as enhancing the performance of the transportation sector. However, the micro-simulation in the present study was conducted only for small segment of UAE's freeways. Therefore, despite the promising results of the study, further detailed simulations should be conducted for a larger study area. In addition, further investigation of the optimum HOT lanes features, such as the toll rate function and the weaving lengths around interchanges should be carried out. The current HOT lanes projects established worldwide must be thoroughly evaluated as they represent a great resource for better implementation of HOT lanes. Additionally, results of a preliminary analysis regarding public perception to HOT lanes showed that there was a big portion of the public that had positive feedback regarding such implementation. However, the toll rate they are willing to pay is relatively small, which indicates that they are not completely interpreting how the HOT lanes can be operated efficiently. Therefore, prior to potential implementation of HOT lanes projects in the UAE, a wider range of the public opinion must be obtained and analyzed in order to depict an idea regarding the best employment of HOT lanes.

5 Conclusion

The High Occupancy Toll (HOT) lanes would provide a great opportunity for enhancing the transportation sector through potential reduction of roadway congestions, reducing the overall sector costs, and mitigating the carbon footprint. The aim of this study is to explore the potential of implementing HOT lanes on the local freeways using micro-simulation approach in the VISSIM software. Multiple parameters for simulating HOT lanes have been acquired from the literature, in addition to current roadway characteristics. The simulation was conducted on a selected segment of one of the UAE's major freeways, the E311. However, public perception to the implementation of HOT lanes in the UAE was acquired through a questionnaire-based survey. The results of the micro-simulation showed that HOT lanes would reduce delay time and travel time by around 19.02 and 23.44%, respectively, while increasing the average travel speed would increase by around 24.98% compared to conventional conditions. Additionally, it was found that HOT lanes would significantly decrease air emissions such as CO and NOx by 19.9 and 15.9% respectively, as a result of reducing fuel consumption by around 12.7%. The results of the study aim to support decision makers in the UAE with future planning for the management and enhancement of the transportation sector. Thus, those results indicate that future implementations of HOT lanes can be an excellent sustainable solution to enhance the performance of the freeways.

References

Abulibdeh, A., Zaidan, E.: Analysis of factors affecting willingness to pay for high-occupancy-toll lanes: Results from stated-preference survey of travelers. J. Transp. Geogr. **66**, 91–105 (2018)

Curry, A.E., Mirman, J.H., Kallan, M.J., Winston, F.K., Durbin, D.R.: Peer passengers: how do they affect teen crashes? J. Adolesc. Health **50**(6), 588–594 (2012)

Alhmoudi, M., Abuzwidah, M., Hamad, K.: Road users' opinion about pedestrian safety in the emirate of Sharjah, UAE- survey results. In: 2017 MATEC Web of Conferences, vol. 120, p. 07006 (2017)

Elawady, A., Khetrish, A., Abuzwidah, M.: Driver behaviors' impacts on traffic safety at the intersections. In: 2020 Advances in Science and Engineering Technology International Conferences, ASET 2020 – 9118291 (2020)

Elbaz, Y., Naeem, M., Abuzwidah, M., Barakat, S.: Effect of drowsiness on driver performance and traffic safety. In: 2020 Advances in Science and Engineering Technology International Conferences, ASET 2020 – 9118242 (2020)

European Commission. Road safety statistics: what is behind the figures? (2019)

Schultz, G.G., Mineer, S.T., Hamblin, C.A.: An analysis of express lanes in Utah. Transp. Res. Procedia **15**, 561–572 (2016)

Mcgwin, G., Brown, D.B.: Characteristics of traffic crashes among young, middle-aged, and older drivers. Accid. Anal. Prev. **31**, 181–198 (1999)

Zhang, G., Ph, D., Yan, S., Wang, Y., Ph, D.: Simulation-based investigation on high-occupancy toll lane operations for Washington state route 167. J. Transp. Eng. **135**(October), 677–686 (2009)

Konishi, H., Mun, S.: Regional science and urban economics carpooling and congestion pricing: HOV and HOT lanes. Reg. Sci. Urban Econ. **40**(4), 173–186 (2010)

Dill, J., Weinstein, A.: How to pay for transportation? a survey of public preferences in California. Transp. Policy **14**, 346–356 (2007)

Finkleman, J.: The HOT solution: an examination of the desirability for high-occupancy/toll (HOT) lanes in the greater Toronto area by (2010)

Rolison, J.J., Regev, S., Moutari, S., Feeney, A.: What are the factors that contribute to road accidents? an assessment of law enforcement views, ordinary drivers' opinions, and road accident records. Accid. Anal. Prev. **115**(March), 11–24 (2018)

Langford, J., Koppel, S.: Epidemiology of older driver crashes – identifying older driver risk factors and exposure patterns. Transp. Res. Part F **9**, 309–321 (2006)

Yuan, J., Abdel-aty, M., Cai, Q., Lee, J.: Investigating drivers' mandatory lane change behavior on the weaving section of freeway with managed lanes: a driving simulator study. Transp. Res. Part F: Psychol. Behav. **62**, 11–32 (2019)

Braitman, K.A., Kirley, B.B., Mccartt, A.T., Chaudhary, N.K.: Crashes of novice teenage drivers: characteristics and contributing factors. J. Safety Res. **39**, 47–54 (2008)

Jang, K., Chung, K., Yeo, H.: A dynamic pricing strategy for high occupancy toll lanes. Transp. Res. Part A **67**, 69–80 (2014)

Podgorski, K.V., Kockelman, K.M.: Public perceptions of toll roads: a survey of the Texas perspective. Transp. Res. 1–20 (2005)

Desertot, M., Lecomte, S., Gransart, C., Delot, T.: Intelligent Transportation Systems. Elsevier Inc. (2013)

Barth, M.J., Wu, G., Boriboonsomsin, K.: Intelligent transportation systems and greenhouse gas reductions. Curr. Sustain./Renew. Energy Rep. **2**(3), 90–97 (2015). https://doi.org/10.1007/s40518-015-0032-y

Saad, M.: Safety, Operational, and Design Analyses of Managed Toll and Connected Vehicles' Lanes (2019)

Burris, M.W., Stockton, B.R.: HOT lanes in Houston — six years of experience. J. Public Transp. **7**(3), 1–21 (2004)

Najah, A., Abuzwidah, M., Khalil, D.: The impact of the rear seat belt use on traffic safety in the UAE. In: 2020 Advances in Science and Engineering Technology International Conferences, ASET 2020 – 9118388 (2020)

OCED, Road Accidents (2017). https://data.oecd.org/transport/road-accidents.htm. Accessed 28 Oct 2019

Guensler, R., et al.: Factors affecting Atlanta commuters' high occupancy toll lane and carpool choices. Int. J. Sustain. Transp. 1–12 (2019)

RTA: Sustainability Report (GRI Standards Report), Dubai (2017)

Boyles, S.D., Gardner, L.M., Bar-gera, H.: Incorporating departure time choice into high-occupancy/toll (HOT) algorithm evaluation. Transp. Res. Procedia **9**, 90–105 (2015)

Sajjadi, S., Kondyli, A.: Science direct macroscopic and microscopic analyses of managed lanes on freeway facilities in South Florida. J. Traffic Transp. Eng. (Engl. Ed.) **4**(1), 61–70 (2017)

Shabib, A., Khalil, M., Abuzwidah, M., Barakat, S.: Public perception and willingness to pay for high-occupancy toll (HOT) lanes in the United Arab Emirates: questionnaire-based survey. In: 2020 Advances in Science and Engineering Technology International Conferences, ASET 2020 – 9118315 (2020)

Toledo, T., Sharif, S.: The effect of information on drivers' toll lane choices and travel times expectations. Transp. Res. Part F: Psychol. Behav. **62**, 149–159 (2019)

Toledo, T., Mansour, O., Haddad, J.: Simulation -based optimization of HOT lane tolls. Transp. Res. Procedia, **6**(June 2014), 189–197 (2015)

United Nations: World Population Prospects (2017)

Ma, X., Sun, S., Cathy, X., Ding, C., Chen, Z., Wang, Y.: A time-varying parameters vector auto-regression model to disentangle the time varying effects between drivers' responses and tolling on high occupancy toll facilities. Transp. Res. Part C **88**(January), 208–226 (2018)

Driver Behavior Strategies

Posted Road Speed Limits in Abu Dhabi: Are They Too High? Should They Have Been Raised? Evidence Based Answers

Francisco Daniel B. Albuquerque[✉]

United Arab Emirates University, Al Ain, Abu Dhabi, UAE
daniel@uaeu.ac.ae

Abstract. Road posted speed limits (PSL) in Abu Dhabi (AD) were raised in mid-2018. Before raising PSL by 20 kph, AD drivers were legally allowed to drive 20 kph above PSL. Thus, there was a speed buffer within which drivers would not be fined for speeding. While it has been stated (through public media) that the elimination of this speed buffer would increase safety, this paper challenges this view. This paper describes how the argument for increased safety through increased PSL quickly breaks down. The paper bases its arguments mainly on basic physics, as well as local research data and findings. More specifically, this paper not only shows how existing road design may not safely accommodate raised PSLs (or even previously lower PSLs), but also how fatal road injuries are more likely to occur on AD roads with higher design speeds. The paper also describes how increased PSL may increase the potential for liability and greenhouse gas emissions. This paper concludes that while speed buffers should have been removed, PSL should not have been increased since higher PSL may negatively affect road safety while decreasing travel times only marginally. That is, there is not much to gain from raising PSLs, but there is potentially a lot to lose. Thus, the author urges policy makers to make decisions based on science and solid evidence. This is relevant as decisions such as increasing PSL on public roads may have a significant negative impact on public health through increased injury/mortality rates and pollution levels.

Keywords: Road Safety/Design · Speed management · Crashes · Injuries

1 Introduction

The Gulf Cooperation Council (GCC) region has been impacted by an excessive number of road crashes and injuries (Al Kuttab and Abdullah 2017; Al Ramahi 2017; Al Shaibany 2017; Ansari et al. 2000; Rohrer 2016; Scott 2014; Toumi 2017a, 2017b). The United Arab Emirates (UAE), in particular, has shown declining road-injury and fatality records; yet, road crashes continue to take an economic toll on the country (Shahbandari 2015).

Unfortunately, the number of detailed, publicly available vehicle crash studies conducted in the GCC region has historically been scarce, making it difficult for researchers to identify road safety problems and assess safety countermeasures. This lack of publicly available, road-safety-related scientific studies may have opened the door to road-safety-related claims with no scientific backing. For example, 3 years ago, in an article published by The National newspaper (shown in Fig. 1), it was announced that road posted speed limits (PSLs) in Abu Dhabi were being raised in order to increase safety (Dajani 2018). In addition, in an article published by Gulf News, it was stated that "A comprehensive study was undertaken before deciding to implement the new system, which will help raise road safety standards" (Zaatari 2018). However, no scientific evidence has been shared demonstrating how safety would be increased by raising PSLs. It has been argued that the raising of PSLs was undertaken in order to remove the 20 kph-speed-limit buffer that was previously in place. That is, up to 2018, Abu Dhabi drivers were allowed to drive at 20 kph above PSLs before they were fined for speeding. Even though it is really not clear why this buffer ever existed in the first place, it has been said that it caused confusion among drivers in regards to which speed they could drive up to, and therefore, it posed a safety problem (Zacharias 2018).

Fig. 1. Article published by the national (Dajani 2018)

Scientific literature refutes the claim that higher PSLs can increase road safety. (Brubacher et al. 2018; Donnell et al. 2016; Farmer 2017, 2019; Haselton et al. 2002; Kockelman and CRA International, Inc 2006; Ledolter and Chan 1996; Nilsson 2004; Renski et al. 1999; Rock 1995; Tarko et al. 2019). In fact, even slight increases in PSLs have been shown to have a significant negative effect on safety (Shirazinejad et al. 2019). On the other hand, scientific literature backs the argument that lower PSLs improve safety (Pauw et al. 2014).

Hence, there is a need to revise the decision of raising PSLs in Abu Dhabi, as higher PSLs may be negatively affecting public health through increased injury/mortality rates and pollution levels.

2 Objectives

The objective of this paper is to challenge the view that higher PSLs can improve road safety in Abu Dhabi.

3 Methods

This paper exposes the risks involved with higher PSLs by raising ten opposing arguments. Some of these arguments are based on scientific research conducted locally, in the Emirate of Abu Dhabi.

4 Arguments Against Raised PSL

4.1 No Backing from High-School Physics

Based on high-school-level physics, it is known that if an object is to be moved, a force must be applied to it. The product of force and displacement is represented by work. Once work has been done, energy is transferred to the object, and the object will then move. This energy is called kinetic energy. As shown in Eq. 1, kinetic energy depends on object mass as well as on the speed of the object squared. Therefore, doubling the speed of an object results in quadrupling kinetic energy (Halliday et al. 2014).

$$E_k = \frac{m.v^2}{2} \tag{1}$$

Thus, even marginal increases in vehicle speed translates into significant increases in the amount of kinetic energy that must be absorbed by a vehicle collision. For example, assuming a car crashes into a fixed object coming to an abrupt stop, the kinetic energy would have to be transferred through heat, sound, and physical deformation. Once this energy transfer occurs, the vehicle would be stopped. The problem is that this process usually occurs within an extremely short period of time, causing decelerations which are not bearable by a human body, potentially resulting in serious organ injuries.

As such, there are two means through which one can make this same collision safer: i) spread the length of the collision over a longer period of time, or ii) lower vehicle speed (Haven 2000). Even though more deformable vehicles and the adoption of roadside safety hardware (e.g., barriers and crash cushions) can extend the period of time between the vehicle's initial impact and its coming to a complete stop (AASHTO 2009), there is just only so much this period of time can be extended by. In addition, even when this period of time is extended, if speeds are too high, this time expansion will not be long enough to save vehicle occupants from suffering seriously. Thus, one is left with the second option for more effective results in terms of reducing road crash injuries: lower vehicle speeds. This can be accomplished via the implementation of lower (not higher) PSLs along with effective speed enforcement measures.

The good news is that implementing lower PSLs is a low-cost safety measure. Indeed, the implementation cost associated with this measure may be negligible. On the other hand, installing roadside safety hardware or improving vehicle design may result in significant installation/maintenance/repair and engineering design costs, respectively.

4.2 Poorer Driver Cognitive Performance and More Expensive Road Design

Higher travel speeds make the task of driving more challenging, as drivers need to process/interpret information and react to it more quickly (Engström et al. 2005; Shinar 2007; Wu and Liu 2006). In addition, higher design speeds result in increases in design variables such as stopping sight, passing sight, underpass sight, and horizontal curve sight distances as well as horizontal curve radii. The increase in these design variables translates into flatter grades, longer curves, and wider right-of-ways, which, in turn, results in higher road project costs (Garber and Hoel 2020).

4.3 More Severe Crashes on Roads with Higher PSLs

There is a large amount of evidence in the scientific literature showing that road crash severity tends to increase on higher design speed roads or on roads with higher PSLs. Thus, this relationship is well established. Nevertheless, the author backs what has been long established by scientific research (i.e., that higher PSLs tend to be detrimental to road safety) with findings from two studies conducted in the Emirate of Abu Dhabi (Albuquerque and Awadalla 2019a; Awadalla and Albuquerque 2021). These studies investigated factors that significantly contributed to increased risk of severe and fatal injuries produced by road crashes that occurred in the Emirate of Abu Dhabi between years 2012 and 2017. These studies found that the likelihood of severe and fatal injuries were higher on roads with higher PSLs.

Table 1 shows the results from one (i.e., Awadalla and Albuquerque 2021) of these two studies. As can be seen from results shown in Table 1, the risk of a fatal injury to occur is 12.22 times higher on roads with PSLs between 80 and 100 kph than that on roads with PSLs no higher than 60 kph, as well as 36.62 times higher on roads with PSLs equal to or higher than 120 kph than that on roads with PSLs no higher than 60 kph. These are striking differences in fatality risk.

Table 1. Multivariate analysis results – all variables (Awadalla and Albuquerque 2021)

Variable	ANOVA (P-value)	Non-baseline category	Baseline category	Odds	P-value	Goodness of fit (P-value)
Time of day	0.00	14:00–21:59	22:00–5:59	1.88	0.00	0.46
		6:00–13:59		0.00	0.92	
Day of week	0.00	Weekend	Weekday	1.44	0.00	
Crash type	0.00	Run-over	Others	907.89	0.00	
Main reason for crash	0.00	Running red light	Others	12.08	0.00	
Posted speed limit (kph)	0.00	80–100	≤60	12.22	0.00	
		≥120		36.62	0.00	
Gender of driver at fault	0.01	Male	Female	1.59	0.01	
Nationality of driver at fault	0.00	Emirati	Others	1.91	0.00	

4.4 More Severe Single-Vehicle, Run-off-Road (SVROR) Crashes on Roads with Higher PSLs

There is a large amount of evidence in the scientific literature showing that road crash severity tends to increase on higher design speed roads or on roads with higher PSLs. Thus, this relationship is well established. Nevertheless, the author backs what has been long established by scientific research (i.e., that higher PSLs tends to be detrimental to road safety) with findings from a study conducted in the Emirate of Abu Dhabi. That is, a research study conducted recently investigated the impact of SVROR crashes on fatality risk (Albuquerque and Awadalla 2020a).

In this study, the authors developed a series of multivariate logistic regression models in order to assess the impact of collisions with different fixed objects on fatality risk, while controlling for important variables such as design speed and seatbelt usage. Models 1, 2, and 3 (shown in Table 2) compare the fatality risk between tree/pole and other hazard (other than barriers) crashes, tree/pole and guardrail crashes, as well as tree/pole and concrete barrier crashes, respectively. All three models control for design speed, which was classified into two categories: ≤80 and ≥100 kph. Due to a much larger number of cases in model 1, the variable seatbelt usage could remain as a second control factor. Results contained in Table 2 show that according to models 1, 2, and 3, the risk of fatal injury occurrence is 1.93, 3.3, and 3.9 times higher on roads with design speeds equal to 100 kph or higher as compared to roads with design speeds equal to 80 kph or lower, respectively.

Table 2. Results from past SVROR research conducted in Abu Dhabi (Albuquerque and Awadalla 2020a)

Model	Variable	Non-baseline category	Baseline category	Total # observations	# Non-baseline observations	Odds	P-value
1	Most harmful object struck	Tree	Others (No barrier)	380	77	1.04	0.92
		Pole			111	1.46	0.28
	Design speed	$\geq$ 100 kph	$\leq$ 80 kph		172	1.93	0.03
	Seatbelt	No	Yes		145	2.21	0.01
2	Most harmful object struck	Tree	Guardrail	299	77	3.1	0.02
		Pole			111	4.7	0.00
	Design speed	$\geq$ 100 kph	$\leq$ 80 kph		189	3.3	0.01
3	Most harmful object struck	Tree	Concrete barrier	239	77	1.1	0.81
		Pole			111	1.9	0.23
	Design speed	$\geq$ 100 kph	$\leq$ 80 kph		118	3.9	0.00

4.5 Roadside Design in AD not Ready for Higher PSLs

There is a large amount of roadside safety-/design-related studies available in the literature (AASHTO 2011), but none of these studies have assessed the level of compliance of citywide roadside design to state-of-the-art guidelines (AASHTO 2011). In the past two years, three studies have been conducted in the Emirate of Abu Dhabi.

The first study assessed the compliance level of the roadside design in the city of Al Ain (i.e., low-density environment) to a selected benchmark (Albuquerque and Awadalla 2019b). After visiting over 100 roadside crash locations and assessing the roadside design compliance of these locations relative to the selected benchmark, this study found that a staggering 80.17% of the SVROR injury crashes investigated occurred at locations where roadside design deviated from the benchmark (i.e., non-compliant). Lack of an adequate clear zone was the main cause of non-compliance. Most SVROR injury crash locations containing roadside design with deviations from the benchmark were located on roads with PSLs of 100 kph or higher. This Al Ain study concluded that significant revision of the existing roadside design not only in the area studied but throughout the UAE is recommended. The authors of this study proposed measures that may be useful in making roadside design in the area studied (i.e., Al Ain) better align with the benchmark requirements such as lowering PSLs. The authors explain that lower PSLs may significantly decrease clear zone requirements in terms of lateral width, while also reducing injury risk at a negligible implementation cost.

The second study assessed the compliance level of the roadside design in the city of Abu Dhabi (i.e., high-density environment) to a selected benchmark (Albuquerque and Awadalla 2020b). As in the Al Ain study, after visiting over 100 roadside crash locations and assessing the roadside design compliance of these locations relative to the selected benchmark, this study found that almost all of the SVROR injury crashes investigated occurred at locations where roadside design deviated from the benchmark (i.e., noncompliant). Lack of the minimum recommended clear zone (CZ) provision was the main cause of non-compliance, while 80% of all locations suffered from barrier misplacement. The authors of this Abu Dhabi city study concluded that "roadside design guidelines have been poorly implemented in the area studied, and findings indicate that more focus on proper, on-site design implementation is warranted".

The third study not only assessed the compliance level of the roadside design in the entire emirate of Abu Dhabi to a selected benchmark, but it also investigated whether noncompliant design as well as hazard and barrier lateral offsets resulted in higher fatality risk (Awadalla and Albuquerque n.d.). After visiting 1,070 roadside crash locations (over 200 of these locations were included in the first two studies) and assessing the roadside design compliance of these locations relative to the selected benchmark, this study found that two-thirds of the studied locations contained non-compliant design. Table 3 shows the results from the second part of this study in which whether roadside design compliance status (RDCS) or hazard lateral offsets had any significant impact on fatality risk was investigated. Design speed was controlled for in all of the models contained in Table 3. As can be seen from the very low p-values associated with the design speed variable, fatality risk statistically increases on roads with higher design speeds. Results shown in Table 3 also show that hazard lateral offsets larger than 12 m tended to lower fatality risk.

Table 3. Results from Past RDCS research conducted in Abu Dhabi (Awadalla and Albuquerque n.d.)

Model	Variable	Non-baseline category	Baseline category	Total # observations	# Non-baseline observations	Odds ratio	P-value
1 (All SVROR crashes)	RDCS	No	Yes	1026	705	1.3	0.24
	Design speed	≥ 100 kph	≤ 80 kph		575	1.8	0.01
	Rollover	Yes	No		380	1.6	0.02
	Seatbelt	No	Yes		325	1.9	0.00
2 (No collision-free events or curb crashes)	Hazard lateral offset	<6	>12	383	316	4.9	0.13
		6–12			45	9.7	0.04
	Design speed	≥ 100	≤ 80		164	2.7	0.00
3 (No collision-free events or curb crashes)	Traveling-Lane-to-Hazard lateral offset	<6	>12	376	206	1.5	0.45
		6.0–12.0			124	1.6	0.33
	Design speed	≥ 100 kph	≤ 80 kph		163	2.6	0.00

In sum, these three studies indicate that i) the majority of the road sections in Abu Dhabi do not contain compliant roadside design and ii) noncompliant roadside design (e.g., excessively narrow clear zone width) may increase fatality risk.

It is important to note that, in the three studies described above, compliance status was based on PSLs before the 20-kph increase that occurred in mid-2018 since data included in these studies included crashes that occurred between years 2013 and 2016. Therefore, it is expected that fatality risk may have increased after the 20-kph PSL increase while significant changes to roadside design have not taken place.

These findings are relevant in the context of what this paper intends to address, as they show that roadside design in Abu Dhabi was not ready for PSL increases. For example, Fig. 2 shows an article published by a newspaper in Abu Dhabi back in 2019. From this article, it can be learned that a vehicle left the roadway and hit an unshielded light pole located quite close to the road edge. This pole did not break away upon impact. A woman lost her three children in this crash: two brothers, aged 11 and 15, and their sister aged 13. The nanny was also killed. The mother survived. This article writes "Initial police investigations revealed that the mother did not have a driver's license and was speeding. Police said she swerved suddenly, lost control of the vehicle, and crashed into a lamp post in Al Falah area". While the mother's driving behaviour might have contributed to the severity of this crash, it is surprising that no attention is brought to the fact that there was a road design problem in the first place. That is, firstly, rigid light poles should never be in such close proximity to roads that have PSLs of 80 kph and above, as it is often the case of Al Falah roads. Secondly, if the pole was to stay where it was, it could have been made collapsible, so that it could break away upon impact, significantly reducing crash severity. Finally, if the pole was to stay where it was and the installation of a breakaway device, or of any other sort of energy absorbing mechanism, was not an option, this pole should have been shielded by a crash-tested roadside barrier (AASHTO 2011). If any of these three safety measures had been adopted, the lives of these three children and their nanny could have been saved despite the fact that the driver might have been driving in an unsafe manner. In other words, proper road design could have "forgiven" the driver for her mistake - a mistake which may haunt her for life, as she may believe she is the one to be blamed for the loss of her children. However, human beings make mistakes. This is natural. What she needs to know is that according to state-of-the-art roadside guidelines (AASHTO 2011), the road she was traveling on was not safe in the first place.

Fig. 2. Fatal light pole crash in AD killed four people

The real-life crash shown in Fig. 2 demonstrates how a flawed road design can affect public health. Unfortunately, according to findings from locally-conducted, previously-described studies (Albuquerque and Awadalla 2019b, 2020b; Awadalla and Albuquerque n.d.), flawed road design seems to be widespread in Abu Dhabi. For example, Fig. 3 shows a multitude of fixed-object hazards (i.e., palm trees and light poles) installed in close proximity of a 120-kph PSL road located in the city of Al Ain. In Fig. 4, while fixed-object hazards have been shielded by WB-beam guardrails, these barriers are not crash-tested for impact speeds of 160 kph (AASHTO 2011). As such, the safety outcome from collisions with these WB-beam guardrails involving impact speeds of 160 kph is unknown.

Fig. 3. Roadside design noncompliance in the city of Al Ain

Fig. 4. Excessively high PSLs not congruent with roadside barrier performance

4.6 Increased PSL Differentials

Another issue related to raised PSLs is the potential for increased PSL differentials. Figure 4 shows PSLs set for small vehicles and trucks at 160 kph and 80 kph, respectively. Three years ago, the PSL set for small vehicles on this road was 140 kph (though a 20-kph speed buffer existed which allowed drivers to drive 20 kph above PSL before they were fined).

The problem with such large PSL differential as shown in Fig. 4 is the potential gap between small vehicle and truck speeds is very large, as small vehicles are allowed to travel as much as twice as fast as trucks. This is relevant as previous studies have found that increased PSL differentials have detrimental effects on safety (Gates et al. 2016), even when PSL differentials are narrow (Tarko et al. 2019). On the other hand, some other studies have found no significant increase in crashes or injuries on roads under different speed limits for trucks and cars as compared to those on roads under uniform speed limits (Garber and Gadiraju 1991; Hall and Dickinson 1974; Harkey and Mera 1994; Idaho Transportation Department Planning Division 2000; Virginia Transportation Research Council 2004). However, speed differentials considered in these studies were extremely narrow when compared to that shown in Fig. 4.

In addition, travel speed differentials between small vehicles can be increased as PSLs are set too high, as shown in Fig. 4. That is, when PSLs are raised to very high speeds (e.g., 160 kph), some drivers may not desire or feel uncomfortable driving at those speeds, making it more likely for them to choose to drive at lower speeds (e.g., perhaps 120–130 kph). As a result, the difference in speeds between these slower moving drivers and the drivers moving at a higher PSL (i.e., 160kph) create a larger speed differential than speed differentials created on roads with lower PSLs (e.g., 140 kph). Thus, in sum, increased PSLs may not only increase speed differentials between small vehicles and trucks, but also among small vehicles.

4.7 Poorer Intersection Safety

Another issue related to raised PSLs is that of intersection safety. That is, Fig. 5 shows a signalized intersection located in the city of Al Ain. Three years ago, PSLs on all roads connected to this intersection were set at 80 kph (though a 20-kph speed buffer existed which allowed drivers to drive 20 kph above PSL before they were fined). Today, PSLs are set at 100 kph. In addition, at that time, instead of a signalized intersection, there was a roundabout at this location. This paper is about the impact of raised PSLs on safety and not about the impact of intersection type (i.e., signalized intersection versus roundabout) on safety. Thus, even though the author believes that safety at this intersection may have deteriorated dramatically due to the combined effect of change in intersection design (Elvik 2003, 2017) and an increase in PSL, this text will restrict itself to the discussion of the impact of raised PSL on safety.

Fig. 5. Signalized Intersections in the city of Al Ain: a. 24.230617304386936, 55.71961094444156 and b. 24.207914245747, 55.68680951698089

Intersection-related crashes have accounted for a significant portion of all road-related crashes and severe injuries worldwide. For instance, intersection crashes have accounted for one-quarter and one-half of all traffic-related fatalities and injuries, respectively, in the United States (Federal Highway Administration (FHWA) 2021). In Abu Dhabi city, running-red-light crashes alone have accounted for almost one-fifth of all traffic-related fatalities. Eighty-five percent of these crashes involved right-angle impacts (which tend to result in severe collisions), while the number of injuries and vehicles involved in these crashes were found to be 27 and 32%, respectively, higher as compared to those from all the crash types combined (Albuquerque 2015).

In addition, according to results shown in Table 1, the risk of a fatal injury to occur is 12 times higher on crashes which had "Running red light" as the main reason for crash as compared to other crashes.

Thus, the raising of PSLs on the roads connected to signalized intersections such as those shown in Fig. 5 significantly increases the potential for severe injuries and deaths resulting from intersection-related crashes, more specifically from running-red-light crashes.

In 2015, the author of this paper recommended the adoption of design options (e.g., roundabouts) with the potential to reduce traveling speeds (i.e., due to curved entry points and center island presence) at intersections in an attempt to improve intersection safety in Abu Dhabi.

4.8 Increased GHG Emissions

There is strong evidence that anthropogenic greenhouse gas (GHG) emissions are the main drivers behind climate change (Gao, Kovats et al. 2018b). Transportation is the fastest-growing major contributor to climate change, and it accounts for approximately 14% of the global GHG emissions (The World Bank 2011). The transportation sector is the second-largest energy consumer after the industrial sector (Gao et al. 2018a). Seventy-two percent of the total GHG emissions produced by the transport sector are road-transport related (Gao et al. 2018a; The World Bank 2011).

While the UAE's total (i.e., in absolute terms) GHG emissions due to road transport are significantly less than those of other countries like China and the United States (US), emissions normalized by national road mileage are alarmingly high. For example, the normalized road-transport-related GHG emissions in the Emirate of Abu Dhabi, the largest emirate in the UAE, are 9 times higher than those of China and 2.5 times higher than those of the US (Albuquerque et al. 2020).

Unfortunately, increased PSLs may only contribute to the already excessively high GHG emissions produced by road transport in AD as fuel consumption (and, therefore vehicle-related GHG emissions) increases in a non-linear fashion with vehicle operating speeds (Haworth and Symmons 2001).

4.9 Increased Potential for Liability

Traffic control devices are the means of communication between authorities and the driving public. PSLs, more specifically, convey the message related to allowable-traveling speeds which, in turn, should be speeds that can be safely accommodated by the road design in place. Therefore, authorities need to be aware that, once they install a PSL sign, they are basically telling the motoring public that it is safe for drivers to travel at the speeds indicated by those PSL signs. However, as discussed in Sect. 4.5, past research shows that this is not necessarily true in the case of Abu Dhabi, where roadside design cannot safely accommodate raised PSLs. In fact, roadside design in AD could not even safely accommodate PSLs before they were raised. Thus, instead of raising PSLs, they should have been lowered, given roadside design remained unchanged.

This discussion leads to the concept of liability. The Cambridge Dictionary describes liability as "the fact that someone is legally responsible for something". In the US, for example, the cost associated with traffic-related liability is very significant (Agent and Turner 2003; Turner et al. 2006). As a result, US Departments of Transportation tend to be very cautious about road-design-related flaws that may make them vulnerable to lawsuits and potential liability-related costs. However, in the UAE, it is not common for the public to hear about traffic-related liability. Nevertheless, as the legal system in the UAE evolves, the concept of traffic-related liability should be more of a concern and, as a result, more accountability for flawed road design and policy should be expected.

4.10 Marginal Mobility Improvement

Raising road PSLs by 20 kph may offer marginal mobility improvement. That is, for example, the traveling distance between Abu Dhabi and Al Ain (the two largest cities in the emirate of Abu Dhabi) is only approximately 130 km. Previously, PSL on highway E22, the main rural freeway connecting Abu Dhabi to Al Ain, was 140 kph. Assuming one would travel at 140 kph from Abu Dhabi all the way to Al Ain, the estimated travel time would be approximately 55 min. On the other hand, assuming one would travel at 160 kph from Abu Dhabi all the way to Al Ain, the estimated travel time would be approximately 49 min. Thus, traffic moving between the two largest cities in the Emirate of Abu Dhabi has little to gain in terms of travel time when PSL is raised by 20 kph. Of course, these travel times are based on uncongested conditions and, therefore, this marginal travel time reduction would likely be even more negligible as volume-to-road-capacity ratio rises.

5 Conclusions

This paper concludes that while speed buffers should have been removed, PSLs should not have been raised in Abu Dhabi. There was little to gain, in terms of mobility improvement (as discussed in Sect. 4.10), by raising PSLs by 20 kph in Abu Dhabi. Meanwhile, road project costs are increased (as discussed in Sect. 4.2), fatality risk is higher (as discussed in Sect. 4.3), roadside design issues became even more problematic (as discussed in Sect. 4.5), speed differential issues were magnified (as discussed in Sect. 4.6), intersection safety may have deteriorated (as discussed in Sect. 4.7), GHG emissions likely became even larger (as discussed in Sect. 4.8), and the potential for liability increased (as discussed in Sect. 4.9). In other words, there is little to gain from raising PSLs, but potentially a lot to lose.

Thus, the author urges for policy makers to make decisions based on science and solid evidence. This is relevant as decisions such as increasing PSLs on public roads may have a significant, negative impact on public health through increased injury/mortality rates and pollution levels.

Acknowledgements. The authors would like to thank the United Arab Emirates University for funding this research effort [grant number 31R202].

References

AASHTO: Manual for assessing safety hardware. Washington, DC, USA: American Association of State Highway and Transportation Officials (2009)
AASHTO: Roadside design guide (4th ed.). Washington, DC, USA: American Association of State Highway and Transportation Officials (2011)
Agent, K.R., Turner, D.S.: Roadway related tort liability and risk management: 5th Edition (No. KTC-03-01/TT(1)-02-1F). Kentucky Transportation Center Research Report (2003). https://uknowledge.uky.edu/ktc_researchreports/1522

Al Kuttab, J., Abdullah, A.: Two people die every day on UAE roads. Khaleej Times. News, 12 March 2017. https://www.khaleejtimes.com/nation/two-people-die-every-day-on-uae-roads

Al Ramahi, N.: Dubai Police: 525 road traffic deaths in 2017. The National News. News, 28 December 2017. https://www.thenationalnews.com/uae/dubai-police-525-road-traffic-deaths-in-2017-1.691319

Al Shaibany, S.: Oman begins traffic safety campaign as road deaths rise. The National News. News, 15 March 2017. https://www.thenationalnews.com/world/oman-begins-traffic-safety-campaign-as-road-deaths-rise-1.75680

Albuquerque, F.D.B., Maraqa, M.A., Chowdhury, R., Mauga, T., Alzard, M.: Greenhouse gas emissions associated with road transport projects: current status, benchmarking, and assessment tools. Transp. Res. Procedia **48**, 2018–2030 (2020). https://doi.org/10.1016/j.trpro.2020.08.261

Albuquerque, F.D.B.: The running-red-light problem in Abu Dhabi. Technical Material, Abu Dhabi Municipality (2015)

Albuquerque, F.D.B., Awadalla, D.M.: Characterization of road crashes in the emirate of Abu Dhabi. In Transportation Research Procedia (Vol. 48, pp. 1095–1110). Presented at the World Conference on Transport Research, Mumbai, India: Elsevier (2019a)

Albuquerque, F.D.B., Awadalla, D.M.: Roadside design assessment in an urban, low-density environment in the Gulf Cooperation Council region. Traffic Inj. Prev. **20**(4), 436–441 (2019b). https://doi.org/10.1080/15389588.2019.1602770

Albuquerque, F.D.B., Awadalla, D.M.: Roadside fixed-object collisions, barrier performance, and fatal injuries in single-vehicle, run-off-road crashes. Safety **6**(2), 27 (2020a). https://doi.org/10.3390/safety6020027

Albuquerque, F.D.B., Awadalla, D.M.: A Benchmark compliance evaluation of the roadside design in an urban, high-density area. In The Built Environment (Vol. 200, pp. 117–129). Presented at the 26th International Conference on Urban Transport and the Environment, Online: Wessex Institute (2020b). https://www.wessex.ac.uk/2011/HBNFE452/UT20-9781784664091.pdf

Ansari, S., Akhdar, F., Mandoorah, M., Moutaery, K.: Causes and effects of road traffic accidents in Saudi Arabia. Public Health **114**(1), 37–39 (2000). https://doi.org/10.1038/sj.ph.1900610

Awadalla, D.M., Albuquerque, F.D.B.: Fatal road crashes in the Emirate of Abu Dhabi: contributing factors and data-driven safety recommendations. Transp. Res. Procedia, **52**, 260–267 (2021). Manuscript under peer-review. https://doi.org/10.1016/j.trpro.2021.01.030

Awadalla, D.M., Albuquerque, F.D.B.: Impact of roadside design compliance and hazard offset on the risk of single-vehicle, run-off-road crash fatalities. Int. J. Inj. Control Saf. Promot. (n. d.). https://doi.org/10.1080/17457300.2021.1942923

Brubacher, J.R., Chan, H., Erdelyi, S., Lovegrove, G., Faghihi, F.: Road safety impact of increased rural highway speed limits in British Columbia Canada. Sustainability **10**, 3555 (2018). https://doi.org/10.3390/su10103555

Dajani, H.: Speed limits to be increased on Abu Dhabi roads to boost safety. The National News. News, 29 November 2018. https://www.thenationalnews.com/uae/transport/speed-limits-to-be-increased-on-abu-dhabi-roads-to-boost-safety-1.797068. Accessed 1 June 2021

Donnell, E.T., Hamadeh, B., Li, L., Wood, J.S.: 70 mph study (No. FHWA-PA-2016-009-PSU WO 13). Commonwealth of Pennsylvania Department of Transportation (2016)

Elvik, R.: Effects on road safety of converting intersections to roundabouts: review of evidence from non-U.S. studies. Transp. Res. Rec. J. **1847**, 1–10 (2003)

Elvik, R.: Road safety effects of roundabouts: a meta-analysis. Accid. Anal. Prev. **99**(Part A), 364–371 (2017). https://doi.org/10.1016/j.aap.2016.12.018

Engström, J., Johansson, E., Östlundb, J.: Effects of visual and cognitive load in real and simulated motorway driving. Transp. Res. F Traffic Psychol. Behav. **8**(2), 97–120 (2005). https://doi.org/10.1016/j.trf.2005.04.012

Farmer, C.M.: Relationship of traffic fatality rates to maximum state speed limits. Traffic Inj. Prev. **18**(4), 375–380 (2017). https://doi.org/10.1080/15389588.2016.1213821

Farmer, C.M.: The effects of higher speed limits on traffic fatalities in the United States, 1993–2017. Insurance Institute for Highway Safety (2019). https://www.iihs.org/topics/bibliography/ref/2188

Federal Highway Administration (FHWA): About intersection safety. FHWA, 9 March 2021. https://safety.fhwa.dot.gov/intersection/about/

Gao, J., Hou, H., Zhai, Y., Woodward, A., Vardoulakis, S., Kovats, S., et al.: Greenhouse gas emissions reduction in different economic sectors: Mitigation measures, health co-benefits, knowledge gaps, and policy implications. Environ. Pollut. **240**, 683–698 (2018). https://doi.org/10.1016/j.envpol.2018.05.011

Gao, J., Kovats, S., Vardoulakis, S., Wilkinson, P., Woodward, A., Li, J., et al.: Public health co-benefits of greenhouse gas emissions reduction: a systematic review. Sci. Total Environ. **627**, 388–402 (2018). https://doi.org/10.1016/j.scitotenv.2018.01.193

Garber, N.J., Gadiraju, R.: Impact of Differential Speed Limits on Highway Speeds and Accidents. University of Virginia, Charlottesville (1991)

Garber, N.J., Hoel, L.A.: Traffic and highway engineering, 5th edn. Cengage, Boston (2020)

Gates, T.J., et al.: Safety and operational impacts of differential speed limits on two-lane rural highways in Montana (No. FHWA/MT-16-006/8224-001). Montana Department of Transportation (2016)

Hall, J.W., Dickinson, L.V.: An operational evaluation of truck speeds on interstate highways. Department of Civil Engineering, University of Maryland (1974)

Halliday, D., Resnick, R., Walker, J.: Principles of Physics (10th ed.). Wiley, Hoboken (2014)

Harkey, D.L., Mera, R.: Safety impacts of different speed limits on cars and trucks (No. FHWA-RD-93-161). Washington, DC, USA: U.S. Department of Transportation, Federal Highway Administration (1994)

Haselton, C.B., Gibby, A.R., Ferrara, T.C.: Methodologies used to analyze collision experience associated with speed limit changes on selected California highways. Transp. Res. Rec. J. Transp. Res. Board **1784**(1), 65–72 (2002). https://doi.org/10.3141/1784-09

Haven, H.D.: Mechanical analysis of survival in falls from heights of fifty to one hundred and fifty feet. Inj. Prev. **6**(1), 62–68 (2000). https://doi.org/10.1136/ip.6.1.62-b

Haworth, N., Symmons, M.: The relationship between fuel economy and safety outcomes (No. 188). Victoria, Australia: Monash University Accident Research Centre (2001)

Idaho Transportation Department Planning Division: Evaluation of the impacts of reducing truck speeds on interstate highways in Idaho, -Phase III. Idaho, US: Idaho Transportation Department Planning Division (2000)

Kockelman, K.: CRA International, Inc. Safety impacts and other implications of raised speed limits on high-speed roads. Washington, DC, USA: The National Academies Press (2006). https://doi.org/10.17226/22048

Ledolter, J., Chan, K.S.: Evaluating the impact of the 65 mph maximum speed limit on Iowa rural interstates. Am. Stat. **50**(1), 79–85 (1996). https://doi.org/10.1080/00031305.1996.10473546

Nilsson, G.: Traffic safety dimensions and the Power Model to describe the effect of speed on safety. Lund, Sweden: Lund Institute of Technology (2004)

Pauw, E.D., Daniels, S., Thierie, M., Brijs, T.: Safety effects of reducing the speed limit from 90 km/h to 70 km/h. Accid. Anal. Prev. **62**, 426–431 (2014). https://doi.org/10.1016/j.aap.2013.05.003

Renski, H., Khattak, A.J., Council, F.M.: Effect of speed limit increases on crash injury severity: analysis of single-vehicle crashes on North Carolina interstate highways. Transp. Res. Rec. J. Transp. Res. Board **1665**(1), 100–108 (1999). https://doi.org/10.3141/1665-14

Rock, S.M.: Impact of the 65 mph speed limit on accidents, deaths, and injuries in Illinois. Accid. Anal. Prev. **27**(2), 207–214 (1995). https://doi.org/10.1016/0001-4575(94)00058-t

Rohrer, W.M.: Road traffic accidents as public health challenge in the Gulf Cooperation Council (GCC) Region. Public Health Open J. **1**(3), e6–e7 (2016). https://doi.org/10.17140/PHOJ-1-e004

Scott, V.: Study: Traffic accidents cost Qatar nearly $5 billion over six years. Doha News. News, 8 April 2014. https://dohanews.co/study-counts-huge-cost-road-accidents-qatar/

Shahbandari, S.: Road crashes cost Dubai Dh1.8b a year. Gulf News. News, 11 March 2015. https://gulfnews.com/news/uae/road-crashes-cost-dubai-dh1-8b-a-year-1.1470277

Shinar, D.: Traffic Safety and Human Behavior. Emerald Group Publishing Limited, Bingley (2007)

Shirazinejad, R.S., Dissanayake, S., Al-Bayati, A.J., York, D.D.: Evaluating the safety impacts of increased speed limits on freeways in Kansas using before-and-after study approach. Sustainability **11**(1), 119 (2019). https://doi.org/10.3390/su11010119

Tarko, A.P., Pineda-Mendez, R., Guo, Q.: Predicting the impact of changing speed limits on traffic safety and mobility on Indiana freeways (No. FHWA/IN/JTRP-2019/12). West Lafayette: Purdue University (2019)

The World Bank: Transport - Greenhouse gas emissions mitigation in road construction and rehabilitation : A toolkit for developing countries (No. 69659), pp. 1–78. The World Bank (2011). http://documents.worldbank.org/curated/en/660861468234281955/Transport-Greenhouse-gas-emissions-mitigation-in-road-construction-and-rehabilitation-A-toolkit-for-developing-countries. Accessed 3 Sep 2017

Toumi, H.: A car accident every 10 minutes in Kuwait. Gulf News. News, 27 July 2017a. https://gulfnews.com/news/gulf/kuwait/a-car-accident-every-10-minutes-in-kuwait-1.2065034

Toumi, H.: One traffic accident per minute in Saudi Arabia. Gulf News. News, 23 November 2017b. https://gulfnews.com/news/gulf/saudi-arabia/one-traffic-accident-per-minute-in-saudi-arabia-1.2129414

Turner, S., Sandt, L., Toole, J., Benz, R., Patten, R.: Federal highway Administration University course on bicycle and pedestrian transportation (No. FHWA-HRT-05-133), p. 374. Texas, US: Texas Transportation Institute (2006)

Virginia Transportation Research Council: The safety impacts of differential speed limits on rural interstate highways (No. FHWA-HRT-04-156). Turner-Fairbank Highway Research Center (2004)

Wu, C., Liu, Y.: Queuing network modeling of driver workload and performance. Proc. Hum. Factors Ergon. Soc. Ann. Meet. **50**(22), 2368–2372 (2006). https://doi.org/10.1177/154193120605002204

Zaatari, S.: Abu Dhabi buffer speed limit removal explained. Gulf News. News, 26 July 2018. https://gulfnews.com/uae/transport/abu-dhabi-buffer-speed-limit-removal-explained-1.2257426. Accessed 8 June 2021

Zacharias, A.: Abu Dhabi 20kph speed limit buffer to be removed in August. The National News. News, 25 July 2018. https://www.thenationalnews.com/uae/abu-dhabi-20kph-speed-limit-buffer-to-be-removed-in-august-1.753875. Accessed 1 June 2021

Transport Responses to the Pandemic

The Impact of the COVID-19 Pandemic on Mobility Behavior in Istanbul After One Year of Pandemic

Ali Atahan[✉] and Lina Alhelo

Civil Engineering Faculty, Istanbul Technical University, Istanbul, Turkey
atahana@itu.edu.tr

Abstract. Worldwide, the COVID-19 pandemic has resulted in significant changes in activity patterns and travel behavior. Some of these changes are due to responses to governmental or personal measures for controlling the spread of the pandemic (e.g. partial lockdowns, remote education, or work). This paper examines changes in human mobility over 1 year period (from March 1st, 2020 to March 30th, 2020) in Istanbul, Turkey, a megacity in developing countries. The study period imposed three distinct waves, 1st wave (March-April), the second wave (Nov/202–Jan/2021), and 3rd wave (March/2021-still continue). The paper includes two parts. The first part is an analytical analysis for traveler mobility trends using data of public transit automatic fare collection system and traffic volume counting sensors along with Istanbul metropolitan. For the 1st wave, mobility changes have a V-shape trend for public transit riders and a smooth U-shape for reported vehicles by highway sensors. Despite the severity level of the COVID-19 pandemic and mobility restriction measures during 2nd wave, there was a smooth drop in mobility indicating a change in human reaction toward pandemic severity. The second part is investigating differences in individuals' trip characteristics for before the COVID-19 pandemic and after one year of pandemic based on data collected from an online survey. Tends to shift to use personal vehicle for daily trips, indicate more effective decisions should be implemented next to staggered working hours and reductions transit capacity to encourage travelers for using the sustainable transportation system.

Keywords: COVID-19 · Human mobility · Travel behavior · Istanbul

1 Introduction

More than 1 year has passed since the first reported case infected by COVID-19 on 31/Dec/2019 in Wuhan/China. On 11th March 2020, the World Health Organization declared that COVID- 19 to be a global pandemic. Worldwide COVID-19 pandemic has significantly affected various facets of our daily life (Ozaras et al. 2020). Among them, one important aspect that captures the attention of researchers and policymakers is the human's mobility. That is because the COVID-19 outbreak is related to people's mobility, so mobility control was the key means to control the spread of the virus (Oztig and Askin 2020).

© The Author(s), under exclusive license to Springer Nature Switzerland AG 2022
A. Akhnoukh et al. (Eds.): IRF 2021, SUCI, pp. 933–949, 2022.
https://doi.org/10.1007/978-3-030-79801-7_65

To encounter the COVID-19 outbreak different measures have been adopted on a local and international scale. Governments have been implemented various measures and regulations to curb the outbreak of infection, by enforcing curfews and imposing restrictions on people's free mobility (Santamaria 2020). Besides the imposed mobility restrictions impact, encouraging staying at home, increasing distance/online activities, even social one, and other actions that impact on mobility requirement; resulted in a reduction in human mobility. COVID-19 pandemic and the governmental measures aiming at overcoming the virus spread may have very large impacts on urban mobility (Kanda 2020). These impacts are either in response to restrictive measures imposed by the Government to slow down the spread of the virus or a person self-commitment to protect them self (Bhaduri 2020).

COVID19 pandemic becomes an interest of investigators, besides medical investigations, to reflect its impact from various approaches like impact on tourism (Abbas et al. 2021; Duro et al. 2021), on social-economic (Gandasari, 2020) and other areas. Growing researches on examining the COVID-19 pandemic effects on human's mobility patterns and travel behaviors exist (e.g., Zho et al. 2020; Santamaria et al. 2020; Haas et al 2020, Willberg et al. 2021; Matson et al. 2021).

Although previous studies provided a useful base for future research, they have some limitations. Previous studies focused on the different stages of the COVID-19 pandemic (e.g., first wave, second wave). Considering that the COVID-19 pandemic in Turkey is still continuing as public health crisis at the time of writing. Investigating the mobility changes in a more comprehensive timeline would provide a meaningful approach in understanding the impacts of the COVID-19 pandemic on human's travel behaviors and mobility patterns.

Thus, this paper aims to fill this gap by investigating changes in human mobility levels in Istanbul metropolitan for 1 full year period (from March 1st, 2020 to March 31st, 2021). This investigation conducted based on analyzing the impact of COVID-19 severity level, mobility restriction measures and policies on mobility pattern based on recorded statistics collected from automatic fare collection for public transit, number of vehicles traveling from traffic counting sensors and daily activity trend gathers from Google COVID-19 community mobility reports.

The second part of this paper includes an analytical analysis for an online survey, conducted with 272 people, to observe the impact of the COVID-19 crisis on their mobility pattern after passing one full year from the 1st case infected by COVID-19 in Turkey.

Survey analysis was the most approach used for evaluating the impact of COVID-19 on mobility (Barbieri et al. 2020; Beck et al. 2020; Shakibaei et al. 2021). This approach provides an insightful explanation for the general view for a certain case. Shakibaei et al. 2021, used to distribute surveys during different phases of the study, 1st phase when the first case appeared in China. Second phase one when border countries infected by the virus, and 3rd phase after appearing 1st case infected by COVID-19 in Turkey. Others performed a comparison on the impact of 1st and 2nd pandemic waves on travel behaviour (Beck 2020) while Barbieri et al. 2020 survey has characterized the frequency of use of all transport modes before and during the enforcement of the restrictions.

1.1 COVID-19 and Measures

To halt the rise of virus spread, a series of regulations and measures being implemented through this crisis even before the first case appearing in Turkey, that was on March 11th 2020 (Ministry of Health/Turkey, COVID 19 information page). Figure 1 and Table 1 summarize the timeline of precaution and restrictions measures that were implemented. The first measures were restricting mobility with boarder infected countries, like Iraq and Iran, and suspending all flights from other infected countries. Definitely disinfection work was implemented in public gathering places, public transportation vehicles and stations, air ports, health centers, etc. After the 1st case appearing, a quick decision to ban all education process was announced to start implementing distance education two weeks later with recommending other people to stay at home.

Later group of mobility restrictions were quickly implemented either by imposing full or partial curfew after March 11th as presented in Fig and Table 1. Part of the measures implemented by Istanbul Metropolitan municipality were reducing the operation hours for public transportation, e.g. for buses, railways, Metro lines to end on 21:00 instead of 00:00, ending sea buses trips at 17:00 instead of 22:00, pause Funicular and Nostalgic Tram up to further notice and operating all public transportation services under 50% of its normal capacity (www.ibb.istanbul). Besides that, group of full curfews were imposed during the first wave, includes travel restrictions for non working group −18 ages and old age group + 65 ages.

In the first wave the daily reported cases in Turkey reached to an initial peak of 4789 daily reported cases in the mid of April month 2020 approximately one month since 1st case appearing in Turkey. Number of cases drops later to reach daily average value around 1000 case per day after 13th May 2020. After this drop, the governmental approached to the gradual relaxation of restrictions after May 29th, based on the underlying principles of maintaining a social distance of 1.5 m from those not in the family unit; regular hygiene and sanitization practices, staying at home if unwell, and a COVID safe plan for workplaces and premises.

This condition continues after risk falling till 1st third of Nov. Later sharp increase in reported cases to be 33198 per day presented. At the beginning of this increment, new measures and restrictions have been re-implemented, by encouraging shifts work system, curfew on + 65 and −18 years old, reducing working time, and implementing curfew on weekends. These restrictions continued till the end of March 2021.

Another measure implemented to reduce the passenger interaction and crowded trips was adopting staggered working hours strategy that is used to spread the travel demand during peak hours for longer time period to reduce the traffic congestion during the peak hours (Mutlu et al. 2020). This strategy is used to reduce passenger density within public transportation system, by specifying different starting working time for organizations stating from 8:00 AM to 10:00 AM, the same for closing time.

2 Impact of COVID-19 Pandemic on Humans' Mobility

2.1 Mobility Trend Using Public Transit

In this section, the impact of the pandemic on humans' mobility is investigated. Public transportation modes operating in Istanbul include road (tire wheels), rail systems, and sea transportation. Road transportation includes journeys made by bus and bus rapid transit (BRT). Rail systems cover journeys made by subway lines and light rail systems (tram, funicular, cable car, nostalgic tram, and historical tunnel) while sea transportation includes trips made by ferry (City Lines) only. Fig. 3 represents the number of passengers made by each mode on an average working daily basis. These numbers were collected from automated fare collection systems for each mode.

As this study's aim is to observe the impact of the pandemic on mobility, weekend's day has been excluded from the analysis, because most of the curfews implemented were on the weekends. After 1st infected case, there is a sudden drop in the number of passengers for all modes of transportation within two weeks. All modes have a similar trend along the period of observation. Moving to Fig. 4, represents the percentage of change in passengers number for all public transportation modes, compared to the average weekday's number of trips prior to March (from Jan 6th–Feb 28th). The percentage change reached to 88% reduction at the first wave and range from 88% to 32% at best condition along 1 year. According to the timeline of health measures taken after 11th Mar, the first action taken on 16th Mar was closing all education and public gathering places, followed by a curfew for + 65 and −18 years old. Within one week, the average of transit trips made has sharply dropped to 81.5% (Fig. 1).

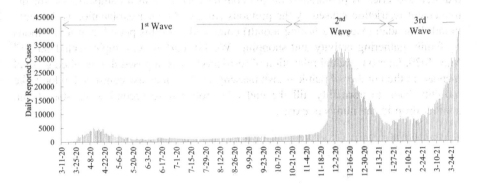

Fig. 1. Daily reported cases in Turkey

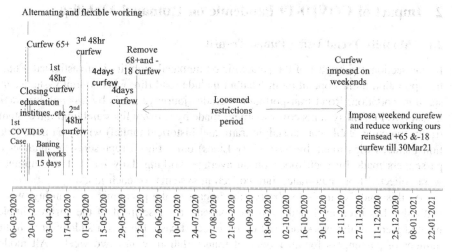

Fig. 2. Ongoing timeline of Curfew imposed

Within the first month for the 1st wave, there was a fluctuation in the number of passengers in the period of 17th April–22nd May between increasing and decreasing in the mobility. The first curfew was implemented on 11th–12th April that was announced 2 h prior to implementation. For coming curfew, the announcements were within 5–3 days prior to implementation. From Fig. 3, after each curfew there is increase in the mobility. This increase could be explained due to public activities in purchasing required groceries as preparation for the coming curfew or as a compensation for the movement restriction caused by the previous curfew. Another explanation is that holy month Ramadan (Muslims fasting month) coincided with this period, and it is famous by family gathering activity and shopping. Weekly curfew was implemented till 31st May 2020; later on gradual relaxation of restrictions was imposed that resulted gradual increase in the number of public transit passengers. This increase continued to look like mobility back to normality, till the end of October were second wave started and another drop in the numbers exists.

Table 1. Summarising key actions in ongoing COVID-19 timeline.

24-Jan	Precautionary measures on the airports included infrared guns, disinfection at all customs gates and the handing out of free surgical masks and instruction leaflets has adopted
3-Feb	Announced a stop of all the flights from China and stop all flights to and from Iran
29-Feb	Announced the termination of all flights to and from Italy, South Korea and Iraq
8-Mar	Carry out massive disinfection work in public places and mass transit vehicles
11-Mar	Country's first corona virus case
14-Mar	Cases rose to five. Stop all flights to and from Germany, France, Spain, Norway, Azerbaijan Etc
16-Mar	Closing all education institutes, public gathering places cafes, gyms, bars, libraries, praying places, etc. excepting shops and increase stopped flight countries
19-Mar	Football, volleyball, basketball and handball leagues were postponed
21-Mar	Curfew for those who are over the age 65 or chronically ill. Banning barbecuing in gardens, parks and promenades
22-Mar	Allow for alternating and flexible schedules and enforce remote working if possible in public institutions and organizations
23-Mar	Banks across the country and limited their working hours to 12:00 pm to 5:00 pm
24-Mar	Markets could serve customers between 9:00 am to 9:00 pm, with 0.5 of its normal capacity. Public transportation vehicles will serve fewer than 50% of their capacity
27-Mar	All overseas flights were terminated. Intercity travel was subject to permission by the state governors
28-Mar	Starting from 28 to 29 March, picnics, fishing at the shores, doing outside physical exercises (including running and walking in the weekends were banned
30-Mar	Sea buses and ferry services that make intercity trips in Istanbul would stop operating from 5:00 pm
31-Mar	Banning all works for at least 15 days except those offering essential goods and services
10-Apr	48-h curfew was declared from 10.04.2020 at 24.00 until 12.04.2020 at 24.00
13-Apr	Announcing curfew that will be imposed from Friday night, 17th April to Sunday night, 19 April/2020
21-Apr	Announcing curfew that will be imposed from 22.04.2020 at 24.00 to 26 April at 24
30-Apr	Curfew that imposed from 30.04.2020 at 24.00 to 3 May at 24.00
11-May	Main automotive factories in the country will start to operate again
19-May	Announcing 4days curfew that will be imposed from 22.05.2020 at 24:00 until 26.05.2020 at 24:00
20-May	Citizens aged 65 and over, whose curfew is restricted, were allowed to go to the settlements they want in one direction, provided that they do not return for at least one month as of 09:00 on Thursday, May 21

(continued)

Table 1. (*continued*)

28- May	Announcing 2days curfew that will be imposed from 29.05.2020 at 24.00 and 31.05.2020 at 24.00
29-May	Loosened restrictions as "controlled social life",
30-May	End travel restriction in 15 provinces as of May 31, 24.00
1-Jun	Domestic flights were resumed and most public spaces were opened, restaurants, beaches, libraries and museums
10-Jun	Remove the restrictions for under 18 and above 65 age citizens
17-Nov	curfew would be imposed on weekends except between 10:00 am–08:00 pm, restaurants would only provide take-away service, and shopping malls and markets would close at 08:00 pm
1-Dec	Ministry of Health reinstated the curfew on people age 65 and older and people twenty and younger. Public institutions and organizations have been determined as between 10:00 and 16:00 Partial curfew from 21:00 to 5:00 am weekdays and from 21:00 on Friday to 5:00amMonday in the weekend
1-Mar-2021	Turkish government eased the restrictions, but the rates of infections rose

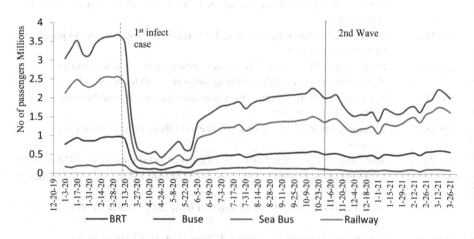

Fig. 3. Average number of passenger for working days

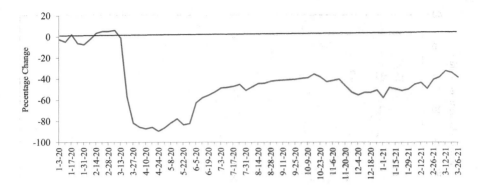

Fig. 4. Percentage change in public transit passengers

Minor reductions exist in public transit passengers from October to November due to weather condition. For 2nd wave, there was mobility decline after the implementing of + 65 and −18 years old curfew, but it is lower than 1st wave despite the existence of mobility restriction policies and the worse condition COVID-19 pandemic in this period as daily reported cases reach 35 thousand case compare to 4500 case in the 1st wave. So restricting people's mobility to control the pandemic may be effective only for a short time period (Kim and Kwan 2021) and other strategies should be implemented. Along this period still the education institutes did not resume face to face education and part of official jobs still performed online.

Similar outcomes could be reported from Fig. 5 that represents the percentage change in basic daily activities. This chart is based on Google Community Mobility Reports data (Google 2021), which aggregates data across Istanbul Metropolitan, and compared to the average time spent value for the corresponding day to the day of the week during the 5-week period Jan 3–Feb 6, 2020 that is specified to be a baseline of zero percent change.

Figure represents total reduction in time spent at grocery, transit station, work location, recreation location since the 1st case appearing in Turkey. Chart show similar reduction rate for all activities except for grocery visits. Time spent at grocery reduction rate is lower than other activities along the period. Its maximum reduction rate was 30% while other activities range from 57% to 72%. While an increase in percent change achieved by the recreation activities, grocery attains higher rate than pre COVID-19 pandemic condition as it is exceeds zero line. This explained that people tend to practice only mandatory activities on this period to avoid as much possible chance of COVID19 infection and people fair from any sudden curfew or movement restrictions resulted in an increase in visiting rate for grocery markets (Beck et al. 2020).

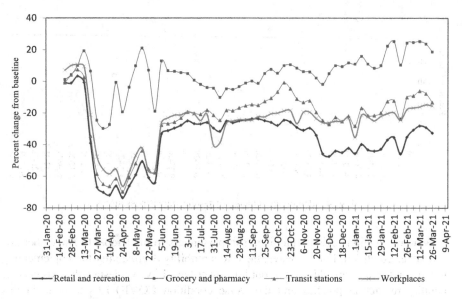

Fig. 5. Google mobility report weekly averages for Istanbul

Easing restriction is one of the causes to increase the aggregate mobility, (Beck and Hensher 2020). This is visible from the all charts at Fig. 3, 4 and 5. Percent in using public transportation, visiting grocery shops, recreation centers and workplaces started to slowly return immediately after the governmental approach to the relaxation of restrictions that is implemented after 31st May as controlled social life. During this period, travel activity has started to slowly return for all transportation modes, in particular by private car and for the purposes of shopping and social or recreational activities.

2.2 Mobility Tend Using Personal Vehicle

Globally there is shift to use personal vehicle on public transportation as precaution step from getting infected by the virus. Because of this it is important to observe how the pandemic impact personal vehicle mobility within Istanbul Metropolitan. Number of registered automobile per 1000 person in Istanbul reduced in period 2018 to 2019 from 192 to 185, while it is increased to 190 in 2020.

In Turkey, after March 10, 2020, a short-lived positive spike in gasoline consumption is observed after the announcing of the 1st case infected by COVID -19 in Turkey followed by a steady downward trend (Gungor 2020). This means the number of moving private cars increased immediately after this date. Figure 6 illustrates the reduction in public transit and private car trips along the period from 1st Jan–6th Nov/2020. Percent reduction in private car trips gathered from sensors data along Istanbul Metropolitan highways from Jan to 10th Nov/2020. It is clear from the diagram that the percent change in private cars does not have the same trend or percent compared to the percent reduction in public transportation ridership. So the reduction in mobility for vehicle owners is not significant compared to public transpiration riders. A significant shift from public transportation to the private cars has being observed in many studies (Jenelius and Cebecauer 2020; De Haas et al. 2020), this may explain the lower reduction in mobility for vehicle owners.

Another spikes present in the first wave period that occurred after each curfew imposed within this period. The same tend observed here, of smooth reduce in the reduction rate that shows an increase in the number of vehicle to be too near from zero percent (0%) reduction line.

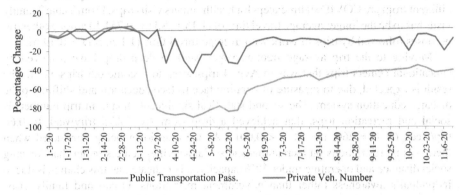

Fig. 6. Percent change in public trans. ridership and private vehicles number

3 Survey Descriptions

Survey was designed and circulated through various online modes to make it suitable to the respondents and also to reach a greater geographical coverage within a short period of time. The survey includes different parts, starting by social-economic information, such as gender, age, income, then working condition, changes in working nature, shopping activity, participating in social life activities like family visits, recreation trips, education centers, etc. and what transportation modes being used for each activity. The major interest here is observing how people have changed their mobility for different trip purposes due to the pandemic.

Number of participants was 275 persons, their characteristics summarized in Fig. 7. Socio-demographics characteristics are based on gender (Female = 150, Male = 122), age (lower than 30 (17–30, n = 180); middle-age (30–60, n = 83); 60 or older, n = 8). Also working condition as student (school or university) = 152, working (employee or employer) n = 91 and not a student nor working (n = 29). Household income classified for 4 groups, (less than 2500Turkish Lira n = 80, 2500–3500 TL n = 62, 3500–4500 n = 31, higher than 4500, n = 56).

Survey contain two parts, question about pre COVID 19 aspects and current day after passing 1 year for COVID-19 pandemic. Current day(after COVID19), answers depends on the last two weeks from answering the questioner, that is conducted from 17th–31st/May/2021, after removing most of mobility restrictions imposed in the previous month, April 2021.

4 Results

Purpose of trips classified for 7 classes as shown in Table 2. Family visits mean relatives visits, while work trips are trips for the sake of job requirements. Health centers trips include trips to hospitals; clinics and any other center provide medical services. Education centers means schools, university, libraries, etc. while recreation trips are trips for recreation purpose like, restaurant, cinema, gym, etc.

For each trip purpose average travel time and frequency per week being observed. Comparing trip frequency before and after, all trip purpose have undergone significant different from pre COVID status except for health centers visits trips. From Table 2 family visits has to be the longer average travel time of (1.15 h, S.D = 0.751 h) among other trip purposes followed by trips to work with average travel time (1.1 h, S.D = 0.59 h).

Moving to the trip average frequency per week, a deep drop have achieved by educational centers trips that was in Avg. 4 trips/week to become 0.5 trips/week. The result is expected, due to measure of banning face to face education and shifting to the distance education system. The second statistical significant change in trip numbers is social and recreation trips, that achieved a drop from Avg. 2.75 trip/week to 0.66 trip/week (at $N = 260$, $t = 10$, $p < 0.001$). As the questioner being conducted when there was no restrictions on recreations and social activities centers rather than keeping social distance and operating under 50% capacity, this indicate that this change is due to individual's awareness rather than government restrictions. Friend and family visits frequency almost the same and significant drop appears for this trips purpose, $p < 0.001$.

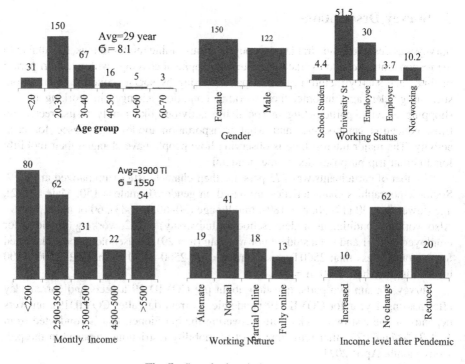

Fig. 7. Sample description summary

For working persons, that are 33% of participants, their average trips to their work is 3.7 day per/week, median = 4.1 day/week. So, it is clear that normal/in office work condition is still active.

For gender categorical classification, male and female, significant difference in the frequency means for friend's visit and recreation trips has being found. For recreation trips mean trip frequency for male is 1.37 trip/week, S.D = 1.7 and 0.58 trip/week, S.D = 0.75 for female, t (270) = 4.6 m, p < 0.001 while for recreation trips male frequency found to be 1.08 trip/week, S.D = 1.71 while females mean frequency 0.4 trip/week, S.D = 0.8 t (270) = 3.6, p < 0.001. We can classify these trips as non mandatory trips as there was a lot of calls to reduce non required movement, so it seems female take self measures for protecting more than males, as there was no statistically significant difference between the means of these purpose for pre COVID19 pandemic. According to the age group and trip purpose, there was no any significant difference between age groups and certain trip purpose frequency nor the income level.

Figure 8 illustrate number of days used to travel per week for general purpose. For different group classification range, average number of travelled day ranges from 4 to 6 days/week before pandemic, while now ranges from 1 to 3.5 days/week. Classification group is based on working condition, income level, age, gender and mode of transportation used.

Excepting working condition, none of the variable had a significant relation with number of trips conducted pre COVID-19 period. A Chi-square to test was used to test any significant relation between the variable and number of average travelled days/week, Chi square (30, N = 272) = 65.7, p < 0.001. This means that number of days travelling depends on working condition, as to be employee, employer, student, or not working, that is acceptable. That is because pre-COVID, nature of work and teaching for most of the sectors required persons to travel to perform his task.

Table 2. Trip purpose statiscal properties

Trip purpose	Family visit	Shopping	Work	Health center	Education cent.	Friend Visits	Social/Recreation
Avg. travel time (hr)	1.15	0.84	1.1	0.77	1.07	1.09	1.05
S.D	0.751	0.39	0.59	0.42	0.67	0.56	0.5264
Avg. weekly Trips Before COVID 19	1.67	2.15	5.20	0.70	4.00	2.84	2.75
S.D	1.61	1.54	1.18	1.30	2.23	1.92	1.90
Avg. weekly Trips Nowadays	0.90	0.87	3.73	0.54	0.50	0.87	0.66
S.D	1.45	1.07	2.18	1.01	1.22	1.18	1.25

Moving to after COVID-19 period, according to transportation mode used, weekly average travelling days for personal vehicle owners found to be higher than other modes users. That is acceptable as opportunity to get infection increases when number of contact person increases. So, mode of transportation to be used is important factor for trip conducting rather than the purpose of trip.

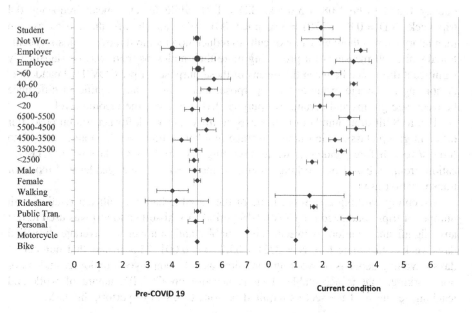

Fig. 8. Average weekly trips based on various socio- demographic groups

For gender variable, the average trips made by males are significantly higher than that made by females (M = 2.99, S.D = 1.97), (M = 1.99, S.D = 1.53); t (272) = 4.6, p < 0.001 respectively. These shows female have taken the situation more responsible than the males. Regarding the income level and age according to t-test, there is significant difference between groups for each category. For income level, there is statistically significant difference between income levels groups and average trips frequency per week (t (272) = 42.1, p < 0.001) similarly for age groups (t (272) = 47.5, p < 0.001). For age group variable start with 1.5 day per week for young groups < 20 years old, and increase to reach 3.5 day per week for 40–50 years old and then reduce back to 2.0 day per week for age group older than 60 year. This is expected result, as < 20 years old group are not working group, and as per trip purpose, higher trip frequency achieved by working group for work trip purpose, the same explanation for the older groups.

Moving to the next part of the survey, to investigate the shift in travel modes, and the causes of this shift. Regarding the changes in most common modes of transport used for daily trips, outcomes are illustrated in Fig. 9. According to the mode of choice, 82% of participants mostly were using public transit vehicles for their daily trips, and

17.9% using personal vehicles before the COVID-19 pandemic. Out of this 82% = 210 person, 22.4% have shifted to a personal vehicle after the pandemic, while 17.14% walk for their daily trips and 8% using shared vehicles like yellow taxis, Uber, etc. Only 51% of people who were using public transportation continue to use it for their daily trips nowadays. For persons who shifted from public transposition to other modes, 90% of them explained the reason to be health safety concerns by avoiding crowded modes. It was clearly observed that there is an increment in the number of those started to use a private car instead of public transportation and this outcome has been observed in the previous section for comparing reduction rate in private vehicle trips compared to public transit trips.

In other hand 17.9% of pre using personal vehicle, 8.7% shifted to public transit. As it is observed this was due to economic condition as being explained in another question for the reason over changing transportation mode. Also 6.5% of personal vehicle users shifted to walking and bike mode for their daily trips (4person), for physical health purposes.

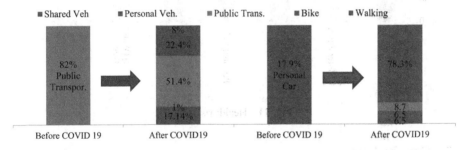

Fig. 9. Shift in trasportaion mode choice

Figure 10 illustrates the different transportation mode used percent for different trip purposes. There is no relatively difference in mode choose and trip purpose along different trips. We can observe that for family visits, passenger vehicle percent used for this purpose of trips is higher than other modes, 45% of family trips were performed using personal vehicle.

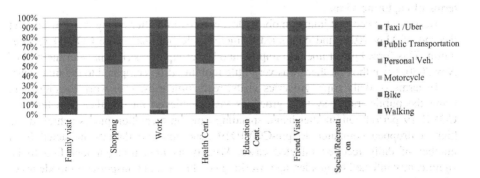

Fig. 10. Mode of Choice according to trip purpose

Last part of the survey includes evaluation for the health risk for different transportation systems that illustrated in Fig. 11 that is classified for no risk, low risk and high risk grades. 75% of the participant believes that road public transportation system is high risk system. In other hand ferry (water transportation) classified as low risk and no risk transportation system. Domestic flight rated as safer than international flight thus while shared vehicle rated to be more low risk system. These results indicate that water transportation system is even safer than domestic flights. This encourages having real investment in water transportation for inner and outer city trips.

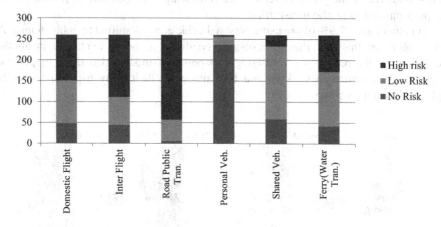

Fig. 11. Health risk rating

5 Conclusions

This research examined changes in people's mobility over 1 year period (from March 1st, 2020 to March 31st, 2021) during the COVID-19 pandemic in Istanbul metropolitan. The results exposed that mobility changes in 1st wave were major, reached an 85% reduction in public transit users and 40% in personal vehicles crossing Istanbul highways. A significant shift from public transportation to personal vehicle observed. This shift explained by finding public transportation is a risky environment in terms of capturing virus.

Results revealed that human mobility levels recovered after easing the restriction at the end of May 2020, and continued to increase despite the severe COVID-19 condition. This forced government to impose another group of restrictions after the mid of Nov 2020. Even though, there was very little mobility drop compared to the first wave.

In terms of daily trips, groceries visits, recreation trips, visit work location, and using the public transit system, groceries visits achieved rates higher than the pre-COVID19 period. On another hand, spending time on recreation places achieved the highest dropped compared to pre-COVID19 condition and directly impacted by a number of daily recoded infected cases. Weekly average trips number find to be significantly infected by gender, age, working condition, and transportation mode used, while trip length and purpose were not.

Further, this study has important implications for COVID-19 control measures specifically that aim on restricting mobility. The results revealed that people's mobility levels quickly recovered after May 2020 despite the severe COVID-19 situation and mobility restriction policies in Turkey. Even though, after each curfew imposed, a spike increase in mobility exists. This indicates that restricting human mobility to control the pandemic may be effective only for a short period, after which mobility restrictions may become ineffective in curbing the spread of the virus.

So mobility management will be a more useful tool in controlling the virus outbreak, like adopted a staggered working hour's strategy that is used to spread the travel demand during peak hours for the longer time period, which reduces the passengers' congestion during the peak hours at public transit systems.

Among shared transportation systems, water transportation evaluated to be the safer after personal vehicle. Also, the drop in sea transportation was minor compared to other public transportation systems. There must be a tendency to increase the use of sea transportation and investments.

On other hand, long-term consequences of the COVID-19 pandemic may cause a reduction in mobility needs due to changes like digitalization of work and other daily activities shopping, education, etc. (Kanda 2020). Working from home is another important strategy in reducing travel and pressure on constraining mobility, which will be an important step towards a more sustainable transport future if contoured to be carried over to post COVID19 pandemic.

References

Abbas, J., Mubeen, R., Iorember, P.T., Raza, S., Mamirkulova, G.: Exploring the impact of COVID-19 on tourism: transformational potential and implications for a sustainable recovery of the travel and leisure industry. Curr. Res. Behav. Sci. **2**, 100033 (2021)

Barbieri, D.M., et al.: A survey dataset to evaluate the changes in mobility and transportation due to COVID-19 travel restrictions in Australia, Brazil, China, Ghana, India, Iran, Italy, Norway, South Africa, United States. Data Brief **33**, 106459 (2020)

Beck, M.J., Hensher, D.A.: Insights into the impact of COVID-19 on household travel and activities in Australia – The early days of easing restrictions. Transp. Policy **99**, 95–119 (2020b)

Bhaduri, E., Manoj, B.S., Wadud, Z., Goswami, A.K., Choudhury, C.F.: Modelling the effects of COVID-19 on travel mode choice behaviour in India. Transp. Res. Interdiscip. Perspect. **8**, 100273 (2020)

De Haas, M., Faber, R., Hamersma, M.: How Covid-19 and the Dutch 'intelligent lockdown' change activities, work and travel behavior: Evidence from longitudinal data in the Netherlands. Transp. Res. Interdisc. Perspect. **6**, 100150 (2020)

Duro, J., Perez-Laborda, A., Turrion-Prats, J., Fernández-Fernández, M.: Covid-19 and tourism vulnerability. Tourism Manage. Perspect. **38**, 100819 (2021)

Gandasari, D., Dwidienawati, D.: Content analysis of social and economic issues in Indonesia during the COVID-19 pandemic. Heliyon **6**(11), e05599 (2020)

Google COVID-19 community mobility reports. https://www.google.com/covid19/mobility. Accessed 20 Apr 2021

Güngor, B., Ertugrul, H., Soytas, U.: Impact of Covid-19 outbreak on Turkish gasoline consumption. Technol. Forecast. Soc. Change **166**, 120637 (2021)

Jenelius, E., Cebecauer, M.: Impacts of COVID-19 on public transport ridership in Sweden: analysis of ticket validations, sales and passenger counts. Transp. Res. Interdiscip. Perspect. J. Urban Econ. **8**, 100242 (2020)

Kanda, W., Kivimaa, P.: What opportunities could the COVID-19 outbreak offer for sustainability transitions research on electricity and mobility? Energy Res. Soc. Sci. **68**, 101666 (2020)

Kim, J., Kwan, M.: The impact of the COVID-19 pandemic on people's mobility: a longitudinal study of the U.S. from March to September of 2020. J. Transp. Geogr. **93**, 103039 (2021)

Oztig, L.I., Askin, O.E.: Human mobility and corona virus disease 2019 (COVID-19): a negative binomial regression analysis. Public Health **185**, 364–367 (2020)

Matson, G., McElroy, S., Lee, Y., Circella, G.: Longitudinal analysis of COVID-19 impacts on mobility: an early snapshot of the emerging changes in travel behavior, UC Davis: 3 Revolutions Future Mobility Program (2021). https://escholarship.org/uc/item/2pg7k2gt

Mutlu, O., Durak, Z., Hasan, A.: Staggered working hours in order to reduce traffic congestion. Pamukkale Univ. J. Eng. Sci. **26**, 730–736 (2020)

Ozaras, R., Leblebicioglu, H.: COVID-19 pandemic and international travel: Turkey's experience. Travel Med. Infect. Dis. **40**, 101972 (2021)

Santamaria, C., Sermi, F., Spyratos, S.: Measuring the impact of COVID-19 confinement measures on human mobility using mobile positioning data. Eur. Reg. Anal. Saf. Sci. **132**, 104925 (2020)

Shakibaei, S., Jong, G.C., Alpkokin, P., Rashidi, T.H.: Impact of the COVID-19 pandemic on travel behavior in Istanbul: a panel data analysis. Sustain. Cities Soc. **65**, 102619 (2021)

Willberg, E., Jarv, O., Vaisanen, T., Toivonen, T.: Escaping from cities during the COVID-19 crisis: using Mobile phone data to trace mobility in Finland. ISPRS Int. J. Geo Inf. **10**, 103 (2021)

Zhou, Y., Xu, R., Hu, D.: Effects of human mobility restrictions on the spread of COVID-19 in Shenzhen, China: a modelling study using mobile phone data (2020). www.thelancet.com/digital-health. 2 Aug 2020

The Evaluation of the Impacts on Traffic of the Countermeasures on Pandemic in Istanbul

Mahmut Esad Ergin[1(✉)], Halit Ozen[2], and Mustafa Ilıcalı[1]

[1] Istanbul Commerce University, Istanbul, Turkey
`meergin@ticaret.edu.tr`
[2] Yildiz Technical University, Istanbul, Turkey

Abstract. Governments have taken various countermeasures to slow down the effect of the Covid-19 virus, which has affected the whole world since the beginning of 2020. This study aims to evaluate the impacts of the countermeasures taken by the government on travel behavior in Istanbul, Turkey, through a large-scale survey (approx. 150.000 respondents), remote traffic microwave sensor (RTMS) and transit system electronic toll collection (ECT) data. The countermeasures have been taken by the governments were all day on weekends and between 9 pm and 5 am on weekdays, closure of the restaurants, cafes except take away, stepwise working hour measure and determination of the working hour between 10 am and 4 pm. The survey was developed to allow electronic surveys to be designed on a word processor, sent to, and conducted on standard entry level mobile phones. As a result of the survey, it is estimated that there is a 9% increase in the use of private vehicles, and the road traffic congestion is expected to be increased accordingly. Despite the stepwise working hour measure of the government, the morning and evening peak hours of the traffic did not change. Also, the number of vehicles before and during the pandemic passing through the Bosporus via two bridges which connect the two continents and are the main transportation corridor of Istanbul, is analyzed. According to the November, 2020 data, the number of the vehicles has decreased by almost 14% on weekdays in comparison with the data of November, 2019 for both bridges.

Keywords: Travel behavior · Measure impacts · Pandemic · Survey study · RTMS

1 Introduction

Coronavirus disease (COVID - 19) data and place of the first case are unknown but it is admitted that the first outbreak started from Wuhan in China in late 2019. As the number of cases increased, the alert level was increased by World Health Organization (WHO). After all, on March 11 of 2020, WHO recognized the outbreak as a pandemic which is named Covid-19 (WHO 2020). At the time of the writing this paper, over 146 million confirmed cases of Covid-19, and over 3 million deaths are reported by WHO (WHO 2021).

COVID – 19 has had a huge effect on our lives and has created an extraordinary situation in which people almost all over the world have been lockdown and enterprises and, institutions have been suspended by governments in order to slow down and decrease the effects of the outbreak (Fatmi 2020). Several countries in Europe, including France and Italy, have introduced national lockdowns, banning all non-essential travel. Other countries, such as Sweden, were less stringent and permitted citizens to go to pubs, restaurants, and schools. The Dutch government imposed its so-called "intelligent lockout". People were encouraged to spend as little time away from their homes as possible and to work from home. Furthermore, pubs, stores, classrooms, gyms, and 'touch occupations' were closed, and visitors to nursing homes were prohibited. Even though people were encouraged to stay at home, they were able to walk about easily as long as they maintained a 1.5 m distance from others (de Hass et al. 2020).

As a result of the measures, daily routines of the people dramatically changed. Telecommuting and online shopping are displacing long-standing habits such as driving to work and shopping in-store and the pandemic significantly accelerated the process which is already being undertaken (Shamshiripour et al. 2020). One of the important changes that stand out is the change of travel behavior and numerous studies are being carried out from all over the world (Shamshiripour et al. 2020; de Hass et al. 2020; Circella 2020; Eisenmann et al. 2021; Beck and Hensher 2020; De Voss 2020; Shakibaei et al. 2021; Fatmi 2020; Parady et al. 2020). Many of the studies focus on the process of the measures and compares the time before the pandemic, present (Eisenmann et al. 2021; de Hass et al. 2020) and the time in future (Circella et al. 2020) by using online surveys (Circella et al. 2020; Beck and Hensher 2020; De Voss 2020; Parady et al. 2020), online platforms (de Hass et al. 2020) and mobile applications (Axhausen 2020).

This study focuses on changes in travel behaviors after the first concrete restrictions imposed by the central government in May 2020 when the first times of widespread of pandemic in Turkey. A survey study was carried out in order to investigate the travel behavior of the people because of the new countermeasures, to determine the tendency of private vehicle use, and to find out the reaction of individuals to the pandemic and countermeasures. Besides the survey analysis, the Intelligent Transportation System (ITS) data were used to evaluate the effects of COVID – 19 on traffic flow, Remote traffic microwave sensor (RTMS) and transit system electronic toll collection (ETC) data on Bosporus which is the border of Europe and Asia and, divided Istanbul into two main zone as a screen line, were examined.

2 Countermeasures

Countermeasures taken by the government for the first wave is briefly given as follow after Turkey's first Covid-19 case was confirmed on March 11, 2020 and showed in Fig. 1:

- Border crossing with countries that, had a high risk for COVID – 19, were closed,
- Face-to-face education was interrupted and then entertainment places were closed on 15 and 16 of March,
- As of March 21, going out of the citizens aged 65 and over was restricted in many provinces,
- Flexible working hour regulation was started with the Presidential Circular as of March 22,
- Markets were allowed to be opened between 09: 00–21: 00 on March 24 and education was suspended until April 30,
- All international flights were suspended on March 27 and intercity travels were subject to permission,
- Weekend curfew was imposed for the first time on 11–12.04.2020 following April 10, and this regulation was implemented in the following weeks,
- On May 4, it was decided a relaxation of the restrictions that were generally spread over the months of May, June and July.

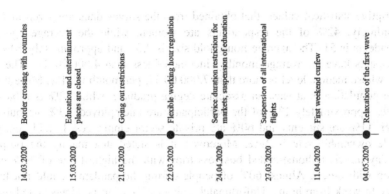

Fig. 1. First countermeasures

3 Survey Data and Questioneries

First of all, it was necessary to determine the method of survey study. In order to examine the impact of the pandemic on people living in Istanbul, a questionnaire is designed. The questionnaire consists of 32 questions, which was prepared to analyze the impact of the Covid-19 outbreak, which negatively affected the whole world, on Istanbul residents and the travel behavior of the individuals. The prepared questionnaire consists of 3 parts: (1) socio-economic characteristics, (2) pre-pandemic, pandemic time and post-pandemic (controlled social life) travel preferences, and (3) stated preferences in public transport service scenarios under pandemic conditions. The pre-pandemic period refers to the situation until the first case is seen, while the period of the pandemic refers to the situation under the restriction decisions taken after the first case, and the controlled social life period refers to the situation where restrictions are gradually lifted, but compliance with cleaning, mask and distance is sought. Since face-to-face surveys were discouraged due to the Covid-19 outbreak, it was planned to perform an online survey as soon as possible, since this would be the most practical method. In a short period of 11 days with the great support of the Istanbul Governorship and Istanbul Metropolitan Municipality, the surveys are conducted. In this context, the support by GSM Operators, that were Turkcell, Vodafone and Turk Telekom, was received and the link of the online survey sent as an operator short message (SMS) to the smart phone users living in Istanbul. Conducting of the surveys were started on 01.06.2020 and ended on 12.06.2020 at 09:00 pm. Random survey links were sent once to the users without distinction of persons' age, employment status, income level, district of residence, etc. Moreover, the distribution of age groups obtained is in accordance with the age group distribution published by the Turkish Statistical Institute (TUIK) for Istanbul in 2019.

4 Analysis of Survey Data

Descriptive statistical values, that obtained from the survey data, are given in Table 1. Accordingly, 42% of the respondents are women, while the average age of the respondents is 51. The average household size is 3.61 and approximately 50% of the participants have an average monthly income of less than 4,000.00 TL. The rate of those whose income level is more than 7,000.00 TL per month is 21%. More than half of the population is at least an associate degree graduate, while 32% is a bachelor's degree. Approximately 13% of the participants are unemployed, 13% are public personnel, 11% are students and 60% are private sector employees. In addition, private vehicle ownership rate is 38%. Moreover, it is stated that mainly the purpose of weekday travels is home-ended business trips with the highest rate of 76% in comparison with others. Almost 60% of people during the pandemic could not find any chance to work from home. Unfortunately, about 27% of respondents stated that their workplace was closed completely or temporarily. Moreover, it was concluded that 38% of the users would not prefer to work from home after the pandemic.

According to the survey data, 34% stated that their starting and ending time of the work were not changed. While the starting time of the work mainly concentrated at around 9:00 am and before (69%), in the duration of the pandemic it was concentrated at around 10:00 am and after (46%). However, there is no significant difference in the end of working hours of pre-pandemic and duration of the pandemic time periods (Fig. 2).

The preferred transportation modes by users before the pandemic, during the pandemic and the controlled social life periods were examined in 3 groups as public transportation (PT), private vehicle (PV) and public transportation + private vehicle (PTPV) (Fig. 3). While the rate of preference of PT for home based work/school/other travels was 65% before the pandemic, this rate decreased to 57% during the pandemic, and according to the data it would increase to 59% in the controlled life period. On the other hand, according to the survey data, the preference rate of the PV mode was 25%, while it was observed that this rate increased to 38% during the pandemic, and this rate will be 34% in the controlled life period. The reasons of the modal shift, respondents highlighted in-vehicle crowding (40%) and hygiene concerns (43%).

Table 1. Descriptive statistics

Criteria	Classification	N	Percentage	Criteria	Classification	N	Percentage
Age	13–24	27,977	19.0%	Education level	No school life	1,059	0.72%
	25–44	91,216	61.8%		Primary school	24,793	16.77%
	45–64	27,650	18.7%		High school	43,052	29.13%
	65+	686	0.5%		Associate degree	18,764	12.69%
Gender	Female	62,247	42.11%		Undergraduate	46,764	31.64%
	Male	85,560	57.89%		Graduate/Master/PhD	13,375	9.05%
HH size	1–4	113,924	77.08%	Occupation	Unemployed	19,329	13.08%
	5–10	33,883	22.92%		Retired	5,061	3.42%
Income	2250	30,690	20.76%		Civil servant	19,148	12.95%
	3250	41,914	28.36%		Student	15,540	10.51%
	4750	27,908	18.88%		Private sector officer	88,729	60.03%
	6250	15,837	10.71%	Willingness for working remotely	No	55,856	37.79%
	7750+	31,458	21.28%		Yes	64,394	43.57%
Private car ownership	No	91,961	62.22%		Undecided	27,557	18.64%
	Yes	55,846	37.78%				

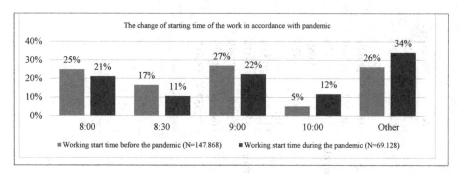

Fig. 2. The change of starting time of the work in accordance with pandemic

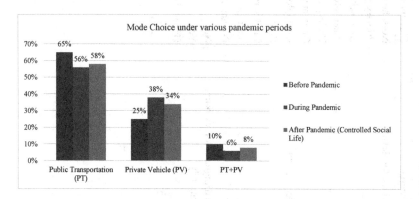

Fig. 3. Modal splits in pre-pandemic, during the pandemic and controlled life periods

As it is seen in Fig. 3, it is expected that the use of PV will be increased (Circella and Dominguez-Faus 2020) by 9% in the controlled life period after the pandemic. Moreover, in the beginning of the pandemic, the use of PV increased dramatically as the use of PT decreased, however after a while, the use of PV decreased and on the other hand the use of PT increased slowly. This fluctuation represents adjusting of the people to the pandemic in terms of transport modes. In the analysis of stated preferences, 39% of respondents would not change their transportation mode which was used daily before the pandemic. It has been stated that 34% of the users will use PV during the controlled life period, and 40% will continue to use PT.

In order to examine the transportation mode choice reaction of the PT users, additional questions were asked. As it is seen in Fig. 4 that the rate of PT users who would not prefer PT in the absence of hand disinfectant in PT is 23%, while the rate of those who will definitely not use PT, it is 34%, the rate of those who will not prefer PT if PT is not suitable for social distance is 24%, and the rate of those who will not absolutely choose 47%. Moreover, if masks are not used in PT, 22% of the participants stated that they would not prefer public transportation and 43% of them would definitely not (Fig. 2). Consequently, approximately 65% of those who used public transportation before the epidemic stated that they would not prefer public transportation if the necessary measures are not taken.

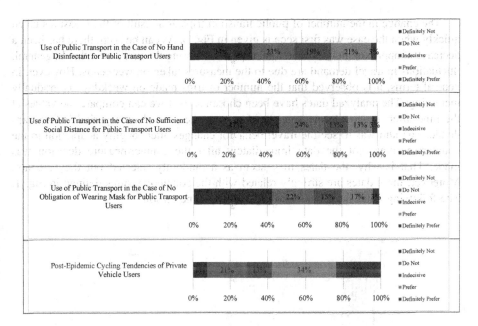

Fig. 4. Reasons for change in mode preference during the controlled life period

According to the statistics of the use of public transport, including the time period in which this survey was conducted, the use of public transport has shown a tendency to decrease since the first case was seen, but there was an increase in the use of public transport after the decision to gradually remove the restrictions in the summer months. Figure 5 presents that the mobility by public transportation modes is almost the same proportions with the rate of pre-pandemic period. However, it should be noted that the use of sea transportation is more than doubled in during controlled social life times. As a result of this, it can be considered that the PT users prefer sea transportation because of the possibility to transport in the open air.

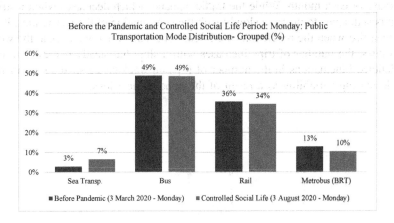

Fig. 5. Public transportation mode distribution on mondays

The change in the number of public transport trips as a result of the measures taken quickly when the case was first seen is given in Fig. 6. As can be seen, there has been a decrease of approximately 80% in total trips made during the first 3 months. In general, fluctuations in travel demand are due to the measures taken on weekends. However, in general terms, it is observed that the number of trips made on weekdays is gradually increasing. The analyzed dates have been chosen so that we can compare the values at the time after the countermeasure decision with the normal time period before pandemic. The aim was to see the travel behavior changes after roughly adaptation to the countermeasures, not the date immediately after the countermeasure decision was made. That is why, the dates may seem as a randomly selected but they are not. Moreover, these dates are strongly related with the countermeasures defined in Fig. 1. The following dates have been chosen with the same reason.

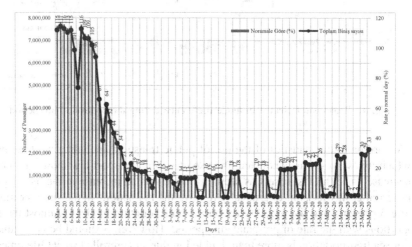

Fig. 6. Immediate effects of the countermeasures in the first 3 months

Figure 7 gives the average number of public transport trips on Mondays and Saturdays of each month. While the traffic demand, which decreases when restrictions are imposed, tends to increase slowly with the continuation of the restrictions, it shows a sudden rise when the restrictions are lifted. Especially when we look at the weekend trip values, the number of trips increases with a slow but steady rise even in case of restrictions. Fluctuations in the number of trips on weekdays also show an increasing and decreasing variability as a result of the political decisions.

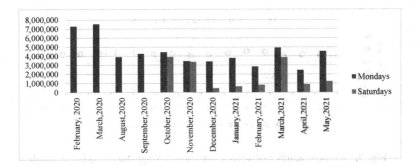

Fig. 7. Public transport trips on Mondays and Saturdays of each month

There was an increase in the use of public transport, especially on Mondays following the weekends when the curfew was imposed. This situation is thought to indicate that people who cannot leave their home for about two days tend to go out more when their freedom comes, even if they do not have to travel.

5 Analysis of RTMS Data

According to the pre-pandemic travel demand, especially in private vehicle journeys, it decreased by approximately 22% during the pandemic process on weekdays, while there was a great decrease due to the restriction decisions taken on weekends. According to the vehicle count data for December 2019 and 2020, this situation works for Fatih Sultan Mehmet Bridge and July 15 Martyrs Bridge and is shown in Fig. 8 and Fig. 9, respectively. It should be noted that pandemic does not affect the weekday travel demand significantly.

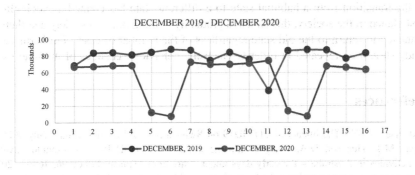

Fig. 8. Average number of vehicle passing through Fatih Sultan Mehmet Bridge

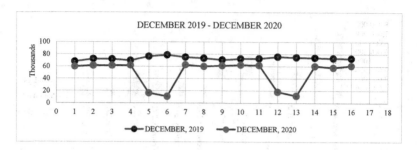

Fig. 9. Average number of vehicles passing through July 15 Martyrs Bridge

6 Conclusions

From the first appearance of the cases, different countermeasures continue to be taken according to the course of the case in various periods. In the process we are in, a constantly changing policy is being implemented. After the countermeasures taken with the first wave, there was a decrease of up to 80% in private vehicle travels. In addition, the number of passengers decreased by half in the pre-pandemic public transport travels, which approached 8 million. However, as can be seen from the graphics, even if there is no change in the measures, there is a slow increase in the number of passengers. Especially in these days when the summer season has come, less care is taken to obey the new rules. Although public transportation users state that they will pay close attention to disinfectant, mask and distance rules, public transportation usage tends to increase despite the measures taken. It is thought that this is due to the fact that users do not have a different alternative. Moreover, it was observed that some measures did not have a significant effect on travel behavior. For example, after the regulation made within the scope of the starting and ending time of the working hours, the peak time of the home-based trips during the morning peak and the non-home-based trips during the evening peak hours did not change. The countermeasures could not change the peak hours of the daily routine of the city. As a result, although the urban memory or the transition from a habitual state to a different state has created a shock effect at first glance in the society, this shock is gradually coming out as a society and there is a tendency to return to the old state regardless of the number of cases. Furthermore, this study should be repeated in order to catch the behavioral changes of the road users.

References

Axhausen, K.W.: The impact of COVID19 on Swiss travel. TU Delft Webinar, July 2020

Beck, M.J., Hensher, D.A.: Insights into the impact of COVID-19 on household travel and activities in Australia – the early days under restrictions. Transp. Policy **96**, 76–93 (2020)

Circella, G., Dominguez-Faus, R.: Impacts of the COVID-19 pandemic on transportation use: updates from uc davis behavioral study. 4th of August, 2020. Webinar: UC Davis Institute of Transport Studies, 3 Revolutions Program (2020). https://its.ucdavis.edu/blog-post/impacts-of-the-covid-19-pandemic-on-transportation-use-updates-from-uc-davis-behavioral-study/

De Hass, M., Faber, R., Hamersma, M.: How COVID-19 and the Dutch 'intelligent lockdown' change activities, work and travel behaviour: evidence from longitudinal data in the Netherlands. Transp. Res. Interdisc. Perspect. **6**, 100150 (2020). https://doi.org/10.1016/j.trip.2020.100150

De Voss, J.: The effect of COVID-19 and subsequent social distancing on travel behavior. Transp. Res. Interdisc. Perspect. **5**, 100121 (2020)

Fatmi, M.R.: COVID – 19 impact on urban mobility. J. Urban Manag. **9**(3), 270–275 (2020)

Parady, G., Taniguchi, A., Takami, K.: Travel behavior changes during the COVID-19 pandemic in Japan: analyzing the effects of risk perception and social influence on going-out self-restriction. Transp. Res. Interdisc. Perspect. **7**, 100181 (2020)

Shakibaei, S., de Jong, G.C., Alpkokin, P., Rashidi, T.H.: Impact of the COVID-19 pandemic on travel behavior in Istanbul: a panel data analysis. Sustain. Cities Soc. **65**, 102619 (2021)

Shamshiripour, A., Rahimi, E., Shabanpour, R., Mohammadian, A.: How is COVID-19 reshaping activity-travel behavior? Evidence from a comprehensive survey in Chicago. Transp. Res. Interdisc. Perspect. **7**, 100216 (2020)

World Health Organization (WHO) (2021). https://covid19.who.int/. Accessed 26 Apr 2021

World Health Organization (WHO): WHO timeline - COVID-19 (2020). https://www.who.int/emergencies/diseases/novel-coronavirus-2019/interactive-timeline. Accessed 26 Apr 2021

New Approaches to Performance Delivery

From Reactive to Proactive Maintenance in Road Asset Management

Timo Saarenketo[1(✉)] and Vesa Männistö[2]

[1] Roadscanners Oy, Rovaniemi, Finland
timo.saarenketo@roadscanners.com
[2] Finnish Transport Infrastructure Agency, Helsinki, Finland

Abstract. In road asset management, reactive pavement maintenance is still used in most countries. In this policy, corrective measures are only initiated after clear pavement distress or other deficiencies in road condition have been identified. However, in the philosophy of proactive maintenance, the root causes of damages are always determined before making any corrective measures. In addition, any measures chosen, focus on eliminating or reducing the causes of the problems. The central theme of proactive maintenance is to extend the service life of pavement in accordance with the following principles: 1) the avoidance of unnecessary repairs, 2) damage must never become an acceptable "normal," and 3) approaching damage is monitored and responded to in a timely manner.

In Finland, a pilot towards proactive asset management started in 2015, in cooperation between The Finnish Transport Infrastructure Agency and Roadscanners. The goal of the pilot was to develop and test new innovative methods to improve the productivity of paved roads maintenance. A further aim was to provide new information on the life cycle costs of various roads and the factors that increase these costs.

The results of piloting have been promising. The weakest links increasing asset management costs have been identified and maintenance measures addressing the real root causes of the problems have been implemented. As a result, calculated annual paving costs have been reduced already by 20–40%. In addition, valuable information has been collected of the LCC of different maintenance measures. This information is used when moving towards more proactive maintenance policies.

Keywords: Road asset management · Pavement management · Proactive maintenance · Life cycle costs · PEHKO project · Road diagnostics

1 Introduction

Road asset management policies have been mainly based globally on reactive maintenance, where corrective measures have been initiated only after clear pavement damage or another types of deficiencies in structural or functional condition have been identified. A good example of this is the fact that roads are only paved after visible pavement distress has been recorded. The problem with this policy is that the cracked asphalt has at that already lost its tensile strength and the service life of the new coating is significantly lower, and the life cycle cost higher, than if the coating had been repaired before damage. Therefore, monitoring pavement distress only measures how late the needed maintenance actions are. Rut depth or roughness limits are also used as trigger tools in reactive maintenance based on the impact on road safety. But a good indicator or parameter to describe life cycle costs have been missing from the asset management toolbox.

The next step forward from reactive maintenance is preventive maintenance, in which the timing of maintenance measures is based on specific time periods, for example repetitive drainage measures at certain intervals. Another good example in winter maintenance is preventative de-icing based on a weather forecast. However, a problem with time-based preventive maintenance on pavements is that road and street structures do not behave homogeneously over time. If the action date is chosen according to the average service life, then 50% of the length of the road section could be treated too late. If the action date is moved earlier, it will be immediately reflected in the annual cost and resources will be wasted at the same time.

In the philosophy of proactive maintenance, prior to repair, the root causes of premature damage are always determined and the chosen corrective actions focus on eliminating, or at least improving, the causes of these problems. The central theme of proactive maintenance is to extend the service life of machines, or roads, in accordance with the following principles. 1) repairs are not carried out if nothing is wrong, i.e. for safety's sake, 2) damage must never become acceptable "normal," and 3) the approaching damage condition is monitored and responded to in a timely manner.

2 Steps Towards Proactive Maintenance

2.1 General

As stated earlier, proactive road maintenance requires first of all a good knowledge of road diagnostics techniques and an understanding of pavement fatigue and permanent deformation and other processes behind road damages. This is made possible with the new road survey technologies currently available that give a comprehensive picture of the functional and structural of the road, and also its surroundings and their interrelationships (Saarenketo et al. 2012a; Herronen et al. 2015; Saarenketo 2016). The implementation of this new technology, after appropriate data analysis techniques, will enable new maintenance operations to be put into practice.

2.2 Technology Steps

Fast developments in digitalization and new innovations in non-destructive techniques (NDT) in road condition monitoring have prepared the path towards proactive road maintenance. Earlier, condition criteria based on the measurement of rutting and roughness, and evaluation of pavement distress, was used to inform reactive maintenance. Over the recent years road condition evaluation has been moving towards comprehensive time series analysis of both pavement surface and road and its surrounding, subsurface structures and their condition as well as stiffness of the pavement structure. Thanks to the development of sensor fusion techniques all the needed data sets can now be collected by one or two road survey vehicles.

One of the major issues in the implementation of Big Data is the development of data processing and data storage capacity When the first pavement management systems (PMS) were developed in the 1980's the data storage capacity was so expensive that the PMS data averaging distance was designed to be 100 m. This was sufficiently accurate for national and network level evaluation for road condition, but not for project level systems or for proactive maintenance, where the length of many critical pavement problems can generally be measured in tens of metres.

Another big technical step forward has been the development of global positioning systems (GPS) that have enabled accurate positioning of the road surveys data and also the development of reliable time series analysis.

The last critical technological development towards proactive maintenance are the 4G and 5G networks, that enable cost effective real time monitoring of the functional condition of the road network.

2.3 Data Analysis Techniques

In reactive maintenance the motivation of the maintenance crew has been a fairly easy task as problems are mainly seen by the human eye. In proactive maintenance however decisions on maintenance measures are mainly made before visual indications of problems can be seen. That is why road condition survey data processing, analysis, diagnostics and especially visualization of the results has to be made in such a way that the whole maintenance organisation can internalize the results and understand the facts behind the maintenance measures (Fig. 1). In addition in pavement management the new technologies, and their precise analysis result based time series, provide new possibilities for forecasting and optimizing the life time of pavement sections (Fig. 2). For this reason road asset condition analysis can be divided into the following areas (Gandomi and Haider 2015; McMahon et al. 2020).

a) descriptive analytics, which include reporting/online analytical processing (OLAP), dashboards/scorecards, and data visualization.
b) diagnostic analytics, which is used for discovery or to determine why something happened;
c) predictive analytics, which can suggest what will occur in the future. The methods and algorithms for predictive analytics include regression analysis, machine learning, and neural networks; and
d) prescriptive analytics, which can identify optimal solutions, often for the allocation of scarce resources. Prescriptive analytics are seen as the future of big data.

Fig. 1. Example of a visualization of a road survey and diagnostic data presented on a point cloud background. The red and green colors present rutting and deformation in the road surface, and the pink and blue colors present high moisture content inside the pavement structure. Together these two datasets demonstrate how drainage problem and moisture in the road structure can lead to pavement permanent deformations.

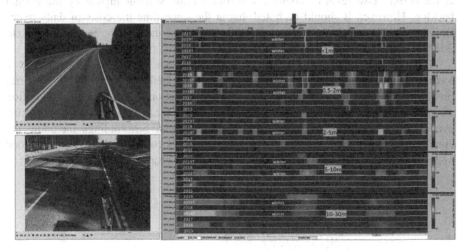

Fig. 2. Time series analysis of road roughness using vertical acceleration wave lengths from road 83 Sinettä – Pello in Finland. Data has been collected from 2015 to 2019 and in 2018 and 2019 roughness has been collected and analysed also in the winter time. The data shows the quick development of transverse cracks causing increasing annoying shocks to cars, and also early phase transverse cracks where the growth could be slowed down or prevented with sealing and another proactive measures.

3 Experiences from PEHKO Pilot Project in Finland

3.1 PEHKO Project Goals

Since 2015 the new technologies and policies, including proactive maintenance in road asset management, have been tested "full scale" in a 10 year PEHKO project in Finland (Fig. 3). The project's goal is to improve practices and policies in paved road maintenance and management and thereby improve the condition of the paved road network, or at least keep it at the current level using less resources. This is being done by focusing on three targets: 1) Improving daily maintenance, especially drainage. 2) Applying new NDT methods in the diagnostics of paved roads. These techniques allow engineers to focus rehabilitation measures exactly on the problem sections and address the root. 3) Changing maintenance policies from reactive to more proactive maintenance, allowing maintenance crews to fix the potential problem sections before serious pavement damages appear (Tapio et al. 2016; Saarenketo et al. 2019). The total length of the paved road network in the PEHKO areas is roughly 2400 km.

The PEHKO project has very ambitious financial goals, based on the ROADEX results and recommendations, (Saarenketo et al. 2019) by requiring that in 2025, the annual pavement costs in the pilot areas should be about 50% lower than the former computational costs (= annual paving budget + annual backlog). This would mean that in 2025 and 2028 the PEHKO roads should be able to be maintained in good condition at the current cost level of about € 130–140 million without increasing the budget backlog. However, it should be noted that at the same time the costs of daily maintenance would be expected to increase slightly from the current level (Saarenketo 2019).

The data collection in the PEHKO pilot has been made using a Road Doctor Survey Vehicle (RDSV), and the continuous Traffic Speed Deflectometer (TSD). This has proven to be a very good combination for the functional and structural condition analysis of roads (Saarenketo 2017). At the same time it has also been possible to map even small changes in the pavement surface and road surroundings including the condition of the road drainage system. RDSV data is collected throughout the road network annually and TSD measurements every fifth year (Saarenketo et al. 2019).

Fig. 3. The first PEHKO project pilot areas in 2015 were located in Finnish Lapland and in Central Finland and from 2018 in Southern Finland.

Efforts have also been made in the PEHKO project to take into account the requirements of future digitalisation and intelligent traffic developments. For this reason digital point cloud models were developed for the entire road network surveyed and all the collected data linked to these models. This makes it possible to gather accurate historical data and to move forward to proactive maintenance based on the monitoring results.

The basic asset management process used in the PEHKO pilots is described in the following (Saarenketo et al. 2019):

1. Surveying the paved road network using new techniques, locating the problem sections and defining their root causes.
2. Identifying road sections that are still in good condition, but are close to the end of their service life, ie before visual damages can be seen. These sections are paved with a new and, if needed, thicker pavement.
3. Identifying road sections suffering from inadequate daily maintenance. These sections are treated by raising the standards of daily maintenance and improving practices and technologies.
4. Locating quickly damaging paved road sections and diagnosing the root causes of these damages. A focused rehabilitation plan is prepared for each section and executed.
5. Continuing annual monitoring of the entire road network. This involves launching proactive maintenance operations immediately if the monitoring results indicate an increased risk of pavement damage. The effectiveness of various treatments and remedies is also monitored.
6. Repeating steps 1–5.

3.2 PEHKO Results, Weakest Links

In the first PEHKO years the project plan was, according to the proactive maintenance basic idea, to focus on finding the root causes for the premature pavement fatigue and deterioration identified. As a result altogether 6 main factors were found that explained the problems behind the road sections with highest life cycle costs (Saarenketo et al. 2019).

1. Heavy trucks on road sections with weak subgrade.
2. Heavy trucks and thin pavements (<150 mm).
3. Pavement quality, including aggregate quality, debonding, creep, remix pavements, poor quality patching, etc. Pavement quality was found to be a problem also in thicker pavement sections.
4. Poor daily drainage maintenance. This applies mainly to private access road junctions, or to poorly performing and shallow side ditches (Fig. 4).
5. Pavement deterioration problems due to inadequate winter maintenance. The main reason identified was delayed snow removal from the road shoulders.
6. The extensive use of de-icing salt on thin and porous pavements. This is a recently discovered cause for increasing rut growth rates.

Once these factors were identified they could be used as a gubasis for planning maintenace and rehabilitation measures and policies towards minimazing the effects of the factors. For instance a special steel grid structure has been used on road sections with weak subgrades. This structure has proven to cut down well permeanet deformations on the these sections.

Also thin pavements have been made thicker, especially when pavement strains have indicated approaching high risk for pavement cracking. These measures follow the third principle of practive maintenance. The solution has proven to be succesful and according to the latest results a new overlay has cut down pavement strains by around 50–80 microstrains, which means 4–10 million new axles in the first phase, and more than 10 million axles in the second phase (see Fig. 5). At the same time remix treatment has not improved the stiffness of the pavement.

Road sections where problems could be related to poorly performing road drainage, have been analysed and their severeness ranked, and guidelines for drainage improvement made. A good way for motivating maintenance organizations for better maintenance actions has been to show the impact of maintenace deficiencies in the local currency.

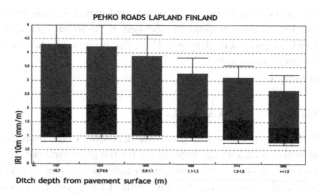

Fig. 4. Distrubution of 10 m IRI value on road sections with different ditch depths on PEHKO roads in Finnish Lapland. Data shows improvement of the median values of IRI when ditches were deeper. But even more important is the information is that high roughness values could be detected only on section with ditch depths of less than 1.1 m.

3.3 Experiences of Proactive Maintenance Actions

In Finland the decision for a new asphalt overlay or pavement remix is based on pavement distress data and/or high rutting or roughness survey results. However, the big problem with this procedure is that asphalt that is already cracked has already lost its strength and the new overlay will continue to have all the shear stresses focused exactly on the points with cracks beneath the new asphalt. According to the theoretical calculations made in the Finnish PEHKO project this can reduce the lifetime of the new asphalt, and calculated annual paving costs can be 60% lower in comparison if the new overlay has been made before cracks have appeared. In addition, microcracks will be expose a pavement to water infiltration, and then the pavement will become susceptible to freeze-thaw related deterioration. This is why new cost-effective proactive pavement asset management requires information concerning early phase microcracking in the asphalt.

The proactive approach has now been tested in several road sections and pavement strains have been calculated from TSD data both in 2015 and 2020 before and after the new overlay has been made (Figs. 5 and 6). The results show that the improvement in the average pavement strain, have been of the order of 70–80 microstrain, and in some cases 100 microstrain (Fig. 6). This has been slightly smaller than the theoretically calculated 150 microstrain but still very significant. If the average initial value was around 270 microstrain the improvement was around 4 million axles, but if the initial average strain value was 220 microstrain the improvement was 15 million standard axles. In the cases of remix pavement there were no improvement of the pavement stiffness. This results clearly show the economical benefits of placing thicker overlays even in the earlier phase of the pavement fatigue, ie. when asphalt strains are around 250 microstrains.

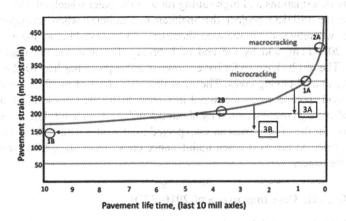

Fig. 5. Modelling and PEHKO results of the development of asphalt pavement strain during the last 10 million axles before pavement failure according to the Finnish asphalt design guide. Figure also shows areas where microcracking (1A) and macrocracking are (2A) initiated. Figure shows the calculated lifetime of a new 40 mm overlay if paving has been done in the microcracking initiation phase (1A, 17 million axles), or if it is done after macro cracks have appeared (2B, 3.5 million axles). Graphs 3A and 3B show also the measured average strains of two cases in the Lapland PEHKO area where the new overlay has been placed when the initial average strain value was 280 microstrain (3A) and 220 microstrain (3B). The improvement in both cases was roughly 70 microstrain.

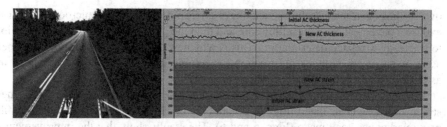

Fig. 6. Example of a PEHKO case in road 81 where pavement thickness was increased from 50 mm to 100 mm and at the same time the pavement strains were improved by roughly 100 microstrain, approximating to 4 million axles.

Another important and successful finding of the benefits of proactive maintenance has been the early removal of the snow from the road shoulders. This has been traditionally done in Finland in March-April when the snow has started to thaw in the road shoulders. However latest research results have shown that this has been leading water infiltration into the base course under the pavement due to cryosuction and further frost heave in the pavement edge. Later in the spring the thawing of these ice lenses lead to permanent deformations and high rutting rates in the outer wheelpath (Vuorimies et al. 2020). In the PEHKO project the maintenance policies where changed with new guidelines whereby snow in the road shoulders was removed before the end of January down to 20–30 cm, and by the end of February the snow ion the road was totally removed. The results have so far have shown that this poicy has had 10–20% impact on cutting the annual paving costs. The most dramatic effect of the failed removal of the snow in road shoulders was measured in winter 2019–2020 in Central-Finland where the annual paving costs increased by about 0.5 million € on a 400 km of road sections. The reason for the failures was an unexpected mild winter with several repeated snow thawing cycles on the road when maintenance crews were always late with the snow removal measures.

3.4 Life Cycle Cost Improvement 2016–2020

In the PEHKO project, life cycle costs of the paved roads, especially on main roads, have been calculated based on annual rut increase because historically in Finland it has been the main factor in triggering the need for the repaving or rehabilitating the road. And on the Finnish paved road network increased rutting rates can be always measured before visual cracks appear. Figure 7 shows the development of annual paving costs in the PEHKO pilot areas on main roads and on the whole road networks. Over the first years the PEHKO proactive maintenance concept had been mainly focusing on the main road as can be seen in the reduced annual paving costs, which are in Lapland 27% and in Uusimaa even after first year already 22%. In the Central Finland pilot area the improvement was fast over the first years and in 2019 the calculated savings were already 38% compared to the 2015–2016 costs, but then the winter 2019–2020 was very mild and winter maintenance operations could not react to the quickly changing weather conditions. This led water from melting snow to infiltrate into the pavement structure and further quicken pavement deformations and rutting. As a result the annual paving costs in 2020 were 5% higher than in 2015–2016 in main roads, and 17% higher on the whole road network.

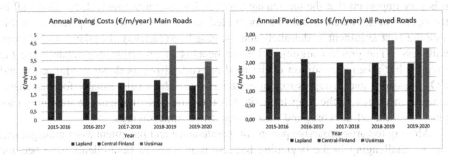

Fig. 7. The development of average annual paving costs for main roads (left), and all roads (right) in the PEHKO road network in Finland from 2015–2016 to 2019–2020.

Presenting average or median values give only a part of the story, because life cycle costs of road sections are mainly defined by the worst 10–20% values of each road section. Figure 8 presents a distribution chart of annual rut increase in different AADT classes in the Uusimaa area in 2018–2019. It shows that in the same AADT class the annual rut increase, and thus life cycle costs, are often more than 3–4 times higher on the worst 10% of the case compared to the best 10% road sections. For instance the annual rut increase on the worst 10% of AADT class <1000 vehicles is higher than the median value of AADT 8000–12,000 vehicles.

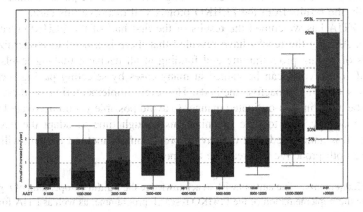

Fig. 8. The distribution of annual rut increase as 10 m values (mm/year) in different AADT classes in the Uusimaa PEHKO pilot area roads in 2018–2019.

4 Conclusions

Digitalization and fast development of new survey technologies have been enabling the movement towards proactive road maintenance. But experiences have already shown that digitalization is only 1/3 new technology, and 2/3 change in mindsets. That is why it is very important that the top management and middle management, but also staff in the operation activities, should be committed to the new modes of operation. In this area the best argument is definitely the increased productivity, in other words "more with less". The greatest argument for the benefits of proactive maintenance are the savings, but a better condition of the road network also cuts down driving costs and CO_2 emissions and provides road users better and safer roads. Finally focused maintenance measures in the long term shorten the time that maintenance measures take place on the road network, and thus lead to travel time savings.

Moving towards proactive pavement maintenance and pavement asset management can be challenging in the beginning because road users as final clients have become used to the fact that engineers only fix poor quality roads. That is why, based on the Finnish experience, the new methods should also be presented to the public, i.e. "the taxpayers", to inform them how much of their money it saves.

One of the most important learning experiences of the PEHKO project has been the fact how big the impact daily maintenance has had on the life cycle cost of the paved road network. The 0,5 million € extra cost due to winter maintenance problem in annual paving costs of 400 km of road network in Central Finland demonstrates well, that good quality proactive maintenance policies should also include winter maintenance.

So far the first proactive maintenance principle, "repairs are avoided if nothing is wrong" has not gained general acceptance in Finland, and long good quality road sections continue to be repaved or remixed within paving projects even though their expected life time could be still more than ten years. The argument has been that productivity is better in paving works if long sections can be paved without stopping. Hopefully the results from the PEHKO project will disprove this case.

The FTIA has welcomed the results of the first half of the PEHKO research programme. Already the results have revealed that there is a need to modernize current methods of paving programming and funding at all decision making levels. Clearly, lower life cycle costs can be gained in many cases by selecting paving sections and treatments utilizing proactive methods. However, implementation of these methods need to be made in a co-ordinated manner to avoid possible yet undiscovered pitfalls in the methodology, and to give organizations enough time to adapt the new way of thinking. It is anticipated, that the paving programming method toolbox will in future comprise both proactive and reactive treatments.

Acknowledgements. The authors wish to acknowledge the fruitful co-operation of regional Ely-centres in the implementation of the PEHKO research programme, as well as FTIA for financing the programme.

References

Gandomi, A., Haider, M.: Beyond the hype: big data concepts, methods, and analytics. Int. J. Inf. Manage. **35**(2), 137–144 (2015)

Herronen, T., Matintupa, A., Saarenketo T.: Experiences of integrated analysis of TSD, GPR and laser scanner data. In: 2015 Proceedings of International Symposium on Non-destructive Testing in Civil Engineering (NDT-CE) (2015)

McMahon, P., Zhang, T., Dwight, R.: Requirements for big data adoption for railway asset management. IEEE Access **PP**(99), 1 (2020)

Saarenketo, T.: Experiences of integrated GPR and Laser scanner analysis – we should not only look down but also around. In: Proceedings of the 17th Nordic Geotechnical Meeting Challenges in Nordic Geotechnic 25th–28th May 2016, Reykjavik (2016)

Saarenketo, T.: What new technologies can provide to intelligent road asset management. In: Proceedings of the 29th International Baltic Road Conference, 28–30 August 2017, Tallin (2017)

Saarenketo, T., Munro, R., Matintupa, A.: PEHKO project, implementing recommendations for rural road asset management in Finland. In: Proceedings of the International Symposium on Cold Regions, ISCORD 2019, 17–19 June 2019, Oulu, Finland (2019)

Saarenketo, T., Matintupa, A., Varin, P.: The use of ground penetrating radar, thermal camera and laser scanner technology in asphalt crack detection and diagnostics. In: Scarpas, A., Kringos, N., Al-Qadi, I.A.L. (eds.) 7th International Conference on Cracking in Pavement. RILEM Book Series, vol. 4, pp 137–145. Springer, Dordrecht (2012a). https://doi.org/10.1007/978-94-007-4566-7_14

Saarenketo, T., Matintupa, A., Kourim, B.: Experiences with new technologies in road problems diagnostics. In: Proceedings of EPAM 2012 Conference, Malmo, Sweden (2012b)

Tapio, R., Lehtinen, J., Ylinampa, J., Saarenketo, T.: PEHKO project 2015–2025, increasing the productivity of paved road management in Finland. In: Proceeding of EAPA Conference, Prague 2016. Digital Object Identifier (2016). https://doi.org/10.14311/EE.2016.144

Vuorimies, N., Kurki, A., Kolisoja, P.: Keskeisiä havaintoja tierakenteen instrumentoinnnista Aurora koekohteilla 2017–2020. Loppuraportti. Väyläviraston julkaisuja (in print) (2020). In Finnish

Observing How Influence of Nature Phenomena Against Inside Tunnel by Air Pressure Information

Kensaku Kawauchi[1(✉)], Yumi Watanabe[2], Yuichi Mizushima[2], and Takeo Hosokai[3]

[1] National University of Singapore, Singapore, Singapore
kensaku@nus.edu.sg
[2] Nexco Engineering Niigata Co. Ltd., Niigata, Japan
[3] East Nippon Expressway Company Limited Niigata Regional Office, Chiyoda City, Japan

Abstract. Monitoring inside the tunnel is important to be safe for road. A long mountain tunnel such as Kan-Etsu tunnel must ventilate the road. Ventilation system for Kan-Etsu tunnel controls the rotation speed of fans installed inside the tunnel to fill the lack of wind volume for ventilation. However, the fans cannot change rotation speed abruptly. Therefore, the ventilation system estimates near future conditions and calculates the necessary rotation speed. The ventilation is influenced by winds which are generated by traffic cars and nature phenomena, so that its estimation needs to recognize cars and nature effecting. Although the existing system measures traffic influence with a traffic counter installed in front of the tunnel, the system cannot measure natural phenomena. In addition, installed wind sensor in the entrance measures wind speed. The measured speed is treated as a ventilation result, but natural wind cannot be extracted from the sensor result. Therefore, the influence from nature wind haven't been clear. To measure the influence is important to improve the ventilation controlling in the tunnel. In this paper, we reported influences of natural phenomena against inside tunnel conditions by monitoring air pressure. By observing pressure movement at each entrance, nature effecting can be measured. In this paper, pressure sensor arrays were installed at entrances and inside roads. By observing sensor behaviors, tunnel condition was monitored. The observation was executed multiple times but different seasons. As a result, this paper showed natural wind from each entrance influences the tunnel inside condition.

Keywords: Environment influence · Ventilation · Kan-Etsu tunnel · Pressure sensor array · Pressure gradient

1 Introduction

Maintaining a highway tunnel is important for drivers to drive at high speed, safely.

Exhaust gas, which the car discharges, influences visibility of the road in the tunnel. For keeping clear conditions, a ventilation system is installed into the tunnel. If the tunnel length is long, the ventilation role is more important. Kan-Etsu tunnel (JianDong 2011; Nakahori 2010) is one of the longest mountain tunnels in Japan. The location is in the mountains where natural phenomena affect the environment inside the tunnel. In addition, the whole area has to be clean for visibility. Therefore, the ventilation controlling is a complex structure because the system ventilates using multi sources to cover the whole area.

Ventilation system in Kan-Etsu tunnel has multiple roles. The main purpose is to discharge dust such as exhaust gas to the outside. As one of other roles, the system controls smoke flow for people to leave to safety space in a fire accident. When a fire accident occurs in a tunnel, smoke spreads into the tunnel. People cannot breathe through the smoke. In addition, people cannot move to a safe space because they cannot see the view through smoke. Ventilation system handles smoke flow not to spread in the tunnel and avoids those situations.

The system in Kan-Etsu tunnel ventilates the tunnel inside with wind, but blowing direction has to be considered. To ventilate the whole area, Kan-Etsu tunnel ventilates inside with winds generated by cars, installed blowers and natural phenomena. By effectively ventilating control with winds, the system ventilates efficiently inside the tunnel. Blowing direction is monitored by installed sensors because the system needs to avoid situations that ventilation wind influences drivability. However, influence from the natural environment against the tunnel has not been monitored, though wind has been monitored. In this paper, we investigated the influence of ventilation effecting from natural phenomena. Therefore, we observed pressure movement by pressure sensors to measure influence from natural phenomena against Kan-Etsu tunnel.

Monitoring pressure movement with pressure sensors is useful to recognize natural phenomena, especially nature wind. From the pressure movement, pressure distribution can be estimated. The pressure distribution shows a summary of air flow on the air. If the relationship between high- and low-pressure positions can be understood, wind direction and volume can be estimated by the position (Wallace 2006; Holton 1973). This knowledge can be adopted into calculating the influence from nature, pressure distribution can be estimated by analyzing data from pressure sensors installed at some places. Tunnel is a simple structure like a tube. Therefore, pressure gradient can be instead of the distribution. Pressure gradient can be shown in a gap of pressure value between two observation areas.

However, pressure sensors receive noise from other phenomena. To measure potential power by nature, those influences have to be removed from pressure value. For example, wind pushes sensor measurement space and pressure values increase. The phenomena, which influences pressure sensor value, can be separated by two things. One is wind generated by a blower or car, and the other is air movement by pressure gradient. Wind generated from pressure gradients is influenced from the natural environment so that wind generated from the others has to be removed to measure the natural influences against the tunnel.

To remove that noise, we adopted a pressure sensor array system to monitor pressure movement. Single pressure sensor reads pressure values, but the origin of change can not be analyzed from transition data. The reason is that the element can not be extracted from the pressure value. Then, comparison with pressure values for measuring the gradient is difficult using normal pressure sensors installed at different observations. Pressure sensor measures air weight on the sensor, and air volume is different by altitude and by temperature. Therefore, each installed position to observe pressure gradient has to be in a similar environment and same weather condition for removing wind influences. It is difficult to prepare the places in the actual environment. On the other hand, a pressure sensor array consists of 25 pressure sensors on a single board. To measure pressure gradient can be realized on one board by comparing with pressure sensors.

One of the measurement tools to evaluate ventilation affection is the wind sensor. Wind sensor measures the volume of air flow and its volume equals ventilation affection. To monitor ventilation, the wind sensor, which measures wind speed and direction with doppler affection, is installed on the ceiling at each entrance in Kan-Etsu tunnel. However, wind sensor value includes total value from all phenomena so that wind sensor cannot extract a potential information generated by nature.

To increase ventilation quality, investigation of influence from the natural environment against tunnels is needed. However, extracting influence from a sensor data is difficult for a wind sensor or single pressure sensor. In this paper, we proposed a method to monitor the influences with a pressure sensor array. Following is our contribution:

- Visualizing influences from around nature environment against Kan-Etsu tunnel
- Proposing a method to visualize tunnel condition with pressure information from sensor array Investigating tunnel condition at more than once.
- Investigating tunnel condition at more than once.

2 Kan-Etsu Tunnel

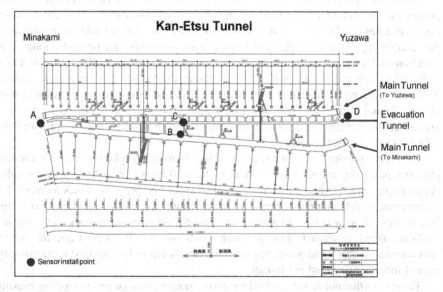

Fig. 1. Kan-Etsu tunnel outline. Kan-Etsu tunnel consists of three tunnels and other tunnels for facility.

Kan-Etsu tunnel (shown in Fig. 1) is located in the high mountains on the border of three prefectures in Japan. This tunnel length is 11 km between entrances. Although wind from the entrance does not reach to the end of the tunnel because of the length problem, ventilation wind has to be provided to the whole area. In addition, the ventilation controlling system has to manage smoke during fire accidents inside the tunnel. To solve the ventilation problem, Kan-Etsu tunnel has a complex ventilation controlling system.

Kan-Etsu tunnel consists of two main tunnels, an evacuation tunnel, connection roads which are joining between main and evacuation tunnels, and tunnels for blowing and discharging air as a ventilation function. Two main tunnels are ventilated, but the evacuation tunnel is not ventilated in normal condition. When a fire accident occurs, the evacuation tunnel is pressurized to prevent smoke intrusion. The connection roads are closed by doors. In emergencies, drivers can go through one of the roads to leave the accident area by opening the doors. Ventilation control is different depending on the situation and is required to control delicate air volume.

In ventilation controlling, blowers at two places around the middle area are installed to fill a requirement to become over a specific wind volume. Ventilation controlling estimates wind volume in near future by traffic volume and current wind volume. When the estimated wind volume does not fill a requirement to ventilate inside the tunnel, the controlling system adjusts wind volume by blowers.

Kan-Etsu tunnel is located at high altitude so that weather conditions change suddenly. Then, the condition is different between summer and winter seasons so that range of weather conditions is wide. For example, snow falls at one entrance, but snow does not fall at the opposite entrance. Different weather conditions influence the inside tunnel. The system has to measure and estimate those natural influences. In addition, the ventilation system considers wind generated from car passing. Currently, natural wind volume is treated as wind sensor value. The sensor value includes other wind, which is generated from the car, so that existing system cannot extract natural wind volume from the installed sensor data. If the system also measures influence from natural phenomena, accuracy of ventilation controlling is increased.

3 Investigations Environment and Measuring Process

For evaluation of this paper, we installed 4 units of pressure sensor array in Kan-Etsu tunnel (shown in Fig. 1 and Fig. 2). To measure outside conditions, two pressure sensor arrays were put in front of each entrance of the evacuation tunnel. Position A is an entrance of the evacuation tunnel on Minakami area. Position D is an opposite entrance of the evacuation tunnel, and is located at Yuzawa. Then, others were installed on the side of the middle point of the evacuation tunnel (Position B) and main tunnel (Position D). The sensors installed outside were set at the same height from the ground. This is not the same height above sea level because the altitude of each tunnel entrance is different. Pressure value was logged over 12 h at least. The examination has been held 5 times. Each monitoring day is a different day. Pressure gradient and normalized pressure value were calculated for investigation. In addition, the difference of pressure value at each entrance was investigated with the same device, as reference of pressure difference by altitude.

To evaluate the relationship between inside tunnel and outside environments, pressure movement is compared with each sensor. The evaluation adopted average pressure value from 25 pressure sensor value as pressure value at the observation place.

The reason about the usage of the average is that the pressure sensor has characteristic noise. Noise from the structure can not be removed, even if the pressure sensor was calibrated. To remove sensor characteristics, the average of 25 pressure sensors was calculated. Then, calculated data was treated as a pressure movement graph in the observing period. The graphs were evaluated by correlation coefficient (Lawrence 1989). Correlation coefficient is one of statistics methods to evaluate similarity of degree activity of graphs. It is difficult to prepare in the same environment, so comparison of pressure values is difficult for us. This paper needs to evaluate the similarity of each graph with the degree of increase or decrease volumes. Therefore, correlation coefficient is an ideal evaluation method. When correlation coefficient value is low, the cause is analysed using pressure gradient. Pressure gradient visualizes air flow around a sensor, so that what factor influences a result can be realized.

Calculation of pressure gradient is following optical flow (Barron 1992; Sun 2008) and related algorithm (Van Roosmalen 1996; Zhang 2007). Pressure sensor was treated as a pixel of the camera. Pressure sensors on the pressure sensor array were treated as pixels of the camera, and gradient was calculated using an algorithm. The calculated result is gradient direction and difference between maximum pressure and minimum pressure values. High pressure value can be treated as light color, and low pressure value can be treated as dark color. Then, the algorithm can calculate pressure sensor data as a 5 by 5 matrix image. Therefore, this algorithm can be adopted into calculation of pressure gradient. This algorithm is one of the most famous methods in computer vision and can detect slight change. Pressure values from pressure sensors, which were installed in a small space, are not very different. In this environment, the algorithm is useful for calculating gradients.

Fig. 2. Scenes of installing sensors that measure pressure for monitoring pressure movement.

4 Evaluation Result

At first, pressure volume was measured at both of two entrances with the same pressure sensor array. The entrance of Minakami side (634.140 m) is lower than Yuzawa side (669.636 m). Average of pressure value was 947 hPa at Minakami side, and another was 942 hPa. The value was calculated by an average of 25 pressure sensors. Pressure value at Minakami was higher than at Yuzawa.

Correlation coefficient result is shown in Table 1. Each sensor data was compared with each other, and the result of correlation coefficient was calculated. Table 1 is the calculated result at each investigation. Cars were going through the main tunnel during monitoring. As a result, wind generated cars influence pressure sensor value. On the other hand, the pressure sensor array installed at each entrance was not affected by the car because the location is only used by maintenance service. Pressure movement in the main tunnel did not tend to be similar against each entrance. The sensor, which is installed inside the evacuation tunnel, tended to show similar behavior of sensor activity installed at the entrance in Minakami.

Pressure gradient at evacuation tunnel shown in Fig. 3. This table represents the average direction of pressure gradient. The direction information is a direction data from higher pressure value area to lower pressure value area than others. Mostly, sensor data in the evacuation tunnel showed that Minakami side has high pressure. The case iii was that correlation coefficient between evacuation tunnel and outsides was lower value. In the case, the direction information pointed at a wall. The wall is an opposite wall where the sensor was installed.

Table 1. Correlation coefficient result

(i)

	(A) Minakami	(B) Main Tunnel	(C) Evacuation Tunnel	(D) Yuzawa
(A) Minakami	1	0.57138972	0.79145795	-0.0993202
(B) Main Tunnel	0.57138972	1	0.77476632	-0.1024529
(C) Evacuation Tunnel	0.79145795	0.77476632	1	-0.3448394
(D) Yuzawa	-0.0993202	-0.1024529	-0.3448394	1

(ii)

	(A) Minakami	(B) Main Tunnel	(C) Evacuation Tunnel	(D) Yuzawa
(A) Minakami	1	0.9454798	0.84909322	0.92426497
(B) Main Tunnel	0.9454798	1	0.84201493	0.86709549
(C) Evacuation Tunnel	0.84909322	0.84201493	1	0.80940694
(D) Yuzawa	0.92426497	0.86709549	0.80940694	1

(iii)

	(A) Minakami	(B) Main Tunnel	(C) Evacuation Tunnel	(D) Yuzawa
(A) Minakami	1	0.22373929	0.11558016	0.13094824
(B) Main Tunnel	0.22373929	1	0.19625015	0.42208201
(C) Evacuation Tunnel	0.11558016	0.19625015	1	0.2048673
(D) Yuzawa	0.13094824	0.42208201	0.2048673	1

(iv)

	(A) Minakami	(B) Main Tunnel	(C) Evacuation Tunnel	(D) Yuzawa
(A) Minakami	1	0.66978388	0.71318156	0.54278237
(B) Main Tunnel	0.66978388	1	0.61642777	0.26581759
(C) Evacuation Tunnel	0.71318156	0.61642777	1	0.66537155
(D) Yuzawa	0.54278237	0.26581759	0.66537155	1

(v)

	(A) Minakami	(B) Main Tunnel	(C) Evacuation Tunnel	(D) Yuzawa
(A) Minakami	1	0.64346088	0.77638655	0.76473445
(B) Main Tunnel	0.64346088	1	0.82582828	0.61136834
(C) Evacuation Tunnel	0.77638655	0.82582828	1	0.8109618
(D) Yuzawa	0.76473445	0.61136834	0.8109618	1

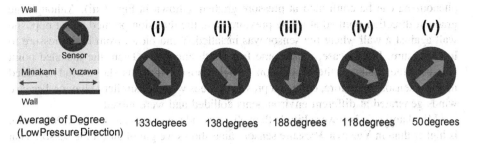

Fig. 3. Install environment of point (C), and each evaluation result of pressure gradient that was measured by pressure sensor array installed at point (C). Vector shows direction from high pressure area to low pressure area. This direction information is the average data of collected direction information that pressure sensor array calculated from 25 pressure sensors.

5 Discussion

We haven't collected enough data in a whole year, but the outside environment influenced the inside Kan-Etsu tunnel was confirmed by the correlation coefficient result. Pressure movement in the evacuation tunnel is similar behavior against either both of entrances or one side. Evacuation tunnel is not ventilated in normal condition, and traffic in the evacuation tunnel is not busy. Cars, which is for maintaining the tunnel, and rescue cars in emergency only use the evacuation tunnel. Monitoring environmental information at the evacuation tunnel is the same as observing influences from outside weather conditions. Correlation coefficient value against Minakami is over 0.7 at four in five examinations. At least, the inside tunnel is influenced by the outside environment. Especially, weather conditions from Minakami influence Kan-Etsu tunnel more than Yuzawa in these evaluations.

In the fifth evaluation examination, correlation coefficient value against Yuzawa is higher than against Minakami. Pressure information from the sensor, which was installed at the evacuation tunnel, was that the sensor near Minakami is higher than near Yuzawa on the board. Under the consideration of installation position and direction, strong wind blew from Minakami to Yuzawa in the evacuation tunnel. We though that the weather conditions in Minakami did not only affect the evacuation tunnel but also Yuzawa in this time.

The result of the third evaluation examination is different from others. The correlation coefficient value is lower than others. Pressure sensor movement at the evacuation tunnel is not similar activity against others. To analyze the situation, pressure gradient provided from pressure sensor array and wind information, which were provided from wind sensor installed at each entrance of evacuation tunnel, were confirmed. Figure 4 and Fig. 5 show wind volume and wind direction. In Minakami side, wind volume is plus at the middle part (shown in Fig. 4). This means the wind blows to Yuzawa. On the other hand, wind volume at Yuzawa side is expressed as minus (shown in Fig. 5). The expression is blowing from outside to the tunnel. In the middle part of those graphs, wind blowing to the tunnel from each entrance was confirmed. Then, those winds, which were coming from outside, collided inside the tunnel. The phenomena can be confirmed at pressure gradient (shown in Fig. 3 iii). Although the gradient direction pointed at a low pressure area, the direction pointed at the opposite wall against a wall where the sensor was installed. Wind blows from high pressure to low pressure. In this case, wind came from both entrances. Then, the installed point was near the center of the evacuation tunnel. Those symptoms show winds collided near the sensor. Therefore, collected pressure values were not similar behavior, because winds generated at different environments collided and were mixed.

Minakami side is lower altitude than Yuzawa side, and pressure value in Minakami is higher than in Yuzawa. Pressure sensor value shows weight of air on the sensor. That means low altitude is more air volume than high altitude if the weather conditions are the same. The phenomena were observed from our evaluation, too. Comparison of pressure values is not a way to measure influence against a tunnel in the actual environment. There is not the same environment and same altitude from the sea level. For this reason, our evaluations are using the tendency of pressure movement, and pressure gradient with a pressure sensor array. As a result, comparison with single pressure sensors cannot be measured, and the benefit of using a pressure sensor array was confirmed.

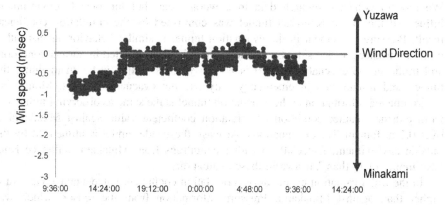

Fig. 4. Wind sensor log at an entrance of evacuation tunnel at Minakami.

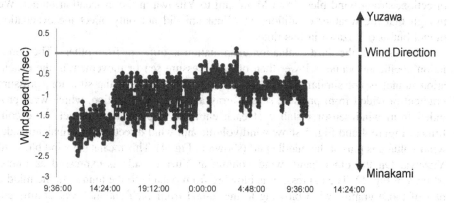

Fig. 5. Wind sensor log at an entrance of evacuation tunnel at Yuzawa.

6 Applications

From investigation in this paper, air flow in the tunnel can be visualized by understanding pressure movement and pressure gradient. For example, wind behavior, such as case iii of the investigation in this paper, was analyzed by pressure gradient. Using pressure information, current conditions can be analyzed in more detail. Following is availability and potential of applications by analyzing pressure information.

Visualizing influences from outer environments is useful to analyze a factor of wind blowing. Current installed sensors can not classify winds. Therefore, the ventilation system can not analyze and recognize current winds, except wind generated from cars. Then, forecasting wind volume after cars pass is difficult. If a sensor can detect different types of wind, the system can recognize current wind volume generated by nature. Then the system can forecast the near future from the current natural wind. On the other hand, using monitoring pressure movement with the pressure sensor array, wind from natural phenomena can be separate from wind blowing in the tunnel. Conversely, wind generated from cars can be extracted. The ventilation system can increase accuracy of estimation and quality of ventilation.

In addition, the pressure sensor array can extract wind generated from the car in more detail. Winds treated as ventilation are classified with three categories: car, nature and blower. Ventilation controlling system manages the blower, so that the system can remove the blower's wind from wind blowing in the tunnel. Then, natural wind also can be extracted, too. Therefore, the system can collect car's wind data in more detail. The Existing system estimated the wind volume from the size of the car. Although the simulator can calculate it, actual wind volume, which each car generates, has not been measured. Wind sensors measure total wind volume but do not measure each different type of wind. Using pressure information, generated wind volume from cars can be calculated. Analyzing performance will be increased, so that the controlling system can work under more efficiency and effectiveness.

7 Conclusion

Kan-Etsu tunnel is one of the longest mountain tunnels in Japan. Its location is at high altitude and around mountains. Although this is a special environment, the inside tunnel has to be ventilated to keep drivability and safety. However, Kan-Etsu tunnel is a long tunnel so that it is difficult to ventilate in a whole area. Therefore, the system ventilates tunnels by multi types of wind. Current ventilation controlling in Kan-Etsu tunnel estimates future conditions inside the tunnel with installed sensors and controls wind volume blowing in the tunnel. When the system understands wind volume as an index of ventilation functioning, wind sensor values are referenced. If the volume does not achieve the requirement to ventilate, the ventilation system supplies winds by blower installed in the tunnel. If the system ventilates under effective and efficient conditions, accuracy of recognizing current conditions is needed. However, wind sensors can not detect an influence of natural phenomena against tunnels.

In this paper, we proposed a method to investigate the influence of the natural environment against tunnels using observing pressure movement and gradient. To observe those phenomena, pressure sensor arrays were installed in four places: two entrances of the evacuation tunnel, evacuation tunnel and main tunnel. As a result, the evacuation tunnel is influenced by the outside with correlation coefficient. Pressure movement in the evacuation tunnel tended to be similar pressure movement in Minakami, which is one of the entrance areas in Kan-Etsu tunnel. Although we haven't evaluated it in a whole year, four out of five evaluation examinations showed the manner. The rest was confirmed by wind sensor information and pressure gradient from the pressure sensor array. As a result, wind was blowing from both entrances to the inside of the tunnel at the same time. Then, the pressure gradient showed that wind collided around the pressure sensor array installed in the evacuation tunnel. Using the pressure sensor array, the system was able to detect natural phenomena to influence the tunnel. By the results of this paper, monitoring pressure values showed a potential to increase quality of ventilation.

References

Barron, J.L., Fleet, D.J., Beauchemin, S.S., Burkitt, T.A.: Performance of optical flow techniques. In: Proceedings 1992 IEEE Computer Society Conference on Computer Vision and Pattern Recognition, pp. 236–242. IEEE, June 1992

Holton, J.R.: An introduction to dynamic meteorology. Am. J. Phys. **41**(5), 752–754 (1973)

JianDong, C., ShihChung, C.: A comparison of planning among long highway tunnel: Guanyin-Gufeng tunnel, Qinling Zhongnanshan tunnel and Kan-Etsu tunnel. Tunnel Constr. S1 (2011)

Lawrence, I., Lin, K.: A concordance correlation coefficient to evaluate reproducibility. Biometrics, 255–268 (1989)

Nakahori, I., Mitani, A., Vardy, A.: Automatic control of two-way-tunnels with simple longitudinal ventilation. na (2010)

Sun, D., Roth, S., Lewis, J.P., Black, M.J.: Learning optical flow. In: Forsyth, D., Torr, P., Zisserman, A. (eds.) ECCV 2008. LNCS, vol. 5304, pp. 83–97. Springer, Heidelberg (2008). https://doi.org/10.1007/978-3-540-88690-7_7

Van Roosmalen, P.M., Westen, S.J., Lagendijk, R.L., Biemond, J.: Noise reduction for image sequences using an oriented pyramid thresholding technique. In: Proceedings of 3rd IEEE International Conference on Image Processing, vol. 1, pp. 375–378. IEEE, September 1996

Wallace, J.M., Hobbs, P.V.: Atmospheric Science: An Introductory Survey, vol. 92. Elsevier, Amsterdam (2006)

Road Asset Management: Innovative Approaches

Soughah Salem Al-Samahi[1] and Fernando Varela Soto[2](✉)

[1] Roads Department, Ministry of Energy and Infrastructure,
Dubai, United Arab Emirates
[2] Polytechnic University of Madrid, Madrid, Spain
fernando.varela@upm.es

Abstract. Maintenance and management of road assets is a crucial factor when it comes to making strategic decisions aimed at achieving optimal and sustainable conservation of the network. While it is widely acknowledged that digitalization and smart data analysis of different processes is on the rise, the Road asset management field is no stranger to this matter. As a case in point of the latter, some of the latest innovations include the collection of real time data using smart sensors, intended to optimize the management of asset's risk of failure. This type of innovative approaches has proved to be highly effective when implemented in the Federal Roads Network of the United Arab Emirates, where real time data of traffic, flood monitoring systems, and environmental parameters are being constantly collected.

Under the project "Supply of technical assistance in the Civil Engineering field for the Assets and Roads Management Systems" and the consultancy services of the Spanish company Rauros Management Consultants, several smart sensors have been installed in strategic and critical locations in order to analyze the evolution of the parameters and ultimately, serve as a baseline for decisions to be made based on the immediate risks and conditions the Road Network may face. Following a strict communication protocol and data synchronization with the expert management system ICARO (developed by Rauros), the Ministry of Energy and Infrastructure has achieved an exceptional level of standards, ranking in the highest positions worldwide in Road Asset Management development and innovation.

Keywords: Asset · Management · Maintenance · Road · Network · Conservation

1 Introduction

Road infrastructure constitutes, in many cases, an indicator of the degree of development of a country, exerting a direct influence on its economy and the social well-being of citizens (World Road Association 2014). The management and maintenance of the assets of this infrastructure is a play an important role for decision-making focused on optimal and sustainable conservation. This is why, more and more, greater efforts are being destined to management, maintenance and prioritization of the actions that must be carried out on these assets.

Generally, in order to know the conservation condition of a road network, different techniques and technological devices must be used in order to establish the evolutionary models of each asset separately, and consequently, its degree of deterioration. It is this degree of deterioration, along with other factors, that will allow the technicians to determine the priority of the actions to be carried out, in addition to establishing a maintenance and operation schedule based on the condition and quality parameters.

According to the Organization for Economic Cooperation and Development, road asset management is defined as a systematic process of maintaining, updating and operating assets, combining engineering activities with sound business practices and economic justification, in order to provide tools to facilitate decision-making through an organized approach (OECD 2001) (Fig. 1).

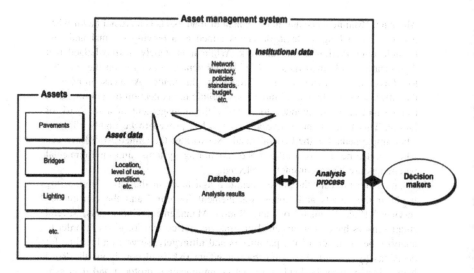

Fig. 1. Road asset management data flow. Source: (OECD 2001)

Once the data has been collected in the field, and using an adequate management system, the condition of the assets and their evolution over the years can be studied, thus having a technical basis on which the corresponding administrations will be based for taking the corresponding decisions.

This article is organized as follows: Sect. 2 describes the objectives to be achieved using an innovative approach in road asset management. Section 3 details the methodology used to obtain the results of Sect. 4. Finally, the conclusions and analysis will be stated in Sect. 5.

2 Objectives

The objective of this article is to bring to attention the different innovative methods currently applied to asset management by presenting a case study carried out in the United Arab Emirates, through the Ministry of Energy and Infrastructure and with the collaboration of the Spanish company RAUROS.

The innovative approach applied in the management of the Federal highways of the United Arab Emirates and its benefits will, in turn, make it possible to extrapolate the results and make other administrations aware of the importance of conserving their roads to reduce public investment.

3 Methodology

Once the general lines of this article have been described, the methodology used in the specific case study mentioned above will be described below.

According to the World Road Association, there are three levels of road asset management (PIARC, n.d.); The basic level, the proficient level and the advanced level (Fig. 2).

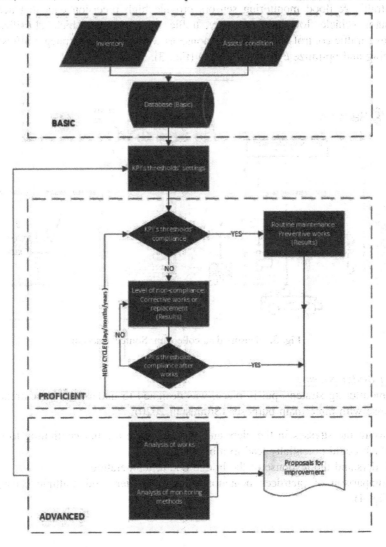

Fig. 2. Maturity levels, PIARC.

The basic level consists of inventorying all the assets present in the road network and organizing this information in a basic database. The proficient level consists of implementing a work methodology, as well as the control of KPI's based on international experience and, finally, the advanced level consists of the analysis of the useful life and risk management based on different indicators.

The process of inventorying assets to know their condition, which is necessary to overcome the basic level, has been adapted in parallel with the availability of new, increasingly advanced technologies, such as in the case of bridges or slopes. For these assets, different devices and sensors were used to know the degree of corrosion, movement under loads and forces, strain, among others.

Traffic and Flood Sensors

The traffic & flood monitoring sensors provide high precision data that accurately measures vehicle flow, traffic density, traffic velocity, and vehicle classification to improve traffic control and disaster response as well as support strategic infrastructure planning and optimize evacuation routes (Fig. 3).

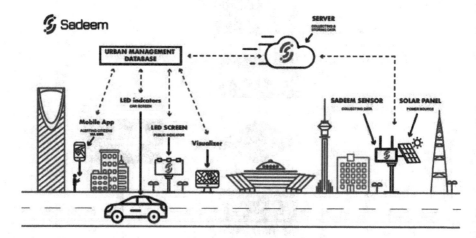

Fig. 3. Sensors data collection. Source: Sadeem

Smart Bridge Sensors

The monitoring strategy put in place, was designed to understand the current bridge response with three main purposes (Strainstall 2020):

1. Ensure the stresses in the piers and superstructure are not continuing to increase which could potentially lead to a bridge collapse.
2. Understand the response of the bridge due to temperature.
3. Comparison of pier/deck behavior at multiple piers and multiple carriageways (Fig. 4).

Fig. 4. Bridge sensor. Source: Strainstall

Corrosion Sensors

The deterioration of reinforced concrete and pre-compressed reinforced concrete is one of the most serious problems for our society and institutions due to its social and economic repercussions.

The aim of installing these sensors is to predict the damage growth curve, specific to the structure investigated. In fact, the continuous monitoring of the conservation status of the concrete, calculating the speed of degradation, can simulate a curve of the actual damage and its deviation from the theoretical one based on the service life of the project, and estimate the current and future costs for repair (Estimated Service Life) (Fig. 5).

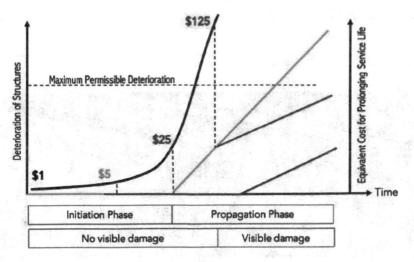

Fig. 5. De Sitter's law of fives

Once all this information is collected, it was entered in the expert management system ICARO for its utilization in different reports and studies, that will serve as a base for the next maturity level.

For the implementation of the work methodology, the key indicators to achieve the optimization and prioritization of resources must be known. In this case study, data from several parameters such as IRI (International Regularity Index), corrosion, flooding, traffic, among others were collected in order to establish unique thresholds appropriate to the country for the risk management. In the following section, the evolution of these indicators over the years of study will be shown.

4 Results

This section will explain how risk management can be optimized by the real-time data mentioned in the previous section, and its use through direct communication protocols in the ICARO comprehensive management system.

Figure 6 shows, in the expert management system, the inspections carried out on the asset under study, based on the data received from the sensors in real time.

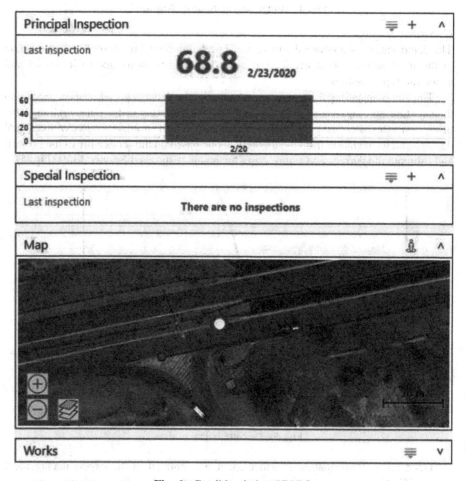

Fig. 6. Condition index, ICARO

In the event that the structure is in poor condition, a communication protocol is opened to proceed to carry out the corresponding action as shown in Fig. 7.

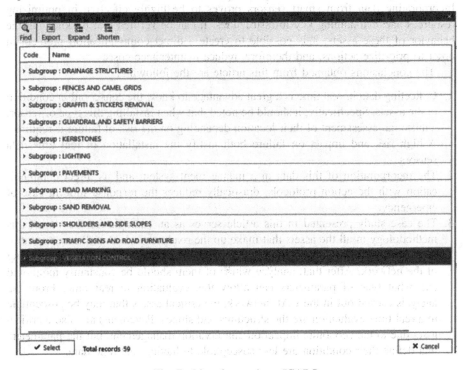

Fig. 7. List of operations, ICARO

After defining the evolution model of all the key assets of the road network, it will de possible to prioritize the mandatory and required actions to be done on each one as per their condition index, constantly updated with real time data (Fig. 8).

Fig. 8. Map of assets, ICARO

5 Analysis and Conclusions

Incorporating data from smart sensors proves to be highly effective in optimizing resources and maintaining key infrastructures. It can be verified that by knowing the behavior of these assets, it is possible to create a direct communication protocol in order to prioritize actions, and therefore, reduce maintenance costs.

The conclusions obtained from this article are the following:

1. Collecting data in real time is a great advantage to know the conservation condition of the assets. Specifically, it should be noted that when installing this type of sensor, a thorough assessment of their location depending on the risk of failure is required. A high risk and impact on failure both justify the installation of real time data sensors.
2. The incorporation of this data in a management system and its direct communication with the action protocols, drastically reduces the response time in case of emergency.
3. The case study presented in this article serves as an example for the use of this methodology in all the assets that make up the road network. A dull assessment of the assets is required, distinguishing those that are critical for the proper functioning of the network. After that, analyze which of them should be constantly monitored and what type of parameters can allow this evaluation in real time. From the analysis carried out in the UAE network, the critical assets that may be susceptible to a real time evaluation are the structures and slopes. Pavements are also a critical asset due to the economic impact on conservation management, but the parameters that define their condition are less susceptible to having real-time data.

References

OECD: Asset Management for the Roads Sector. OECD Publications Service, Paris (2001)
PIARC: Maturity levels RAMS (s.f.). https://road-asset.piarc.org/en/planning-programming/maturity-levels
Strainstall: Installation report for smart bridge trial. Dubai (2020)
World Road Association: Importancia de la conservación de carreteras. World Road Association, La défense (2014)
El-Gioshy, S., Osman, O., El-Desouky, A., Fatalla, S.: Guidelines for maintenance of airport pavements case study: Aswan international airport (2006)
Atoh-Okine, N., Adarkwa, O.: Pavement condition surveys-overview of current practices. Panagiotis Anastasopoulos. Infrastructure asset management: A case study on pavement rehabilitation. Ralph Haas and W Roanld Hudson. Pavement asset management
World Road Association et al.: The piarc asset management manual

Traffic Planning and Forecasting – 2

Evaluating the Efficiency of Constructability Review Meetings for Highway Department Projects

Amin K. Akhnoukh[1]([✉]), Minerva Bonilla[2], Nicolas Norboge[3], Daniel Findley[4], William Rasdorf[2], and Clare Fullerton[5]

[1] East Carolina University, Greenville, NC, USA
akhnoukha17@ecu.edu
[2] North Carolina State University, Raleigh, USA
[3] NC Office of State Budget and Management, Raleigh, USA
[4] Institute for Transportation Research and Education, Raleigh, USA
[5] North Carolina Department of Transportation, Raleigh, USA

Abstract. Constructability review (CR) meetings are held before the completion of the project design to define potential site conflicts; reduce design errors, delays, and cost overruns; and improve construction site safety. The main objective of this research is to assess current practices associated with conducting CR meetings and provide specific guidelines that increase CR meeting efficiency. The research methodology included evaluating current CR meeting practices by different state DOTs and conducting interviews with multiple construction project stakeholders to assess different parameters affecting CR meeting success. The research findings showed that CR meetings should be scheduled at 60% to 70% of design completion. At least 3 contractors should be included, and a formal checklist of items should be provided by the constructability review manager for discussion throughout the meeting. Successful CR meetings result in reduced site conflicts, cost overruns, schedule delays, and they improve site safety.

Keywords: Constructability review · Safety · Cost overrun · Schedule overrun · Constructability

1 Introduction

The term "Constructability" (often referred to as buildability) refers to the extent to which a facility design provides for ease of construction yet meets the overall requirements of the project. Similarly, a Constructability Review (CR) is defined as a project management technique to review the construction processes from start to finish during preconstruction. It is used to uncover and discuss obstacles before a project is built to reduce or prevent errors, delays, and cost overruns (Anderson and Fisher 1997 and AASHTO 2000). Constructability reviews ensure that a construction project is biddable, buildable, and cost-effective. Specific CR meetings objectives include the following:

1. Reduce potential conflicts during the construction process
2. Improve the efficiency of construction site activities
3. Increase the overall safety of the construction site
4. Avoid issues and conflicts that might result in arbitration or litigation
5. Reduce the number of changes to the original design.

CR meetings are usually held upon the completion of a specific percentage of the project design phase (and can be held multiple times throughout the design process). Some state and federal agencies have standard policies and procedures for organizing CR meetings, including the timing of conducting the meeting, the format of the meeting, who is to be invited, and a checklist for all the items to be included in the CR meeting. Moreover, some agencies require conducting multiple follow-up meetings to monitor the outcome of the original CR meeting and to assess the meeting's outcomes. A typical CR meeting is attended by different project stakeholders, such as the project managers, project engineers, multiple contractors, safety engineers, and material suppliers.

The afore-mentioned stakeholder composition may vary depending on project type and characteristics, project owner, and the project delivery method. Despite the dynamic nature of construction projects, common parameters are always present in highway department CR meetings including utilities, right of way, permits, traffic control, and detours.

The advantages of effective CR meetings include improved project planning for construction projects, a substantial reduction in project design and construction costs, fewer site conflicts, and a shortened schedule. These advantages stem from the ability of the project team to resolve conflicts and eliminate their negative consequences in design rather than during construction. Thus, early detection of potential conflicts results in significant project cost savings. Recent research outcomes show that the early stages of a project design (preconstruction), where expenditures are minimal, have a significant influence on project cost savings. Later phases (construction) require increased expenditures and reactive corrective actions. The pre-construction phases, including conceptual and schematic designs, design development, and production of construction documents and project specifications, provide project stakeholders with an adequate window for possible revisions that provides significant project savings at minimal expenditure (AASHTO 2000).

2 Literature Review

The highway construction project industry is globally renowned for the lack of coordination between design (including the development of design drawings and specifications) and construction (including the development of construction drawings). In the last three decades, the idea of integrating design and construction has been investigated and implemented in multiple ways to avoid site issues, schedule delays, site safety violations, and more (Stamatiadis 2017 and Bakti 2012).

Several research investigations have suggested that the lack of CR meetings results in an increased number of change orders, contracts being settled through arbitration and litigation, extensions to construction schedules, and an increased number of safety violations (as indicated by OSHA citations) that result in an increasing number of site accidents, injuries, and fatalities (Salleh 2009; Goodrum et al. 2009). Recent research projects considered the success factors of construction projects, with special emphasis on infrastructure projects, due to the magnitude of problems associated with these projects (Gambatese et al. 2007). Variables affecting project success are highly dependent on project type, mission, size, and complexity (Othman 2011; Dunston et al. 2002).

Significant benefits are attained as a direct result of conducting CR meetings in the presence of different project stakeholders (Hancher et al. 2003). Attained benefits are maximized when a pre-determined format is prescribed for the CR meetings and formal follow-up procedures are mandated (Stamatiadis et al. 2013). Today's highway construction projects are characterized by increased complexity, complicated designs, the use of new products, and the execution of projects using multiple contractors. In addition, fast-track projects are increasingly adopted in the local market to minimize construction duration. The modeling of constructability knowledge and the inclusion of lessons learned from previous projects in a detailed database is crucial for the success of new CR meetings. However, the creation of a constructability database is dependent on the successful categorization of construction problems and on providing an adequate description of the problem given the project size, budget, location, and project delivery method.

Due to the increased importance of CR meetings and concerns about developing constructible plans and specifications, a National Cooperative Highway Research Program (NCHRP) study was conducted in the 1990s to provide a detailed format for a successful CR meeting. The NCHRP study provided an evaluation of the losses associated with ignoring the provided CR rigid format. Because of the comprehensive checklist provided in the NCHRP study, state DoTs described the implementation of the study recommendation as being laborious and resource-consuming. Hence, even partial implementation of the study was adopted by few DoTs among the United States.

The American Association of State Highway and Transportation Officials (AASHTO) expressed concern about the applicability and buildability of different types of construction projects. AASHTO's main concern was the inability of some AASHTO members to execute construction projects according to the construction plans. AASHTO surveyed its members about conducting CR meetings and compiled the attained benefits associated with the meetings. The survey results showed that less than 30% of the AASHTO members conduct CR meetings using a pre-structured format, and only 8 states have a written formalized procedure for the CR meetings. According to the AASHTO, an effective CR meeting should accomplish multiple goals that are beneficial to any transportation agency when conducting a construction project. The CR meeting's main goals are to determine whether:

1. The project plans and specifications are executable using standard construction materials and methods.

2. The project documents provide the general contractor with precise and clear instructions to be used in preparing a successful bid.
3. The resulting project can be maintained by the owner cost-effectively during the project life span.

Based on its surveys, AASHTO developed a "Best Practice Guide" to provide information for its members regarding the best practices to hold successful CR meetings, expected CR meetings outcomes, and potential issues to consider when conducting the meetings. The AASHTO guide does not require AASHTO members to fully utilize the guide. Instead, agencies were advised to use the provisions that suit their project type, size, location, and specific parameters as per the agency conditions and resources. According to the AASHTO guide, it is advisable to maintain a responsible key person to request and conduct the CR meetings. The key person could be the design project manager or construction, office manager. Few states (New Jersey is one) establish a construction team that is solely responsible for the CR meetings. According to the AASHTO guide, a successful CR team may include:

1. Internal Members: many states require in-house CR meetings. Thus, engineers and construction personnel from their design and construction offices take full responsibility for the meetings and provide the project design engineer with their recommendations. New Jersey, California, and Washington are among the states that conduct internal CR meetings. Florida also conducts CR meetings internally and determines the project duration through their CR team. Connecticut is among the states that start the CR meeting internally, but they invite industry professionals towards the end of the CR meetings to provide external feedback.
2. Construction Professionals: are invited and assist internal engineers and project managers through their feedback. External members are mostly drawn from lists of contractors approved and certified by the agency conducting the CR meeting. The North Carolina Department of Transportation, for example, invites a list of certified AGC contractors to its CR meetings for feedback. The Pennsylvania DoT employs a retired contractor for the specific purpose of assisting with CR meetings. In Kansas, a joint task force comprised of representatives from the Kansas Contractors Association, Heavy Constructors Association, and Kansas Department of Transportation (KDoT) conducts the CR meetings. Maine DoT conducts its CR meetings using internal DoT personnel and invited industry professionals actively involved in Maine construction market.
3. Consultants: state highway agencies may contact consultants to join the CR meetings.
4. Regulatory: representatives of federal and state agencies may be invited to the CR meetings.
5. Utility Personnel: representing utility companies to be affected by the construction project may be invited to the CR meeting for further discussions.
6. Vendors and Material Suppliers: are sometimes invited for a CR meeting in case the design includes construction materials of an unusual or non-standard nature.

To date, agencies conducting CRs recommend that they be conducted during design. Early CR meetings have the greatest potential to provide meaningful benefits to the review team; and can substantially affect the final cost and time savings of the project. On the other hand, a CR meeting after 90% of the design will not significantly improve a project's schedule and budget. Currently, the timing of most CR meetings is dependent on the design engineer. The lead designer determines if a CR meeting is necessary and specifies the items to be discussed during the meeting.

3 Research Objectives and Methodology

The overall objective of this paper is to present CR meeting best practices for transportation agency projects including highway, tunnel, and bridge construction. Specific objectives include the following:

1. Investigate and assess successful CR meeting format, timing, duration, attendees, location, and the development of a checklist for meeting items.
2. Provide transportation agencies with a list of recommendations to increase CR meeting efficiency and enhance its potential benefits.

The research methodology included a two-phase research program. Phase I consisted of surveying different state DOT's to determine their CR meeting practices and assessing meeting outcomes. Twenty state Departments of Transportation were surveyed using a standard research questionnaire, in addition to conducting a one-hour follow-up virtual interview. Virtual interviews were conducted with NCDOTs head of research, and value management office (VMO) chief engineer in attendance.

Phase II consisted of attending different CR meetings held by the NCDOT for various projects and interviewing different construction professionals within the state of North Carolina to better understand their perceptions of the effect of the CR meetings on construction projects. These effects were documented, and a checklist was developed to guide further CR meetings.

4 Research Results

Phase I: Current State DOTs Practices

Current state DOT practices prescribe how the CR meetings are held. The literature search and the DOT personnel interviews showed that all DOTs conduct CR meetings. However, as shown in Fig. 1 only 26% of state DOTs have a formal process for the meeting.

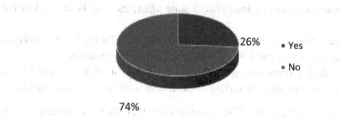

Fig. 1. Formal process for DOT CR meetings

All state DOTs reported that CR meetings are held during the project design phase to avoid or eliminate issues and problems during construction. As shown in Fig. 2, feedback from DOT's construction offices shows that the majority of CR meetings are held prior to 90% of design completion.

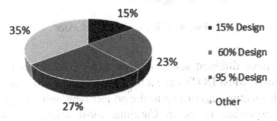

Fig. 2. Timing of CR meetings

Projects that often require CR meetings include projects with traffic mitigation, right-of-way problems, water-crossings, culverts or bridges, bridges with very high span-to-depth ratio, and/or projects where potential safety hazards are prevalent. As shown in Fig. 3, data collected from state DOTs shows that 59% of view CR meetings as a requirement for all construction projects, 22% only consider major urban interstate projects for CR meetings, and 13% require CR meetings for only bridge construction projects.

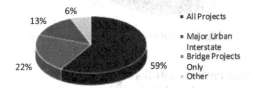

Fig. 3. Types of projects included in CR meetings

Constructability review meetings necessitate the representation of a broad range of project stakeholders to provide comprehensive and insightful feedback regarding different obstacles and potential problems that may obstruct project activities during the construction phase. Possible stakeholders to attend CR meetings include the project owner, engineer, consultant, general contractors, subcontractors, and material suppliers. Data collected from different state DOTs and illustrated in Fig. 4 shows that transportation agency personnel, project consultants, and contractors represent the main expertise providing input during CR meetings.

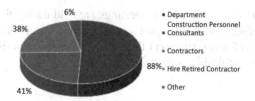

Fig. 4. CR meeting stakeholder

State DOTs have different selection criteria for contractors to provide feedback during CR meetings. Most DOTs have a list of pre-certified contractors to be invited to participate in CR meetings according to the project type, job size, geographic location, and contractor expertise. In North Carolina, most selected contractors are AGC members with relevant qualifications. On average, three different contractors are invited to attend every CR meeting. Most state DOTs also invite contractors who have a presence within the geographic location of the new project (see Fig. 5).

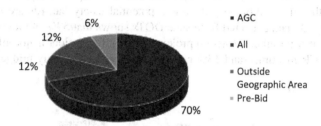

Fig. 5. Source of CR meeting contractors

Phase II: Construction Professionals Survey

A questionnaire was developed for investigating current NCDOT CR practices and their effectiveness. Thirty-five interviews were held, including a) 19 interviews for NCDOT personnel in highway construction division, bridge construction division, and value management office, and b) 16 interviews for industry personnel, primarily AGC certified contractors. The outcomes of the questionnaires are presented in the following subsections. Each table presented below lists the number of interviews (out of 35) that identified the specified stakeholders, limitations, and activities that are a part of the CR process.

Participants in CR Meetings

Effective CR meetings require the participation of a diverse group of relevant construction project stakeholders. Table 1 shows the survey feedback regarding the attendees at current CR meetings and the recommendation for the possible inclusion of different stakeholders in future projects to increase the efficiency of CR meetings (see Table 1).

Table 1. Participation of stakeholders in CR meetings

Stakeholders	Survey recommendations
Owners	9
Contractors	32
NCDOT engineers	32
Construction managers	30
Utility representatives	13
Railroad representatives	1
Municipality representatives	1
Right of way personnel	1
Suppliers	1

Limitations/Disadvantages of Current CR Meetings Practices

Table 2 articulates a list of possible limitations and disadvantages for various aspects of the CR meeting format and regulations. These concerns present major impediments to the success and efficiency of the CR meetings.

Table 2. Limitations/disadvantages of current CR meetings practices

Limitations/Disadvantages	Survey feedback
Time for scheduling CR meetings	5
Current CR process is not well communicated	2
Project information should be provided to stakeholders far ahead of meeting time	2
Lack of incentives for participating contractors to share ideas	1
CR meetings do not follow a strict format	1
No specific concerns are relayed to contractors ahead of the meeting	2

Indicators/Triggers for CR Meetings

CR meetings benefit different types of construction activities within a given project. The increased complexity of construction projects and the evolution of new techniques provide project stakeholders with sufficient challenges that define specific construction activities and/or project parameters that will ultimately require and benefit from CR meetings. A comprehensive list of CR meeting indicators is attained as a result of the construction professionals survey is shown in Table 3.

Table 3. Indicators of the need for CR meeting

Project activity	Survey recommendations
Traffic management	9
Structures issues, long spans, shallow sections	6
Right of way issues	6
Construction phasing	4
Wetland proximity	2
Proximity to historical areas	1
Geotechnical issues, very high-water table	2
Multiple permits	1
Environmental aspects, water streams proximity, endangered species	3

Optimum Timing for Scheduling CR Meetings

Surveyed professionals prefer construction meetings with a duration ranging from 2 h to no more than 3 h in length. The survey results showed that successful CR meetings are optimally held during project design. For increased efficiency, CR meetings should be conducted no later than 60%–70% of the design phase completion.

5 Conclusions

CR meetings significantly contribute to the success of construction projects, especially for projects with a complex nature, including urban interstate construction, bridge construction, traffic intersections, and projects requiring significant utility relocation. The efficiency of CR meetings is affected by their scheduled time, participating stakeholders, and timely communications and exchange of information among different project parties. Successful meetings result in better compliance with the project budget and schedule, in addition to improving overall site safety and minimizing work conflicts. CR meetings conducted at 60% to 70% design completion, in the presence of project owner, architect, design engineer, project manager, utility representatives, and multiple contractors with sufficient project information attained sufficiently enough ahead of the CR meeting, would provide construction stakeholders with optimum outcomes.

Acknowledgements. The authors would like to acknowledge Engineer Clare Fullerton at the Value Management Office of the North Carolina Department of Transportation for her continuous support of the research activities resulting in the results presented herein.

References

AASHTO: Constructability review best practices. american association of state highway and transportation officials, Washington D.C. (2000)

Anderson, S., Fisher, D.: Constructability review process for transportation facilities. NCHRP Report 390, Transportation Research Board, Washington D.C. (1997)

Anderson, D.K., Merna, T.: Project management strategy—project management represented a process-based set of management domains and the consequences for project management strategy. Int. J. Proj. Manag. **21**, 387–393 (2003)

Bakti, E.S.: Constructability database knowledge model and guidelines of industrial plant and construction stage. A dissertation. Technological University of Malaysia (2012)

Dunston, P., McManus, J., Gambatese, J.: Cost/benefits for constructability reviews. Research Project 20-7, American Association of State Highway and Transportation Officials (AASHTO), Washington D.C (2002)

Gambatese, J.A., Pocock, J.B., Dunston, P.S.: Constructability Concepts and Practices. American Society of Civil Engineering, ASCE Book Series (2007)

Goodrum, P.M., Taylor, T.R.: Change orders and lessons learned. Research Report KTC-10-8, Kentucky Transportation Center, Lexington, KY (2009)

Hancher, D.E., Goodrum, P.M., Thozhal, J.J.: Constructability issues on KyTC projects. Research Report: KTC-03-17, Kentucky Transportation Center, Lexington, KY (2003)

Othman, A.A.: Improving building performance through integrating constructability in the design process. Organ. Technol. Manag. Constr. Int. J. **3**(2), 333–347 (2011)

Salleh, R.: Critical success factors of project management for brunei construction projects: improving project performance. a dissertation. Queensland University of Technology, Brisbane (2009)

Stamatiadis, N., Goodrum, P., Shocklee, E., Wang, C.: Quantitative analysis of state transportation agency's experience with constructability reviews. ASCE J. Constr. Eng. Manag. **140**(2), 04013041 (2013)

Stamatiadis, N., Sturgill, R., Amiridis, K., Taylor, T.: Estimating constructability review benefits for highway projects. In: Proceedings of the Lean and Computing in Construction Congress. Joint Conference on Computing in Construction (2017)

A Novel Method for Aggregate Tour-Based Modeling with Empirical Evidence

Yanling Xiang[1(✉)], Shiying She[2], Meng Zheng[2], Heng Liu[1], and Huanyu Lei[1]

[1] Shenzhen Urban Transport Planning Center CO. Ltd.,
Shenzhen, Guangdong, China
xiangyanling@sutpc.com
[2] Wuhan Transportation Development Strategy Institute, Wuhan, Hubei, China

Abstract. The use of agent-based or activity-based model (ABM) is growing to overcome the limitations of traditional 4-step models. Aiming at capturing complex behaviours, over the last 30 years of development, ABM behaviour realism is still limited, let alone vast costs and poorness in application, especially for major schemes and government interventions. This paper presents a state-of-art tour-based mechanism in aggregation, based on Department for Transport UK advice and PTV Visum tour-based models. We address their limitations by presenting novel solutions, and mitigate the challenge in traffic forecasting by an alternative framework. Featured with ABM properties (considering the linkage of trip chains, variable activity durations, tour main/sub mode choice, tour main destination/intermediate-stop choice), the aggregate design focuses on demand-supply consistency, built-in realism, inherent sensitivities to transport policies. Complex tours with intermediate stops are considered, joint travel with parents escorting children to school and half-tours to external areas are approximated with assumptions. The aggregate tour framework is subject to rigorous tests on model system convergence, achieved by MSA feedbacks with overnight model runs. A gap function will be shown to measure the equilibrium status between demand and assignment models (small number of ABM feedbacks, typically fixed, prone to errors with sensitivity issues). Outturn elasticities for realism tests (fuel cost, parking cost, transit fare) will be shown in comparison to established values, and model sensitivity will be shown by transit fare strategy tests. A case study of Wuhan City will be shown as empirical evidence, carried out successfully funded by the World Bank.

Keywords: Travel forecasting · Aggregate demand model · Tour-based model · Equilibrium convergence · Gap function · Transport elasticity · Sensitivity test

1 Introduction

Traditional four step models (FSM) have been used in transport planning for more than seventy years since 1950s (McNally 2000; Boyce and Williams 2015). Originally applied in highway forecasting context, the traditional aggregate approach has been enhanced and expanded over the years with established software. They are considered as practical tools for transport system studies operating at transport analysis zone (TAZ) level efficient in runtimes. However, the FSM suffers with fundamental limitations that all trips are independent of each other with trip chains breaking into disconnected trips, classified into home-based (HB) and non-home-based (NHB). The FSM has no concept of activities, nor tour start/end times and durations. Mode choice and trip distribution are modeled independently with no relations in-between. These are clearly against the reality of how people travel. The weaknesses of FSM are inherent due to the lack of time/space consistency and poor in accessibility measure utilization, thus inapplicable for behaviour-based studies and emerging TDM policies. It is paramount to develop advanced transport models with behaviour realism, to support complex decision making thus meet the challenges of the 21st century transport planning.

The activity-based model (ABM) is developed to overcome the limitations of FSM (Castiglione et al. 2015). Aiming at capturing complex behaviours and reducing aggregate bias, the ABM modeling has made substantial contributions to transport research especially on maintaining spatial/temporal/modal consistency for individual travelers. with industrial implementations on real life applications in the USA (Bradley and Bowman 2004). However, over the last 30 years of developments, the applications of ABM are still limited outside the USA and the objectives and goals of the original ABM developments are open to challenge (Tajaddini et al. 2020). The ABM implementation suffers setbacks in travel forecasting especially for traditional planning tasks and government interventions, not least because of abundant development cost, intensive data needs, complex model structure, daunting calibration/validation tasks, excessive model runtimes, and micro-zone spatial resolutions burdening for future years (Bernardin 2015). Further, the ABM uncertainty and the level of accuracy are difficult to quantify in scenario testing and evaluation. These unsolved obstacles make ABM models less applicable, if not acceptable, for transport interventions and business case appraisals that require detailed economic benefit analysis.

The aggregate tour-based models, trying to bridge the gulf between the ABM and the FSM by incorporating strength and mitigating weaknesses, have been used in many European and Middle East cities from late 1990s, which are mainly based on the following two development approaches:

- The pseudo-tour approach (DfT-UK 2020c), and that for the Greater Bristol Modeling Framework (Xiang et al. 2009). In a similar incremental-logit structure, they consider simple HB tours only focused on home-work-home (HWH), home-Employer's Business-home (HBH), and home-other-home (HOH), with NHB tours and longer chains treated as NHB trips. The hybrid approach in both tours and trips, are simple and straightforward by production-attraction pairs on pseudo trip linkage for tours, using the gap function in (1) below to control the MSA feedback, with parking costs embedded within tour durations. However, the pseudo-tour approach is too restrictive for real life applications allowing only highway and public transport main mode modelling. The modeling methods cannot deal with NHB

tours, multi-stop complex tours, nor allowing mode switching between outward and return trips of tours;
- The aggregate Visum tour modeling approach, namely "Activity chain based model (tour-based model)" (PTV 2020). This formulation considers multimodal HB/NHB tours generically no matter simple or complex trip chains. However, the Visum tour implementation does not consider tour durations explicitly as such parking costs etc. cannot be addressed appropriately. The feedback to Visum highway and transit assignment models does not utilize any form of equilibrium measures, but its COM interface allows user to define own criteria for model convergence. The mode switching between exchangeable modes are allowed for trip chains, however, no clear mathematic description is given transparently on how the exchangeable modes are considered in conjunction with the destination choice. Another fundamental issue with this approach is that, only trip results are available as model output despite tour information are specified as Visum input. The tour demand modeling procedure is largely black-boxed for users, it is thus impossible to digest model output results in terms of trip chain characteristics for understanding tour patterns.

We present a new aggregate tour-based model (ATBM) for travel forecasting, which solves the issues in the above two approaches with innovations. In the ATBM, each component of the nested logit structure is defined mathematically, and the accessibility logsum is used for vertical consistency for demand hierarchy. HB/NHB tours are addressed consistently from simple to complex with tour durations spanning multiple time periods. While seeking behaviour realism and model sensitivities, duration-based parking, mode switching between exchangeable modes, and intermediate stop location choices etc. are considered explicitly in the formulation. To reduce aggregate bias or model noise, the same gap function defined in (DfT-UK 2020a) is used for the convergence between supply and demand models for market equilibrium. The data requirements and the number of parameters to be tuned are kept at minimum in model designs, indeed at a similar level to that of the FSM. The ATBM is aimed for major scheme bids and transport interventions with efficiency and accuracy, with transparent model structure and consistent tour representation throughout fit-for-purpose for generic tour based modeling.

The case study for Wuhan city in China will be presented as empirical evidence, a major study carried out successfully as a part of the project "Wuhan Integrated Transport Development" funded by the World Bank. The application of the ATBM will be shown fit-for-purpose in calibration/validation. Based on the data collected from Household Interview (HIS) Surveys and Stated Preference (SP) surveys, we shall focus below on the aspects of demand-supply model convergence, realism tests for outturn elasticities, sensitivity tests for model uncertainty etc., due to the space allowed in page restrictions.

The paper is divided into the following sections: Sect. 2 is for the overview of the ATBM; Sect. 3 is for the mathematic formulation for basic tours in terms of nested logic structure; Sect. 4 is for complex tours based on Sect. 3; Sect. 5 is for the case study of Wuhan City; Sect. 6 is for conclusion and further development.

2 Aggregate Tour-Based Modeling: A Novel Design

Figure 1 presents our aggregate tour-based model (ATBM) in the overall framework together with supply modeling. The ATBM takes basic tours by person group and rates, land use by purpose, highway and transit skims etc. as input, and output tours and trips after system convergence. The classification of basic tours in Fig. 1 is indicative only, showing compulsory HB tours (work/education), other HB tours, office-based tours, other NHB tours (e.g. hotel based, interchange-hub based). The number of tours by type is parameterized specified by users.

Figure 1 is transparent and clear, self-explaining in five parts where Part 1 and 2 are for model input and tour generations, Part 4 is for highway/transit supply modeling by multicore parallel assignments and skims (for Wuhan it is built on Emme 4 with its Modeler interface connected to the demand model ATMB), Part 5 is for final tour/trip outputs and housekeeping. The Part 3 is focused on the nested choice modeling, with the rightmost branch for the accessibility logsum (bottom-up) and the middle branch for demand calculation (top-down), both of which are presented in Sect. 3 for basic tour modeling. The intermediate stop location choice per half-tour is shown in the leftmost branch of Part 3, to be discussed in Sect. 4 below.

The ABM simulates each person in turn taking excessively long model runtimes, with calculations done often in days. In comparison, the ATBM is aggregate and deterministic in terms of person groups with matrix-based manipulations. Thanks for the advanced matrix technologies, the Python implementation of the ATBM modules is computationally efficient with model runs achieved in affordable runtimes. For the framework shown in Fig. 1, the efficiency and speed in aggregation provide the practical basis for the market equilibrium between demand and supply models, which is critical for intensive scenario testing in nowadays transport planning, since it is increasingly combined with infrastructure developments and government interventions.

The convergence Gap%, between Part 3 and 4 of Fig. 1, is a measure of how far the current flow patterns are away from the equilibrium status with supply-demand consistency. We use the measure to monitor the MSA feedback loops between the ATBM and highway/transit assignment models. The gap function (1) is defined in (DfT-UK 2020a), where "X" and "C" are for demand and skimmed costs respectively, and "D" for the demands calculated for the next loop. The subscripts "$ijpcmT$" represent matrix cells in multiple dimensions, in particular, referring to origin i, destination j, purpose p, person group c, trip mode m, and time period T, respectively.

$$\text{Gap\%} = \frac{\sum_{ijpcmT} C(X_{ijpcmT}) |D(C(X_{ijpcmT})) - X_{ijpcmT}|}{\sum_{ijpcmT} C(X_{ijpcmT}) X_{ijpcmT}} * 100 \qquad (1)$$

A Novel Method for Aggregate Tour-Based Modeling 1011

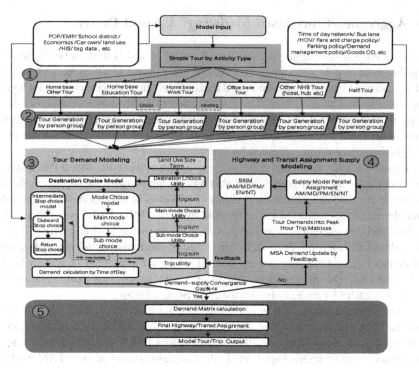

Fig. 1. Aggregate tour-based modeling framework

The convergence gap (1) is evaluated after the ATBM in Part 3 of Fig. 1 per iteration, and the feedback continues until an ε target, being 0.2 by (DfT-UK 2020a), is satisfied. Section 5 will show how the gap is diminishing by iteration in computer runtime while reaching the equilibrium convergence. This is in a clear comparison to the common practice of the ABM and FSM on demand-supply feedbacks:

- Most ABM runs finish with a pre-fixed small number of iterations (often, ≤ 5) in order to reduce model runtimes, which is prone to errors resulting instability, let alone that the ABM is realized by Monte Carlo simulations in random draws with inherent uncertainty. This is one of main issues for applying the current ABM practice to transport assessments for example major scheme bids, for which the cost-benefit analysis is essential for economic benefit evaluation;
- Many FSM implementations use link flow differences (e.g. the maximum changes in absolute flows between consecutive loops) as the measure to control the feedback loops. This practice assumes that stable solutions or some stability statistics imply the consistency between supply and demand models. This type of check for convergence is inferior, generally not indicating of how close that the solution is to the equilibrium point (DfT-UK 2020a).

Please note that many formulas are used below in Sect. 3 and 4 – we shall omit repeated explanations for the notations throughout the paper, for example, "*ijpcmT*" in (1) will be used below with no further comments.

3 Basic Tour Modeling

A basic tour is a round chain starting/ending at the same location with a single destination, with two half-tours inter-linked via the destination, referred as the outward trip and the return trip, respectively. Basic tours take more than 85% of all travel patterns in Chinese cities regardless HB or NHB, demonstrated by the HIS and GPS data collected (see Sect. 5 for Wuhan situation). Many Chinese tend to work more than eight hours a day and kids stay longer in schools (often with lunchtime naps in work/school places). Comparing to other parts of the World especially in the West, their travelling are dull with much simpler patterns. The basic tour formulation fits into the environment as the core engine of the ATBM, supported by empirical datasets in China.

For clarity, this section is divided into the following four sub Sects. 1 to 4, first giving the tour specifications by time periods and mode combinations, then the definitions for utilities by logsums from the bottom trip level up to the top main tour level, then the calculations for discrete choice probabilities top-down, finally the demand calculation for tour and trips, respectively.

1) Basic tour Specification (for average weekdays in 24-h)

Table 1 presents the 15 tours arranged in the top-right triangle by the time period combination of outward and return trips, with each of the two linked half-tours realized either in the same time period or two different time periods, regardless HB or NHB:

- T = 1, 2, 3, 4, 5, daily time period index, refers to AM (06:00–09:00), MD (09:00–16:00), PM (16:00–19:00), EN (19:00–22:00), and NT (22:00–06:00), respectively (the number of time periods may be expanded if required);
- t = 1, 2, ..., 14, 15, tour index (ID) for the 15 tours in Table 1, with each in the paired combinations of time periods i.e. $t = (t^o, t^r)$, where $t^o = T$ for the time period ID of outward trips, and $t^r = T$ for the time period ID of return trips; and, the pairs satisfy the time-period forwarding convention i.e. $t^o \leq t^r$;
- Tour duration per tour is the time difference between the two middle points of time periods (t^r & t^o), except when $t^r = t^o$ and the NT with simple assumptions;
- The percentages in Table 1 next to the tour ID, are example parameters W_{pct} used in Eq. (3) etc., calibrated from HIS by purpose and person group, with all 15% summing to 100%.

Table 1. Basic tour specification by paired time periods of outward and return trips

tour t = 1, 2, ..., 14, 15		Return time period (t^r)				
		AM (t^r = 1)	MD (t^r = 2)	PM (t^r = 3)	EN (t^r = 4)	NT (t^r = 5)
Outward time period (t^o)	AM (t^o = 1)	*1* (0.29%)	*2* (2.70%)	*3* (72.58%)	*4* (14.47%)	*5* (0.87%)
	MD (t^o = 2)		*6* (0.38%)	*7* (1.96%)	*8* (5.13%)	*9* (1.18%)
	PM (t^o = 3)			*10* (0.00%)	*11* (0.01%)	*12* (0.11%)
	EN (t^o = 4)				*13* (0.00%)	*14* (0.00%)
	NT (t^o = 5)					*15* (0.32%)

The following modes are distinguished in the ATBM based on the data available currently (the number of modes can be added if required):

- main mode M = {car driver (CD), car passenger (CP), transit, shared drive (taxi/TNCs), slow mode}
- exchangeable mode: *mix* = {CP, transit, shared drive, slow mode}
- mechanized mode = {CD, CP, transit, shared drive};
- slow mode = {walk, bike, eBike}.

Table 2 presents the basic tours in terms of paired mode combinations of (m, m^r) for outward and return trips, respectively, where m ∈ M and m^r ∈ M. When an outward mode uses an exchangeable mode in *mix*, all other modes shown off-diagonal in *black* are allowed to use by switching. However, this is not the case if an outward mode uses car as driver - the CD shown in red is the single non-exchangeable mode for the inter-linked trips. In summary, for basic tours, the pairs of modes (m, m^r) satisfy the condition: if m = CD, m^r = m = CD; if m ∈ *mix*, m^r ≠ m & m^r ∈ *mix*.

Table 2. Basic tour specification by pair of modes for outward and return trips

		Mode of return trip (m^r)				
		car driver	car passenger	transit	shared drive	slow mode
Mode of outward trip (m)	car driver (CD)	√				
	car passenger (CP)		√	√	√	√
	transit		√	√	√	√
	shared drive		√	√	√	√
	slow mode		√	√	√	√

For the convenience, the mode and destination of outward trips in the ATBM are also named as the main mode and main destination of basic tours, respectively (to be consistent to the complex tours presented in Sect. 4).

2) Utility (U) Definitions for Mode and Destination Choice (bottom-up from trip to tour)

Trip-level: the utility function for trip mode choice is defined by (2) with notations in Table 3 below.

$$U^T_{ijpcm} = \beta^1_{pm} T^{ivt}_{ijmT} + \beta^2_{pm} T^{a+e}_{ijmT} + \beta^5_{pcm} C_{ijmT} + \beta^3_{pcm} \ln\frac{D_{ijmT}}{\beta^4_{pm}} + \beta^7_{pm} T^{wait}_{ijmT} + \beta^8_{pm} N^{trnsf}_{ijmT} + \beta^6_{pcm} \tag{2}$$

Table 3. Notations for trip-level utility function for mode choice

T^{ivt}_{ijmT}	Skimmed in-vehicle time for mechanized mode or total travel time for slow mode, in minutes
T^{a+e}_{ijmT}	Sum of access time and egress time in minutes, for mechanized modes only, from assignment skims
C_{ijmT}, D_{ijmT}	Trip cost in Chinese money (Yuan), and trip distance in meter, respectively, from assignment skims
T^{wait}_{ijmT}, N^{trnsf}_{ijmT}	Total wait time and the no of exchanges for linked transit trips, respectively, from assignment skims
β^1_{pm} – β^8_{pm}	8 coefficients correspond to each of variables in Eq. (2) explained in the above and one for mode alternative constant ASC, calibrated from SP survey results, shown in Sect. 5

Tour-level: if outward and return trips use the same non-exchangeable mode CD, the daily tour utility by (3) for the main mode choice is defined as the weighted sum of utilities of the 15 tours by Table 1, with each tour utility being the sum of outward and return trip utilities by (2) explained in Table 4:

$$U_{ijpcm} = \sum_t W_{pct} * U^t_{ijpcm} = \sum_t W_{pct} * (U^{t^o}_{ijpcm} + U^{t^r}_{jipcm}) * 0.5 \quad (3)$$

Table 4. Notations for tour utility definition based on trip utilities

$U^{t^o}_{ijpcm}$, $U^{t^r}_{jipcm}$	Outward trip and return trip utilities of basic tour, respectively
Pair (α_1, α_2)	for (m, m^r) in Table 2: α_1 the proportion for using the same mode for return as outward, and α_2 the proportion for using other exchangeable modes on return. When $m \in mix$, Eq. (4) is used with $\alpha_1 + \alpha_2 = 1$ (e.g. $\alpha_1 = 0.9, \alpha_2 = 0.1$)

However, if the outward trip is an exchangeable mode (m ∈ *mix*), as shown by (m, m^r) in Table 2, complication occurs as travellers may use exchangeable modes in *black* on return, tour utility (4) is then expanded from utility (3) by introducing a new term while cooperating the pair of parameters (α_1, α_2), explained in Table 4.

$$U_{ijpcm} = \sum_t W_{pct}[U^{t^o}_{ijpcm} + (\alpha_1 U^{t^r}_{jipcm} + \alpha_2 U^{t^r}_{jipcm*})] * 0.5 \quad (4)$$

The added term weighted by α_2 in (4), is the logsum defined by (5) below for the expected maximum utility between the *black* modes in Table 2, where λ^m_{pc} is the scaling parameter for return-trip sub mode choice:

$$U^{t^r}_{jipcm*} = \ln \sum_{\substack{m^r \in mix \\ m^r \neq m}} e^{\lambda^m_{pc} U^{t^r}_{jipcm^r}} \quad (5)$$

The (main) destination choice in the ATBM is placed above the main mode choice as shown in Fig. 1. The utility function is by (6) in a standard definition, consisting of:

- the logsum of main mode choice utilities with scale parameter β_{pc}^d (being 1 currently)
- the logged size term by purpose, with each sub land use type weighted by σ_{jpk}, obtained from calibration

$$U_{ijpc} = \beta_{pc}^d \ln(\sum_m e^{U_{ijpcm}}) + \ln(\sum_k \sigma_{jpk} * SizeTerm_{jpk}) \quad (6)$$

3) Choice Probability Definitions for Destination and Mode Choice (top-down from tour to trip).

The probability definition for the main destination choice is given by (7), where $\theta_{pc}^d \leq 1$ is the destination choice structure parameter; B_{jp} are the column balance factors for the doubly-constrained destination choice, applied to purpose work/education only, estimated from the converged Iterative Proportional Fitting procedure. For all other purposes, the probability is by singly-constrained destination choice where $B_{jp} = 1$ for all j per p.

$$P_{ijpc} = \frac{B_{jp} e^{\theta_{pc}^d U_{ijpc}}}{\sum_k B_{kp} e^{\theta_{pc}^d U_{ikpc}}} \quad (7)$$

The probability for the main mode choice is given by (8), where λ_{pc}^m are scale parameters for multinomial mode choice, estimated from the stated preference (SP) surveys between modes.

$$P_{ijpcm} = \frac{e^{\lambda_{pc}^m U_{ijpcm}}}{\sum_k e^{\lambda_{pc}^m U_{ijpck}}} \quad (8)$$

The probability (9) below is for the return-trip sub mode choice between *black exchangeable* modes, defined by trip utility (2) among all black modes in Table 2, by assuming the same scale parameter λ_{pc}^m as the main mode choice.

$$P_{jipcm^r t^r} = \frac{e^{\lambda_{pc}^m U_{jipcm^r t^r}}}{\ln \sum_{\substack{m^k \in mix \\ m^k \neq m}} e^{\lambda_{pc}^m U_{jipcm^k t^r}}} \quad (9)$$

4) Tour and Trip Demand Calculation Based on Model Inputs, Utilities and Probabilities

Daily tour demands (10) are calculated by appling probability functions with (7) first for destination choice and then (8) for the main mode choice, before splitting into 15 tours by W_{pct} shown in Table 2. Table 5 explains the population group by H_{ic} and the tour rate by R_{ipc}, as input to the ATBM:

$$TourD_{ijpcmt} = H_{ic} * R_{ipc} * P_{ijpc} * P_{ijpcm} * W_{pct} \qquad (10)$$

Table 5. Population and tour rates in person groups and tour purpose

H_{ic}	The number of persons by segmented population groups for >6yr olds for each TAZ. The segmentation of population groups are by user input; flexibly specified by the combination of car availability, income, gender, and age etc. for HB tours; but for NHB tours, they are calibrated from jobs, hotel rooms, interchange hub services per TAZ, as well as big data collected
R_{ipc}	Tour generation rates for each TAZ by purpose calibrated from variety of data sources. The no of purposes are user input, including work, education, personal/employer's business, entertainment, eating, and leisure etc. See Sect. 5 for examples

The trip demands are collected from relevant tour demands by time period. Outward trip demands by (11) are straightforward, by summing all tour demands that the outward trips are realized in time period T.

$$TripD_{ijpcmT} = \sum_{t^o=T} TourD_{ijpcmt} \qquad (11)$$

Return trip demands by CD are collected by (12), by summing the tour demands that return trips are realized in time period T (in matrix terms, by the summation of transposed tour demands):

$$TripD_{jipcmT} = \sum_{t^r=T} (TourD_{ijpcmt})' \qquad (12)$$

Return trip demands by exchangeable mode in time period t^r are by (13), calculated by sub mode choice i.e. applying probabilities by (9) to the transposed of basic tour demands in matrix terms:

$$TripD_{jipcm^r t^r} = \alpha_2 * (TourD_{ijpcmt})' * P_{jipcm^r t^r} \qquad (13)$$

Total return trip demands for an exchangeable outward mode are by (14), including a) the return trips by the same outward mode, and b) those return trips by *black* exchangeable modes in Table 2:

$$D_{jipcmT} = \alpha_1 \sum_{t^r=T} (TourD_{ijpcmt})' + \sum_{\substack{t^r=T \\ m^r \neq m}} (TourD_{jipcm^r t^r}) \qquad (14)$$

4 Complex Tour Modeling

The complex tours are any tours that cannot be represented by basic tours regardless HB or NHB. ATMB intermediate stops are inserted per outward or return trip, shown in the leftmost of Part 3 in Fig. 1. The ATMB allows maximum three stops per trip chain including the main destination, covering almost 99% of patterns in China. The following briefs the ATMB methodologies for complex tour modeling, based on the basic tours presented in Sect. 3.

4.1 Parents Escorting Children to School en-Route to Work (via Commuting HWH and Education HEH Demands)

Intermediate stops for parents escorting children to school are approximated by relating commuting demands to school education demands by CP: the HCWH demands (Home-EscortChildren-Work-Home by car driver CD) are estimated from the basic tour demands HWH and HEH already available in ATBM by Sect. 3. Further, school-stop insertions for HCWH parents are achieved via two knowns factors N_{is}^{CP} and N_{ij}^{CD}, for no of HEH pupils by CP from home TAZ_i to school TAZ_s and no of HWH parents by CD from TAZ_i to work TAZ_j who drop kids, respectively.

- Get the proportion of pupils from home TAZ_i to school TAZ_s over all pupils from TAZ_i by CP:

$$P_{is}^{CP} = N_{is}^{CP} \Big/ \sum_i N_{is}^{CP} \quad (15)$$

- Apply proportion (15) to car drivers from TAZ_i to work TAZ_j via stop TAZ_s regardless different TAZ_j, by utilizing proportionality method (Bar-Gera et al. 2012) for most-likely route flow patterns, i.e. by assuming that HCWH demands $(i \rightarrow s \rightarrow j)$ are proportional to HWH parents via *unknown* school-specific factor x_s:

$$N_{isj}^{CD} = x_s * (P_{is}^{CP} * N_{ij}^{CD}) \quad (16)$$

- Associate HEH and HWH parents by *known* car occupancy factors at school (collected from school education surveys), since that the average car occupancy (O_s) at TAZ_s can also be viewed as ratio (17) by the total number of pupils by CP over the total number of HWH parents via TAZ_s by CD, also school-specific:

$$O_s = \sum_i N_{is}^{CP} \Big/ \sum_i N_{is}^{CD} \quad (17)$$

- Obtain HCWH demands by solving the x_s in (16) via (17) (omitting detailed steps for page restriction):

$$N_{isj}^{CD} = \left[\sum_i N_{is}^{CP} \Big/ O_s * \sum_{i,j} (P_{is}^{CP} * N_{ij}^{CD})\right] * (P_{is}^{CP} * N_{ij}^{CD}) \tag{18}$$

HWCH/HCWCH, parents escorting children in return from work, are simplified by adjusting HCWH demands.

4.2 Rubber Banding Method for Intermediate Stop Insertion

In the ATBM work/education stops are defined as the main destinations of complex tours; for other tours that consist of trips without the two compulsory purposes, the second stop of long chains is defined as the main destination.

The intermediate stop location is estimated by the rubber-banding method (TMIP 2012) after the outward/return trips have been collected from the basic tours in the ATBM, ready from the main destination output by Sect. 3, as shown by arrows in Fig. 1. Without loss of generality, insertion is implemented by the catchment area assumption, with the maximal distance restricted to be 10-times to the main destination in length from the anchor locations per half-tour. The detailed rubber-banding methods are widely applied for complex tour modelling, omitted for space.

4.3 Half-Tour Modeling with Assumptions Reflecting the Reality

Half-tours regardless HB- or NHB- based, are round basic tours in the ATBM that either outward or return trip is not realized in the 24-h weekday, classified by the internal half-tours within the study area boundaries and the external half-tours from/to the study area (the latter cover trips by certain surface-access modes from/to interchange hubs such as airports, rail stations and river ports etc.).

The outward and return trips per half-tours of basic tours are assumed to be symmetric in the ATBM: the number of from-origins trips are the same as the number of to-origin trips at TAZ level in aggregation by segmentation, regardless anchored at homes, offices, hotels, or interchange hubs. The proxy is close to the real-life traveling in China, verified by daily hotel/hub surveys and mass transit supplies (well-balanced from/to the cities). As such, the basic tours in Sect. 3 are utilized to approximate half-tour demands with adjustments. Special half-tour generation rates for the ATBM are estimated from available big data at the TAZ level for all person groups by purpose. Note that the output half-tour demands take advantage of the aggregate method with outward/return trips realized by different persons within each demand segment, whereas the basic tours in Sect. 3 are realized by the same persons for both legs of basic tours.

5 The Case Study for Wuhan City

This section presents the empirical evidence by applying the framework in Fig. 1 to Wuhan city as a case study, funded by the World Bank as part of the project "Wuhan Integrated Transport Development". With almost 12 million residents covering 8500 km^2 in central China, Wuhan is the mega city boosted with advanced multimodal transport systems, integrated with freeways/Yangtze-River bridges/highway roads, Rail/Metro/Tram/BRT/Bus/Ferry, high speed trains and an international airport. Behaviour-based surveys, traditional count data and big data collections have been undertaken for the case study.

5.1 Data Analysis and Calibrated Parameters

An activity-based household interview survey (HIS) was undertaken in December 2020 with 0.5% sample rate, with 15000 sample sizes collected in total. Table 6 and 7 present summary statistics for the HIS tours, with tour type and purpose code given in Table 7 below. Table 6 presents the number of tours undertaken by residents during an average weekday:

- The number of independent daily tours: 90% of residents take only one tour, 9% take two tours, <1% take more than two; these clearly verify that basic tours are the core concerns in the ATBM.
- The number of intermediate stops on tours: 87% of tours are basic in one stop only, 12% with two stops, and about 1% with more than two stops; our simplified approaches for complex tours in Sect. 4 are clearly evidence based.

Table 6. Household survey summary on no of tours and stops

No of independent tours	No of samples	Percentage	Daily tour examples	Daily activity chain examples
1	24463	90.2%	HWH	HWH
2	2547	9.4%	HWOH-HSH	HWOHSH
3	106	0.4%	HWH-HCH-HWH	HWCHWH
4 +	10	0.0%	HWH-HCH-HWH-HTH	HWHCHWHTH
No of intermediate stops	No of samples	Percentage	Example tour types	Basic tour/Complex tour
1	25984	86.9%	HWH, HEH	Basic tour
2	3617	12.1%	HWSH, HCWH	2-stop Complex tour
3	273	0.9%	HSWTH, HCWCH	3-stop Complex tour
4 +	43	0.1%	HCTSAH	4 + stop Complex tour

Table 7 lists the most frequent basic tour types from raw HIS samples collected. For each of them, the average tour duration per main destinations is also given by activity purpose, where column "Code" represents purposes. Note that the number of records is more than the number of samples per row, because that the former contains complex tours with the same purpose at main destinations. Only one example NHB basic tour is shown in Table 7 – special adjustments have been made to uplift the NHB tours by GPS data collected (detailed explanations omitted).

Table 7. Most frequent tour types by activity purpose in tour durations

No	Basic tour type	Samples Total	Samples Percentage	Activity	Code	Average tour duration (hour)	No of records	Percentage
1	HWH	17435	58.3%	Work	W	9.7	18446	19.6%
2	HEH	1658	5.5%	Primary school	E	9.1	1703	1.8%
3	HMH	1295	4.3%	Secondary sch	M	11.3	1367	1.5%
4	HCH	857	2.9%	Escorting	C	1.3	1500	1.6%
5	HUH	142	0.5%	University	U	8.5	149	0.2%
6	HSH	4119	13.8%	Shopping	S	1.7	6399	6.8%
7	HTH	2880	9.6%	Entertainment	T	2.9	3558	3.8%
8	HQH	436	1.5%	Visiting	Q	3.7	753	0.8%
9	HYH	423	1.4%	Hospital	Y	2.4	519	0.6%
10	HOH	416	1.4%	Other	O	5.1	1109	1.2%
11	HBH	115	0.4%	Business	B	2.7	340	0.4%
12	FOF (NHB)	141	0.5%	Home	H	10.8	58276	61.9%

Table 8 presents example tour generation rates for HWH commuters in three area types, with person groups classified from the combination of household incomes (low/medium/home) and car availabilities (CA or NCA).

Table 8. Example daily tour rates for HWH calibrated from HIS

Person group	R1 (CBD area)	R2 (Suburb Area)	R3 (rural area)
Car available, high income	0.80	0.63	0.54
Car available, medium income	0.77	0.61	0.52
No Car available, medium income	0.81	0.66	0.56
Car available, low income	0.60	0.51	0.43
No Car available, low income	0.59	0.44	0.38

For mode choice, a stated preference (SP) survey has been undertaken for sensitivity parameters i.e. the coefficients of utility function (2) in Sect. 3. We designed and implemented the SP surveys with 1500 questionnaires collected, and Biogeme software is used for multinomial mode choice after data cleaning. Table 9 presents the SP results for parameters calibrated, using medium income as an example:

- The numbers in red are volume of time (VoT, yuan/hour): 44.78/40.08 and 19.50/16.64 by purpose and CA/NCA, respectively. These VoTs are compared nicely to the values estimated by a World Bank estimation method (VTPI 2020), which are 41.03 and 17.87 by purpose, respectively;
- The values of walk-time and wait-time ranges 1–2.3 and 1.37–3.75 times in-vehicle time (IVT), which are comparable to 1.5–2.0 and 1.5–2.5 times IVT by (DfT-UK 2020b), respectively. The range of number of interchanges is 3.62–6.98 min of IVT per interchange, whilst (DfT-UK 2020b) is 5–10 min of IVT (Note that the sign for CA commuter is positive in Table 9, for which we failed to estimate the parameter).

Table 9. Example SP calibration results for Utility functions by Biogeme

Parameters	Commuting/Other					Business				
	Value	Std err	t-test	p-value	Ratio	Value	Std err	t-test	p-value	Ratio
Car Available	Sample size: 132					Sample size: 307				
Time (minute)	-0.56	0.56	-1.01	3.11E-10		-1.74	0.70	-2.50	1.24E-02	
Cost (yuan)	-1.74	0.60	-2.88	4.03E-03	19.50	-2.34	0.70	-3.33	8.72E-04	44.78
Walk time (minute)	-1.05	0.64	-1.65	9.97E-02	1.87	-4.01	0.83	-4.83	1.00E-06	2.30
Wait time (minute)	-2.12	2.13	-1.00	3.20E-05	3.75	-2.93	1.66	-1.76	7.84E-02	1.68
No of nterchanges	2.79	4.69	0.60	5.51E-03	-	-6.31	5.19	-1.22	2.24E-01	3.62
No Car Available	Sample size: 224					Sample size: 387				
Time (minute)	-2.48	0.50	-4.93	8.36E-07		-2.13	0.23	-9.12	0.00E+00	
Cost (yuan)	-8.96	1.33	-6.74	1.56E-11	16.64	-3.19	0.63	-5.11	3.27E-07	40.08
Walk time (minute)	-4.32	0.82	-5.29	1.21E-07	1.74	-2.13	0.34	-6.32	2.65E-10	1.00
Wait time (minute)	-7.00	1.90	-3.68	2.32E-04	2.82	-2.92	0.76	-3.86	1.12E-04	1.37
No of nterchanges	-17.35	3.73	-4.65	3.30E-06	6.98	-8.29	1.86	-4.47	7.81E-06	3.89

5.2 Model Convergence, Realism and Sensibility

It is of crucial importance to demonstrate that the demand-supply models converge to a satisfactory degree, in order to have confidence that the model results are as free from error and noise as possible. Model accuracy is a prerequisite for transport modeling and appraisal (Hartgen 2013). Figure 2 shows the evolution of Gap% defined in (1) by iteration and runtime for Wuhan in 3877 TAZs. The required convergence <0.2% (DfT-UK 2020a) is reached with 13 demand-supply loops in 26.5 h, with five peak-hour (AM/MD/PM/EN/NT) assignments/skimming running on parallel in Emme 4 which take most of model runtimes.

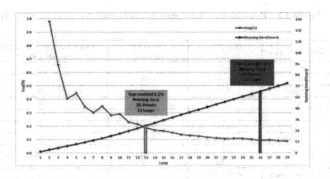

Fig. 2. Gap% by demand-supply iteration and runtimes

Sensitivity tests on model accuracy by Gap% are undertaken in the same laptop with 128gb of RAM. Figure 2 shows that, for the finest convergence of Gap% < 0.1, 26 loops are needed in 66 h with 2.5 times more runtimes. Convergence tests for smaller number of TAZs (e.g. 2500 zones) have also been carried out: reaching the equilibrium status of Gap% < 0.2 takes typically less than 20 h by the MSA method. For scenario model runs in the Wuhan study, the convergence Gap% < 0.2 is used throughout for model accuracies.

The realism tests are undertaken to ensure that the ATMB responds rationally with a broad range of transport changes, i.e. testing demand responses with a modest change to cost components for acceptable elasticities. The elasticity form $\eta = \frac{\log D_2 - \log D_1}{\log C_2 - \log C_1}$ is used where C_2 is for the 10% changing from the reference case cost C_1, and D_2 and D_1 are demands corresponding to costs C_2 and C_1, respectively. Table 10 gives preliminary results for Wuhan including five HB basic tours and three NHB basic tours as examples, where the estimated trip elasticities of fuel cost, transit fare and duration-based parking cost are given by tour type, respectively. The recommended transport elasticity values by (Litman 2021), derived mainly from the studies in developed countries, are presented for comparison (few published elasticity values are found from our extensive internet search for developing countries). Table 10 demonstrates that the trip elasticities derived for Wuhan are comparable to the reference ranges by tour type with purposes, but with outliers (e.g. the HCH in red with few trips from HIS samples). Further studies on the realism test are needed to consolidate the range of elasticity values for Wuhan and beyond in model development.

Table 10. Realism test with outturn elasticities for trips by tour type

Basic tour type	Fuel price elasticity	Transit fare elasticity	Parking elasticity in activity durations for CBD
HWH	-0.28	-0.52	-0.74
HBH	-0.06	-0.48	-0.21
HSH	-0.07	-0.56	-0.89
HOH	-0.08	-0.85	-0.24
HCH	-0.16	-2.48	-1.51
FAF	-0.04	-0.37	-0.71
FBF	-0.05	-0.36	-0.78
FOF	-0.05	-0.34	-0.81
Reference elasticities by (Litman 2021)			
Recommended	-0.05 to -0.55	-0.4 to -1.0	-0.541

A series of sensitivity tests shall be done comprehensively in model validation with scenarios in different population growth, land use and supplies etc. Preliminary sensitivity tests have been undertaken to understand model uncertainties for Wuhan. Table 11 presents transit fare sensitivity examples with three fare strategies, for encouraging multimodal transit linkage among Metro, tram, BRT, bus, and ferry services. Everything else the same, the sensitivity tests for the similar plans on transit demands are shown from model outputs. The transit mode shares among main modes in Table 11 are increased from the reference case 24% to 26.18%, 26.45%, and 26.04%, respectively; the increase of transit demands over the reference case is 8%, 9%, and 7% respectively, with minor differences anticipated in design.

Table 11. Sensitivity test for transit fare strategies for uncertainty

	Plan1	Plan 2	Plan 3
Transit sub mode change numbers and time restriction	Maximum three changes allowed within 90 min		
Discount per change among Tram, BRT, Bus, Ferry	50% off	60% off	30% off
Discount per change between Tram/BRT/Bus/Ferry and Metro	Per Metro entry, Metro fare discount 1 Yuan	Per Metro entry, Metro fare discount 0.8 Yuan	Lower fare on change is free max discount 2 Yuan
Reference case transit fare: Metro fare distance-based, Tram/BRT/Bus/Ferry 2 yuan per boarding	% of transit share (% of transit demand) over reference case		
Reference case transit mode share: 24%	26.18% (8%)	26.45% (9%)	26.04% (7%)

6 Conclusion and Further Development

We presented a new aggregate tour-based model for travel forecasting, addressing the issues in state-of-practice approaches with innovations. In our ATBM, each component of the nested logit structure is defined clearly with the accessibility logsum utilized for vertical demand hierarchy. HB/NHB basic tours are addressed in time/space consistency with tour durations spanning multiple time periods. While seeking behaviour realism, exchangeable mode switching, intermediate stops for complex tours, duration-based parking etc. are explicitly considered in the framework. To reduce aggregate bias and minimize errors, a gap function is used explicitly to monitor the convergence between the supply and demand models. Computer runtimes are reasonable for stable results often achieved by overnight model runs.

The ATBM is generic fit-for-purpose for strategic transport planning and government interventions. Being aggregate in nature and deterministic, the ATBM data requirement is at a similar level to that of the traditional 4-step models. Calibration parameters are kept at minimum in system design. The case study for Wuhan city has been presented as empirical evidence, with preliminary test results given for convergence, realism and sensitivity tests, respectively.

The ATBM is under continuous developments with phased version updating on pipeline. Another case study will be presented in a near future for Chinese city Wuxi (medium-sized with the highest GDP per capita in China by the 7th Census in 2020). The near-term developments for Wuxi include:

- Add more main modes than the five in Table 2 and implement sub mode choice per main mode;
- Allow more time periods than the five in Table 1 and add adjustments for shift work patterns within the 24h time span (e.g. basic tours with outward/return trips occurred in the NT/AM time periods, respectively);
- Add tour frequency modeling at the top of nested choice hierarchy allowing tour rates being variables, based on the main destination choice logsums during the demand-supply feedback.

Robust big data are gradually available consolidating the potential application of the ATBM in other Chinese cities and beyond. Flexible, transparent and modular structures shall be provided for user customization on system input/output, parameter/function updating, memory/runtime management etc. – this is particular important for state-of-art demand modeling with ever increasing uncertainties in travel forecasting.

References

Bar-Gera, H., Boyce, D., Nie, Y.: User-equilibrium route flows and the condition of proportionality. Transp. Res. Part B Methodol. **46**(3), 440–462 (2012)

Bernardin, V.: Why one size doesn't fit all activity-based vs trip- based models and everything in between, RSG, USA (2015). https://tnmugfiles.utk.edu/presentations/2015Dec/OneSizeModelDoesNotFitAll.pdf

Bradley, M., Bowman, J.: Activity-based travel forecasting models in the united states: progress since 1995 and prospects for the future. Progress in Activity-Based Analysis (2004). https://trid.trb.org/view/759300

Boyce, D., Williams, H.: Forecasting Urban Travel: Past Present and Future. Edward Elgar Publishing, London (2015)

Castiglione, J., Bradley M., John, G.: Activity-Based Travel Demand Models A Primer. Transportation Research Board, USA (2015). https://www.nap.edu/catalog/22357/activity-based-travel-demand-models-a-primer

DfT-UK: Variable demand modelling. TAG UNIT M2.1, Transport Analysis Guidance (TAG), Department for Transport, UK (2020a). https://www.gov.uk/government/publications/tag-unit-m2-1-variable-demand-modelling

DfT-UK: Public Transport Assignment. TAG UNIT M3.2, Transport Analysis Guidance (TAG), Department for Transport, UK (2020b). https://assets.publishing.service.gov.uk/government/uploads/system/uploads/attachment_data/file/938870/tag-m3-2-public-transport-assignment.pdf

DfT-UK: Dynamic Integrated Assignment and Demand Modeling (DIADEM) User Manual v7.0. Department for Transport, UK (2020c). https://www.gov.uk/government/publications/diadem-software

Hartgen, D.T.: Hubris or humility? Accuracy issues for the next 50 years of travel demand modeling. Transportation **40**(6), 1133–1157 (2013). https://doi.org/10.1007/s11116-013-9497-y

Litman, T.: Understanding transport demands and elasticities - how prices and other factors affect travel behavior. Victoria Transport Policy Institute, Canada (2021). https://www.vtpi.org/elasticities.pdf

McNally, M.: The four step model. Institute of Transportation Studies, University of California, Irvine (2000). https://escholarship.org/uc/item/7j0003j0

Tajaddini, A., Rose, G., Kockelman, K., Vu, H.: Recent progress in activity-based travel demand modeling: rising data and applicability. IntechOpen (2020). https://www.intechopen.com/chapters/73240

PTV: PTV Visum 2020 Manual, PTV AG, Karlsruhe, Germany (2020)

TMIP: Activity-Based Modeling Tour Mode-Intermediate Stop Location-Trip Mode Session 10, FHWA, USA (2012). https://tmip.org/sites/freightmodelimprovementprogram.localhost/files/webinars/2012/TMIP_ABM_Webinars/Session_10/Webinar10_TourTripModeStopDest_Slides_Only.pdf

VTPI: Transportation Cost and Benefit Analysis II – Travel Time Costs. Victoria Transport Policy Institute, Canada (2020). https://www.vtpi.org/tca/tca0502.pdf

Xiang, Y., Wright, I., Meehan, T.: A novel design and implementation of an aggregate UK-based transport model. In: European Transport Conference 2009, The Netherlands (2009). https://aetransport.org/public/downloads/uD5hd/4019-514ec5cdcdad0.pdf

Multilayer Perceptron Modelling of Travelers Towards Park-and-Ride Service in Karachi

Irfan Ahmed Memon[1(✉)], Ubedullah Soomro[1], Sabeen Qureshi[2], Imtiaz Ahmed Chandio[1], Mir Aftab Hussain Talpur[1], and Madzlan Napiah[3]

[1] Department of City and Regional Planning, Mehran University of Engineering and Technology, Jamshoro 76062, Sindh, Pakistan
irfanahmed04crp26@gmail.com
[2] Department of Architecture, Mehran University of Engineering and Technology, Jamshoro 76062, Sindh, Pakistan
[3] Department of Civil and Environmental Engineering, Universiti Teknologi PETRONAS, 32610 Seri Iskandar, Perak Darul Ridzuan, Malaysia

Abstract. The imbalance between public and private transport causes congestion. Currently, congestion is due to individuals driving their automobiles to work in Karachi central business districts (CBDs). Therefore, the park and ride (P&R) service has been utilized widely in several countries as part of travel demand management (TDM). Consequently, P&R has proved successfully reduced congestion and difficulties to locate parking spots in the urban center. Travelers cannot be persuaded to adopt P&R without knowing their travel pattern. Accordingly, a travel behavioral survey was conducted, to eliminate imbalances between public and private mobility. Therefore, modal choice models were to identify the variables influencing the decision to accept P&R service of single-occupant vehicles (SOV). Data were collected by an adapted self-administered questionnaire through a survey approach. Mode choice models developed through multilayer perceptron (MLP) of artificial neural network (ANN) approach by using statistical package for the social sciences (SPSS) version 22. The research findings were more towards the socio-demographic factors. Furthermore, travel time, travel expenses, environmental protection, avoid mental stress, parking problem, vehicle sharing, and travel directly from home to office were found significant variables. In conclusion, the SOV users may be encouraged to move into P&R services by overcoming these influencing elements and balance push and pull measures of TDM. Thus, policymakers can benefit from study results and provide a base for future studies on sustainable modes of public transportation.

Keywords: Park-and-ride · Travel behavior · Artificial neural network (ANN) · Multilayer perceptron (MLP) · Karachi · Mode choice

1 Introduction

Mobility in developing world cities is a very peculiar challenge because it is different from health, education, or housing. Mobility tends to become worse as societies become richer and become an unsustainable model. Mobility, like most other developing country's problems, more than a matter of money or technology, is a question of equality and equity. The inequality in developing countries makes it difficult to observe. For instance, regarding transport "a developed country is not a place where the poor have cars; it is where the rich use public transport" (Kun 2016). Such as, in Amsterdam, 30% of the population use bicycle, although the Netherland has a high income per capita than the United States (U.S) (Peñalosa 2013).

Traffic jams and road rage are one of the eminent problems faced by most of the population around the world. The issue of traffic jams in the main urban centers around the world, such as Karachi, is a common occurrence (Ali et al. 2014; Brohi et al. 2021). Traffic congestion tends to draw out, impede, and worsen nonproductive monetary activities in urban communities. Different studies on traffic-related problems are underway in different clusters, including economic and financial problems (Hamid et al. 2008; Irfan Ahmed et al. 2021). Economical loss because of traffic congestion was observed in different cities of the world. It has been revealed that, in 2018, Karachi's traffic jam cost surpassed 1 billion USD per year, which equals a 2% share of the total gross domestic production (GDP) of Karachi (Ali et al. 2014; MEMON 2018). That cost is ascertained based on the value of time, fuel cost, and consumption of fuel. Karachi is the capital of Pakistan's province of Sindh. Karachi offers transit services along the Arab Sea and is home to Pakistan's two major maritime ports: the Port of Karachi, the Port of Bin Qasim, and Pakistan's most busy airport. (Earth.Esta.Int.). Karachi is one of the busiest and most influential cities in the region of South Asia. The rapid expansion of the city in recent decades increased the demand for city transit infrastructure. Currently, 15 million residents with a density of 24,000 people per square kilometer are living in the city (Statistics 2017). Overall, 60% of the population of Karachi is relying excessively on private transport creating enormous traffic congestion and painful air pollution (ITDP 2015; Indus 2016). Karachi is the only functional port city in Pakistan and the major industrial and economic hub of the region. All major banks in Pakistan are headquartered in Karachi. The Pakistan Stock Exchange formerly, Karachi Stock Exchange is the country's largest having an annual turnover of Rs 436 million (US$7.2 million) is located in Karachi. Almost, Sindh province generates almost 70% of the revenue and 62% of the sales tax charged by Pakistan's government and, 94% of this is generated in Karachi. However, Sindh's share in federal government revenue transfers is just 23.3% (Arif Hasan 2020; Memon et al. 2021).

The provision of public transportation cannot help to reduce the long-term traffic jams. Travel demand management (TDM) strategies executed efficiently with a light rail or bus service integrated with park-and-ride (P&R) can reduce traffic intensity in the long term (Institute 2014; Memon et al. 2016a, b; Memon et al. 2014). Different factors impact the usage of P&R infrastructure, such as public transport service, parking lot circumstances, and parking area accessibility (Turnbull 1995; Shaharyar et al. 2021). Other considerations include choice, time, and cost savings for the whole journey in comparison with travel by car. It is consistent with previous findings that car passengers are likely to use P&R if the travel time of the P&R is minimum (Memon 2010; Memon et al. 2016a, b). The provision of dedicated bus routes in the city center and complex changes can help to reduce travel time by the P&R service. Furthermore,

Hamid et al. (Hamid 2009) described that in comparison with other transport modes, it is important to save money and time to use the P&R service.

Whitfield and Cooper (Whitfield and COOPER 1998-9) focused on their findings to increase attraction on P&R services would be as useful as traveling in a private car (with switching times). Therefore, regularity of service and bus preference techniques should be considered. Similarly, a study conducted in the United Kingdom discovers that 81% of potential bus travelers considered that a 'turn up and go' level of frequency at every 10 min was encouraging them to adopt the service (TAS 2001). Individual lifestyle and status have a significant influence on the travel pattern (Memon 2010; De Vos et al. 2012). Consequently, potential explanations on lower public transit not only include increasing quality but also considering the travel patterns of travelers.

The purpose of this research is to examine the variables affecting the decision and willingness of travelers to move towards P&R service. In this research, an ANN's MLP approach was adopted to develop a mode choice model, and the use of the model was clarified using a case study. Therefore, it is critical to investigate their travel habits and readiness to switch towards P&R if given positive encouragement. The results are beneficial which provides some awareness into policy makings for stakeholders.

2 Literature Review

2.1 Park and Ride Service

Park-and-ride service includes parking lots, provided at commute stations, bus stops, and highway onramps. Majorly, P&R service is usually available at the fringes of the city, to accelerate the use of transit and rideshare. Parking at P&R stations is usually free or at a nominal price than city centers (Institute 2010a, b).

P&R services can improve travel and carpool (Turnbull 1995). Turnbull discussed the effects of park-and-ride services on transport systems. Morrall and Bolger (Morrall and Bolger 1996; Morrall and Bolger 1997), researched that P&R service provision has a major influence on the city center trips. Parkhurst (Parkhurst 2000), discovered that though P&R services reduce urban congestion, that may increase urban vehicles periphery traffic as private transport users are unaware of reaching facilities or unnecessary trips, in certain instances, moving from non-motorized to motorized journey (Parkhurst 2000). The quality of P&R service depends upon real factors, the availability of attractions such as preferential High Occupancy Vehicles (HOV), financial reasons for passengers, and individual conditions like employment dispersion. Various transit companies use modern parking management systems to highlight empty parking places (SAIC 2008).

Karachi is Pakistan's major urban and economic hub. Having an open phase of fast urbanization and motorization (Qureshi and Lu 2007). Karachi's southern district specifically Saddar Town is the administrative hub of Karachi and is known as the economic backbone of the country. The increased urbanization and economic growth in the city have put tremendous pressure on travel demands (Qureshi and Lu 2007). Rising demand has quickly filled the roadway infrastructure as about 33% of all motorized vehicles in the country multitude on its roads and expressways (Qureshi and

Lu 2007). Figure 1 shows that the vehicle fleet is dominated by cars and motorcycles, which account for 87% of the vehicles as compared to 7% for Paratransit; taxis, and rickshaws, and 1% for public transport (Ali 2012). This rapid rise in personal vehicle ownership and the lack of economic instruments, such as charged parking and road pricing, has led to enormous congestion especially in the central part of the city which increases the average commute travel time in Karachi by over 45 min (Qureshi and Lu 2007). Figure 1 shows the increasing number of different modes of transport from 2006 to 2011, it is reflecting that private cars and motorcycles dominantly increased.

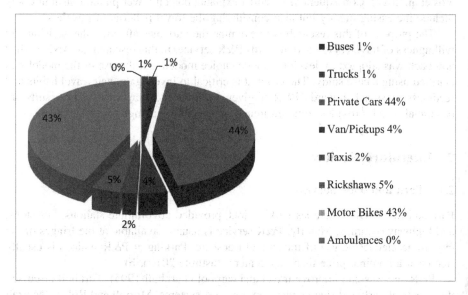

Fig. 1. The proportion of vehicles fleets in Karachi (Memon et al. 2016a, b)

Moreover, the theme of the research is towards investigating the factors that affect travel behavior and modal splits such as environment, trip, individual, transport, quality, and uncertainty specific factors, influencing private and public transport consumers' modal split. To find the results, the researcher must follow specific research techniques. As per conversations, the below research questions contribute to improved knowledge and action. This can encourage private vehicle users to P&R services for future P&R development and planning.

- What are significant factors to develop a mode choice model to affect individual travelers' willingness to adopt the park and ride service?

2.2 Ann Approach in Transportation Research

Transportation information is customarily modeled through two diverse methodologies such as; statistical approach and computational intelligence (CI). Firstly, the statistical approach, is the arithmetic of gathering, arranging, and analyzing numerical information, especially at the point when this information concern the investigation of populace qualities by selecting from sampling (Glymour et al. 1997; Karlaftis and Vlahogianni 2011). Statistics have robust and broadly acknowledged numerical establishments and can give in-depth knowledge on the systems making the information. Statistics may fail while managing complex and exceedingly nonlinear information (the curse of dimensionality). Secondly, computational intelligence CI joins components of learning, adjustment, advancement, and fuzzy logic to make models that are "intelligent" as in structure rises out of an unstructured starting (the information) (Sadek et al. 2003; Engelbrecht 2007).

Neural Networks (NN), a significant and famous class of computational intelligence models, has been broadly adopted to different transportation issues, somewhat they are exceptionally exclusive, precise, and appropriate scientific models ready to reenact numerical model mechanisms efficiently. They have the inborn affinity for putting away exact learning and can utilize as a part of any of three fundamental behaviors (Haykin 1998); firstly, as a model of natural nervous systems. Secondly, as on-time versatile signal processors/controllers. Thirdly, as analytical information techniques. In transportation, NN is primarily utilized as analytical data techniques on account of their capacity to work with extensive measures of multidimensional information, displaying adaptability, learning and speculation capacity, flexibility, and high prescient ability.

2.3 MLP and Mode Choice Modelling

Shmueli et al. conducted mode choice modeling research in which simple multilayer perceptron (MLP) compared with nonlinear classification and regression trees (CART) and proved that both procedures perform similarly well in demonstrating travel behavior (Shmueli et al. 1996). Sayed and Razavi in 2000 (Sayed and Razavi 2000) proved that fuzzy neural network has a comparative same predictive approach as do logit and probit models, whereas Mohammadian and Miller in 2002 (Mohammadian and Miller 2002) revealed that the MLP has a substantial advantage in the nested logit model regarding the cases efficiently characterized in demonstrating household mode choice. In 2006, Celikoglu demonstrated that simple MLP might beat the utility function alignment in mode choice modeling (Celikoglu 2006). Whereas, Rao et al. in 1998 (Rao et al. 1998), Hensher and Ton in 2000 (Hensher and Button 2000), Xie et al. in 2003 (Xie et al. 2003), Adrande et al. in 2006 (Karlaftis and Vlahogianni 2011), and Zhang and Xie in 2008 (Zhang and Xie 2008) revealed MLP's functioning capability is better than multinomial and nested logit models. Vythoulkas and Koutsopoulos in 2003 discussed that outcomes from fuzzy NN in travel choice modeling contrast positively to the logit model (Vythoulkas and Koutsopoulos 2003).

This research extracted the management of traffic solutions of congestion problems by using P&R services in urban centers (Institute for Transport Studies, Seik 1997; CPRE 1998; Institute 2010a, b; Qin et al. 2013). There is a prospect to expand the P&R service to promote SOV users to switch from car to alternative mode choice (Arup and Accent 2012). In the developing world, there was less focus on attracting car travelers towards P&R service (Seik 1997; Qin et al. 2013). Several studies were carried out on urban transport equity, sustainable transport strategies, and pedestrian in Karachi (Khan et al. 1999; Qureshi and Lu 2007; Ahmed et al. 2008), hence, less consideration has been given to P&R service. Putrajaya's modal split is turned opposite and the current modal split is 15:85 between public and private transport (Nor Ghani Md. Nor and Nor 2006). Therefore, factors that impact travelers' mode choice must be investigated.

3 Research Implementation and Methods

This study mainly covers private transport travelers. The research concentrates on the employees or working people. Particularly, those whose places of employment reside in the core economic district or the study center. Specifically, private transport owners who drive their vehicles to work are targeted. The responders have no such age constraint; because all are workers, they must be above 18 and independent. Working people produce their journeys on specific timings while going and return from their workplaces. Therefore, in those specific peak hours, traffic mostly increases and creating traffic congestion in the city center. A non-probability purposive sampling technique has been adopted in this study to present the targeted population. Purposive sampling is also known as subjective, judgmental, or selective sampling technique (Dissertation.laerd.com 2017; Memon et al. 2020). It contains a specific purpose and the approach which usually select to comprise the people relevant to the targeted area and leaving the unmatched with the sample of the research. This method of sampling is chosen because the population of study areas is known, but the sample frame or probability of sample selection is unknown. In other words, it is hard to know the population of buildings, departments, or offices, particularly in study areas. Furthermore, there are different purposive sampling approaches in which a homogeneous sampling approach is suitable for this study. The homogeneous sampling technique is used to focus on a sample that shares the same characteristics or traits to achieve a homogenous sample.

The survey from the non-users of P&R service was carried out and established the significance and willingness of certain criteria to adopt P&R service. The study also collects data on travel choices made by passengers known as revealed preference and hypothetical data is referred to as stated preference. The preferences data are assessed to determine whether or not the traveler is prepared to utilize P&R.

This research includes 1000 samples according to the well-known method for determining the sample size of Krejcie and Morgan (Krejcie and Morgan 1970). In 1996, Ortuzar and Willumsen suggested not less than 250 samples for the mode choice research. According to a recent study at Leeds, England, the sample should be large enough for each group to have a sample size of a minimum of 100 when divided into groups (Pathan and Faisal 2010).

Self-administered questionnaires were delivered in the city centers (CBD) of Karachi to the ministries, the government, and private offices. The respondents of the survey were working in the offices mentioned above. Data was collected through questionnaires and analyzed by IBM Statistical Package for Social Sciences (SPSS) version 22.

4 Model Development

Multilayer perceptron (MLP) is the type of feed forwards network (FFN), and FFN is one of the types of artificial neural network (ANN). Multilayer perceptron uses different algorithm techniques in which the back-propagation algorithm is one of the most popular techniques. In the back-propagation technique, the outputted values are compared with the corrected answer to compare the values of some predefined error. Multilayer perceptron approach applied on the variables which were fitted in Mode Choice Model known as Putrajaya Model. In which it was tested and validated all variables such as gender, monthly income, travel directly, vehicle sharing, parking problem, avoid mental stress, protect the environment, comfort preference, and travel time per trip.

The artificial neural network has two steps of data analyzing: testing and training. Initially, in the training stage, 702 sample sizes were selected for training purposes which were 70.2% of the total sample size, and the remaining 298 samples were held out as shown in Table 1. This section presents the training and testing of the Final Model of Karachi, detailed output results generated by SPSS.

Table 1. Training case processing summary of the Model

		No	Percent
Sample	Training	702	70.2%
	Holdout	298	29.8%
Valid		1000	100.0%
Excluded		0	
Total		1000	

The overall model summary during a training session was satisfactory; there were 0% incorrect predictions. The cross-entropy error was 0.360. The classification table indicates during training; the model correctly classifies 100% of training cases. The sensitivity was 100%. Specifically, 100%, the model's positive predictive values were 100%, and the negative predictive value is 100%.

Similarly, after training, the testing phase was initiated to validate the model. In this section of the modeling, 544 samples were trained, and 158 samples were tested. Total valid cases which are selected by the model are 702 out of 1000; the remaining 298 cases were a holdout. A further detailed summary of case processing is presented in Table 2.

Table 2. Testing case processing summary of the model

		N	Percent
Sample	Training	544	54.4%
	Testing	158	15.8%
	Holdout	298	29.8%
Valid		1000	100.0%
Excluded		0	
Total		1000	

The model summary in the testing session presents both training and testing. The training cross-entropy error is 17.59; the incorrect predictions percentage is 0.9%. The testing's cross-entropy error is 9.59, and the incorrect percent of predictions is 1.9%. Furthermore, in this session, the training model correctly classifies 99.1% of the cases, and the testing model classifies 98.1% of corrected cases. The training model sensitivity was 98.3%, specifically was 99.3%, model's negative values are 97.54%, and positive values are 99.52%. Consequently, the testing model's sensitivity was 84.9% and specifically was 99.2%, the model's negative value was 94.87% and the positive predictive value was 98.33%. Further detail of the testing model is shown in Fig. 2. It is proved that willingness to adopt P&R service is nearer to baseline.

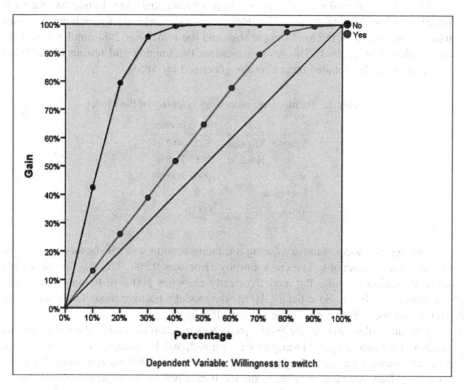

Fig. 2. Cumulative gain chart of the model

Moreover, the independence variables significance which is included in the model is presented in Fig. 3. In this process, as discussed above, the importance of the predictor variable is measured to know how much the network's model-predicted changes for different values of the predictor variable. Therefore, the most important variable in the network was avoided mental stress (Yes) and then protect the environment (Yes), vehicle sharing (Yes), and the least important was private transport ownership (motorbike) as shown in Fig. 3. Thus, it is validated that the variable, which was signified by the logit model, is significant in the multilayer perceptron approach of ANN. Furthermore, logit modeling outputs and MLP results of the Model are compared. Hence, it is validated that the factors with higher odds (β) in logit modeling also have higher importance and normalized importance in MLP modeling which can be observed in Table 3.

Table 3. Logit modelling and MLP output comparative list of model.

Final model's variables	Logit model		MLP model	
	β	p-value	Importance	Normalized importance
Ownership (M. Bike)	5.485	.055	.071	15.3%
Protect environment (Yes)	17.35	.007	.378	80.8%
Vehicle sharing (Yes)	7.129	.058	.082	17.6%
Avoid mental stress (Yes)	21.57	.006	.468	100.0%

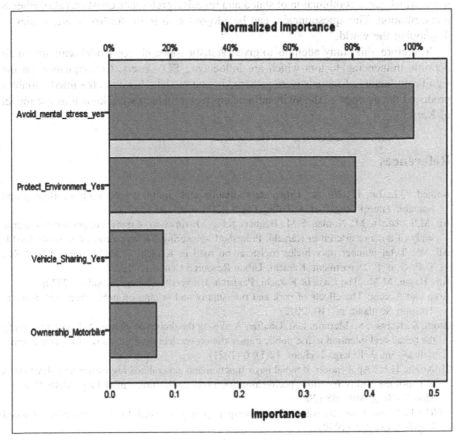

Fig. 3. Importance of predictor variable of model

5 Conclusions

Karachi model consists of eight significant predictive variables; gender, household size, income per month, private transport ownership, protect the environment, frequency of bus, transport share, and avoid mental stress traveling were fitted in the final model. In conclusion, it was found that the people of Karachi are willing to adopt the P&R service if stakeholders are going to provide it. It is predicted that motorbike owners are willing to adopt P&R services to reduce their mental stress and to protect the environment.

Moreover, this research also contributes to the methodology of mode choice modelling. The synthesized factors were modelled through logit modelling. Meanwhile, the factors which were signified through BLR were tested on ANN's MLP approach. MLP technique applied to ensure and further understand the significant factors through computer intelligence techniques. This hybrid methodological approach such as a combination of conventional and computer intelligence techniques contributes to developing world research. Additionally, the development of a questionnaire, which has a combination of stated and revealed preference questions altogether is a contribution. This questionnaire can be adopted as it is for further future studies in developing the world.

Therefore, this study attempts to give a holistic mode choice model with almost all possible influencing factors which are influencing SOV users. Consequently, all the significant factors of models were gathered in one model. However, the fitted variables produced the essence of the swift influencing factors which were known as the model of Karachi.

References

Ahmed, Q.I., Lu, H., Ye, S.: Urban transportation and equity: a case study of Beijing and Karachi. Transp. Res. Part A: Policy Pract. **42**(1), 125–139 (2008)

Ali, M.S., Adnan, M., Noman, S.M., Baqueri, S.F.A.: Estimation of traffic congestion cost-a case study of a major arterial in Karachi. Procedia Engineering **77**(Supplement C), 37–44 (2014)

Ali, W.: Total number of vehicles registered/on road in Karachi – 2011 compiled by URC. T. P. G. o. S. Department. Karachi, Urban Resource Centre (2012)

Arif Hasan, M.M.: The Case of Karachi, Pakistan. University College, London (2020)

Arup and Accent: The effects of park and ride supply and pricing on public transport demand. Trnsport Scotland, p. 110 (2012)

Brohi, S., Kalwar, S., Memon, I.A., Ghaffar, A.: Using the theory of planned behavior to identify the behavioral intention to use public transportation service: the case study of Karachi circular railway. Int. J. Emerg. Technol. **12**(1), 6 (2021)

Celikoglu, H.B.: Application of radial basis function and generalized regression neural networks in non-linear utility function specification for travel mode choice modelling. Math. Comput. Model. **44**(7), 640–658 (2006)

CPRE: Park and Ride: Its role in local transport policy. Council for the Protection of Rural England, London (1998)

De Vos, J., Derudder, B., Van Acker, V., Witlox, F.: Reducing car use: changing attitudes or relocating? The influence of residential dissonance on travel behavior. J. Transp. Geogr. **22**, 1–9 (2012)

Dissertation.laerd.com: Purposive sampling | Lærd Dissertation (2017). http://dissertation.laerd.com/purposive-sampling.php. Accessed 15 May 2017

Earth.Esta.Int.: Karachi, Pakistan. Historical Views. https://earth.esa.int/web/earth-watching/historical-views/content/-/article/karachi-pakistan#:~:text=Situated%20on%20the%20Arabian%20Sea,the%20busiest%20airport%20in%20Pakistan. Accessed 09 Apr 2021

Engelbrecht, A.P.: Computational Intelligence: An Introduction. Wiley, Hoboken (2007)

Glymour, C., Madigan, D., Pregibon, D., Smyth, P.: Statistical themes and lessons for data mining. Data Min. Knowl. Disc. **1**(1), 11–28 (1997)

Hamid, N.: Utilization patterns of park and ride facilities among Kuala Lumpur commuters. Transportation **36**(3), 295–307 (2009)

Hamid, N.A., Mohamad, J., Karim, M.R.: Travel behaviour of the park and ride users and the factors influencing the demand for the use of the park and ride facility. In: EASTS International Symposium on Sustainable Transportation incorporating Malaysian Universities Transport Research Forum conference 2008 (MUTRFC08). Universiti Teknologi Malaysia, UTM, Skudai (2008)

Haykin, S.: Neural Networks: A Comprehensive Foundation. McMillan Publ. Co., New York (1998)

Hensher, D., Button, K.: Handbook of Transportation Modelling. Pergamon (2000)

Indus, T.: Research on traffic congestion (2016). http://www.toyota-indus.com/toyota-research-on-traffic-congestion-2012-13/. Accessed 20 Oct 2017

Institute for Transport Studies, U. o. L.: Park and ride. KonSULT, KNOWLEDGEBASE on SUSTAINABLE URBAN LAND USE AND TRANSPORT. http://www.konsult.leeds.ac.uk/private/level2/instruments/instrument035/l2_035a.htm. Accessed 26 Oct 2013

Victoria Transport Policy Institute: Park & ride convenient parking for transit users. Online TDM Encyclopedia. Victoria, BC, Canada, Victoria Transport Policy Institute (2010a)

Victoria Transport Policy Institute: Park &ride – convenient parking for transit users. TDM Encyclopedia (2010b)

Victoria Transport Policy Institute: Park & ride convenient parking for transit users. Online TDM Encyclopedia. Victoria, BC, Canada, Victoria Transport Policy Institute (2014)

Irfan Ahmed, M., Saima, K., Noman, S., Madzlan, B.N.: Factors that influence travelers' willingness to use or not park-and-ride service in Putrajaya and Karachi CBD. PalArch's J. Archaeol. Egypt/Egyptol. **18**(2), 720–734 (2021)

ITDP: Addressing Congestion on the Streets of Karachi (2015). https://www.itdp.org/addressing-congestion-streets-karachi/. Accessed 21 Oct 2017

Karlaftis, M.G., Vlahogianni, E.I.: Statistical methods versus neural networks in transportation research: differences, similarities and some insights. Transp. Res. Part C: Emerg. Technol. **19**(3), 387–399 (2011)

Khan, F.M., Jawaid, M., Chotani, H., Luby, S.: Pedestrian environment and behavior in Karachi, Pakistan. Accid. Anal. Prev. **31**(4), 335–339 (1999)

Krejcie, V.R., Morgan, W.D.: Determining sample size for research activities. Educ. Psychol. Meas. **30**, 607–610 (1970)

Kun, J.: Where the rich use public transport.... (2016). http://newyork.thecityatlas.org/lifestyle/developed-area-rich-public-transport-ways-city/. Accessed 14 Nov 2016

Memon, I.A.: Factors influencing travel behaviour and mode choice among Universiti Teknologi Malaysia employees. Master of Science in Transportation Planning. Universiti Teknologi Malaysia, Faculty of Built Environment (2010)

Memon, I.A.: Mode Choice modelling to shift car travelers towards park and ride service in the CBD of Putrajaya and Karachi. Ph.D., Universiti Teknologi PETRONAS (2018)

Memon, I.A., Kalwar, S., Sahito, N., Qureshi, S., Memon, N.: Average index modelling of campus safety and walkability: the case study of University of Sindh. Sukkur IBA J. Comput. Math. Sci. **4**(1), 37–44 (2020)

Memon, I.A., et al.: Mode choice modeling to shift car travelers towards park and ride service in the city centre of Karachi. Sustainability **13**(10), 5638 (2021)

Memon, I.A., Madzlan, N., Talpur, M.A.H., Hakro, M.R., Chandio, I.A.: A review on the Factors Influencing the Park-and-Ride Traffic Management Method. Applied Mechanics and Materials. Trans Tech Publications, Ltd., Kuala Lumpur (2014)

Memon, I.A., Napiah, M., Hussain, M.A., Hakro, M.R.: Influence of factors to shift private transport users to Park-and-Ride service in Putrajaya. In: Zawawi, N.A.W.A. (ed.) Engineering Challenges for Sustainable Future: Proceedings of the 3rd International Conference on Civil, Offshore and Environmental Engineering (ICCOEE 2016), Kuala Lumpur, Malaysia, 15–17 August 2016, p. 566. CRC Press, London (2016a)

Memon, I.A., Napiah, M., Talpur, M.A.H., Hakro, M.R.: Mode choice modelling method to shift car travelers towards park and ride service. ARPN J. Eng. Appl. Sci. **11**(6), 3677–3683 (2016b)

Mohammadian, A., Miller, E.: Nested logit models and artificial neural networks for predicting household automobile choices: comparison of performance. Transp. Res. Rec. J. Transp. Res. Board **1807**, 92–100 (2002)

Morrall, J., Bolger, D.: The relationship between downtown parking supply and transit use. ITE J. 32–36 (1996)

Morrall, J., Bolger, D.: Park-and-ride: canada's most effective TDM strategy cost effectiveness through innovation. In: Proceedings of the 1996 TAC Annual Conference, Charlottetown, Prince Edward Island, Canada (1997)

Nor, N.G.M., Nor, A.R.M.: Predicting the impact of demand- and supply-side measures on bus ridership in Putrajaya, Malaysia. J. Public Transp **9**(5) (2006)

Parkhurst, G.: Influence of bus-based park and ride facilities on users' car traffic. Transp. Policy **7**(2), 159–172 (2000)

Pathan, H., Faisal, A.: Modelling travellers' choice of information sources and of mode. University of Leeds (2010)

Peñalosa, E.: Why buses represent democracy in action. TEDCity2.0 Dream me. Build me. Make me real., TimesCenter, Manhattan, New York (2013). Ted.com

Qin, H., Guan, H., Wu, Y.-J.: Analysis of park-and-ride decision behavior based on decision field theory. Transport. Res. F: Traffic Psychol. Behav. **18**, 199–212 (2013)

Qureshi, I.A., Lu, H.: Urban transport and sustainable transport strategies: a case study of Karachi, Pakistan. Tsinghua Sci. Technol. **12**(3), 309–317 (2007)

Rao, P.S., Sikdar, P., Rao, K.K., Dhingra, S.: Another insight into artificial neural networks through behavioural analysis of access mode choice. Comput. Environ. Urban Syst. **22**(5), 485–496 (1998)

Sadek, A., Spring, G., Smith, B.: Toward more effective transportation applications of computational intelligence paradigms. Transp. Res. Rec. J. Transp. Res. Board **1836**, 57–63 (2003)

SAIC: Evaluation of Transit Applications of Advanced Parking Management Systems Final Evaluation Report. Research and Innovative Technology Administration (RITA) (2008)

Sayed, T., Razavi, A.: Comparison of neural and conventional approaches to mode choice analysis. J. Comput. Civ. Eng. **14**(1), 23–30 (2000)

Seik, F.T.: Experiences from Singapore's park-and-ride scheme (1975–1996). Habitat Int. **21**(4), 427–443 (1997)

Shaharyar, B., Irfan Ahmed, M., Saima, K., Noman, S.: Predicting the use of public transportation service: the case study of Karachi circular railway. PalArch's J. Archaeol. Egypt/Egyptol. **18**(3), 4736–4748 (2021)

Shmueli, D., Salomon, I., Shefer, D.: Neural network analysis of travel behavior: evaluating tools for prediction. Transp. Res. Part C: Emerg. Technol. **4**(3), 151–166 (1996)

Statistics, M.o.: Census Quick Results (2017). http://www.statistics.gov.pk/census-results.html. Accessed 6 Dec 2017

TAS, P.L.: Quality bus partnerships: good practice guide. Version 03.05.01. TAS Publications & Events Limited (2001)

Turnbull, K.F.: Effective use of park-and-ride facilities. NCHRP Synthesis Highway Pract. **213** (1995)

Vythoulkas, P.C., Koutsopoulos, H.N.: Modeling discrete choice behavior using concepts from fuzzy set theory, approximate reasoning and neural networks. Transp. Res. Part C: Emerg. Technol. **11**(1), 51–73 (2003)

Whitfield, S., Cooper, B.: The travel effects of park and ride. Public Transport Planning and Operations, European Transport, Loughborough university, England (1998–9)

Xie, C., Lu, J., Parkany, E.: Work travel mode choice modeling with data mining: decision trees and neural networks. Transp. Res. Rec. J. Transp. Res. Board **1854**, 50–61 (2003)

Zhang, Y., Xie, Y.: Travel mode choice modeling with support vector machines. Transp. Res. Rec. J. Transp. Res. Board **2076**, 141–150 (2008)

Congestion on Canada's Busiest Highway, 401 Problems, Causes, and Mitigation Strategies

Abdul Basith Siddiqui[✉]

Carleton University, Ottawa, ON, Canada
abdulbsiddiqui@cmail.carleton.ca

Abstract. The problem of congestion on highways is frustrating and reflects the inefficiency of the transportation network. Growing cities and economies have resulted in high population density regions across the Greater Toronto Area (GTA). The study area of this paper will be a worst-affected section on the Highway ON-401 which runs through dense residential communities from North York till Mississauga. There is a very major bottlenecking problem in the study area which results in very significant delays for commuters and reduces confidence in the transportation infrastructure across the province. To tackle this urban congestion, the Government of Canada has invested more than $8 billion in over 2,700 infrastructure projects across Ontario. The province of Ontario plans to invest more than $7.3 billion in public transit infrastructure over the next ten years (MTO, 2020). This paper aims to identify affected areas, causes, and potential solutions.

Keywords: Congestion mitigation · Trips to CBD · Policy changes · Highway congestion · Congestion solutions

1 Introduction

1.1 Study Area

As seen in Fig. 1 below, the stretch of HWY 401 from Don Valley Parkway (DVP) in the East towards HWY 409 to the west is highlighted. The green areas represent where traffic flows are fastest, yellow areas represent where traffic flows are very slow and red areas represent extremely slow traffic flows. For a commuter travelling from DVP towards HWY 409, the travel time due to this congestion is increased by 1 h and 45 min at the minimum (Google Maps, 2020). This problem of bottlenecking results in high emissions, increased travel times, delays, and inconvenience. The am peak is around 8 AM in the morning and pm peak is around 4 PM in the evening (Google Maps, 2020). This is the rush hour because people leave offices and go to their places of residence and return in the evening. Although, several infrastructure developments such as the 407-toll route extension, increasing of lanes in several regions along the 401 have been implemented, new connected highways to the 401 have been constructed they have not shown any significant improvement in reducing congestion along the study area.

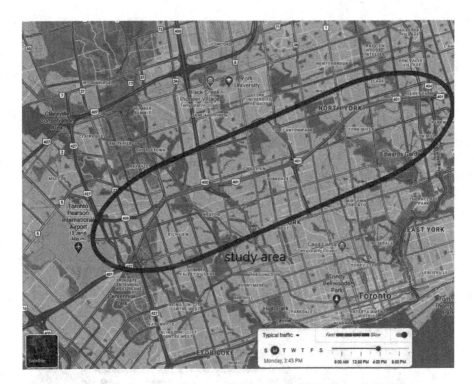

Fig. 1. Study area peak hour traffic on a typical Monday at 3:45 PM

The study area is the section of the Highway 401 starting from the north of Toronto, from Don Valley parkway towards 401-W and ends at the exit for Highway 409-W.

Since the space available for infrastructure expansion is always limited along with the costs associated with it, it is necessary to revisit existing infrastructure which is operating at levels higher than their capacity and identify scope for improvement of existing networks and affected areas. There are many ways to do this such as demand management, providing alternatives such as light-rail transit systems, rapid bus networks, encouraging ridesharing, carpooling and urban policy changes which will be discussed in the following sections.

1.2 Problem Definition

According to the FHWA, between 1980 and 2003, annual vehicle-miles increased by 89%, while total road miles grew by just 3%. There is a similar trend for Ontario. It is necessary to understand why there is so much traffic on the 401, from the figure below, the trips originate from Toronto Downtown, people take the DVP and start heading West towards Mississauga and other residential communities as can be seen in Fig. 2 below.

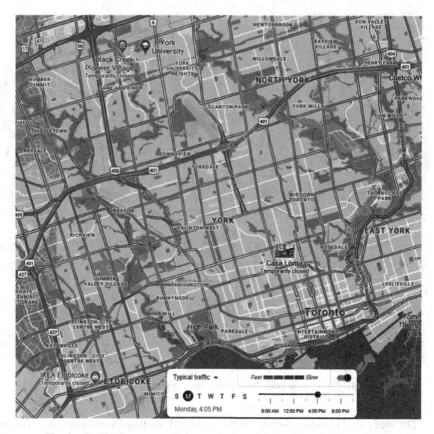

Fig. 2. Traffic flow route during typical peak hour (Google Maps, 2020)

Due to rapid growth of population, urban sprawl, and the inability of Toronto to handle this population, new urban communities such as Mississauga, Brampton, Milton etc. were created with long but reliable transit connections from these communities to the Central Business District (CBD). The transit connections include GO trains, public busses which take the HWY 407 ETR and reach Toronto, Markham, Scarborough etc. Highway 401 starting along the North York of the City of Toronto is Canada's worst highway bottleneck and costs commuters 3 million hours annually (Canadian Pacific Consulting Services 2018). This is corroborated from the 2016 Provincial highways AADT, the problem in the study area is confirmed as it has the highest AADT of 411,600 included in the study area. A bottleneck can be defined as a condition where demand exceeds maximum available capacity. This is generally seen during peak hours. Special events such as harsh weather, merging lanes, on-and-off ramps also contribute to the congestion. Congestion can be defined as the condition in which the travelling speed is far lower than a commuter would expect to drive on. For example, somebody travelling 60 km/h on a highway that is designed for vehicles travelling at 100 km/h.

The main reason for the 401 to come to a halt is because the local road networks are not sufficient to hold the traffic coming from the highway onto the local arterials. The off-ramps which feed traffic to the highway are also very congested because of the traffic in the local networks. When they merge onto the highways, they cause a decrease in speed of the whole platoon behind. These merges and diverges have are main contributors to the delays due to congestion. There are many ways to tackle this such as ramp-metering and weaving sections (Lagerev et al. 2017). Unlike other fields in Civil Engineering, transportation problems are complex, and no simulations or solutions will completely overcome the problem unless a combination of congestion-reduction methods and policy changes to reduce the number of trips to the CBD is done.

2 Methodology

The solution for traffic congestion for a transportation network as congested as the study area of the 401 cannot be achieved as a single entity. There needs to be a combination of congestion-reduction methods such as ramp-metering, improved traffic management of local arterial networks, encouraging public transportation and reducing commuter trips to effectively solve this problem.

The results and measures of effectiveness reported in the references were analyzed to come up with suggested improvements to the existing network. The rationale is that a one-hit solution is not possible for solving a problem as complex as congestion in the study area. To identify the worst-affected section of the 401, a route analysis was performed in the online resource Google Maps to look at the worst affected section of the 401 in the GTA region and the chosen study area was analyzed at peak hour traffic. The congestion on the highway was compared to the congestion levels on the local arterial networks. Although there are various research papers discussing different traffic-improvement and congestion-mitigation techniques, there is a lack of research for a feasible implementation across a highway segment or a region. This study aims to discuss possible congestion-mitigation methods on the worst-affected sections which will help solve the congestion problem on a micro-scale.

3 Data Analysis

The total length of the study area is 23.1 km. The worst affected section is Weston Rd to Islington Ave with an AADT of 411,600 (Highway Standards Branch 2017). This is because, this highly congested stretch is responsible for the subsequent platoons and delays that are experienced by commuters.

From the AADT volumes present, the pattern can be traced as follows and is corroborated with the Google Maps data:

- Don Valley Parkway takes the traffic and feeds it to the 401 Westbound
- A large no. of merging and diverging takes place in the study area due to presence of residential areas around it

- The highway is congested till the HWY 409 exit towards areas of Rexdale and the Toronto Airport
- There is also traffic fed onto the 401 Eastbound from HWY 427 (coming from Burlington, Hamilton etc.)

The main section which is congested is between Weston Rd. to the Islington Avenue exit which is 1.3 km long (Highway Standards Branch 2017). The AADT is 411,600 which is the highest for any section on the 401 (Highway Standards Branch 2017) This is because there is a huge traffic flow from on-ramp vehicles. The presence of dense residential communities near the highway is a reason for the increased congestion (Fig. 3).

Fig. 3. Residential communities along the Weston to Islington stretch on the 401 (Google Maps, 2020)

As seen in the figure above, the study area of the 401 is highly congested whereas the supporting local arterial networks are also in yellow, meaning they are experiencing significant delays. From this, it can be concluded that since the arterial networks are not able to handle such traffic volumes, there is a shockwave effect on the highway and a huge platoon is created. It is calculated that the average AADT volumes for the sections of 401 in the study area is 361,000.

The intersections which are included in the study area and their lengths and AADT volumes are shown in the last two columns of the Table 1 below:

Table 1. AADT volumes on different sections (Highway Standards Branch 2017)

401	HWY 404 IC-375-DON VALLEY PKWY-NORTH YORK	LESLIE ST IC-373-NORTH YORK	2.0	348,000
401	LESLIE ST IC-373-NORTH YORK	BAYVIEW AV IC-371-NORTH YORK	1.9	332,000
401	BAYVIEW AV IC-371-NORTH YORK	HWY 11 IC-369-YONGE ST-NORTH YORK	2.0	341,500
401	HWY 11 IC-369-YONGE ST-NORTH YORK	AVENUE RD IC-367-NORTH YORK	1.7	332,000
401	AVENUE RD IC-367-NORTH YORK	BATHURST ST IC-366-NORTH YORK	1.0	343,000
401	BATHURST ST IC-366-NORTH YORK	ALLEN RD IC-365-NORTH YORK	1.4	350,000
401	ALLEN RD IC-365-NORTH YORK	DUFFERIN ST IC-364-NORTH YORK	0.7	368,000
401	DUFFERIN ST IC-364-NORTH YORK	KEELE ST IC-362-NORTH YORK	1.9	387,700
401	KEELE ST IC-362-NORTH YORK	HWY 400 IC-359-NORTH YORK	3.0	397,100
401	HWY 400 IC-359-NORTH YORK	WESTON RD IC-357-NORTH YORK	1.4	416,500
401	WESTON RD IC-357-NORTH YORK	ISLINGTON AV IC-356-ETOBICOKE	1.3	411,600
401	ISLINGTON AV IC-356-ETOBICOKE	DIXON RD IC-354-ETOBICOKE	2.4	390,700
401	DIXON RD IC-354-ETOBICOKE	HWY 427 IC 352	2.4	275,000

This congestion results in backed-up traffic along the arterial network which adds to the highway congestion woes. A case study of Minnesota showed that the bottleneck reduction programs they implemented resulted in savings of 1.3 million hours (CPCS 2018) which would otherwise have been spent in traffic jams. This saving in time also translates to a high commuter satisfaction, reduced emissions, and efficient transport systems.

4 Discussion

Since infrastructure expansion is costly and causes delays to the existing networks, low capital investment strategies such as actuated traffic signal control systems which ensure that the arterial network is synchronized can be used to avoid a gridlock. These systems must also include a monitoring mechanism to recognize any accidents or incidents on the arterial network and accordingly re-direct or adjust signal timings to reduce congestions.

Congestion on Canada's Busiest Highway, 401 Problems, Causes   1045

The exits worst affected identified are Weston Road Exit to Islington Avenue exit. This is because the supporting arterial network is not able to handle this congestion which creates congestion on the highways. The 3 major causes identified during this project are discussed briefly below. All these causes along with other external factors contribute to creating the congestion.

4.1 High Influx of Traffic from DVP

The Don Valley Parkway which connects the Toronto traffic to the 401, creates a huge influx of vehicles on the eastbound 401 towards other areas of the Greater Toronto Area (GTA) such as North York, Mississauga, Brampton, Oakville etc. The reason for this high traffic volumes on the DVP can be attributed to the fact that Toronto is an economic hub and attracts a huge number of commuters and therefore generates a lot of trips.

4.2 Inefficient Local Arterial Networks

The supporting arterial networks along the exits on the 401 are also not able to efficiently manage the traffic influx from the 401. Due to high traffic growth, these networks are not able to handle these peak hour volumes. These local networks can be optimized to minimize delays and reduce congestion. The below figure shows Google Maps data which shows that the local arterial networks are moving slow in orange color, which means that there are significant delays (Fig. 4).

Fig. 4. Reduced flow in arterial networks (Google Maps, 2020)

It is also noted that suburban traffic during rush hours is increased as greater volume of vehicles arrive at the intersections (Ran et al. 1996).

4.3 Recent Infrastructure Developments for Reducing Congestion

There have been several infrastructure expansion projects launched in Ontario to tackle this congestion with no or little focus to introduce or adopt the suggested improvements made in the following sections. Most recently in May 2020, it was decided to widen the HWY 401 from six to ten lanes in the Cambridge area (Highway Services Board, 2020). The new addition will also include High Occupancy Vehicle (HOV) lanes. Also, to improve traffic congestion across York, Peel and Halton regions, creation of a new highway collectively known as GTA West corridor (between Kirby Road and King-Vaughan Road) with connections to the HWY 401 and 407 interchanges is at the impact assessment stage. Whether this will be a toll route or not is yet to be determined. The HWY 407 toll route which runs parallel to the 401, was extended and increased connectivity which reflects that there is a demand for uncongested infrastructure across the GTA. Constructing toll routes may not be in the best interest of social equity due to the favoring of providing commuters who can afford the toll routes instead of investing in remediation of the existing networks (Litman 2019). Such expansion projects are measures that provide temporary relief and are soon filled by induced demand and so cannot be considered solutions in isolation (Leigh 2016).

5 Suggested Improvements and Its Feasibility

A comprehensive and holistic approach must be taken to tackle the ever-growing problem of congestion on the 401. Using advanced planning and traffic control concepts, the traffic flow can be optimized to reduce congestion. Some methods that can prove to be useful are discussed in the following sections.

5.1 Ramp-Metering Systems (RMS)

Vehicle lane changes on freeways during merging or diverging and in weaving areas adds to the problem of the bottlenecking (Guo et al. 2020). These merges and diverges momentarily decrease the capacity of the highway (Lagerev et al. 2017). RMS will allow controlled access to the freeway and increase overall corridor mobility. RMS uses traffic signal heads which are coordinated with the oncoming traffic for volume detection on the merging sections and gives a signal to the driver to proceed to merge or wait on the ramp. (Intelligent Transportation Systems 2020).

A practical implementation of an RMS can be seen in the figure below (Fig. 5).

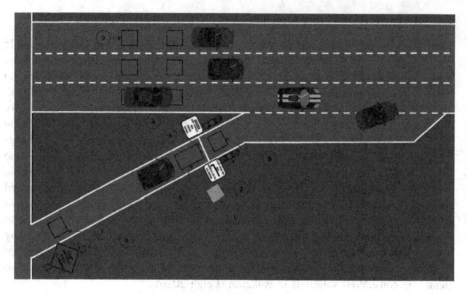

Fig. 5. RMS in action (ITS 2020)

Such a system must be used with an intent to address existing operational issues. Ramp-metering would surely decrease the mainline congestion, but the platoon effected created during peak hours must also be accounted for (Federal Highway Authority 2014).

5.2 Intelligent Transportation Systems (ITS) for Information

Such information systems can be used to advise travelers planning to take the 401 or a particular exit for increased travel times. These can be in the form of display signs, radio broadcasting and mobile applications. This would enable commuters to effectively plan their travel routes and schedules. Traffic can be managed in real-time and can also adopt to changing conditions throughout the day (Pasquale et al. 2017). Driver behavior also plays an important role as many people tend to use engine brakes if they feel they are not far enough from the car in the front, thus creating a shockwave. If people reduce their speeds, instead of applying brakes, it will be easier for them to accelerate to their initial speeds, such a feature can be included in the newer- generation vehicles with their smart technology. This is something which can only be thought of and hard to achieve, but it is a potential solution to avoid stopping. The cars would still be moving in a coordinated way, by going on in reduced speeds but not braking.

5.3 Improvements in Local Arterial Networks

Urban congestion is a serious problem which creates gridlocks (Ran et al. 1996). From the analysis of data, if the supporting arterial networks are efficiently managed and optimized through advanced control methods, the delay due to congestion on the 401 could be reduced significantly. This can be performed in future studies related to this paper and would serve as a very important improvement if the solution is feasible to be adopted by the province. Examples include advanced signal control systems which are also called Traffic Adaptive Control (TAC) systems such as SCOOT can be used. The use of SCOOT systems showed a 17 to 53% reduction in delays (Khan et al. 2001). Also, signal coordination systems which ensure that gridlocks do not occur must also be used. Improved traffic signals reduce congestion and increases reliability and reduces the delay by up to 40%. (CAA, 2018). Roundabouts are slowly gaining popularity across the province and have proven results for an increase in traffic flow and decrease in collisions. (CAA, 2018). As the government has done recently, investments in alternate modes of public transport such as light-rail transit and bus-rapid transit can also prove to be beneficial for reducing car usage.

5.4 Temporary Shoulder Usage

The shoulders must be designed in a way with signage (either actuated for Fixed-Time-Of- Day usage permitted) on the worst-affected sections which permit its usage when needed. This would reduce costs and delays due to construction for any additional lanes and can be easily implemented even on existing highways. Such a measure would provide temporary relief without any significant investments in the form of delaying the start of congestion, increased reliability of transit service and increased road safety (PB Americas et al. 2007).

5.5 Policy Changes

Even though the province is pushing for a transit-oriented development which is evident due to the current huge infrastructure investments the province has made in the Light-Rail Transit (LRT) projects across the most populous cities in the province, more needs to be done. Policy changes must also include developing areas which are far from the highway. The activities which generate many trips must be conveniently away from the highway which would mean the road networks have increased capacity and so the flow on these networks is smooth. A policy change must be initiated to slowly drift away from the CBD concept and instead explore other methods of urban planning. The zoning rules of the Greater Toronto Area must provide flexibility and incentives to establishments to set up workspaces in areas far from already congested areas.

Due to the recent pandemic situation, everything was forced to be switched to remote operations which proved that for most of the work, there is no need for a physical space all the time. Although both methods have their pros and cons, it is very clear that for a commuter living in Brampton and working in Toronto who spent at least 3 h in commute each day can now save that time and make better use of their days and schedules. A comprehensive analysis based on a survey of the people travelling to and from Downton Toronto must be performed based on which the establishments who can switch to remote operations must be encouraged to do so to avoid the burden on the province's highway networks. All these factors must be considered and be used to devise policy which tackles urban congestion problems.

6 Conclusions

Congestion is a major problem facing almost all metropolitan cities. The governments spend billions of dollars on infrastructure, but their effects are not resonated on the worst-affected sections of highways as seen in the case of Highway 401. This is partly due to a one-solution implementation approach whereas there is no perfect strategy which can combat the congestion problems being faced in Toronto, this problem can only be overcome by adopting a combination of congestion-mitigation methods, some of which are proposed in the study. When the traffic offshoots from the Highway's exits, the local arterial network is not equipped to handle such volumes, and thus must be improved. The existing grid-like system in the arterial networks in the study area of roads ensures no cul-de-sacs and good flow of traffic, it also introduces crossing movements of cars at intersections, causing delays.

It was also observed that traffic control methods such as constructing bypass roads, providing large on and off ramps and an aggressive implementation of actuated traffic control systems for the local arterial networks have not yet been adopted in the study area and can prove to be feasible.

Future works could also include collection of survey data based on the above suggested improvements, which can then subsequently be modelled to determine user's perception and behavior for providing policy and strategic direction. A combination of the improvements suggested in this paper can also be implemented as a pilot, which will establish feasibility of implementation and open doors for future research.

References

Alkadry, A., Khan, A.: Methodology for off-line assessment of advanced traffic signal control systems. Can. J. Civ. Eng. **28**, 111–119 (2001)

Canadian Pacific Consulting Services: Grinding to a Halt: Evaluating Canada's worst bottlenecks. Canadian Automobile Association (2017)

Canadian Pacific Consulting Services: Breaking the Bottlenecks: Congestion Solutions for Canada. Canadian Automobile Association (2018)

Leigh, E.: An executive summary: urban congestion inquiry, smarter Cambridge transport (2016)

Federal Highway Administration: Ramp metering: a proven, cost-effective operational strategy. U.S. Department of Transportation (2014)

Guo, Y., Lee, C., Ma, J., Hale, D.K., Bared, J.: Evaluating operational efficiency of split, merge, diverge, and weaving solutions for reducing freeway bottleneck congestion. Transp. Lett. J. **13**, 282–294 (2020)

Highway Standards Branch: Provincial Highways Traffic Volumes 2016 (AADT only). Ministry of Transportation Ontario (2017)

Intelligent Transport Systems Service: ITS903 Freeway Ramp Metering. Intelligent Transport Systems (ITS) (2020)

Lagerev, R., Kapski, D., Burinskiene, M., Barauskas, A.: Reducing a possibility of transport congestion on freeways using ramp control management. Transp. J. **32**, 314–320 (2017)

Pasquale, C., Sacone, S., Siri, S., De Schutter, B.: A multi-class model-based control scheme for reducing congestion and emissions in freeway networks by combining ramp-metering and guidance. Transp. Res. Part C **80**, 384–408 (2017)

PB Americas, Carter + Burgess, EarthTech, Inc., and Telvent Farradyne: Active Traffic Management Feasibility Study. Report to Washington State Department of Transportation, Urban Corridors Office. Seattle, WA (2007)

Ran, B., Huang, W., Leight, S.: Some solution strategies for automated highway exit bottleneck problems. Transp. Res. Part C **4**, 167–179 (1996)

Litman, T.: Smart congestion relief. Victoria Transport Policy Institute (2019)

Multi-Stakeholder Transportation Strategies

A Holistic Approach for the Road Sector in Sub-Saharan Africa

Tim Lukas Kornprobst[1,2], Ulrich Thüer[1,2], and Yana Tumakova[3,4(✉)]

[1] Deutsche Gesellschaft für Internationale Zusammenarbeit (GIZ), Bonn, Germany
[2] Deutsche Gesellschaft für Internationale Zusammenarbeit (GIZ), Amman, Jordan
[3] Eastern Alliance for Safe and Sustainable Transport, Tenterden, UK
yana@easst.co.uk
[4] Eastern Alliance for Safe and Sustainable Transport, Berlin, Germany

Abstract. The road sector is increasingly neglected in development aid. This is particularly visible in Sub-Saharan Africa where road networks are deteriorating rapidly, especially in rural regions, whilst basic connectivity to all-weather roads still lags behind other parts of the world. Major concerns expressed by donors relate to persistent maintenance problem, the associated decline in the value of investments, and the carbon footprint of road projects. Using a case study in Liberia, this paper will recall the value of effective transport linkages for rural regions as a necessary condition for substantial poverty reduction, while also proposing alternative means to develop them through governance reform, an increased role for the private sector, as well as new technical solutions such as the use of low-cost polymer-based roads and low-cost motorbike paths. Taken together, these instruments will give the rural population better access to roads, thus to markets, health and education facilities, and at the same time keep the carbon footprint significantly lower than with conventional technologies.

Keywords: Sub-Saharan Africa · Comprehensive approzach · New road technologies · Polymer roads · Motorcycle paths · Road maintenance

1 The Need for a New Approach in the Road Sector in Sub-Saharan Africa

Infrastructure development in Sub-Saharan Africa is crucial for social and economic progress. In Agenda 2063, the members of the African Union list infrastructure development as one of the greatest priorities of their member states (AU 2020). This is also reflected in various national development strategies, which aim to expand road networks and thus improve connectivity for the population. At the same time, there is an overwhelming amount of scientific empirical evidence for developing countries that access to the road network and its maintenance is of enormous importance for inhabitants of rural regions. Also, through studies by the Federal Ministry for Economic Cooperation and Development of the Federal Republic of Germany (BMZ) and the Gesellschaft für Internationale Zusammenarbeit (GIZ). A connection to the road network significantly reduces transport costs for people and goods. Only then, for example, can many villages sell their agricultural products and thus generate money for their children's school attendance or medical treatment. This is especially true for the rural communities of many Sub-Saharan African countries. However, the majority have access only by a long walk (of more than two kilometres) to an often desolate, unpaved road network. It is hardly passable during the long rainy season in many regions. This has fatal consequences for those affected.

Nevertheless, donors are investing less and less in the road sector. This is because, contrary to the advice of donor organisations, many African governments do not maintain their roads enough. Moreover, this sector is also unattractive in that conventional asphalt concrete roads result in many tonnes of CO_2 emissions and, without suitable alternatives, can lead to more private motorised transport. The strategic focus of many donors in Africa's transport sector is on sustainable urban development, renewable energy, energy efficiency and e-mobility. Such solutions may work for some parts of the world where acceptable road infrastructure exists. However, such solutions are far from the reality of life in rural regions of Sub-Saharan Africa, where the basic infrastructure is not even close to being in place. This approach contradicts the logical step of mobility promotion defined by the *Global Roadmap of Action Toward Sustainable Mobility* 2019. Basic access to the road network must first be ensured before we should think about expanding public transport or even e-mobility in rural areas of Sub-Saharan Africa (SMFA 2019: 7).

It is therefore imperative that donors re-engage with the sector and adopt a new approach that is able to close the infrastructure gap and, in parallel, boost the local economy and minimise environmental impact. Such an approach consists of a comprehensive programme that covers all levels of the sector. This programme must consist of projects in the respective partner country that address both new road construction, road maintenance, tendering procedures, financing of enterprises in the construction sector through loans, vocational training, road safety and the components on good governance. The programme must also build on close coordination between active donors in the sector to maximise leverage and negotiating power. This insight is based on the background of the road sector and the experience gained in the project "Capacity Building in the Transport Sector in Liberia" implemented by the GIZ.

The aim of this advisory paper is therefore to clarify for partner and donor decision-makers how strong the empirical evidence is in favour of investing more in the road sector and, above all, what reasonable measures and technical solutions are available in African partner countries. To this end, the first step will be to show, on the basis of the current state of research, what success the expansion and better maintenance of roads has had for rural regions in developing countries and emerging economies. Then some examples of technical solutions and components of a possible programme will be presented. These include more cost-effective as well as more ecological techniques for road construction such as low-cost roads and motorbike paths, proposals for road maintenance through decentralised, efficient governance structures with a technical road construction administration and a separate financial administration through road maintenance funds. At the same time, the technical and commercial competence of local road construction companies must be developed and access to capital for these companies must be ensured, for example through small loans.

Furthermore, the approaches of the *Construction Sector Transparency Initiative* (CoST Initiative) will be described, which have already contributed to more transparency and government accountability in the sector in various countries. In addition, the experience of GIZ Liberia will be presented, which has been able to generate good results in the sector through various pilot projects of its own and projects of other donors in the country. A forecast of the employment potential and a cost calculation are intended to demonstrate the economic usefulness of the approaches in Liberia as an example. At the same time, however, the difficulties of implementation will make clear the challenges of working with a partner with low capacity.

2   The Importance of Roads and Road Maintenance for Low- and Middle-Income Countries: Findings from Research and Practice

Roads are a necessary condition for economic growth, sustainable development and poverty reduction. The World Road Association (PIARC) estimates that at least 15% of a country's GDP is attributable to the quantity and, above all, the quality of its roads (PIARC 2014: 7). For example, the World Bank was able to demonstrate that there is a clear statistical correlation between the population's access to a road network in low- and middle-income countries (LMICs) and GDP (SMFA 2019). However, it is not only the economic importance of roads that needs to be considered. In rural areas of LMICs, roads and streets are crucial for many aspects of life. The social benefits have been emphasised in many academic papers and reports by international organisations for several decades (World Bank 1992). No access to all-weather roads means significant disadvantages for rural people. High transport costs make the sale of agricultural goods unprofitable. Innovations, such as new knowledge about cultivation methods and new techniques, are hardly accessible to farmers, for example, because they would have to invest a lot of time and, for their income, unreasonably high costs for travelling. Pupils and teachers are more often absent or stay away from school altogether. Clinics are difficult to reach and cannot be visited in time in case of emergency, resulting in higher mortality, especially among expectant mothers and young children. Health workers cannot reach people in remote villages for treatment. People without adequate transport are much more likely to stay away from democratic processes such as elections because travel is too costly. Women in particular suffer from the high cost of transport and the long distances they have to walk for everyday errands. These and other negative consequences have been widely documented in different country contexts (e.g. Donnges et al. 2007: 15–22; Wales and Wild 2012: 5 f., PIARC 2013: 8; Porter 2014 and Allen and Sieber 2016). The people of Sub-Saharan Africa are particularly affected by these negative effects, as only about a third live within a two-kilometre radius of an all-weather road (Iimi et al. 2016 and Hassan 2018). This makes sustainable development and efficient poverty reduction difficult (Berg et al. 2018).

In their joint publication "*The Contribution of Transport to Rural Development*", GIZ and KfW have provided an overview of the damage caused by poor road access and transport facilities. For example, the report states that in Sub-Saharan African countries 10–20% of the harvest is lost due to poor transport because of inadequate infrastructure. This means not only a severe loss of income for farmers, but also a considerable threat to the already fragile food security of many African countries (GIZ and KfW 2013: 10). At the same time, about 75% of maternal mortality could be prevented by faster medical aid in LMICs, mainly due to the lack of transport routes (ibid.: 2). This fact has been demonstrated in various African countries, where child mortality correlates with the condition of roads (e.g. Nigeria, see Porter 2014: 7). Roads therefore contribute to the achievement of many SDGs, for example Goals 3, 4, 8, 9, 11, 13 and 16 (PIARC 2016: 13). In particular, however, they are essential for poverty reduction and thus for the achievement of SDG 1: no poverty. In summary, roads are an important - if not the most important - social infrastructure for rural areas.

Despite this evident key function of rural roads and tracks, their maintenance is massively neglected in LMICs, especially in Sub-Saharan Africa. Road maintenance remains chronically underfunded. It is often not known exactly how much rural infrastructure there even is and who is supposed to maintain it (World Bank 2011: 49 f.). Local institutions lack the resources and knowledge to maintain these roads. The condition of the road network is therefore increasingly deteriorating instead of improving (Donnges et al. 2007: 22 and 38 f.). African countries that have established a road fund invested an average of 60% in maintaining their primary road network in 2007, while only 20% was used to maintain rural transport routes (Gwilliam and Bofinger 2011: 66). Several countries in Sub-Saharan Africa did not finance the maintenance of their rural roads at all in some years (Baril et al. 2013: 12). This is fatal, as in many African countries the rural non-asphalted road network accounts for up to 90% of the existing roads, such as in Tanzania (PIARC 2014: 14).

As a result, the deterioration of Sub-Saharan Africa's roads is progressing rapidly and many people still do not have direct access to a reliable road. Between 1970 and 2010, Sub-Saharan Africa's road network was expanded from approximately 78,000 to 186,000 kms. This was done through the respective governments' own efforts and the support of donor organisations. But maintenance was neglected immediately after construction, leading to a sharp decline in value and high rehabilitation costs (Berg et al. 2018: 257). In 2003, for example, there were calculations by the World Bank that for every kilometre of road renewed in Sub-Saharan Africa, at least three kilometres deteriorate so badly that they would also need to be rehabilitated (World Bank 2003). As a result, a steady deterioration of the road network is progressing, which is estimated to have led to a loss of value of US$150 billion in Sub-Saharan Africa between 1975 and 1995 alone, which could have been prevented by significantly lower maintenance investments (PIARC 2016: 4). Estimates suggest that at least £30 billion (approx. €34.65 billion) needs to be invested in road rehabilitation in Sub-Saharan Africa to stem the acute deterioration through rehabilitation measures (Salih et al. 2016). And the cost of rehabilitation continues will rise exponentially with each passing year, as shown, for example, by the experience of South Africa, where the cost of repair increased 18-fold after 5 years of deferral (Burningham and Stankevick 2005). The multiple increased costs due to lack of maintenance have been pointed out several times (e.g. Levik 2001 and Baril et al. 2013). This means that there is an urgent need for action to sustainably save roads in rural areas in particular from destruction.

The shortcomings of the transport sector in most African countries have strong negative effects on economic development. A poor road network has a serious impact and hampers the entire economy. The rule of thumb is that for every dollar saved in maintenance measures, there is a threat of a loss of 3 US dollars due to increased transport costs, vehicle damage and the like (PIARC 2014: 12). This is the main reason why the cost of transporting goods in Sub-Saharan Africa is the highest of all world regions (Teravaninthorn and Raballand 2009). In many months of the year (especially the rainy season), the affected communities are almost completely isolated due to impassable roads, which is why they can hardly be supplied and the prices for basic supplies rise (Kraybill 2013: 277 f.). At the same time, fertile land can only be used to a limited extent, as the transport infrastructure for efficient cultivation and profitable sale of agricultural products is lacking (Berg et al. 2018: 856). In addition, costs are increased by rampant corruption on transport routes. The state security forces, i.e. traffic police or the gendarmerie, for example, are the most frequent perpetrators, who regularly extort bribes from drivers. Behind this are mostly distribution networks that are driven by the management level of the police or the responsible ministries (Wales and Wild 2012). This significantly increases the costs and time required for transport. For example, studies showed that the costs of corruption in Nigeria and Ghana alone accounted for 29% of total transport costs (GIZ and KfW 2013: 2 and 14 f.).

Despite the difficult situation facing the road sector of LMICs in Sub-Saharan Africa and other regions of the world, various initiatives by governments and donor agencies show how enormous the impacts can be through rural transport projects. Starkey and Hine (2014), after reviewing 360 research papers across different regions, find that rural roads in LMICs clearly promote prosperity in various sectors and reduce extreme poverty. All-weather roads ensure access to service providers and markets, which increases income, consumption and use of basic services, e.g. in health and education (Starkey and Hine 2014). This result is also confirmed by the study by Siebers and Allen (2016), who analyse the successes of development cooperation road projects by various donors. The Independent Evaluations Group comes to a similar conclusion in its evaluation of World Bank projects: Worldwide, 90% of the projects from 2003 to 2013 for rural roads are considered a success (Baril et al. 2013: 36).

A number of examples and studies on the rehabilitation, maintenance and construction of rural roads demonstrate their impact. World Bank projects can demonstrate clear successes. In a project in Bhutan, transport time decreased by 75%, the prosperity of the population increased and the income of farmers from the sale of agricultural products increased by 64%. Based on a project in India, it was found that for every US$ 22,000 invested in roads, 163 people escaped poverty through new economic opportunities. There was also an increase in literacy rates, higher agricultural production and significant increases in school and doctor visits. In Vietnam, the rule was confirmed that for every dong invested, there was more than three dong of agricultural value added. Increased value added also occurred in other sectors of the economy (ibid.: 36 ff.). The effectiveness of roads against poverty in Vietnam is also verified by a study by Glewwe et al. (2002): Households connected to paved roads are 67% more likely to escape poverty. The examples of rural regions in China, Thailand and India showed particularly clearly how important roads are as an highly effective investment in poverty reduction (Fan et al. 2000 and Cook et al. 2005).

The effectiveness of road projects against poverty has also been confirmed many times on the African continent. Morocco rehabilitated and modernised 27,000 kms of rural roads from 1995 to 2012. Afterwards, instead of 43%, 79% of the population now had year-round access to a road. The population benefited in various areas of life as a result. Travel times decreased, school enrolment rates increased by 5.8% and for girls by 7.4%, passenger fares decreased by 26% and freight costs by 15%. Later studies assumed a 20% reduction in freight costs. Employment in agriculture increased by 24% and attendance at medical facilities increased by 26% (SMfA 2019: 45 f.). In Ethiopia, government programmes and donor agency projects improved the condition of and access to the road network. In this context, it was found that child mortality and poverty were significantly reduced in the vicinity of roads (ibid: 6). Dercon et al. (2009) shows from his study of households in Ethiopia that access to an all-weather road increased the income of beneficiaries, reducing poverty rates by 6.7% and increasing consumption by 16%. They therefore confirm that roads are one of the important factors in poverty reduction in Ethiopia. The new or improved roads and tracks gave farmers on average 25% more profit for their products. Basic services for transport, health and education were used much more frequently (SMFA 2019: 6, 24 and 46). Therefore, the desire for road connectivity is high among rural Ethiopians (Wales and Wald 2012: 10). For Uganda, as for Ethiopia, it was found that proximity to a road decreases the likelihood of living in poverty (Sieber and Allen 2016: 31). Thus, in the course of an African Development Bank project in Uganda, about 3000 kms of road network were rehabilitated and supplemented with market places and storage facilities. As a result, the production of agricultural products increased by 7.5% and farmers received 36% more profits on the sale of their goods, which led to an increase in farmers' income of 40% on average. At the same time, travel time was reduced by 67%. Therefore, the rule established itself that for every million Ugandan shillings invested (at today's exchange rate: about 238 euros), 27 Ugandans escaped poverty (Sieber and Allen 2016: 26). It has also been shown through further research in Cameroon, Uganda and Burkina Faso that there is a causality between poverty and distance to all-weather roads (Raballand et al. 2010). These examples of the numerous existing studies thus show that the positive influence of roads on poverty reduction and the use of basic services in Sub-Saharan Africa and beyond has been extensively documented.

German Financial and Technical Cooperation have also achieved clear successes in their projects for roads in rural regions, which will be shown with three examples. The cooperation project "Rural Infrastructure Programme II" of KfW Entwicklungsbank showed strong effects for the target group in Bangladesh through the improvement of roads: an average increase of 197% in annual household income as well as a reduction in annual transport costs by 37% with a reduction in travel and transport time of 56%. Furthermore, school attendance increased by 16% at the lower secondary level and by 26% at the upper secondary level. Significantly more people used health services. Agricultural production increased significantly, by 23% on average for vegetables and by as much as 146% for some crops such as cassava. In line with these achievements, almost three quarters of the respondents said that the improved roads had helped them market their products and improved the flow of goods to the villages (KfW 2013 and Siebers and Allen 2016: 25). KfW's and GIZ's engagement in rural road improvement in Laos has resulted in a 9–10% increase in school attendance and a 22% drop in poverty rates in the target region. In Mali, farmers were supported through small-scale irrigation projects by the German Federal Ministry for Economic Cooperation and Development (BMZ), KfW and GIZ. In addition to the positive effects resulting from better irrigation of the fields, strong impacts were achieved through additional rehabilitation of rural roads. For example, the repair of several hundred kilometres has ensured that health stations can be reached in a fraction of the time it used to take, and sales prices have quadrupled for farmers in the region for some products (such as sweet potato) (GIZ and KfW 2013: 36 ff.). These successes speak for the fact that Development Cooperation can be extremely successful through rural road projects.

Even critical research contributions do not doubt the causality between poverty reduction and access to roads. They mainly mention risks and unused potentials that can occur when improving the rural road network. The risks mainly relate to environmental damage and poor maintenance. For example, poor planning can threaten nature reserves and wildlife populations. At the same time, it is noted that the construction of at least conventional asphalt concrete roads has high CO_2 emissions. Also, the successes of improved road access risk being undermined by lack of maintenance. In addition, there is a risk that the poorest segments of the population will not automatically benefit from roads, as their poverty means that they do not have the necessary means for transport, vehicles or capital for new business opportunities (Porter 2014). This latter assumption is confirmed by Starkey and Hine (2014) in their comprehensive study on the effects of roads in rural areas: people with more resources are better able to take advantage of the new opportunities offered by better connectivity. For road projects to have the greatest possible success, they must be part of a rural spatial planning strategy that takes into account the socio-economic needs of the target group and should be complemented by other development aid projects. This means, for example, that poverty rates of regions must be considered when prioritising new road projects. Important routes, e.g. to hospitals and markets, should be prioritised. Transport subsidies for sparsely populated areas and other support for the local economy and an improvement in the quality of public services should complement road projects (GIZ and KfW 2013). Accordingly, the risks and unused potentials can be avoided through forward-looking planning. The CO2 balance can be improved by new techniques, which will be presented in the next chapter. At the same time, the maintenance of the infrastructure created must be ensured through appropriate maintenance systems. This must be a focus in every road project.

In summary, this means that roads are not a sufficient, but a necessary condition for effective poverty reduction and economic development in rural areas (Hettige 2006 and Sieber andAllen 2016). They alone are sometimes not sufficient if the target group has too few resources at its disposal. Nevertheless, they are the crucial, fundamental basis for prosperity in rural areas. It can thus be deduced from the findings of research and practice that roads are a key social infrastructure against poverty. They are essential for the accessibility of services and for participation in many areas of life. It is therefore advisable for donors to contribute to providing the rural population with access to reliable roads and paths.

3 Components of a Holistic Programme

An approach that supports African partners in the road sector must be holistic. In order to drive truly lasting change, a comprehensive programme is needed that can permanently guarantee the maintenance of African roads and universal access to the road network. For this, the authors propose a programme that can handle such a mammoth project. These can be flexibly assigned to different development aid project teams and provided with different financial volumes depending on the country context. The individual components are described below. These are based on the state of research just described and the active projects of development cooperation, as well as the extensive empirical values of the project in Liberia. The authors have designed the model of such a programme with the intention of achieving the greatest possible impact in terms of poverty reduction and economic development through synergy effects of the individual components. In short: more sustainable successes are to be achieved with less money (*value for money*). Such a programme must include close cooperation and coordination between the other donor organisations. In addition, synergies from other development aid projects in rural areas should be used and joint spatial planning concepts achieved that compliment the new connection, for example, through new health stations or innovation centres, as explained in the previous chapter.

Experiences from Liberia and other African country contexts have shown that government agencies have great difficulties in fulfilling their tasks due to a lack of capacity and, in some cases, political will. For example, public procurement processes are non-transparent, quality control of construction projects is inadequate or non-existent, and financial management is desolately organised. Poor tendering processes lead to higher costs, poorer quality in road construction and facilitate the misappropriation of funds. This means that bad roads are delivered for a lot of money, thanks to a lack of quality control and no compliance with construction standards, which in addition are hardly maintained. The financial management and accounting of many responsible authorities have difficulties in managing the money. As a result, salaries are paid several months late and maintenance work (if any) is carried out only after massive delays. This contributes to further benefit taking and the deterioration of the road network, as some employees take bribes out of financial need and the delayed repair work results in growing damage. At the same time, most African countries do not have an independent road agency that can adequately allocate its resources to maintenance without political interference. This is urgently needed, as it has been proven that African countries with such an independent authority provide for a much better maintained road network with the available resources (Gwilliam et al. 2008). Accordingly, the project should promote the establishment of a functioning financial administration and road authority so that the already very scarce resources are not wasted even more massively through corruption and mismanagement. Tenders must meet international quality standards and procurement procedures must be transparent in order to minimise corruption risks. These and other measures have already been successfully tested in various African countries through the CoST initiative mentioned at the beginning (CoST 2020a). The initiative offers a promising approach to reduce cooperation risks, costs can be lowered, and the quality of road infrastructure can be significantly increased. At the same time, the executing agencies must be trained in maintenance and quality control to ensure necessary standards.

Research shows that inadequate maintenance of rural transport infrastructure in Sub-Saharan Africa is a huge risk. National authorities are hardly able to identify the needs of the local road network from headquarters and take timely action. Therefore, local branches of the road authorities or municipal administrative units must be empowered to fulfil this task. At the same time, routine maintenance in particular should be handed over to small and medium-sized enterprises or local communities in order to promote the local economy and employment. Such models have proven very successful in Latin America (Baril et al.: 39 ff.) In turn, the project could support the establishment of transparent tendering processes and other technical issues. It is important for the implementation of the component that, for example, transfer payments from the central government arrive reliably, which in turn emphasises the need for the upper component in terms of financial management.

The poor quality of roads is also due to the poor training systems in most African partner countries. All too often, it is not standardised by the state. Workers are semi-skilled and fail to provide the necessary quality in road construction due to a lack of knowledge. Nevertheless, there are hardly any projects in the entire portfolio of German development aid that promote training in the sector. And those that do train future skilled workers focus mainly on the university education of engineers (as in Namibia, for example). This is not enough, because in Sub-Saharan Africa there is not so much a shortage of theoretically skilled engineers, but rather of capable and experienced foremen in construction. For example, there has been no formal training for road construction in Liberia since the civil war. The GIZ project in Liberia focuses on training road builders with a specially created curriculum. For this purpose, a practical textbook was created, which is being further elaborated. In this way, the project is developing the capacities of the youth and contributing to the sector through much-needed skilled workers. This could serve as a model for further pilot projects in other African countries. Donors can usefully occupy this important niche in development aid through its experience with vocational training. In addition, the practice-oriented training for companies on business management skills and new, innovative techniques (such as low-cost roads) can be offered in order to increase quality in construction. This has already been tested in Liberia by GIZ with positive results.

In LMICs, road accidents are the number one cause of death for people between the ages of 15 and 30 - and the trend is rising. This is mainly due to poor vehicle safety standards, lack of usage of protective gears (such as seatbelts and helmets), inadequate safety infrastructure (guardrails, crosswalks, bicycle lanes, traffic lights, etc.), drivers without training and insufficient enforcement of applicable safety regulations by the executive (World Bank and GRSF 2020). At the same time, accident victims are threatened with a life of physical limitations and poverty if they become unable to work as a result of the injury (WHO 2020). Particularly in Sub-Saharan Africa, serious accidents are on the rise as urban and rural motorisation increases. Planning authorities lack resources and skills. Therefore, experts are needed to accompany the improvement of road safety. They are needed to advise on necessary legal safety standards and safety infrastructure planning. Also, publicity campaigns for more road safety should be conducted and training for drivers in the informal transport sector (e.g. motorbike taxi drivers) should be implemented. The traffic police must be better trained and given the necessary equipment to enforce current laws. These and other fields need to be addressed in order not to undermine the benefits of mobility through an increase in serious accidents and traffic deaths. Due to the many fields of action, it makes sense to set up a separate project team on road safety. The safety initiative could be supported by a fund, which should ensure the provision of safety infrastructure, equipment for the police and training for stakeholders.

In order to provide the necessary resources for improving road safety, a fund should support the implementation of the project. This should provide money for new safety infrastructure (such as lanes for vulnerable road users like cyclists and motorcyclists) and other road safety measures (e.g. equipment and training for traffic police). The fund could be financed by different donors and by the partner country's own contribution sourced by fines, parking fees and percentages of road construction projects.

Furthermore, empirical evidence has shown that corruption in the transport of goods results in a significant increase in the cost of goods in Sub-Saharan Africa. In particular, poor transport users are heavily burdened financially by the bribe payments. In order to reduce transport costs and protect financially disadvantaged groups, effective anti-corruption measures must therefore be supported by donors. There are various successful approaches around the world, such as using more female traffic police, introducing anonymous complaints bodies and increasing both police wages and penalties for bribes (Karapetrovic 2016). These reforms should be initiated and accompanied by the programme.

Infrastructure projects are not transparent in many LMICs. Data on the costs and condition of roads is not published and decisions on tenders are made closed to the public. The affected population of a road construction project is not consulted about their interests. This encourages misappropriation of funds and ensures that infrastructure projects do not meet the needs of local people. Therefore, the CoST Initiative's approach is to accompany countries in publishing their tenders and data transparently and to involve all stakeholders in the decision-making process. This enables civil society to actively participate and, through control, to obtain better quality and more cost-effective infrastructure in the long term. It creates a fairer business environment in which the private sector can better develop and thus increase employment. The initiative has already achieved relevant successes in Uganda and Malawi, saving millions of euros from being wasted (CoST 2017a and CoST 2017b). Therefore, as a donor organisation, it is advisable to implement CoST's approaches with African partners and ideally convince them to join the initiative as an official partner (CoST 2020b).

Sub-Saharan African countries face the challenge of providing their populations with access to an all-weather road network in order to advance rural development. At the same time, they must prevent the deterioration of their wider road network. The latter can be achieved mainly through regular maintenance, secure financing (such as with the introduction of road maintenance funds) and an independent road agency. The former is a huge challenge with conventional techniques and the limited resources of many African countries. Asphalt concrete roads are very expensive to build (in Sub-Saharan African countries: between 400,000 and more than 1,000,000 US dollars per km). The same applies to maintenance (2–4% of the construction price per year for good maintenance). At the same time, conventional asphalt roads have high CO_2

emissions, which make them unattractive in view of the climate crisis (about 50 tonnes of CO_2 emissions per km). The common alternative is unpaved soil or gravel roads, which are practically impassable during the prolonged rainy season in many African countries and therefore unsuitable for the subtropics. Based on research, IMC Worldwide emphasises that gravel and earth roads are completely unsuitable for regions with more than 2000 mm of rainfall per year due to their susceptibility to water (IMC Worldwide 2017). This is true for a large part of the countries of West and Central Africa. But even in the dry season, they are unsuitable for most transport routes: Gravel roads have a high abrasion rate, which means they often have to be renewed at great expense. The dust created by driving on them increases the risk of accidents and leads to lung diseases among residents (Overby and Pinard 2013).

Low-cost seals refer to various sealing techniques based on bitumen emulsions or cold asphalt. These are mixed with grit (aggregates of rock or chippings) and sprayed or rolled onto the road to be sealed (chip seal and otta seal). These techniques have been used on rural roads since the 1960s in Norway, the USA, Canada (ibid.), but also increasingly in Southern African countries such as Namibia or Botswana (SATCC 2003). Over the decades, these techniques have been improved and perfected through the use of organic polymers. The admixture of polymers in the load-bearing soil layer and the asphalted surface layer ensure that the road becomes significantly more load-bearing and durable due to this soil stabilisation. At the same time, this technique requires much less costly material to build the road. The advantages of this technology: a massive reduction of CO_2 emissions of over 90% during construction and a cost reduction of 60–70% per km compared to conventional asphalt roads (see IMC Worldwide 2017 and Polyroads 2020). They can be constructed with virtually no capital-intensive special machinery and no asphalt mixing plants at all. At the same time, local materials (such as laterite) can be used for construction, which also reduces costs. Construction can be labour-intensive with comparatively simple equipment. This enables local companies to carry out the contracts instead of international companies the contracts usually are awarded to. In this way, the local economy and employment are strengthened significantly more. GIZ Liberia's calculations show that this labour-intensive method creates 10 times more employment compared to capital-intensive construction methods of normal asphalt roads (measured in man-years). Furthermore, the polymer roads are weather resilient. Trials in the rainy region of Africa and also in Liberia show that the roads are still in very good condition after years of use (even without sufficient maintenance). This is neither the case with gravel roads nor conventional asphalt roads. These positive experiences with low-cost roads have been confirmed by pilot projects of GIZ and other donors in Liberia (IMC Worldwide 2017). This not only reduces life-cycle costs through lower maintenance requirements (by at least two-thirds compared to gravel roads), but also guarantees passability at all times of the year. Cold asphalt with polymer additive can also be mixed by hand and can be easily stored in depots. This can then be used to repair road damage easily and cheaply, which can also facilitate maintenance. This work could be carried out in an employment-generating way by local SMEs. Therefore, Haule (2015) argues for the use of low-cost roads primarily as a rational investment for maintaining the road network. The introduction of these techniques should be promoted by and brought closer to national companies, authorities and donors.

Nevertheless, this method has hardly been used on the continent outside of southern Africa. Here, donors could take on a pioneering role for sustainable and innovative improvement of the road network. Soil stabilisation is already being applied by some companies. German companies such as Alphasoil or Powersoil have been testing soil stabilisation methods for years. The South African company Polyroads has already successfully implemented projects with cold asphalt and polymers in 18 countries in Sub-Saharan Africa. In Liberia, there is also the Bleco Engineering Group, which works with similar soil stabilisation techniques. Therefore, there are enough cooperation partners for pilot projects in different partner countries.

Motorbikes are of growing importance for mobility in rural Africa. In the last 30 years their number has increased a hundredfold or a thousandfold in most African countries. Motorbike taxis can be found everywhere in Sub-Saharan Africa (Starkey 2016). They are often the only means of motorised transport in rural areas. As the only means of transport, they guarantee reliable access to distant medical facilities or the transport of goods to the nearest market. In addition, motorbike taxis offer a flexible source of income for young people in the region (GIZ 2012). In some African countries such as Cameroon, more than almost three-quarters of goods and more than 80% of passenger transport is carried out by motorbikes (Starkey 2016). Because of this frequent use, the *Global Roadmap of Action Toward Sustainable Mobility* advocates for motorbike routes in developing countries with little access to roads so that affected villages have reliable connectivity (SMFA 2019: 7 and 15). These are convincing in terms of their utility and cost-effectiveness. Starkey and Hine (2014) found in their extensive study that the combination of paved trails and all-weather roads had the greatest positive effects for rural populations in Ghana, Nepal, Uganda and other countries. These clear effects of motorbike tracks can be substantiated by a GIZ pilot project in Liberia (Peters et al. 2018). At the same time, they are very inexpensive to build (between 1000 to 5000 US dollars per km) and can be constructed by the benefiting communities under guidance and subsequently maintained independently. Maintenance can also be undertaken by the local communities themselves with the necessary equipment and training. This in turn creates jobs. The relevant donor organisation can support the partners in the construction of these roads and develop standards for their construction.

A look at the research has shown that the poorest sections of the population are less likely to benefit from better routes and roads than other income groups, as they have poorer access to transport. Therefore, accompanying measures should ensure that all members of the target communities can use the roads efficiently. Here, for example, the promotion of sharing systems that provide motorbikes or cargo bikes in the villages is an opportunity. These should be accompanied by training for vehicle maintenance. Consideration should also be given to establishing an ambulance motorbike with a couch trailer in this sharing model, as this enables rapid patient transport despite the lack of an ambulance (GIZ and KfW 2013).

A funding offer should be made to partners to co-finance road maintenance funds and sustainable improvement of rural transport infrastructure. Funding for the latter should be disbursed as a grant. This is because, as empirical evidence has shown above, roads and paths are social infrastructure and therefore worthy of subsidy. In addition, an incentive model must be designed that is advantageous for the partners. Highly indebted African states have little interest in accepting a large loan (even with ODA conditions). The African partners should therefore be offered the prospect of a large grant by donors if they are simultaneously prepared to undertake the reforms explained above (conditionality). The main argument here should be that low-cost roads are an excellent way of fulfilling election promises of new, paved roads, which is often the incentive for costly new construction (Wild and Wales 2012). This is because it allows about three times as many sealed roads to be built for the same amount of money. In this way, development cooperation can break through the fatal political economy in the road sector in Sub-Saharan Africa and ensure good transport infrastructure in the long term.

In order to enable the financing and prosperity of the private construction sector, financial institutions are needed to provide loans to entrepreneurs. In most countries in Sub-Saharan Africa, these loans are hardly granted and are unsuitable for the sector, as was emphasised in the introduction. KfW Development Bank has recognised this and created the PTA Bank in Nigeria, for example, which has been successfully financing companies for years. Similar models should also be established in other African countries. Here, cooperative banks or cooperatives are conceivable, in which several companies share costly machines.

4 Conclusion and Final Recommendations

In summary, it has been comprehensively demonstrated that rural transport routes and their maintenance are essential for all areas of life. It has also become clear how disastrous the conditions of the road sector are in Sub-Saharan Africa. The current donor commitment is not sufficient to halt the deterioration of vital infrastructure. The authors appeal to decision-makers to revitalise the road sector of development cooperation in Africa through an overall unified approach and to reform it through innovative techniques. Donors and African partners could thereby become pioneers of an eco-sustainable road infrastructure that is long-lasting and has a significantly lower environmental footprint. At the same time, the respective donor organisation would be able to present clear successes through these projects in terms of poverty reduction and economic growth. Otherwise, many of the United Nations' Sustainable Development Goals, above all poverty elimination (SDG 1), will hardly be achievable in the rural regions of Sub-Saharan Africa.

References

African Union (AU): National & RECs development priorities (2020). https://au.int/agenda2063/priorities. Accessed 20 June 2020

Allen, H., Sieber, N.: Impacts of rural roads on poverty and equity (2016). https://www.unescap.org/sites/default/files/2.%20Impacts%20of%20rural%20roads%20on%20poverty%20and%20equity.pdf. Accessed 23 June 2020

Baril, M., Chamorro, A., Crispino, M.: Best practices for the sustainable maintenance of rural roads in developing countries. World Road Association (PIARC), France (2013)

Berg, C., Blankespoor, B., Selod, H.: Roads and rural development in Sub-Saharan Africa. J. Dev. Stud. **54**(5), 856–874 (2018)

Burningham, S., Stankevich, N.: Why road maintenance is important and how to get it done (2005). https://www.semanticscholar.org/paper/Why-road-maintenance-is-important-and-how-to-get-it-Burningham-Stankevich/68256853d3395c398b4b99c3d07c1cf61579c6f6?p2df. Accessed 20 June 2020

Cook, C., Tyrrell D., Somchai, J., Sharma, A., Wu, G.: Assessing the impact of transport and energy infrastructure on poverty reduction. Asian Development Bank, Manila (2005)

CoST: CoST Uganda – delivering better value public infrastructure (2017a). http://infrastructuretransparency.org/resource/cost-uganda-delivering-better-value-public-infrastructure/. Accessed 21 July 2020

CoST: Engaging citizens to enhance transparency and accountability in public infrastructure (2017b). http://infrastructuretransparency.org/wp-content/uploads/2018/06/3212_Malawi-case-study.pdf. Accessed 21 July 2020

CoST: About us (2020a). http://infrastructuretransparency.org/about-us/. Accessed 23 July 2020

CoST: CoST impact stories (2020b). http://infrastructuretransparency.org/costimpact/. Accessed 22 July 2020

Dercon, S., Gilligan, D., Hoddinott, J., Woldehanna, T.: The impact of agricultural extension and roads on poverty and consumption growth in fifteen Ethiopian villages. Am. J. Agr. Econ. **91**(4), 1007–1021 (2009)

Donnges, C., Edmonds, G., Johannessen, B.: Rural road maintenance. Sustaining the benefits of improved access, SETP 19, International Labor Organization, Bangkok (2007)

Fan, S., Linxiu, Z., Xiaobo, Z.: Growth, inequality, and poverty in rural China. The role of public investments, Research Report 125, International Food Policy Research Institute, Washington D.C. (2000)

GIZ & KfW: Improving the accessibility of rural areas. The Contribution of Transport to Development, Deutsche Gesellschaft für Internationale Zusammenarbeit (GIZ) and KfW Entwicklungsbank, Eschborn, Frankfurt (2013)

GIZ: Challenges of informal motorcycle transport in Liberia. Deutsche Gesellschaft für Internationale Zusammenarbeit (GIZ), Eschborn (2012)

Glewwe, P., Gragnolati, M., Zaman, H.: Who gained from Viet Nam's boom in the 1990s. Econ. Dev. Cult. Change **50**(4), 773–792 (2002)

Gwilliam, K., Foster, V., Archondo-Callao, R., Briceno-Garmendia, C.M., Nogales, A., Sethi, K.: The burden of maintenance. Roads in Sub-Saharan Africa. Background paper 14 (phase I), Africa Infrastructure Country Diagnostic, World Bank, Washington D.C. (2008)

Gwilliam, K., Bofinger, H.: Africa's transport infrastructure. Mainstreaming maintenance and management. World Bank, Washington D.C. (2011)

Hassan, M.: Road maintenance in Africa. Approaches and perspectives. In: E3S Web Conference, vol. 38, p. S.1005-10 (2018)

Haule, J.: Road funds in Africa (2015). https://www.h-a-d.hr/pubfile.php?id=876. Accessed 21 July 2020

Hettige, H.: When do rural roads benefit the poor and how? An in-depth analysis based on case studies. Asian Development Bank, Mandaluyong City (2006)

Iimi, A., Ahmed, F., Anderson, E.C., Diehl, A., Stone, L., Peralta-Quiros, T., Rao, K.S.: New Rural Access Index. World Bank, Washington D.C. (2016)

IMC Worldwide: Applied research on low-cost sealting of roads. Final report. October 2015 - October 2017. European Delegation to the Republic of Liberia, Monrovia (2017)

Karapetrovic, J.: Enhancement of current anticorruption practices to improve traffic safety. New Mexico State University, New Mexico (2016)

KfW: German financial cooperation with Cambodia. Rural infrastructure programme (RIP) II. Ex-post social impact assessment report. KfW Entwicklungsbank, Frankfurt am Main (2013)

Kraybill, D.: Rural development in Sub-Saharan Africa. In: Green, G.P. (ed.) Handbook of Rural Development. Edward Elgar Publishing, Cheltenham (2013)

Levik, K.: How to sell the message of "road maintenance is necessary" to decision makers. In: First Road Transportation Technology Transfer Conference in Africa Tanzania Ministry of Works (2001). https://trid.trb.org/view.aspx?id=688272. Accessed 22 June 2020

Overby, C., Pinard, M.: Otta seal surfacing. Transp. Res. Rec. **2349**(1), 136–144 (2013)

Peters, K., Jenkins, Jack, M., Esther; R., Paul Johnson, T.: Upgrading footpaths to motorcycle taxi accessible tracks. Accelerating socio-economic development in rural Sub-Saharan Africa (2018). http://www.research4cap.org/Library/Petersetal-SwanseaUni-2018-UpgradingFootpathstoMotorcycleTaxiaccesibleTracks-PolicyBrief-181217.pdf. Accessed 21 July 2020

PIARC: The importance of road maintenance. World Road Association, France (2014)

PIARC: Preserve your country's roads to drive development. World Road Association, France (2016)

Polyroads: Environment (2020). http://polyroads.com/community/. Accessed 01 July 2020

Porter, G.: Transport services and their impact on poverty and growth in rural Sub-Saharan Africa. A review of recent research and future research needs. Transp. Rev. **34**(1), 25–45 (2014)

Salih, J., Edum-Fotwe, F., Price, A.: Investigating the road maintenance performance in developing countries. Int. J. Civ. Environ. Eng. **10**(4), 472–476 (2016)

SATCC: Guideline Low-Volume-Sealed Roads. Southern African Development Community, Gaborone (2003)

SMFA: Global Roadmap of Action Toward Sustainable Mobility (GRA). World Bank, Washington D.C. (2019)

Starkey, P.: The benefits and challenges of increasing motorcycle use for rural access (2016). https://www.researchgate.net/publication/332413878_The_benefits_and_challenges_of_increasing_motorcycle_use_for_rural_access. Accessed 21 July 2020

Starkey, P., Hine, J.: Poverty and sustainable transport. How transport affects poor people with policy implications for poverty reduction. Overseas Development Institute, London (2014)

Teravaninthorn, S., Raballand, G.: Transport Prices and Costs in Africa: A Review of the International Corridors. World Bank, Washington D.C. (2009)

Wales, J., Wild, L.: The political economy of roads: an overview and analysis of existing literature (2012). https://www.odi.org/publications/7178-political-economy-roads-overview-and-analysis-existing-literature. Accessed 21 July 2020

WHO: Road Traffic Injuries (2020). https://www.who.int/news-room/fact-sheets/detail/road-traffic-injuries. Accessed 21 July 2020

World Bank: Rural transport and the village. The World Bank, Washington D.C. (1992)

World Bank: Rural transport in multi sectoral and community driven projects. World Bank, Washington D.C. (2003)

World Bank/Global Road Safety Facility (GRSF): Guide for road safety opportunities and challenges. Low- and middle-income country profiles. The World Bank, Washington D.C. (2020)

Innovations in Road Materials

The Introduction of Micro - & Nanodispersed Fillers into the Bitumen Binders for the Effective Microwave Absorption (for the Road, Airfield & Bridge Pavements)

Stanislav Mamulat[1,2,3(✉)], Igor Burmistrov[4,5], Yuriy Mamulat[6], Dmitry Metlenkin[5], and Svetlana Shekhovtsova[7]

[1] The Siberian State Automobile and Highway University, Omsk, Russia
slmamulat@mail.ru
[2] Research and Innovation Center under EC of CTM CIS, Moscow, Russia
[3] Belt and Road International Transport Alliance (BRITA), Beijing, China
[4] National University of Science and Technology MISiS, Moscow, Russia
[5] Plekhanov Russian University of Economics, Moscow, Russia
[6] SpEcoMix Co. Ltd., Moscow, Russia
[7] National Research Moscow State University of Civil Engineering, Moscow, Russia

Abstract. In this paper, the means of improving the efficiency of heating bituminous binders modified with micro- and nanodispersed fillers under the influence of microwave is described. The carbon nanomaterials (multi-walled carbon nanotubesis in particular) were selected upon the analytical review of the effectiveness of various electrically conductive and magnetically sensitive fillers in micro- and ultradispersed form to drive the efficiency of heating a bitumen binder in the microwave field.

The experimental of the study included the preparation of a bitumen binder samples with the addition of electrically conductive and magnetically sensitive fillers, and specifically those selected on the basis of theoretical studies of multi-walled carbon nanotubes and various types of ferrite waste from metallurgy. The influence of modification of bitumen samples on the rate of their heating by electromagnetic microwave radiation is analyzed upon the obtained thermal heating coefficients. It is shown that the combination of small carbon nanotubes and magnetosusceptible waste additives provide the fastest and most uniform heating, and the optimal ratio of the components in the bitumen binder and mastic to ensure effective absorption of microwave energy coatings.

Keywords: Bitumen binders · Nanodispersed filler · Microdispersed fillers · Carbon nanotubes · Ultra-high frequency irradiation (UHF) · Electromagnetic absorption · Induction healing pavements

1 Introduction

Bituminous binders are the main structure-forming agent in the asphalt block, which quality directly determines the reliability and durability of pavings during the entire service life of the road (Shekhovtsova et al. 2019). At the moment, the oxidized bitumen depleted in the oil fraction used in the Russian Federation do not correspond to the ever-increasing loads on the track, which leads to premature destruction of the paving and entails an increase in the cost of maintaining the road. This determines the special relevance of the development of technologies for modifying bituminous binders to improve the quality of their operational and technological properties in the composition of asphalt concrete at all stages of the life cycle of road coatings, as well as the search and testing of new techno-economically effective ways of timely and high-quality repair of road pavements.

Ultra-high-frequency (UHF) electromagnetic radiation is a promising tool for providing volumetric heating and plasticizing of asphalt concrete without local overheating, which can be effectively used both for prompt high-quality repairs with low labor costs and at various stages of asphalt production technology for heating, reducing viscosity during pumping and even volumetric heating during melting. With microwave-induced heating, there is a kind of healing of open microcracks observed in asphalt concrete when the material turning into a plastic state under the action of surface tension forces, and macrocracks are restored with a further decrease in viscosity and mechanical compaction. This is possible when the temperature of the asphalt concrete is high enough so that the bitumen can behave like a Newtonian liquid, which will initiate capillary phenomena at these points of contact and propagate them through the crack, ensuring its filling. In this case, it is very important to evenly heat the bitumen and avoid local overheating, leading to aging and embrittlement.

It is known that "pure" bitumen is characterized by a low ability to absorb microwave radiation. Therefore, the main component of the effective of this technology is the implementing of additives into bitumen that improve both its properties and its microwave susceptibility, that is, they additionally perform the function of channels for the conduction of electromagnetic radiation in the volume of bitumen.

Bituminous binders are polydispersed colloidal systems, and therefore the operational and technical properties and "performance" of these materials in asphalt concrete, usually in the form of a film up to 10 microns thick on the surface of mineral fillers (Shekhovtsova et al. 2019), largely depends on the physical properties bitumen at the micro and nano levels.

Due to the possibility of achieving unique properties when reaching the nanoscale level of these materials, the scientific community has seen a sharp increase in the number of studies of various nanomaterials, methods of their synthesis, study of properties and applications in recent decades (Burmistrov et al. 2017; Cheng and Li 2016; Lee et al. 2018; Kumar et al. 2014). But, despite the wide range of existing methods and materials for modifying bituminous binders, the use of modern advances in nanotechnology in this area currently remains limited, and therefore research in the development of nanomodified materials that improve the properties of asphalt concrete prepared on their basis, are considered promising scientific direction (Ilyin et al. 2014; Mamulat et al. 2015).

The use of microwave (UHF) electromagnetic radiation for heating asphalt concrete mixtures modified with metal particles was tested in a number of works (Tabaković and Schlangen 2015; He et al. 2017), as a result of which it was shown that this direction is promising for increasing the efficiency of "induction healing" asphalt concrete coatings, taking into account the entire set of life cycle costs.

For road pavings, the very first developments of this kind were carried out at the Delft University of Technology in Holland. At the same time, to ensure the acceleration of induction heating of the mixture under the action of microwave radiation, microcapsules with organic "rejuvenators"—bitumen rejuvenators and fragments of stainless wire 0.5–1.5 mm in size were added to the asphalt concrete mixture (Tabaković and Schlangen 2015). Such a composition made it possible to re-compact or level the paving by induction heating in place using a mobile microwave installation after the formation of microcracks and ruts in the paving (Fig. 1).

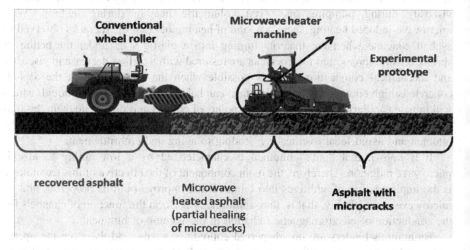

Fig. 1. Technology of microwave-induced recovery of microcracks in asphalt concrete in the road surface.

Along with the traditionally used magnetic and metal fillers to ensure an increase in the absorption of energy of electromagnetic microwave radiation (He et al. 2017), studies began to appear indicating the possibility of obtaining a high absorption coefficient of microwave radiation by composites filled with various types of other fillers, including carbon-containing materials, such as crumb rubber and graphite (Dingbang 2018; Yin 2018).

In the works carried out jointly by National University of Science and Technology MISiS and Tambov State Technological University (Mamulat 2017), the technology of introducing multi-walled carbon nanotubes (MWCNT) into a bitumen matrix was developed, a schematic representation listed in Fig. 2.

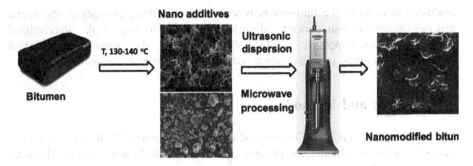

Fig. 2. Technology of implementing multi-walled carbon nanotubes (MWCNTs) into a bitumen matrix

Later, the authors of (Мамулат et al. 2019) established the efficiency of using MCNTs and graphite particles together with microdispersed iron-containing fillers based on industrial waste as induction-heated bitumen modifiers. The principle of road surface restoration without replacing the wear layer using induction-sensitive technology is listed at Fig. 3.

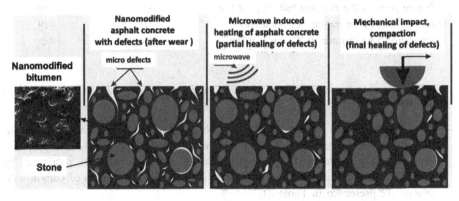

Fig. 3. Reconstruction of the road surface without replacing the wear layer by means of induction-sensitive technology

The aim of this work is to study the possibility of increasing the efficiency of heating bituminous binders in the field of microwave electromagnetic radiation due to the introduction of micro- and nanodispersed electrically conductive and magnetically sensitive fillers. This technology has got high prospects for heating bituminous binder and asphalt concrete mixture at the stages of materials preparation, construction and repair of road pavings. The achievement of the effect declared is ensured through the complex modification of bituminous binders with various nano- and microdispersed particles of carbon materials and iron-containing waste of metallurgical production. Considering the volumetric characteristics of asphalt concrete, it is worth noting that the micro- and nanoscale nature of the particles under consideration will provide the modification of only the bituminous binder without making changes to the frame structure of the asphalt concrete, designed according to one of the standardized methodology like Superpave or Marshall, respectively, the proposed modification method will not entail upon additional costs for asphalt concrete manufacturers.

2 Materials and Methods

In this paper, bitumen BND 70/100 produced by the Moscow Oil Refinery was considered as a bitumen binder, the physical and mechanical characteristics of which are presented in Table 1.

Table 1. Physical and mechanical properties of bituminous binder BND 70/100

Indicator	Standard requirements	Actual value
Penetration depth of the needle, at a temperature of 25 °C, mm^{-1}	71–100	73
Penetration depth of the needle, at a temperature of 0 °C, mm^{-1}, not less	21	25
Softening point of the ring and ball, °C, not less	+47	+50
Fraas brittleness temperature, °C, not higher	−18	−19
Extensibility at temperature 0 °C, cm, not less	3,7	4,2

The following were considered as nano- and microdispersed particles of carbon materials:

– multi-walled carbon nanotubes (MWCNT) of the Taunit-M series, which are quasi-one-dimensional nanoscale, filamentary formations of polycrystalline graphite, predominantly cylindrical in shape with an internal channel, produced by NanoTechCentre.LLC of Tambov (the physicochemical properties of carbon nanotubes are presented in Table 2);
– micrographite, LLC NPP Graphite of the Ukraine, which is a kind of natural graphite obtained from natural material and has a pronounced crystalline form of grains, consisting of 99% carbon and 1% ash;
– rubber crumb from devulcanized tires of a heavy-duty dump truck BelAZ, obtained by grinding in a complex of five stages of grinding, shredding installations, which provides at the output of rubber crumb cleaned of impurities and sorted into various fractions, the work was considered a fraction of 1–3 mm;
– iron-containing waste of PJSC Severstal (oily scale) is a mixture of oxides Fe_3O_4, FeO and Fe_2O_3, and consists of two layers that can be easily separated from each other (the inner layer is porous, black-gray in color, the outer layer is dense and with a reddish tint, both layers are fragile and have ferromagnetic properties; the composition of the iron scale is unstable and depends on the production conditions).

Table 2. Physicochemical characteristics of carbon nanotubes of the Taunit-M series

Characteristic	Taunit-M
External diameter, nm	8–15
Inner diameter, nm	4–8
Length, micron	≥2
Total impurities, % - initial - after cleaning	≤5 ≤1
Specific surface area, m²/g	≥300
Bulk density, g/cm³	0,003–0,005

Preparing Modified Bitumen Samples
The bitumen was heated to a mobile state at a temperature of 130–140 °C in a drying oven. The heated bitumen was poured in portions of 50 g into glass vessels. Fillers were poured into each vessel in specified concentrations: Taunit-M, GAG-2 graphite, rubber crumb, forge scale, and then they were mixed to a homogeneous condition. Then the samples were treated with ultrasound (1 kW, 22 Hz) for 3 min. Then the resulting modified binders were mixed using a paddle mixer for 30–90 min, depending on the modifying component used, and cooled to room temperature for at least 3 h.

Microwave Oven Test
The modified bitumen samples, in glass cups with a wide top, were alternately placed in a microwave oven (Fig. 4) with a power consumption of 800 W for a specified time, the temperature was recorded using a DT-880 remote thermometer.

Fig. 4. The appearance of the microwave oven.

Varying the percentage of nano- and microdispersed particles of carbon materials and iron-containing metallurgical waste in bitumen is based on previous scientific works in this area (Ilyin et al. 2014; Mamulat et al. 2015; Mamulat 2017; Мамулат et al. 2019).

3 Results and Discussion

The process microwave electromagnetic radiation absorption is accompanied by the raise of temperature of the modified bitumen sample, which was measured for the samples under study at different times of microwave exposure with an IR thermometer.

The temperature dependences of bitumen samples modified with various types of electrically conductive and magnetically sensitive fillers on the time of microwave treatment with electromagnetic radiation are shown in Figs. 5, 6, 7, 8 and 9.

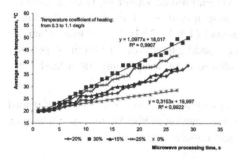

Fig. 5. Influence of scale on the rate of heating of bitumen in a microwave oven

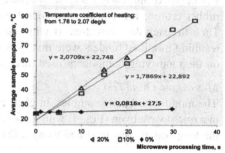

Fig. 6. Influence of rubber crumb on the heating rate of bitumen in a microwave oven

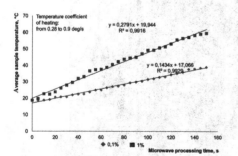

Fig. 7. Influence of micrographite on the heating rate of bitumen in a microwave oven

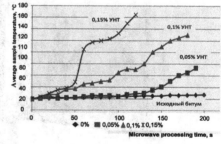

Fig. 8. Influence of MWCNT on the rate of heating bitumen in a microwave oven

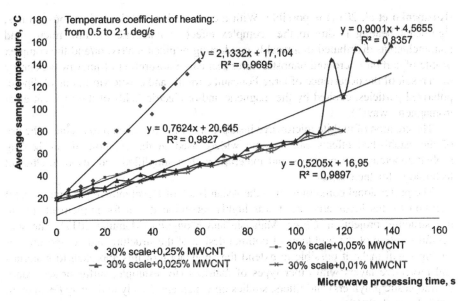

Fig. 9. The combined influence of scale and MWCNT on the heating rate of bitumen in a microwave oven

In the process of analyzing the research results presented in Figs. 5, 6, 7, 8 and 9, it was revealed that the rubber crumb and a combination of the "scale" and MWCNT modifiers provide a thermal heating coefficient of up to 2.1°/s. So, in bitumen containing MWCNTs in an amount of 0.01 wt. % the heating process occurs according to a dependence close to linear, and in samples with a content of 0.025 wt. % and more, a sharp jump in temperature is observed, and then the heating rate decreases (Fig. 9). Presumably, the actual temperature of local areas with an increased content of MCNTs under such conditions exceeds the melting point of bitumen. In this case, the bitumen goes into a molten state and continues to heat up more intensively as a result of a decrease in viscosity. Heat transfer improves in the bulk of the material, and the heating rate decreases due to the inclusion of local areas with a lower concentration of MCNTs, which also begin to melt and absorb some of the heat.

The use of multi-wall carbon nanotubes and nanodisperse materials using the Taunit technology (www.nanotc.ru) without additional cleaning allows you to obtain cost-effective materials for wide use (the cost of nano-modifiers is less than 20 USD per 1 ton of bitumen).

Depending on the content of electrically conductive fillers, several heating mechanisms are possible: in the pre-percolation concentration range, polarization effects occur in the filler particles and energy absorption occurs according to losses during the hopping conduction mechanism; during the formation of percolation clusters in the near-percolation concentration ranges, as well as at concentrations above the percolation threshold, large Foucault currents arise and a rapid rapid heating is observed.

With the introduction of magnetically sensitive fillers, the resonant magnetostrictive mechanism of absorption of microwave energy with the release of heat, described in

(Kuzmenko et al. 2014) is possible. With a combination of fillers of various natures, significant heating is due to the "complex effect of magnetic metallic micro- and nanoinclusions distributed in a weakly conducting bitumen matrix, or/and the complex profile of carbon micro and nanostructures that cause absorption of microwave energy (as a result of the occurrence of large Foucault currents and elastic vibrations of dipole-polarized particles, excited by the magnetic and/or electric fields of the incident electromagnetic wave").

The creation of mobile microwave installations (Fig. 10) taken into account, the use of the established effects can become widely used in the creation of self-healing asphalt-concrete road paving's and materials with a simplified and resource-efficient technology for their repair.

The professional community and the Asian Bank of Development paid rather great attention to this focus area, so it was highly recommended it for implementation in infrastructure projects in China, Malaysia and Mongolia (Mamulat 2017). The successful solution of the problems of optimal design of the structure and composition of paving's will make it possible to extend the application of this approach to materials and protective layers with other types of matrices (for example, sulfur or secondary thermoplastic polymers) and fillers, studies on which are already underway (Wan et al. 2018, Li et al. 2018).

Fig. 10. Suspended microwave installation, created by LLC "Center-Novatsiya", Moscow.

4 Conclusions

In this research, it has been shown that the implementing of electrically conductive carbon or iron-containing fillers can significantly increase the intensity of heating of bituminous binders under the influence of electromagnetic microwave radiation.

It was found that the least effect on the heating rate is exerted by iron-containing waste - scale and micrographite; the coefficient of temperature heating of materials based on them is no more than 1.1°/s.

It was revealed that rubber crumb from devulcated tires of a heavy-duty dump truck and a complex of MWCNT modifiers can provide a coefficient of temperature heating of bitumen up to 2.1°/s.

The most intense heating is observed when MWCNT is added to bitumen. It was also found that in a sample containing MWCNTs in an amount of 0.01 mass. % the heating process occurs according to a dependence close to linear, and in samples with a content of 0.025 mass. % and more, a sharp jump in temperature is observed, and then the heating rate decreases. This is possibly due to the fact that the actual temperature of local areas with an increased content of MCNTs in such conditions exceeds the melting point of bitumen. In this case, the bitumen goes into a molten state, and continues to heat up more intensively as a result of a decrease in viscosity.

Two possible heating mechanisms have been proposed, depending on the content of electrically conductive fillers: in the pre-percolation range of concentrations, polarization effects occur in the filler particles and energy absorption occurs according to losses in the hopping conduction mechanism; during the formation of percolation clusters in the near-percolation concentration ranges, as well as at concentrations above the percolation threshold, large Foucault currents arise and a rapid rapid heating is observed.

The authors plan the scientific and applied development of this area within the framework of a specialized consortium of Russian innovative organizations, and research in the field of bituminous binders with crumb rubber and secondary polymers modified by waste of edible natural oils - within the framework of a joint project of the Faculty of Civil Engineering of Chongqing University (China) under the leadership of Prof. Dr. Ruikun Dong (Dong 2019) and the Department of Building Materials and Materials Science of the National Research University MGSU (Moscow State University of Civil Engineering) (Russia).

The introduction of microwave-susceptible fillers makes it possible to create composites characterized by effective volumetric heating, not only bituminous binder-based, but also in terms of a wide range materials with low absorption of microwaves, for example, polymer waste. A global problem that can be solved using the stated approach is the creation of a microwave heating technology for mould pressing road materials based on unsorted polymer waste that pollutes the environment of many countries (Fig. 11), including on the basis of collection and processing of already accumulated waste.

Fig. 11. Floating islands are already forming in the world's oceans containing millions of tons of polymer waste. Source: https://theoceancleanup.com/media-gallery/#&gid=1&pid=8

References

Burmistrov, I., et al.: Mechanical and electrical properties of ethylene-1-octene and polypropylene composites filled with carbon nanotubes. Compos. Sci. Technol. **147**, 71–77 (2017)

Cheng, L.Q., Li, J.F.: A review on one dimensional perovskite nanocrystals for piezoelectric applications. J. Materiom. **2**, 25–36 (2016)

Dingbang, W.: Surface area and microstructure of microwave activated crumb rubber and its influence on CRM binders. Report at World Transport Convention – 2018, Highway Engineering, Beijing, China, p. 194 (2018)

Dong, R.: Characterization of crumb rubber pre-desulfurized in waste cooking oil. Report at MaxConference "Organic Binders in Road Construction", Moscow, Russia (2019)

Ilyin, S.O., Arinina, M.P., Mamulat, Y.S., Malkin, A.Y., Kulichikhin, V.G.: Rheological properties of road bitumens modified with polymer and solid nanosized additives. Colloid J. **76**(4), 425–434 (2014). https://doi.org/10.1134/S1061933X1404005X

Kumar, R., et al.: Improved microwave absorption in lightweight resin-based carbon foam by decorating with magnetic and dielectric nanoparticles. R. Soc. Chem. Adv. **4**, 23476–23484 (2014)

Kuzmenko, A., et al.: Influence of structural features and physico-chemical properties of metal-carbon nanocomposites with ferromagnetic metal inclusions on microwave radiation. J. Nano Electron. Phys. **6**(3), 03024-1–03024-5 (2014)

Lee, Y., Kim, E., Park, Y., Kim, J., Ryu, W., Rho, J., et al.: Photodeposited metal-semiconductor nanocomposites and their applications. J. Materiom. **4**(2018), 83–94 (2018)

Li, C., et al.: Enhanced heat release and self-healing properties of steel slag filler based asphalt materials under microwave irradiation. Constr. Build. Mater. **193**(2018), 32–41 (2018)

He, L., Zhao, L., Ling, T., Quantao, L.: Research on induction heating of cracks in dense graded asphalt mixture. China J. Highway Transp. **30**(1), 17–24 (2017)

Mamulat, S., Anshin, S., Kuznetsov, D.: Result of application of carbon nanotubes for the modification of polymer-bitumen binders for road construction. Abstracts of 1-st International Scientific-Applied Conference "Graphene and Relative Structures: Synthesis, Production and Application", Tambov, Russian Federation, pp. 203–205 (2015). ISBN 978-5-905724-56-5

Mamulat, S.L.: Induction-healing asphalts, modified by nano- and micro dispersive additives – the new/green era for road's maintenance and international technological cooperation. Report at World Transport Convention 2017, Belt & Road International Transport Alliance Symposium, Beijing, China (2017)

Мамулат, С.Л., Мамулат, Ю.С., Бурмистров, И.Н.: О подходах к модификации битумных вяжущих. Журнал «Мир дорог» **117**, 41–46 (2019)

Shekhovtsova, S., Korolev, E.V., Inozemtcev, S.S., Yu, J., Yu, H.: Method of forecasting the strength and thermal sensitive asphalt concrete. Mag. Civ. Eng. **85**(05), 129–140 (2019)

Tabaković, A., Schlangen, E.: Self-healing technology for asphalt pavements. In: Hager, M., van der Zwaag, S., Schubert, U. (eds.) Self-healing Materials. Advances in Polymer Science, vol. 273, pp. 285–306. Springer, Cham (2015). https://doi.org/10.1007/12_2015_335

Wan, J., Xiao, Y., Song, W.: Self-healing property of ultra-thin wearing courses by induction heating. Materials **11**, 1392 (2018). www.mdpi.com/journal/materials

Yin, W.S.: Research on the Rutting Maintenance with Microwave Heating Technology. Report at World Transport Convention – 2018, Highway Engineering, Beijing, China, p. 193 (2018)

Rheological Properties of Rubber Modified Asphalt Binder in the UAE

Mohammed Ismail[1(✉)], Waleed A. Zeiada[1,2], Ghazi Al-Khateeb[1,3], and Helal Ezzat[1,2]

[1] University of Sharjah, Sharjah, UAE
U18105518@sharjah.ac.ae
[2] Mansoura University, Mansoura, Egypt
[3] Jordan University of Science and Technology, Irbid, Jordan

Abstract. In the 21st century, United Arab Emirates (UAE) is one of many countries that started focusing on building a sustainable eco-friendly system based on recycling waste materials rather than depositing them into landfills. Rubber generated from tires waste is one of those materials that have been proven to be a beneficial material in several applications especially in pavement engineering. This study aims at investigating the rheological properties and performance of Crumb Rubber-modified (CRM) asphalt binders that contain different crumb rubber contents by weight of the binder (5%, 10%, 15%, 20%, and 25%). The testing plan includes conventional tests, namely: penetration, softening point, and Rotational Viscosity (RV) tests and Superpave tests using Dynamic Shear Rheometer (DSR). Moreover, the AASHTOWare Pavement ME Design software was utilized to simulate the effect of using CRM asphalt binders on the pavement performance under local climatic conditions using two different pavement structures (thick and thin). Results showed that increasing the Crumb Rubber (CR) content increased the softening point, RV and rutting parameter (G*/Sinδ) significantly indicating an increased resistance to permanent deformation. The simulated pavement performance showed that modifying the binder with the CR in the thick and thin pavement structures has enhanced the pavement service life remarkably. The control asphalt binder failed due to rutting for thick and thin pavement structures at 12 years and 13 years, respectively. The modified asphalt binder with 25% CR content enhanced the pavement service life by up to 65% and 53% for thick and thin pavement structures, respectively.

Keywords: Rubber modified asphalt · Asphalt binder · Pavement performance · DSR · Rheological properties · AASHTOWare · Rutting

1 Introduction

Asphalt binder can be defined as a viscoelastic material, where its property is highly affected by temperature changes (Somé et al. 2015). It shows a brittle-viscoelastic characteristics at low and intermediate temperatures and fluid characteristics at high temperatures (Batista et al. 2018). As the asphalt pavements are subjected to several external constrains of traffic and climate changes that shorten their expected life span, numerous researchers have been exploring different technologies and additives to enhance the pavement performance against these inevitable conditions (Somé et al. 2015). In UAE, the weather conditions are harsh with extremely high temperatures that dominate most of the year accompanied with heavy and high traffic volumes.

Researchers have tested the effect of modifying the asphalt pavement with several additives that perform better against common pavement distresses such as rutting, fatigue and thermal cracking (Geng et al. 2014; Ghuzlan and Al Assi 2016; Ji et al. 2017; Nega et al. 2014). Multiple modifiers have been used to improve the characteristics of the asphalt binders in terms of stiffness, deformation resistance and stripping (Bulatovic et al. 2012; Mazumder et al. 2016; McNally 2011; Wegan and Brule 1999; Yildirim 2007). One of these modifiers that have been wasted in the landfills and has a great potential in the enhancement process of the asphalt binder is the Crumb Rubber (CR).

There are billions of tires that have been buried in the landfills after reaching their end-of-life stage. Several reports have predicted that by 2030, the annual production of end of useful life tires will reach 1.2 billion because of the continuous increase in population (Narani et al. 2019). Moreover, the presence of this enormous number of toxic-generating wastes in the landfill can affect the soil and groundwater. The re-use process of the tire rubber can be differentiated based on the type of treatment. Two forms of CR can be generated using the chemical and mechanical treatment process, namely: activated and crumb rubber, respectively. CR has shown a high elastic behavior with a significant absorbance of the plastic energy. Moreover, several researchers have tested the characteristics of the material and found out that the CR has high loading resistance, low shrinkage, significant thermal and sound insulation (Brasileiro et al. 2019; Diab and You 2017; Kim et al. 2010; Taylor et al. 2007; Wang et al. 2013). The sizes of the particles can determine the flexibility of the mixture and its performance in overcoming distress.

Different laboratory research tests have expressed the significance of CR content on the asphalt performance. The studies explained how rubber crumb content significantly influenced the rheological properties and performance of asphalt binders (Mashaan et al. 2011, 2012; Mashaan et al. 2014). Xu et al (2016) confirmed that asphalt source and polymer type determine rheological features of the modified asphalt binders. Increasing the CR content in the asphalt binder from 4% to 20% increases the ductility, softening point, viscosity, rutting parameter, and elastic recovery. This aspect is related to the rubber particle absorption, especially those with smaller fraction of binder. Also, such conditions lead to rubber particles increasing during the swelling in the blending procedure. However, higher CR contents of 16% to 20% may increase the Brookfield viscosity value above the 3 Pa.s specification limit. (Al-Khateeb and Ramadan 2015) investigated the effect of modifying the asphalt binder with six rubber percentages on the rheological properties such as Superpave rutting parameter, fatigue parameter, and storage modulus. The study found out that the addition of rubber enhanced the complex shear modulus at all the tested temperatures. Moreover, the rutting parameter, and the fatigue parameter increased significantly with the increase of CR content.

Previous studies evaluated the mechanical properties of dry mixed rubberized Asphalt Concrete (AC) mixtures in terms of temperature susceptibility, moisture sensitivity, permanent deformation and fatigue behavior (Ahmed et al. 2012; Lee et al. 2008; Mohamed and Zumrawi 2017; Thodesen et al. 2009; Kaloush et al. 2009; Kaloush et al. 2012; Zeiada et al. 2014). Generally, both laboratory and field results showed that dry process CRM AC mixture exhibit poor performance or little improvement compared to wet process or conventional AC mixtures. Several laboratory studies have been conducted to determine an appropriate aggregate gradation, design binder content or mixture preparation procedure capable of improving the consistently and performance of a dry process originated mix (Al-Omari et al. 2017; Lee et al. 2007). These studies found that the mechanical properties of the mixtures formed through the dry process method are very sensitive to changes in rubber content.

The AASHTOWare Pavement ME Design software was built on by the assessment of the National Cooperative Highway Research Program (NCHRP) and sponsored by the American Association of State Highway and Transportation Officials (AASHTO) (AASHTO 1995). They relied on the most specialized pavement experts with the help of the huge database retrieved from the long-term pavement performance (LTPP) stations. The purpose of the Pavement ME Design software is to simulate and evaluate the stresses, strains, and deformations occurring in the pavement. Those parameters are influenced by repeated traffic loading and environmental conditions involving climate behavior and temperature changes (Hoerner et al. 2007). The simulation aim is to assess the laboratory tested asphalt pavement modifications on the real-life conditions taken into consideration the aging process in the presence of certain localized traffic and climate conditions (Dokku et al. 2020; Islam et al. 2019; Kocak and Kutay 2020).

This study aims at investigating the effect of modifying local asphalt binder with different CR contents using conventional and Superpave tests at unaged and short-term aged conditions. Moreover, the AASHTOWare Pavement ME Design software have been utilized to simulate the performance of the CRM asphalt binders against rutting under local climate conditions in the UAE.

2 Materials and Experimental Plan

In this study, the utilized CR was grinded using the mechanical process and obtained from Bee'ah. The Bee'ah waste management center is the largest facility in the Middle East with an outstanding rate of waste recovery and waste-to-energy transformation (Beea'h-UAE 2018). Moreover, they have been recycling rubber products into jogging tracks, rubber tiles by grinding the tires rubber into crumb rubber. Table 1 illustrates the sieve analysis results of the CR. 99.9% of the CR particles passed sieve #30.

Table 1. Sieve analysis

Sieve number	Diameter (mm)	Percentage passing (%)
#20	0.840	100.0
#30	0.600	99.9
#40	0.420	82.8
#50	0.300	35.7
#80	0.180	12.5
#100	0.150	6.1
#200	0.075	7.3
Pan	–	0.0

A 60/70 penetration grade asphalt binder obtained from a local supplier was used as a control asphalt binder in this study. The study focuses on investigating the rheological properties of the CRM asphalt binders using five different contents by weight of the asphalt binder. High shear mixer was used to mix the required amount of asphalt binder and CR. The properties of the CRM asphalt binders were evaluated using softening point test, penetration test and rotational viscosity using unaged condition. The Rolling Thin Film Oven (RTFO) was utilized to age the CRM asphalt binders for short term aging condition. The DSR was then utilized to evaluate the complex shear modulus (G*) and phase angle (δ) at five high temperatures. Each test was carried out on two/three replicates. Additionally, the AASHTOWare software was used to simulate the pavement performance using different CR contents against rutting under local climate conditions using climatic database from the Sharjah International Airport, UAE. Table 2 shows the rheological properties of unmodified asphalt binder and the corresponding test standard.

Table 2. Properties of control asphalt binder

Test	Value			Test standard
Penetration	60.3 mm			(ASTM D5/DSM-20 2020)
Softening point	49.62 °C			(ASTM D36-95 2000)
Rotational viscosity (cP)	At 135 °C	At 150 °C	At 165 °C	(ASTM D4402 2015)
	387.1	176	108.2	
PG original	At 64 °C	At 70 °C	At 76 °C	(ASTM D7175-15 2015)
G* (Pa)	1539.6	723.869	362.317	
δ (°)	88.297	88.981	89.414	
PG RTFO	At 64 °C	At 70 °C	At 76 °C	(ASTM D7175-15 2015)
G*/sin δ (kPa)	3.5278	1.5665	0.7378	

The summarized testing methodology of the study is shown in Fig. 1.

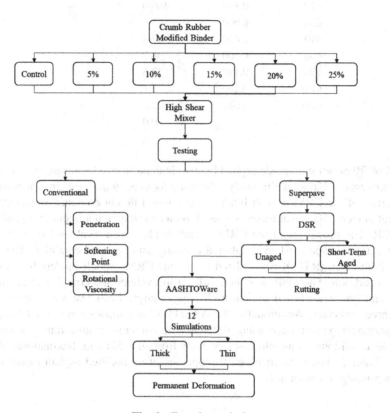

Fig. 1. Experimental plan

3 AASHTOWare Simulations

The AASHTOWare Pavement ME Design software requires a set of traffic, climate, material and structural inputs to carry out an accurate performance simulation. There are three levels of inputs that the Pavement ME Design software offers based on the level of availability of data for the required simulation. The most accurate level of inputs is referred to as level 1, which requires a laboratory testing to be carried out to obtain the required testing inputs. Input level 1 has two testing inputs for the asphalt binder namely: Superpave PG and Penetration/Viscosity grade. The Superpave PG requires the G^* and the δ at different temperatures. However, the Penetration/Viscosity grade needs testing inputs from softening point, penetration, and rotational viscosity tests (National Cooperative Highway Research Program 1999).

3.1 Design Inputs

Table 3 illustrates the summary of the implemented design inputs in the AASHTOWare software. The climate data were obtained from the weather station located in the Sharjah International Airport at a latitude, longitude, and an elevation of 25.330 ft, 55.516 ft, and 34.000 ft, respectively. Moreover, Arizona State calibration factors were utilized in the simulations as shown in Table 3 because of the similarity in the climate conditions between Arizona State and the UAE (Darter et al. 2014).

Table 3. AASHTOWare pavement ME design inputs

\multicolumn{7}{c}{Structural Inputs}
Material	Type	Thickness	Air Void %	Effective Binder Content	Aggregate Gradation
Asphalt Concrete	Flexible	Thick (6 inches)	7%	10.7% / Thick	3/4in (95.02) 3/8in (74.89)
A-2-4 A-7-5	Non-Stabilized Subgrade	Thin (3 inches)		11.7% / Thin	#4 (48.96) #200 (2.8)
\multicolumn{7}{c}{Traffic Inputs}
Section	Two-way AADTT	No. of Lanes	K_d	K_L	Operational Speed (mph)	Growth Rate
Thick	78000	4	60	0.9	70	5%
Thin	50000	4	60	0.9	70	5%
\multicolumn{7}{c}{Climate Inputs}
Station	Mean Annual Air Temperature (°F)	Mean Annual Precipitation (in)	Freezing Index (°F-days)	Average Annual Freeze/Thaw Cycles
Sharjah Airport	80.93	2.07	0	0.03
\multicolumn{7}{c}{Rutting Calibration (Arizona State)}
BR1	BR2	BR3	K1	K2	K3
0.69	1	1	-3.35412	1.5606	0.4791

4 Results and Discussions

The conventional tests were carried out for the control asphalt binder (pen. 60/70) and the five-CRM asphalt binders according to the ASTM testing standards. Figure 2 illustrates both penetration and softening point test results. As shown in Fig. 2, the increase of RC content has increased the ability of the tested asphalt binders to resist the applied load by the penetration needle and therefore lower penetration values. The 15% rubber managed to change the penetration grade to 40/50 according to ASTM-D5. For the softening point test, the 3.5 g balls managed to travel the standard (25 mm) on the control binder at 49.62 °C (ASTM D5/DSM-20 2020). CRM samples showed a high resistance to softening as CR content increases. The softening point increased up to 61.6 °C at 25% CR.

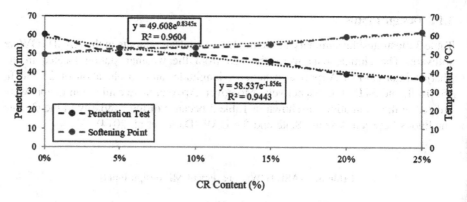

Fig. 2. Penetration and softening point test results

The RV test measures the viscosity at elevated temperatures (110 °C, 135 °C, 150 °C, and 165 °C). The viscosity of the binder tends to decrease with the increase in temperature. The control asphalt binder showed the least viscosity at 135 °C, 150 °C, and 165 °C, with values of 387.1 cP, 176 cP, and 108.2 cP, respectively. The viscosity of the 25% CRM asphalt binder increased significantly at 135 °C and 165 °C up to 6216.63 cP and 2383.88 cP, respectively. It should be noted that at 135 °C, the 20% and 25% CRMA binders failed the maximum allowable viscosity limit (3 Pa.s) according to AASHTO M332 (AASHTO M332 2020) (Table 4).

Table 4. Rotational viscosity test results

Rotational viscosity, (cP)				
Temperature	110 °C	135 °C	150 °C	165 °C
Control 60/70	1640.67	387.10	176.00	108.20
5%	3312.50	729.65	390.50	261.70
10%	4668.07	1165.88	770.25	642.19
15%	8456.50	2176.00	1336.56	1114.50
20%	–	*3598.00*	2096.50	1590.00
25%	–	*6216.63*	3592.38	2383.88

As the increase in the viscosity is expected to affect the workability of asphalt binder, a further analysis was conducted to determine the effect of the CR content on the mixing and compaction temperature ranges. Figure 3 shows the temperature-viscosity relationship and the effect of the CR content on the temperature susceptibility. The slope of the obtained relationship (VTS) represents the level of temperature susceptibility of the Asphalt binder and the range of the mixing and compaction temperatures. It is observed that increasing the CR content increases the mixing and compaction temperature ranges as shown in Table 5. Moreover, asphalt binders with higher CR content is expected to experience a better rutting performance in summer seasons with elevated temperatures due to corresponding higher.

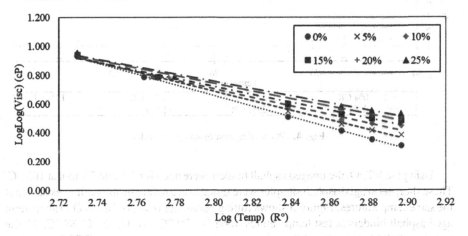

Fig. 3. Temperature-viscosity relationships of tested asphalt binders

Table 5. Upper and lower limits of the compacting/mixing temperatures

Range	0%	5%	10%	15%	20%	25%
Upper	143.5/154.5	160.2/172.9	176.6/191.4	192.3/208.8	206.0/223.6	221.3/240.7
Lower	139.2/149.2	155.3/166.8	170.9/184.3	186.0/200.9	199.2/215.2	213.9/231.4

The Superpave tests were carried out using the DSR to determine G^* and δ. The rutting parameter, $G^*/\text{Sin } \delta$, is defined as the resistibility of the binder against rutting distress which occurs at high temperatures (Kim et al. 2010). In this study, the unaged and the short-term aged asphalt binders were tested to evaluate the effect of the CR content on the asphalt binder resistance to permanent deformation. The rutting parameter for the unaged binder is limited to $G^*/\sin \delta$ of 1 kPa. However, the RTFO-aged rutting parameter is limited to a $G^*/\sin \delta$ of 2.2 kPa. Figure 4 shows the summary results of the tested asphalt binders at 5 different temperatures (64 °C, 70 °C, 76 °C, 82 °C, 88 °C). All of the modified unaged asphalt binders passed the minimum rutting limit at both 64 °C, and 70 °C. Only the 15%, 20%, and the 25% passed the minimum rutting limit at 76 °C. At 82 °C, the rutting parameter of the 20% and the 25% CRM asphalt binders was 1.19 kPa and 1.651 kPa, respectively. However, only the 25% CRM asphalt binder passed the limit of 1 kPa with a $G^*/\sin \delta$ value of 1.022 kPa. The unaged control binder passed minimum rutting limit at 64 °C only, with a $G^*/\sin \delta$ value of 1.746 kPa.

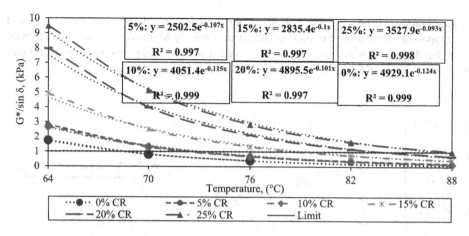

Fig. 4. PG testing results-unaged binder

Using the RTFO, the unaged asphalt binders were aged for 1 h and 25 min at 165 °C. These short-term aged asphalt samples were tested also to evaluate the rutting resistance at the same temperatures. Figure 5 shows rutting parameter results of the RTFO short-term aged asphalt binders at test temperatures of 64 °C, 70 °C, 76 °C, 82 °C, 88 °C. All the tested asphalt binders passed the minimum rutting parameter limit of RTFO condition, 2.2 kPa, at 64 °C. The 20% and 25% CRM asphalt binders passed the rutting parameter limit at all the tested temperatures except 88 °C. However, the 5%, 10% and 15% CRM asphalt binders only passed at 64 °C and 76 °C.

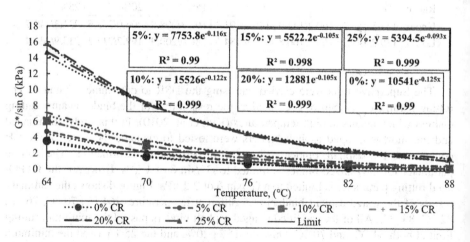

Fig. 5. PG testing results-short-term aged binder

The True PG of each binder needs is determined to identify the maximum real temperature (T_{max}) that the binder can handle as shown in Table 6. The 0% and 5% RTFO CRM asphalt binders can handle a temperature of 68.06 °C and 69.94 °C, respectively. Also, the 10% and 15% RTFO CRM asphalt binders showed almost the same T_{max} with a value of 74.2 °C and 74.7 °C, respectively. The highest T_{max} of the tested RTFO asphalt binders was obtained for 25% RTFO asphalt binder (84.54 °C).

Table 6. True PG of the RTFO-binder

Binder	RTFO-PG	True PG
0%	64	68.06
5%	64	69.94
10%	70	74.19
15%	70	74.77
20%	82	83.32
25%	82	84.54

The AASHTOWare Pavement ME Design software was utilized to simulate 12 runs to target the total rutting performance of two pavement structures, thick and thin, under the local conditions of the UAE. Figure 6 illustrates the total rutting for the thick (a) and thin (b) pavement structures. The control asphalt binder achieved the 0.75-in. minimum rutting requirements in the thick and the thin pavement structures at 12 years and 13 years, respectively.

For the CRM asphalt binders, the corresponding pavement lives due to rutting are observed to be higher for thick pavement structure, the 5%, 10%, 15%, 20%, and 25% CRM asphalt binders failed due to rutting at 14.1, 14.7 16.2, 18.8, and 19.8 years, respectively. The same trend was observed for the thin pavement structure, where the CRM asphalt binders showed an increase in the pavement life as the CR content increases (14.9, 14.9, 16.8, 19.0, 19.9 years for 5%, 10%, 15%, 20%, and 25% CRM asphalt binders respectively).

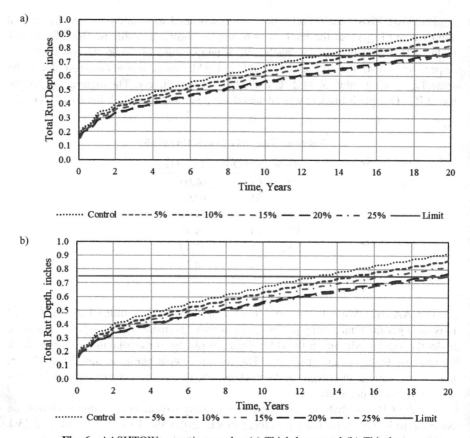

Fig. 6. AASHTOWare rutting results: (a) Thick layer; and (b) Thin layer

The percent pavement life increase due to the use of different CR contents was also calculated by comparing the failing year of each CRM asphalt binder to that of the control asphalt binder as presented in Fig. 7. For the thick pavement structure, modifying the asphalt binder with 5%, 10%, 15%, 20%, and 25% CR contents extended the life of the asphalt pavement by 17.7%, 22.9%, 34.7%, 56.9% and 64.6%, respectively. For the thin pavement structure, the enhancement pavement life against rutting was almost the same for the 5% and 10% CRM asphalt binders. The increase in pavement life of 5% and 10% CRM asphalt binders compared to the control was 14.74%. The addition of the 15%, 20%, and 25% CR contents increased the pavement life against rutting 28.8%, 46.2%, and 53.2%, respectively.

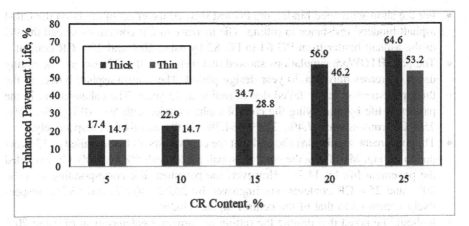

Fig. 7. Enhanced pavement life percent against rutting for thick and thin pavement structures

In general, this study found out that with the increase in the CR content in the asphalt binder, the performance of the CRM asphalt binders exhibited an enhanced performance against rutting which is the main distress in the UAE. It should be noted that the enhancement in pavement performance didn't improve remarkably for the 20% and 25% CRM asphalt binders.

5 Conclusions

This study aimed at evaluating the CR modification of the asphalt binder using 5 different CR contents (5%, 10%, 15%, 20%, 25%) against the most common distresses in the UAE, which is rutting using conventional and Superpave DSR test. 12 pavement performance simulations were carried out using the AASHTOWare Pavement ME Design software to investigate the effect of the CR content on the rutting performance of asphalt pavement during a 20-year period under local UAE climate conditions. The main conclusions and findings of the study are summarized below:

- The conducted conventional tests showed that with the increase of the CR content, the viscosity of the CRM asphalt binder increases significantly. Modifying the binder with 5% and 10% CR almost showed no noticeable changes in results of the penetration and softening point tests.
- The rotational viscosity test showed a remarkable increase in the viscosity with the increase of CR content. However, increasing the CR beyond 15% fails to achieve the viscosity requirements of maximum 3 Pa.s.
- The PG testing results for the unaged binder showed that with the CR content increase, the G* increased remarkably compared to the control asphalt binder. Moreover, the phase angle decreased with the increase in rubber content. Therefore, higher rubber content binders showed a higher rutting parameter value and a significant enhancement in the rutting susceptibility.

- For the short-term aged binder, the PG test showed the effect of aging on the CRM asphalt binders' resistance to rutting. The increase of CR content enhanced the PG of the asphalt binder from PG 64 to PG 82 by using 20% and 25% CR content.
- The AASHTOWare simulations showed that increasing the CR content, the rutting depth decreases during a 20 year design period. The control asphalt binder in the thick pavement structure failed due to rutting at 12 years. The enhancement in the pavement life by modifying the control asphalt binder with 5%, 10%, 15%, 20%, 25% CR content was 17.4%, 22.9%, 34.7%, 56.9% and 64.6%, respectively.
- Thin pavement simulations showed that the control asphalt binder failed at 13 years due to rutting. Modifying the control asphalt binder with 5% and 10% CR extended the pavement life by 14.7%. However, the pavement life corresponding to 15%, 20%, and 25% CR contents was improved by 28.8%, 46.2%, and 53.2%, respectively compared to that of the control asphalt binder.
- It should be noted that despite the rutting performance enhancement of using 20% and 25% CR contents, these CRM asphalt binders showed low workability as their viscosity values were relatively high at mixing and compacting temperatures.

References

AASHTO: AASHTO Provisional Standards. Washington DC (1995)

AASHTO M332: Standard Specification for Performance-Graded Asphalt Binder Multiple Stress Creep Recovery (MSCR), 7 (2020)

Ahmed, E.I., Hesp, S.A.M., Kumar, S., Samy, P., Rubab, S.D., Warburton, G.: Effect of warm mix additives and dispersants on asphalt rheological, aging, and failure properties. Constr. Build. Mater. 37, 493–498 (2012). https://doi.org/10.1016/j.conbuildmat.2012.07.091

Al-Khateeb, G.G., Ramadan, K.Z.: Investigation of the effect of rubber on rheological properties of asphalt binders using superpave DSR. KSCE J. Civ. Eng. 19(1), 127–135 (2015). https://doi.org/10.1007/s12205-012-0629-2

Al-Omari, A.A., Khedaywi, T.S., Khasawneh, M.A.: Laboratory characterization of asphalt binders modified with waste vegetable oil using SuperPave specifications. Int. J. Pavement Res. Technol. 11, 68–76 (2017). https://doi.org/10.1016/j.ijprt.2017.09.004

ASTM D36-95: Standard Test Method for Softening Point of Bitumen (Ring-and-Ball Apparatus). ASTM International (2000). Accessed www.astm.org

ASTM D4402: Standard Test Method for Viscosity Determination of Asphalt at Elevated Temperatures Using a Rotational Viscometer (2015)

ASTM D5/DSM-20: Standard Test Method for Penetration of Bituminous Materials (2020)

ASTM D7175-15: Standard Test MEthod for Determining the Rheological Properties of Asphalt Binder Using a Dynamic Shear Rheometer (2015)

Batista, K.B., et al.: High-temperature, low-temperature and weathering aging performance of lignin modified asphalt binders. Ind. Crops Prod. 111, 107–116 (2018). https://doi.org/10.1016/j.indcrop.2017.10.010

Beea'h-UAE: Bee'ah's Ultra-Efficient Material Recovery Facility Becomes World-Leading Producer of Recovered Plastics (2018)

Brasileiro, L., Moreno-Navarro, F., Tauste-Martínez, R., Matos, J., del Carmen Rubio-Gámez, M.: Reclaimed polymers as asphalt binder modifiers for more sustainable roads: a review. Sustain. (Switz.) 11(3), 1–20 (2019)

Bulatovic, V.O., Rek, V., Markovic, K.J.: Rheological properties and stability of ethylene vinyl acetate polymer-modified bitumen. Polym. Eng. Sci. **47**(22), 935068 (2012)

Darter, M.I., Von Quintus, H., Bhattacharya, B.B., Mallela, J.: Calibration and Implementation of the AASHTO Mechanistic-Empirical Pavement Design Guide in Arizona, September (2014)

Diab, A., You, Z.: Small and large strain rheological characterizations of polymer- and crumb rubber-modified asphalt binders. Constr. Build. Mater. **144**, 168–177 (2017). https://doi.org/10.1016/j.conbuildmat.2017.03.175

Dokku, B., Savio, D., Nivitha, M.R., Krishnan, J.M.: Development of rutting model for Indian highways based on rut depth simulations from AASHTOWare pavement ME design software. J. Transp. Eng. **146**(2), 1–11 (2020). https://doi.org/10.1061/JPEODX.0000160

Geng, J., Li, H., Sheng, Y.: Changing regularity of SBS in the aging process of polymer modified asphalt binder based on GPC analysis. Int. J. Pavement Res. Technol. **7**(1), 77–82 (2014)

Ghuzlan, K.A., Al Assi, M.O.: Predicting the complex modulus for PAV aged asphalt binder using a master curve approach for sasobit modified asphalt binder. Jordan J. Civ. Eng. **10**(3), 390–402 (2016)

Hoerner, T.E., Zimmerman, K.A., Smith, K.D., Cooley Jr., L.A.: Mechanistic-Emprirical Pavement Design Guide Implementation Plan (2007)

Islam, S., Hossain, M., Jones, C.A., Bose, A., Barrett, R., Velasquez, N.: Implementation of AASHTOWare pavement ME design software for asphalt pavements in kansas (2019). https://doi.org/10.1177/0361198119835540

Ji, J., et al.: Preparation and properties of asphalt binders modified by THFS extracted from direct coal liquefaction residue. Appl. Sci. **7**(11), 1155 (2017)

Kaloush, K.E., Biligiri, K.P., Rodezno, M.C., Zeiada, W.A., Souliman, M.I., Reed, J.X.: Performance evaluation of asphalt rubber mixtures in Arizona—Lake Havasu project. Arizona Dept. of Transportation, Materials Group, Phoenix (2009)

Kaloush, K.E., et al.: Laboratory evaluation of asphalt-rubber gap graded mixtures constructed on Stockholm highway in Sweden. In: Asphalt Rubber Conference, Munich, Germany, pp. 1–21 (2012)

Kim, H., Lee, S., Amirkhanian, S.: Rheology investigation of crumb rubber modified asphalt binders. KSCE J. Civ. Eng. **14**, 839–843 (2010). https://doi.org/10.1007/s12205-010-1020-9

Kocak, S., Kutay, M.E.: Fatigue performance assessment of recycled tire rubber modified asphalt mixtures using viscoelastic continuum damage analysis and AASHTOWare pavement ME design. Constr. Build. Mater. **248**, 118658 (2020). https://doi.org/10.1016/j.conbuildmat.2020.118658

Lee, S., Amirkhanian, S.N., Putman, B.J., Kim, K.W.: Laboratory study of the effects of compaction on the volumetric and rutting properties of CRM asphalt mixtures, December, pp. 1079–1089 (2007)

Lee, S.J., Akisetty, C.K., Amirkhanian, S.N.: The effect of crumb rubber modifier (CRM) on the performance properties of rubberized binders in HMA pavements. Constr. Build. Mater. **22**(7), 1368–1376 (2008). https://doi.org/10.1016/j.conbuildmat.2007.04.010

Mashaan, N.S., Ali, A.H., Karim, M.R., Abdelaziz, M.: Effect of blending time and crumb rubber content on properties of crumb rubber modified asphalt binder. Int. J. Phys. Sci. **6**(9), 2189–2193 (2011)

Mashaan, N.S., Ali, A.H., Karim, M.R., Abdelaziz, M.: An overview of crumb rubber modified asphalt. Int. J. Phys. Sci. **7**(2), 166–170 (2012). https://doi.org/10.5897/IJPSX11.007

Mashaan, N.S., Ali, A.H., Karim, M.R., Abdelaziz, M.: A review on using crumb rubber in reinforcement of asphalt pavement, 2014(i) (2014)

Mazumder, M., Kim, H., Lee, S.: ScienceDirect Performance properties of polymer modified asphalt binders containing wax additives. Int. J. Pavement Res. Technol. **9**(2), 128–139 (2016)

McNally, T.: Polymer Modified Bitumen: Properties and Characterisation. Woodhead Publishing, Sawston (2011). https://doi.org/10.1179/1433075X11Y.0000000021

Mohamed, M., Zumrawi, E.: Effect of crumb rubber modifiers (CRM) on characteristics of asphalt binders in Sudan. Mater. Sci. Appl. **6**, 1–6 (2017). https://doi.org/10.11648/j.ijmsa.s.2017060201.11

Narani, S.S., Abbaspour, M., Hosseini, S.M.M.M., Aflaki, E., Nejad, F.M.: Sustainable reuse of waste tire textile fibers (WTTFs) as reinforcement materials for expansive soils: with a special focus on landfill liners/covers. J. Clean. Prod. Elsevier B.V. (2019). https://doi.org/10.1016/j.jclepro.2019.119151

National Cooperative Highway Research Program: Guide for Mechanistic-Empirical Design APPENDIX AA (1999)

Nega, A., Ghadimi, B., Nikraz, H.: Developing master curves, binder viscosity and predicting dynamic modulus of polymer-modified asphalt mixtures. Int. J. Eng. Technol. **7**(3), 190–197 (2014)

Somé, S.C., Gaudefroy, V., Pavoine, A.: Viscoelastic behavior of fluxed asphalt binders and mixes. In: 22ème Congrès Français de Mécanique (2015)

Taylor, P., Shen, J., Amirkhanian, S.: The influence of crumb rubber modifier (CRM) microstructures on the high temperature properties of CRM binders, December 2012, pp. 37–41 (2007). https://doi.org/10.1080/10298430500373336

Thodesen, C., Shatanawi, K., Amirkhanian, S.: Effect of crumb rubber characteristics on crumb rubber modified (CRM) binder viscosity. Constr. Build. Mater. **23**(1), 295–303 (2009). https://doi.org/10.1016/j.conbuildmat.2007.12.007

Wang, H., Dang, Z., Li, L., You, Z.: Analysis on fatigue crack growth laws for crumb rubber modified (CRM) asphalt mixture. Constr. Build. Mater. **47**, 1342–1349 (2013). https://doi.org/10.1016/j.conbuildmat.2013.06.014

Wegan, V., Brule, B.: The structure of polymer modified binders and corresponding asphalt mixtures (1999)

Xu, O., Xiao, F., Han, S., Amirkhanian, S.N., Wang, Z.: High temperature rheological properties of crumb rubber modified asphalt binders with various modifiers. Constr. Build. Mater. **112**, 49–58 (2016). https://doi.org/10.1016/j.conbuildmat.2016.02.069

Yildirim, Y.: Polymer modified asphalt binders. Constr. Build. Mater. **21**, 66–72 (2007). https://doi.org/10.1016/j.conbuildmat.2005.07.007

Zeiada, W.A., Underwood, B.S., Pourshams, T., Stempihar, J., Kaloush, K.E.: Road Materials and Pavement Design Comparison of conventional, polymer, and rubber asphalt mixtures using viscoelastic continuum damage model, October, pp. 37–41 (2014). https://doi.org/10.1080/14680629.2014.914965

Recycling Waste Rubber Tires in Pervious Concrete Evaluation of Hydrological and Strength Characteristics

Sahil Surehali, Avishreshth Singh[✉], and Krishna Prapoorna Biligiri

Department of Civil and Environmental Engineering, Indian Institute of Technology Tirupati, Tirupati 517619, India
{ce18d001,bkp}@iittp.ac.in

Abstract. An alarming increase in the number of waste rubber tires (WRT) generated annually calls for the need to identify sustainable waste management practices as they pose serious threats to the quality-of-life. One such waste disposal strategy is recycling of WRT in pavement materials. Therefore, the objective of this research study was to investigate the effect of inclusion of WRT derived aggregates on properties of pervious concrete (PC). The size of recycled rubber aggregates (RA) varied from 4.75–2.36 mm, which were added to PC mix at 5 and 10% by weight of coarse aggregates; the new mix was called rubber-modified pervious concrete (RMPC). The porosity and permeability increased with increasing proportions of RA with reduction in the density and compressive strength. At 5% RA content; the magnitude of porosity and permeability increased by 6.9 and 53.1%, respectively, while the density and compressive strength reduced by 7.4 and 63.7%, respectively, compared to the control mix. Similarly, at 10% RA content; density and compressive strength decreased by 12.5 and 77.7%, respectively, while porosity and permeability increased by 15.2 and 63.4%, respectively, compared to the control mix. Owing to lower compressive strength (∼5–8 MPa), RA as filler may be utilized in PC materials with non-vehicular traffic sections such as sidewalks and footpaths. Due to higher porosity and permeability, RMPC can help harness the stormwater benefits associated with PC, while simultaneously allowing for sustainable disposal of WRT.

Keywords: Recycling · Waste rubber tires · Pervious concrete · Porosity · Permeability · Compressive strength · Sustainability

1 Introduction

The number of waste rubber tires (WRT) accumulating every year is soaring at an alarming rate. It is estimated that by 2030, this number will reach about 1.2 billion (Mondal and Biligiri 2018). WRT being generated in large proportions, poses major environmental, economic, and social threats (Azizian et al. 2003; Dong et al. 2013; Singh et al. 2015), while urging authorities at the global level to impose stringent rules and regulations to control landfill operations and promote alternative measures of WRT disposal (Lekkas 2013). Recycling and reusing of tire-derived products are adjudged as the top solutions for WRT disposal, followed by energy recovery methods such as incineration and landfilling as the most unfavorable strategies (Birgisdóttir et al. 2007). An innovative technique to dispose WRT in the infrastructure industry is to utilize them as additives in asphalt and/or cement concrete mixtures. Rubber particles and asphalt binder have exhibited excellent compatibility, which makes the use of crumb rubber (CR) in asphalt paving mixtures a successful practice, and further improves the performance characteristics of rubberized asphalt pavements (Bressi et al. 2019; Heitzman 1992; Lo Presti 2013; Miknis and Michon 1998; Nanjegowda and Biligiri 2020; Venudharan et al. 2017). Researchers reported that CR modified concrete mixtures improved freeze-thaw resistance (Paine and Dhir 2010; Richardson et al. 2011, 2012; Savas et al. 1997), thereby enhancing durability compared to the conventional concrete mixtures and supporting the sustainability credentials.

However, the impervious surface wearing courses of traditional asphalt and concrete pavements do not allow for stormwater infiltration, resulting in problems such as increased surface runoff, flashfloods, urban heat islands (UHI), and increased pollutant load at the downstream end (Chandrappa and Biligiri 2016a; Singh et al. 2020a). To mitigate the problems associated with traditionally paved impervious roadway systems, one of the approaches is to develop pervious concrete pavement (PCP) systems. PCP is characterized by an interconnected porous network structure achieved by using little or no fine aggregates. The aggregate gradation for pervious concrete (PC) mixtures typically consists of either single-sized coarse aggregates or a binary mixture of coarse aggregates; blended with cement such that a small fillet of cement paste holds the matrix together. The water-to-cement (w/c) and aggregate-to-cement (a/c) ratios vary between 0.28–0.40 and 3:1 to 6:1, respectively (Chandrappa and Biligiri 2016a, 2016b; Deo and Neithalath 2010; Singh et al. 2020a; Torres et al. 2015; Zhou et al. 2019). Research suggests that PCP could be a suitable alternative for conventional concrete pavements as parking lots because of lower embodied energy and greenhouse gas emissions (Singh et al. 2020b). Additionally, the other benefits of using PCP include mitigation of UHI effect and stormwater runoff as well as conservation of depleting natural resources. However, the field implementation of this innovative pavement mixture is limited due to the absence of standard design guidelines at the global level.

There is a significant need to develop pavement materials that can alleviate the problems associated with the traditional impervious pavement systems and acknowledge the inclusion of rubber aggregates (RA) in the mix matrix, promoting WRT recycling, and consequentially contributing to low impact development. Several researchers have investigated the effect of inclusion of different rubber types and proportions in PC, and reported the properties of the modified composites (Bonicelli et al. 2017; Gesoğlu et al. 2014a, 2014b; Mondal and Biligiri 2018). Bonicelli et al. (2017) investigated the properties of rubber-modified pervious concrete (RMPC) prepared by utilizing two different CR sizes (0.08–1 and 0.6–2.5 mm) as partial replacement (5 and 10% by volume) of coarse aggregates. At 5% replacement level, RMPC with coarser CR showed a slight improvement in the tensile strength compared to PC, while fine CR reduced the tensile strength significantly. Gesoğlu *et al.* (2014a) reported an increase in the reduction of density and permeability of RMPC with increasing size of rubber particles. The compressive strength of RMPC decreased with increasing rubber content and size, while splitting tensile strength improved with increasing rubber size. In another study by the same research group, it was reported that the inclusion of RA resulted in reduction of the flexural strength of the PC mix, whereas the freezing-thawing and abrasion resistance improved significantly (Gesoğlu et al. 2014b). The authors suggested that rubber size influenced the freezing-thawing resistance and abrasion resistance more than the rubber content. Mondal and Biligiri (2018) found a slight reduction in the porosity and permeability of PC mixtures modified by partial replacement of coarse aggregates with CR. However, the compressive strength of RMPC increased with increasing rubber content entailing in the reduction of void content in the mix. Further, the CR particles arrested the crack propagation, thereby restricting the early failure of the RMPC mixtures and ultimately enhancing their compressive strength.

Although the studies in the past have characterized various RMPC mixtures with rubber being utilized as partial replacement of aggregates, none of the studies investigated the effect of utilizing RA as a filler material in PC. In addition, majority of the researchers added either fine CR or used a broad range of RA in PC. Therefore, the main objective of this study was to develop RMPC by utilizing single-sized RA (4.75–2.36 mm) as filler material in the mix and investigate the properties of RMPC while comparing those with traditional PC materials. RMPC mixtures were prepared by including RA as filler material (5 and 10% dosage by weight of natural aggregates) in PC. Three replicate specimens, each of one control and two RMPC mixtures were prepared. The scope of the effort included measurement of hardened density and porosity, permeability, and compressive strength of control and RMPC mixtures, along with recommending the road class on which the RMPC could be applied comfortably.

2 Experimental Program

2.1 Materials

Ordinary Portland cement 53-grade conforming to IS:12269 (Bureau of Indian Standards 2013) was used. The specific gravity of cement was 3.08, and the initial and final setting times were 270 and 510 min, respectively. Single-sized coarse aggregates (passing 10 mm sieve and retained on 6.7 mm sieve) with specific gravity and water absorption capacity of 2.59 and 0.33% were used. The gradation curve for the coarse aggregates is shown in Fig. 1. RA within the size range of 4.75 to 2.36 mm were procured from Tinna Rubber and Infrastructure Limited, Panipat, India. The RA used in the study are shown in Fig. 2, and their gradation curve is presented in Fig. 3. The specific gravity and water absorption capacity of RA were 0.89 and 5.42%, respectively. It was noted that the specific gravity of RA was about 65% lower than that of coarse aggregates, while the water absorption capacity was 94% higher. A polycarboxylic ether-based superplasticizer (SP) conforming to the American Society for Testing and Materials (ASTM) International C494 Type F (ASTM 2019) was used at a dosage of 0.25% by dry mass of cement.

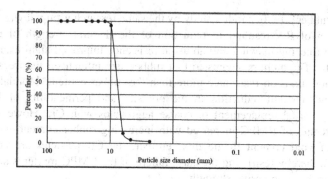

Fig. 1. Gradation curve of coarse aggregates

Fig. 2. Rubber aggregates used in the study

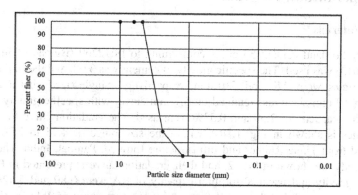

Fig. 3. Gradation curve of rubber aggregates

2.2 Mix Design

Due to the absence of a standard mix design methodology for PC production, trial mixtures were prepared based on the proportions adopted by researchers in the past (Putman and Neptune 2011; Rao et al. 2020; Zhang et al. 2020). The trial PC mixtures were prepared at an a/c ratio of 3.75 and cement content of 325 kg/m³ using three distinct w/c ratios, as presented in Table 1.

Table 1. Trial mix proportions

Sample ID	Coarse aggregates size (mm)	w/c ratio	a/c ratio	Cement (kg/m³)	Coarse aggregates (kg/m³)	Water (kg/m³)	SP (kg/m³)
A	10–6.7	0.27	3.75	325	1218.75	87.75	0.81
B	10–6.7	0.30	3.75	325	1218.75	97.50	0.81
C	10–6.7	0.33	3.75	325	1218.75	107.25	0.81

Based on the various test results summarized in Table 2, the mixture (designated as C) corresponding to the highest compressive strength while qualifying the minimum porosity criteria of 15% mentioned in (ACI 2010; NRMCA 2004) was adopted as the control mix for this study.

Table 2. Properties of trial mixture

Sample ID	Hardened density (kg/m³)	Hardened porosity (%)	Permeability (cm/s)	Compressive strength (MPa)
A	1756.49	30.48	0.91	12.15
B	1798.06	26.81	0.72	13.84
C	1927.45	20.68	0.29	21.37

RMPC mixtures were prepared by blending RA at 5 and 10% dosages by weight of coarse aggregates in the control PC mix. The final mixture designation and mix proportions are presented in Table 3.

Table 3. Mixture designation and mix proportions

Sample ID	Coarse aggregates size (mm)	w/c ratio	a/c ratio	Cement (kg/m³)	Coarse aggregates (kg/m³)	RA (kg/m³)	Water (kg/m³)	SP (kg/m³)
Control PC	10–6.7	0.33	3.75	325	1218.75	–	107.25	0.81
RMPC-5	10–6.7	0.33	3.75	325	1218.75	60.94	107.25	0.81
RMPC-10	10–6.7	0.33	3.75	325	1218.75	121.88	107.25	0.81

2.3 Test Methods

The PC and RMPC were tested for hardened density and porosity, permeability, and compressive strength. The test methods used during this study are discussed in the subsequent sections.

2.3.1 Hardened Density and Porosity

The hardened density and porosity tests were performed in accordance with ASTM C1754 (ASTM 2012). The test setup used for porosity measurements of PC and RMPC cylindrical test specimens is shown in Fig. 4.

Fig. 4. Hardened porosity test setup for cylindrical specimens

2.3.2 Permeability

Permeability test on the cylindrical specimens was performed based on the falling head permeability test principle using an in-house fabricated falling head permeameter (Fig. 5). The permeability coefficients were calculated using Eq. (1), where K is the permeability coefficient (cm/s), a is the cross-sectional area of the acrylic tube (cm^2), A is the cross-sectional area of the test specimen (cm^2), L is the average height of the test specimen (cm), t is the time taken by water to fall from an initial head to final head (s), and h_1 and h_2 are the initial and final head across the test specimen (cm), respectively.

$$K = \frac{al}{At} \ln\left(\frac{h_1}{h_2}\right) \tag{1}$$

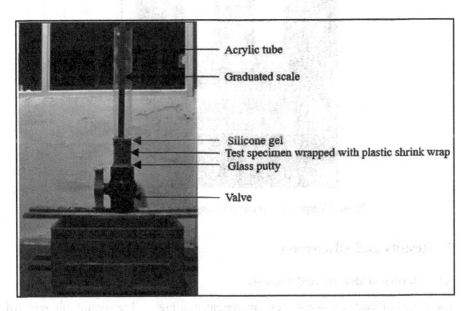

Fig. 5. Permeability test setup

2.3.3 Compressive Strength

The compressive strength of RMPC and PC mixtures was measured in accordance with ASTM C39 (ASTM 2020). However, the standard loading rate of 0.25 MPa/s as given in ASTM C39 was modified to 0.10 MPa/min considering the inherent porous network structure of PC and RMPC to minimize any detrimental impacts of shock loading, which may lead to sudden failure of the materials (Singh et al. 2019). Figure 6 presents the compressive strength test setup of the capped PC specimen.

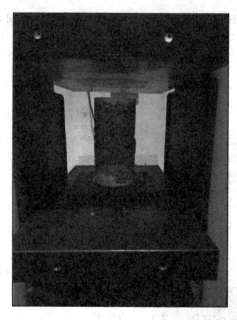

Fig. 6. Compressive strength testing setup of PC specimen

3 Results and Discussion

3.1 Hardened Density and Porosity

The results of hardened density test are presented in Fig. 7. The control mix reported the highest density of 1927.45 kg/m^3. The addition of RA decreased the density of the mix, which is in agreement with the observations reported in a previous study (Gesoğlu et al. 2014a). The reduction in density may be attributed to the lower specific gravity of RA compared to that of the coarse aggregates. It was observed that the density of RMPC decreased with increasing rubber content in the mix. At 5 and 10% rubber dosages, the density of RMPC was about 7 to 12% lower than that of the control PC mix.

Further, the porosity of all the mixtures was higher than the recommended minimum porosity level of 15% (ACI 2010; NRMCA 2004). As shown in Fig. 8, the porosity of the control PC mix increased with the inclusion of RA, which is in contradiction with the findings of past studies (Gesoğlu et al. 2014b; Mondal and Biligiri 2018). This may be attributed to the fact that due to the size of RA used in this study, RA replaced some coarser aggregates in the matrix instead of occupying the open pores, resulting in an increase in porosity. The porosity of RMPC-5 and RMPC-10 mixtures increased by about 7 and 15%, respectively, compared to that of the control PC mix.

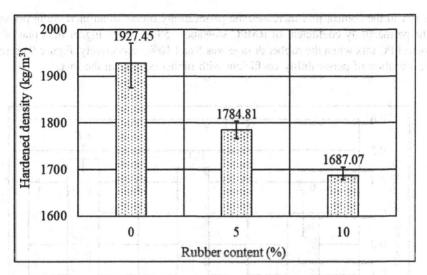

Fig. 7. Hardened density vs. rubber content

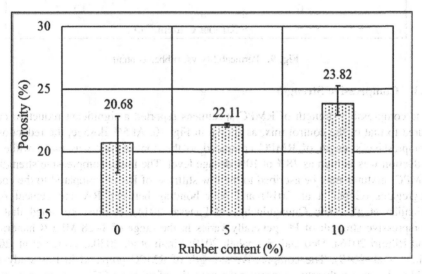

Fig. 8. Porosity vs. rubber content

3.2 Permeability

The permeability coefficient of PC typically varies in the range of 0.20–0.54 cm/s (ACI 2010; Singh et al. 2020a, 2019), although higher values have also been reported by other researchers (Chandrappa et al. 2018; Ibrahim et al. 2014; Singh et al. 2019; Vaddy et al. 2021; Yeih and Chang 2019; Zhou et al. 2019). The permeability coefficient of the PC and RMPC mixtures was between 0.29 and 0.47 cm/s. The inclusion of RA in the control mix increased the permeability owing to an increase in porosity. The permeability coefficient of RMPC was about 53 and 63% higher than that of the control PC mix when the rubber dosage was 5 and 10%, respectively. Figure 9 presents the variation of permeability coefficient with rubber content in the mix.

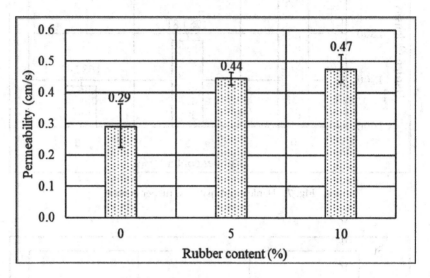

Fig. 9. Permeability vs. rubber content

3.3 Compressive Strength

The compressive strength of RMPC specimens reported a significant reduction compared to that of the control mix, as shown in Fig. 10. At 5% dosage, the reduction in compressive strength of RMPC compared to the control mix was 64%, while the reduction was as high as 78% at 10% dosage level. The lower compressive strength of RMPC mixtures may be ascribed to the low stiffness of RA as compared to the coarse aggregates (Gupta et al. 2014) and poor bonding between RA and cement paste (Ganjian et al. 2009; Onuaguluchi and Panesar 2014). It must be noted that the compressive strength of PC generally varies in the range of 1–28 MPa (Chandrappa and Biligiri 2016a; Deo and Neithalath 2010; Singh et al. 2020a; Torres et al. 2015; Zhou et al. 2019). The compressive strength of RMPC prepared in this study was within this range, thereby presenting the potential of utilizing PC as an efficient source for disposal of WRT through their utilization in the mix.

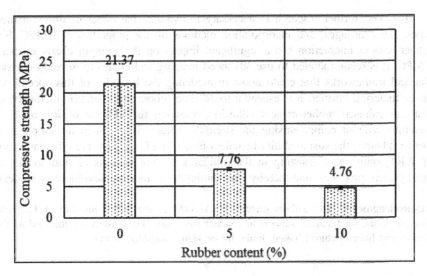

Fig. 10. Compressive strength vs. rubber content

4 Conclusions and Recommendations

Based on the laboratory tests performed during this study, the following conclusions were drawn:

- Incorporating RA in the PC decreased the density of the mix, attributed to the lower specific gravity of RA in comparison with coarse aggregates. The density of RMPC decreased with increasing rubber content in the mix while the porosity increased. The porosity of all the mixtures prepared in this study varied between 20 and 24%.
- The permeability coefficients were in the range of 0.29 to 0.47 cm/s. The permeability values of the RMPC increased with increasing rubber content, indicating a potential increase in the interconnected porous network in the matrix.
- In comparison with the control mix, the compressive strengths of RMPC were 64 and 78% lower at 5 and 10% rubber dosage, respectively. The lower compressive strength of RMPC was due to the lower stiffness of RA and poor adhesion between rubber and cement paste.

This study reported results of the experimental program to characterize RMPC encompassing single-sized RA as filler material at two dosages. The incorporation of RA increased the porosity and permeability of PC while the density and compressive strength decreased. It must be noted that the compressive strength of RMPC fell within the general range observed in the literature. Therefore, there is a potential of utilizing RA as a filler in PC mixtures that are suitable for application in construction of sidewalks and pathways, i.e., low-volume roads. Such an approach may serve as a simple and efficient strategy for WRT disposal in low-strength paving mixtures that are not subjected to vehicular traffic during their design lives and also contribute to low impact development.

However, further research is necessary to evaluate the effect of different rubber types, sizes, dosages, and incorporation methods on the properties of RMPC. Since rubber-cement interaction has a significant impact on the strength characteristics of RMPC, it is recommended to use advanced imaging techniques to develop micromechanical frameworks that could assist in modeling the behavior of this special composite material. Further, it is essential to develop rubber incorporation methodologies that can enhance rubber-cement adhesion, allowing for addition of RA in the PC materials without compromising the strength characteristics. Such an approach will further advance the sustainability benefits offered by PC by allowing efficient disposal of WRT, while also assisting in the mitigation of the impacts of traditional waste management practices, and thereby contributing to sustainable development strategies.

Acknowledgments. The authors gratefully acknowledge the Indian Institute of Technology Tirupati, India, for financial support to conduct this study. The authors are thankful to Tinna Rubber and Infrastructure Limited, India, for providing shredded rubber.

References

ACI: ACI 522R-10 Report on Pervious Concrete, American Concrete Institute (2010). (Reapproved 2011)

ASTM: ASTM C39/C39M − 20: Test Method for Compressive Strength of Cylindrical Concrete Specimens (2020). https://doi.org/10.1520/C0039_C0039M-20

ASTM: ASTM C494/C494M - 19: Specification for Chemical Admixtures for Concrete (2019). https://doi.org/10.1520/C0494_C0494M-19

ASTM: ASTM C1754/C1754M - 12: Test Method for Density and Void Content of Hardened Pervious Concrete (2012). https://doi.org/10.1520/C1754_C1754M-12

Azizian, M.F., Nelson, P.O., Thayumanavan, P., Williamson, K.J.: Environmental impact of highway construction and repair materials on surface and ground waters: case study: crumb rubber asphalt concrete. Waste Manag. **23**, 719–728 (2003). https://doi.org/10.1016/S0956-053X(03)00024-2

Birgisdóttir, H., Bhander, G., Hauschild, M.Z., Christensen, T.H.: Life cycle assessment of disposal of residues from municipal solid waste incineration: recycling of bottom ash in road construction or landfilling in Denmark evaluated in the ROAD-RES model. Waste Manag. Life Cycle Assess. Waste Manag. **27**, S75–S84 (2007). https://doi.org/10.1016/j.wasman.2007.02.016

Bonicelli, A., Fuentes, L.G., Bermejo, I.K.D.: Laboratory investigation on the effects of natural fine aggregates and recycled waste tire rubber in pervious concrete to develop more sustainable pavement materials. IOP Conf. Ser.: Mater. Sci. Eng. **245**, 032081 (2017). https://doi.org/10.1088/1757-899X/245/3/032081

Bressi, S., Fiorentini, N., Huang, J., Losa, M.: Crumb rubber modifier in road asphalt pavements: state of the art and statistics. Coatings **9**, 384 (2019). https://doi.org/10.3390/coatings9060384

Bureau of Indian Standards: IS 12269: Ordinary Portland cement, 53 grade (2013)

Chandrappa, A.K., Biligiri, K.P.: Pervious concrete as a sustainable pavement material – research findings and future prospects: a state-of-the-art review. Constr. Build. Mater. **111**, 262–274 (2016). https://doi.org/10.1016/j.conbuildmat.2016.02.054

Chandrappa, A.K., Biligiri, K.P.: Comprehensive investigation of permeability characteristics of pervious concrete: a hydrodynamic approach. Constr. Build. Mater. **123**, 627–637 (2016). https://doi.org/10.1016/j.conbuildmat.2016.07.035

Chandrappa, A.K., Maurya, R., Biligiri, K.P., Rao, J.S., Nath, S.: Laboratory investigations and field implementation of pervious concrete paving mixtures. Adv. Civ. Eng. Matls. **7**, 20180039 (2018). https://doi.org/10.1520/ACEM20180039

Deo, O., Neithalath, N.: Compressive behavior of pervious concretes and a quantification of the influence of random pore structure features. Mater. Sci. Eng. A, Spec. Top. Sect.: Local Near Surface Struct. Diffr. **528**, 402–412 (2010). https://doi.org/10.1016/j.msea.2010.09.024

Dong, Q., Huang, B., Shu, X.: Rubber modified concrete improved by chemically active coating and silane coupling agent. Constr. Build. Mater. **48**, 116–123 (2013). https://doi.org/10.1016/j.conbuildmat.2013.06.072

Ganjian, E., Khorami, M., Maghsoudi, A.A.: Scrap-tyre-rubber replacement for aggregate and filler in concrete. Constr. Build. Mater. Compat. Plasters Renders Salt Loaded Substrates **23**, 1828–1836 (2009). https://doi.org/10.1016/j.conbuildmat.2008.09.020

Gesoğlu, M., Güneyisi, E., Khoshnaw, G., İpek, S.: Investigating properties of pervious concretes containing waste tire rubbers. Constr. Build. Mater. **63**, 206–213 (2014). https://doi.org/10.1016/j.conbuildmat.2014.04.046

Gesoğlu, M., Güneyisi, E., Khoshnaw, G., İpek, S.: Abrasion and freezing–thawing resistance of pervious concretes containing waste rubbers. Constr. Build. Mater. **73**, 19–24 (2014). https://doi.org/10.1016/j.conbuildmat.2014.09.047

Gupta, T., Chaudhary, S., Sharma, R.K.: Assessment of mechanical and durability properties of concrete containing waste rubber tire as fine aggregate. Constr. Build. Mater. **73**, 562–574 (2014). https://doi.org/10.1016/j.conbuildmat.2014.09.102

Heitzman, M.: Design and construction of asphalt paving materials with crumb rubber modifier. Transportation Research Record (1992)

Ibrahim, A., Mahmoud, E., Yamin, M., Patibandla, V.C.: Experimental study on Portland cement pervious concrete mechanical and hydrological properties. Constr. Build. Mater. **50**, 524–529 (2014). https://doi.org/10.1016/j.conbuildmat.2013.09.022

Lekkas, P.T.: Discarded tyre rubber as concrete aggregate: a possible outlet for used tyres [WWW Document] (2013). https://journal.gnest.org/publication/617. Accessed 20 Sept 2020

Lo Presti, D.: Recycled Tyre Rubber Modified Bitumens for road asphalt mixtures: a literature review. Constr. Build. Mater. **49**, 863–881 (2013). https://doi.org/10.1016/j.conbuildmat.2013.09.007

Miknis, F.P., Michon, L.C.: Some applications of nuclear magnetic resonance imaging to crumb rubber modified asphalts. Fuel **77**, 393–397 (1998). https://doi.org/10.1016/S0016-2361(98)80029-X

Mondal, S., Biligiri, K.P.: Crumb rubber and silica fume inclusions in pervious concrete pavement systems: evaluation of hydrological, functional, and structural properties. JTE **46**, 892–905 (2018). https://doi.org/10.1520/JTE20170032

Nanjegowda, V.H., Biligiri, K.P.: Recyclability of rubber in asphalt roadway systems: a review of applied research and advancement in technology. Resour. Conserv. Recycl. **155**, 104655 (2020). https://doi.org/10.1016/j.resconrec.2019.104655

NRMCA: NRMCA-Concrete in Practice-38 (CIP-38), National Ready Mix Concrete Association (NRMCA) (2004)

Onuaguluchi, O., Panesar, D.K.: Hardened properties of concrete mixtures containing pre-coated crumb rubber and silica fume. J. Clean. Prod. **82**, 125–131 (2014). https://doi.org/10.1016/j.jclepro.2014.06.068

Paine, K.A., Dhir, R.K.: Research on new applications for granulated rubber in concrete. Proc. Inst. Civ. Eng. – Constr. Mater. **163**, 7–17 (2010). https://doi.org/10.1680/coma.2010.163.1.7

Putman, B.J., Neptune, A.I.: Comparison of test specimen preparation techniques for pervious concrete pavements. Constr. Build. Mater. **25**, 3480–3485 (2011). https://doi.org/10.1016/j.conbuildmat.2011.03.039

Rao, Y., Ding, Y., Sarmah, A.K., Liu, D., Pan, B.: Vertical distribution of pore-aggregate-cement paste in statically compacted pervious concrete. Constr. Build. Mater. **237**, 117605 (2020). https://doi.org/10.1016/j.conbuildmat.2019.117605

Richardson, A., Coventry, K., Dave, U., Pienaar, J.: Freeze/thaw performance of concrete using granulated rubber crumb. J. Green Build. **6**, 83–92 (2011). https://doi.org/10.3992/jgb.6.1.83

Richardson, A.E., Coventry, K.A., Ward, G.: Freeze/thaw protection of concrete with optimum rubber crumb content. J. Clean. Prod. **23**, 96–103 (2012). https://doi.org/10.1016/j.jclepro.2011.10.013

Savas, B.Z., Ahmad, S., Fedroff, D.: Freeze-thaw durability of concrete with ground waste tire rubber. Transp. Res. Rec. **1574**, 80–88 (1997). https://doi.org/10.3141/1574-11

Singh, A., Jagadeesh, G.S., Sampath, P.V., Biligiri, K.P.: Rational approach for characterizing in situ infiltration parameters of two-layered pervious concrete pavement systems. J. Mater. Civ. Eng. **31**, 04019258 (2019). https://doi.org/10.1061/(ASCE)MT.1943-5533.0002898

Singh, A., Sampath, P.V., Biligiri, K.P.: A review of sustainable pervious concrete systems: emphasis on clogging, material characterization, and environmental aspects. Constr. Build. Mater. **261**, 120491 (2020). https://doi.org/10.1016/j.conbuildmat.2020.120491

Singh, A., et al.: Uncontrolled combustion of shredded tires in a landfill – Part 2: population exposure, public health response, and an air quality index for urban fires. Atmos. Environ. **104**, 273–283 (2015). https://doi.org/10.1016/j.atmosenv.2015.01.002

Singh, A., Vaddy, P., Biligiri, K.P.: Quantification of embodied energy and carbon footprint of pervious concrete pavements through a methodical lifecycle assessment framework. Resour. Conserv. Recycl. **161**, 104953 (2020). https://doi.org/10.1016/j.resconrec.2020.104953

Torres, A., Hu, J., Ramos, A.: The effect of the cementitious paste thickness on the performance of pervious concrete. Constr. Build. Mater. **95**, 850–859 (2015). https://doi.org/10.1016/j.conbuildmat.2015.07.187

Vaddy, P., Singh, A., Sampath, P.V., Biligiri, K.P.: Multi scale in situ investigation of infiltration parameter in pervious concrete pavements. J. Test. Eval. **49**, 9 (2021)

Venudharan, V., Biligiri, K.P., Sousa, J.B., Way, G.B.: Asphalt-rubber gap-graded mixture design practices: a state-of-the-art research review and future perspective. Road Mater. Pavement Des. **18**, 730–752 (2017). https://doi.org/10.1080/14680629.2016.1182060

Yeih, W., Chang, J.J.: The influences of cement type and curing condition on properties of pervious concrete made with electric arc furnace slag as aggregates. Constr. Build. Mater. **197**, 813–820 (2019). https://doi.org/10.1016/j.conbuildmat.2018.08.178

Zhang, Y., Li, H., Abdelhady, A., Yang, J.: Comparative laboratory measurement of pervious concrete permeability using constant-head and falling-head permeameter methods. Constr. Build. Mater. **263**, 120614 (2020). https://doi.org/10.1016/j.conbuildmat.2020.120614

Zhou, H., Li, H., Abdelhady, A., Liang, X., Wang, H., Yang, B.: Experimental investigation on the effect of pore characteristics on clogging risk of pervious concrete based on CT scanning. Constr. Build. Mater. **212**, 130–139 (2019). https://doi.org/10.1016/j.conbuildmat.2019.03.310

Incorporation of CFRP and GFRP Composite Wastes in Pervious Concrete Pavements

Akhil Charak[1], Avishreshth Singh[1], Krishna Prapoorna Biligiri[1(✉)], and Venkataraman Pandurangan[2]

[1] Department of Civil and Environmental Engineering, Indian Institute of Technology, Tirupati, Tirupati, India
bkp@iittp.ac.in

[2] Department of Mechanical Engineering, Indian Institute of Technology, Tirupati, Tirupati, India
raman@iittp.ac.in

Abstract. Over the past decade, utilization of carbon fiber reinforced polymer (CFRP) and glass fiber reinforced polymer (GFRP) composites in aerospace, automotive, and wind energy sectors has led to generation of huge quantities of composite wastes (CW) at their end-of-life. This poses serious threat to environment and society, calling for identification of eco-friendly and cost-effective strategies to recycle CW into various applications. Therefore, the objective of this study was to recycle CFRP-CW and GFRP-CW in pervious concrete (PC) paving mixtures. The scope of the effort included addition of CFRP-CW and GFRP-CW into the PC mix at 0.33, 0.65, and 1% by volume to investigate their effects on hydrological and mechanical characteristics of the material. For GFRP-CW, porosity and permeability increased at the highest dosage by a magnitude of about 1 and 8.5%, respectively. However, for CFRP-CW, 1% increment in porosity was recorded at the lowest dosage, while permeability decreased by about 5–10% at all dosages. Furthermore, compressive strength increased by about 10–23% at all dosages of GFRP-CW, whereas it reduced by about 5% at lowest and highest dosages of CFRP-CW. Based on the test results, the optimum dosages for GFRP-CW and CFRP-CW were found to be 0.33 and 0.65%, respectively. Further, the properties of CW modified PC mixtures were within the general range for PC, reported in the literature. This study demonstrated that utilization of CFRP-CW and GFRP-CW in PC would serve as an environment-friendly and cost-effective strategy to recycle the CW in paving applications. Also, it is envisaged that this study will help formulate a framework for development of CW reinforced pavement materials from a sustainability perspective.

Keywords: Pervious concrete · Carbon fiber reinforced polymer · Glass fiber reinforced polymer · Composite wastes · Recycling · Porosity · Permeability · Compressive strength

1 Introduction

The increased utilization of composite materials in aerospace and wind energy sectors has led to the generation of a large proportion of composite wastes (CW) after their end-of-life (EOL) (Gopalraj and Kärki 2020; Monteiro et al. 2018; Yang et al. 2020; Alves et al. 2017). Studies have shown that maximum amount of CW belong to the families of glass fiber reinforced polymers (GFRP) and carbon fiber reinforced polymers (CFRP), which are highly stiff and durable materials attributed to their high strength-to-weight ratio and resistance against chemicals (Alves et al. 2017; Bank et al. 2018). These CFRP-CW and GFRP-CW are usually discarded into landfills leading to ecological imbalance (Bledzki and Gassan 1999; Kabir et al. 2012). Hence, there is a significant need to formulate eco-friendly solutions for disposal of CW. An innovative strategy could be recycling CW in pavement materials such as pervious concrete (PC), which is a heterogeneous mixture composed of aggregates, cement, and water with little or no fines providing a porous internal structure to the material. The porous nature of material renders several benefits such as reduction in surface runoff, groundwater recharge, mitigation of noise generation due to interaction between tires and pavement surface, and reduction in the urban heat island effects (UHI) (Haselbach et al. 2011; Kevern et al. 2012; Marolf et al. 2004; Singh et al. 2020). Despite its enormous environmental benefits, the exclusion of fine aggregates from PC mix results in lower strength compared to conventional concrete. In the past, various researchers have utilized different materials and additives such as fibers, rice husk ash, cured carbon fiber composite materials (CCFCM), and silica fume to improve the properties of PC mix (Hesami et al. 2014; Ibrahim et al. 2019; Kevern et al. 2008, 2015; Nassiri et al. 2021; Rangelov et al. 2016; Rehder et al. 2014; Rodin et al. 2018).

Rangelov et al. (2016) utilized CCFCM in PC up to 1.5% by volume of concrete mixture and reported that the 28-day compressive strength improved by 4–11%. In addition, there was an increase in the infiltration rate along with lower porosity and a significant improvement in the workability of the modified PC mixtures over the control ones. In another study, using post-industrial CCFCM in the proportion of 3, 4, and 5% by volume (Rodin et al. 2018) showed an improvement in the flexural strength (36–65%) and toughness index (41–54%) after 28 days over the control mix. It was reported that the addition of 4% CCFCM led to the optimal improvement in the properties of PC. Shareedah et al. (2019) studied the feasibility of implementing recycled CCFCM obtained from aerospace industry in PC mixtures. The test section consisted of two PC lanes of 30.6 m × 3.7 m × 0.115 m constructed with three mixtures: a control mixture (M1), PC mixture reinforced with 0.27% CF (M2), and PC mixture reinforced with 0.4% CF by volume of the mix (M3). Compression test carried out on field cores of M1, M2, and M3 mixes showed 14–53%, 6–52%, and 33–61% lower compressive strength in comparison with the laboratory specimens, due to the differences in their compaction methods. The tensile test results showed that the specimens constructed with CCFCM mixture achieved around 77% strength of control specimen at 28 days with increase in porosity of 20% and 25% for M2 and M3 mixes, respectively. The modulus of rupture of all the beam specimens were approximately similar ranging from 1.6–1.8 MPa with an increase of 3% and 15% in porosity of M2 and M3 mix in comparison with the control mix. The comparative results of infiltration test on field and laboratory specimens showed that the infiltration rate was 9.1, 26.3, and 34.4% higher in M1, M2, and M3 mixes relative to the laboratory specimens due to high porosity. Further, the average surface deflection of M2 and M3 mixes was reduced by 30.8 and 7.7%, respectively compared to M1 mix when tested using Light Weight Deflectometer (LWD).

A recent study examined the mechanical and durability characteristics of PC after replacing natural coarse aggregates (NCA) with CCFCM in the proportion of 0.5, 1, and 2% by volume (Nassiri et al. 2021). CCFCMs passing 2 mm size sieve and retaining on 0.84 mm size sieve were utilized in the study. An increase in the porosity with increase in the proportion of CCFCMs was reported. However, the compressive strength was found to improve at the proportion of 0.5% and then further decreased with increase in CCFCMs. Importantly, the mix with 2% CCFCMs showed excellent performance against freeze-thaw, which was attributed to the additional tensile strength provided by CCFCMs.

Sawant et al. (2017) studied the effect of glass fibers on the compressive strength of PC after the addition of glass fibers as a replacement of cement for volume fraction between 1 and 2%. The compressive strength was evaluated for different proportions of glass fibers at 7, 14, and 28 days, which indicated that strength increased with addition of glass fibers and the optimum dosage was found to be 1.5–2%. In another study, Shende et al. (2019) studied the permeability and compressive strength characteristics of PC after incorporating rice husk ash and glass fibers. The rice husk ash was incorporated in the proportion of 8 and 10%, while glass fibers were incorporated in the proportion of 0.2, 0.4, and 0.6%. The glass fibers were 40–50 mm long and 0.15 mm in diameter. The results showed that permeability decreased with increasing glass fiber content, while the compressive strength increased with increasing proportion of rice husk ash while decreased with the increase in the glass fiber content.

Ibrahim et al. (2019) studied the porosity, hydraulic conductivity, and mechanical strength (compressive, split tensile, and flexural) characteristics of PC mix after incorporating recycled concrete aggregates (RCA), silica fume, and glass fibers. RCA was incorporated in the proportion of 25, 50, 75, and 100% by weight of NCA, silica fume in the proportion of 2.5, 5, 10, and 12% by weight of cement, and glass fibers in the proportion of 0.05, 0.1, 0.15, and 0.2% by volume. The results showed that the hydrological properties improved due to the replacement of NCA by RCA and improved further with the increase in RCA content. On the other hand, mechanical strength improved only up to 50% replacement of natural aggregates. Further, porosity increased with addition of glass fibers while hydraulic conductivity decreased.

Based on several investigations, it was understood that past studies focused on the development of PC mixtures by extracting fibers from composites, which is both energy intensive and expensive. Therefore, there is a need to formulate alternative ways to incorporate composite waste in PC. Also, it is essential to understand if the incorporation of mechanically crushed CFRP-CW and GFRP-CW in PC mixtures without extracting the fibers will promote their recycling prowess without degrading the quality of PC. Therefore, the objective of this research was to utilize mechanically crushed CFRP-CW and GFRP-CW in PC mixtures and study their compatibility in the PC material. In this direction, CFRP/GFRP-CW each were incorporated in three different volumetric proportions: 0.33, 0.65, and 1% by volume of the PC mix. Three replicate specimens were prepared, each of one control and six CW modified PC (CW-PC) mixtures. The scope of experimentation included estimations of hardened porosity, permeability and compressive strength characteristics of CW-PC.

2 Experimental Program

2.1 Materials and Mix Proportions

CFRP-CW and GFRP-CW were obtained in the form of plates from National Aerospace Laboratories, Bangalore, India, and shredded to the size ranging from 10–4.75 mm, as shown in Figs. 1(a) through 1(f). The specific gravity (per IS:2386 – Fig. 1(g)) of CFRP-CW and GFRP-CW were found to be 1.49 and 1.30, respectively with water absorption capacities (per IS:2386 – Fig. 1(g)) of 10.1 and 8.6%, respectively.

Fig. 1. CW from NAL: (a) GFRP-CW; (b) CFRP-CW; (c) shredded GFRP-CW; (d) shredded CFRP-CW; (e) sieve analysis GFRP-CW; (f) sieve analysis CFRP-CW; (g) specific gravity and water absorption test

Coarse aggregates were procured from a local quarry in Tirupati, Andhra Pradesh, India with size varying between 12.5 and 4.75 mm. The aggregate gradation curve is shown in Fig. 2. The specific gravity and water absorption of coarse aggregates as per IS:2386 were 2.67, and 0.18%, respectively. Portland cement 53-grade conforming to IS:12269 was used. The specific gravity of cement was 3.12, whereas the initial and final setting times were 130 and 320 min, respectively. In addition, sulfonated naphthalene formaldehyde-based superplasticizer was used as an admixture at a dosage of 0.4% by mass of cement to reduce the water requirement during production.

The PC mixtures were prepared using a water-to-cement ratio of 0.30 at an aggregate-to-cement content of 3.75. Based on the trial mixtures, the cement content was fixed at 350 kg/m^3. Further, CFRP-CW and GFRP-CW were added in proportions of 0.33, 0.65, and 1% by volume of the mix representing low, intermediate, and high

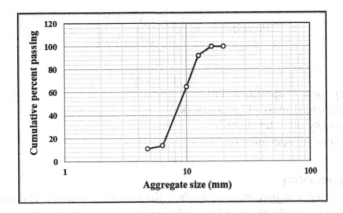

Fig. 2. Aggregate gradation chart

dosages, respectively. In total, three samples of control PC and three specimens each for every dosage of CFRP-CW and GFRP-CW were prepared, and the porosity, permeability, and compressive strength were measured. In total, 21 cylindrical specimens of control PC and CW-PC mixtures with 100 mm diameter and 200 mm height were prepared for testing and evaluation of different properties.

2.2 Methodology

2.2.1 Porosity

The porosity of specimens was evaluated in accordance with ASTM C1754 by using Eq. (1). First, the PC specimens were submerged in water for 2 h and tapped gently with rubber mallet to remove the entrapped air from the pores. The submerged mass of the specimens was recorded as shown in Fig. 3. Later, the specimens were kept in an oven at 38 °C and weighed every 24 h until the difference in the weight between two successive readings was less than 0.5%. The porosity of the specimen was computed using Eq. (1).

Fig. 3. Submerged cylindrical specimen

$$\text{Porosity}\,(\%) = \left[1 - \left\{\frac{A-B}{\rho_w * V}\right\}\right] * 100 \qquad (1)$$

where:

A = dry mass of specimen (kg)
B = submerged mass of specimen (kg)
ρ_w = density of water (kg/m^3)
V = volume of specimen (m^3).

2.2.2 Permeability

Permeability test was performed using the falling head test method. The test specimens were wrapped with shrink wrap from sides and held firmly with a duct tape to ensure vertical flow of water. The PC specimens were saturated before testing to eliminate any air within the pores and attain true permeability rates. Glass putty and silicone gel were applied on the interface of test specimens and the test setup to avoid water leakage followed by pouring of water on the specimen to make it air free. Further, an acrylic sheet with a steel ruler attached to it was fixed over PC specimens and the interface was sealed with silicone gel. The wrapped PC specimens used for permeability tests and the permeability test setup are shown in Fig. 4. The time taken by the water to flow from an initial head of 26 cm to a final head of 1 cm was recorded, and the permeability was calculated using Eq. (2).

Fig. 4. Permeability (a) wrapped test specimens; (b) permeability test setup

$$K = \left(\frac{2.3 \times a \times L}{A \times t}\right) * \log_{10}\left(\frac{H}{h}\right) \qquad (2)$$

where:

K = permeability (mm/s)
a = area of standpipe (mm^2)
L = length of concrete specimen (mm)

A = cross-sectional area of concrete specimen (mm^2)
t = time required to flow the water through concrete specimen (s)
H = initial water head in standpipe using same water head reference (mm)
h = final water head in standpipe using same water head reference (mm).

2.2.3 Compressive Strength

Compressive strength test was performed on cylindrical PC specimens as per ASTM C39. Prior to the test, the PC specimens were capped with gypsum (in accordance with ASTM C617) on top and bottom faces to ensure uniform distribution of load. The capped test specimens, specimen placed in the testing machine, and specimen subjected to compressive forces are shown in Fig. 5. The test was carried out at a loading rate of 0.1 MPa/s and the compressive strength of PC specimens was computed using Eq. (3).

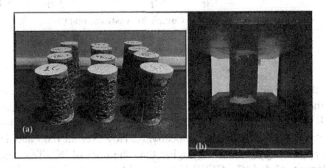

Fig. 5. Compressive strength (a) capped specimens; (b) specimen placed in the testing machine; (c) specimen subjected to compressive forces

$$f_c(\text{MPa}) = \frac{4000 \times P}{\pi d^2} \quad (3)$$

where:

P = maximum applied load (kN)
d = diameter of cylindrical specimen (mm).

3 Results and Discussion

3.1 Porosity

As observed from Fig. 6, the average porosity of control PC (28.36%) and CFRP/GFRP-CW (~27–29%) modified mixtures was found to be higher than 15% as suggested by the American Concrete Institute (ACI) (ACI. 522R-10 2010).

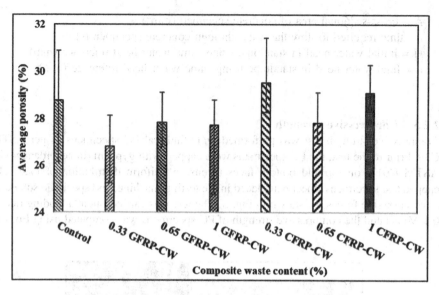

Fig. 6. CW content versus average porosity

Clearly, the porosity increased with increase in GFRP-CW content, but it was still lower than the control mix, except at the highest dosage where an increase in porosity of about 1% was observed. Further, the porosity was about 4 and 1% lower at 0.33% and 0.65% dosages of GFRP-CW, respectively. However, at 0.65 and 1% dosages of CFRP-CW, porosity was about 1% higher than the control PC. Test results indicated that at a dosage of 0.33%, GFRP-CW probably occupied the voids between the aggregates resulting in lower porosity. However, at higher contents, the porosity increased, which may be attributed to the fact that GFRP-CW due to (a) their shape replaced some proportion of coarse aggregates instead of occupying the voids, and (b) their high strength-to-weight ratio offered more resistance to compaction. These results corroborate with the findings reported by (Ibrahim et al. 2019), where the porosity increased with increasing proportion of glass fibers. The respective increase in porosity was reported as 1.1 and 1.09% at lowest and highest dosages of CFRP-CW. The different variations in porosity due to the incorporation of CFRP/GFRP-CW could be attributed to the difference in the structural morphology of CFRP-CW and GFRP-CW. CFRP-CW was found to be in filament form with a diameter between 2 and 4 mm while GFRP-CW was found to be in the form of chips with diameter ranging from 5–8 mm. The variation in the dispersion of CFRP/GFRP-CW due to their different internal structures may be another reason for variation in the porosity.

3.2 Permeability

Figure 7 shows that the permeability of CFRP/GFRP-CW modified mixes was lower than control mix at all dosages and types of CW, except at the highest dosage GFRP-CW. In general, the permeability for control (0.91 cm/s) and CW modified PC (0.8–

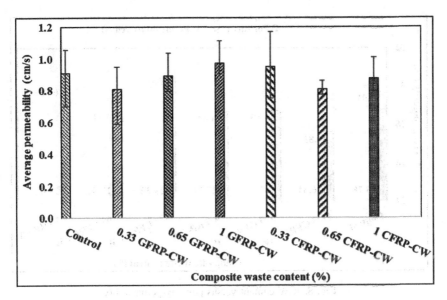

Fig. 7. CW content versus permeability

1 cm/s) mixes was higher than the common range of 0.2–0.54 cm/s, as reported in the literature (ACI. 522R-10 2010).

Permeability increased with increasing GFRP-CW content as the porosity also increased. A similar trend was also observed in the literature (Rangelov et al. 2016; Rodin et al. 2018). In a previous study by Ibrahim et al. (2019), it was found that the hydraulic conductivity of PC decreased with the increase in glass fiber content. However, an opposite trend was observed in this study due to the difference in the diameter of the reinforcing elements. The glass fiber used in that study was reported to be 0.1 mm while the diameter of GFRP-CW used in this study was around 5–8 mm. The change in permeability due to incorporation of CFRP-CW also followed a similar trend to the change in porosity. The porosity could be increased if CFRP-CW can be incorporated in transverse direction, but in this case, they acted as a barrier to the vertical flow of water leading to the reduction in permeability. Thus, the dispersion of CW in PC mixes and interconnectivity of voids also affect the permeability characteristics. Furthermore, it can also be inferred that GFRP-CW was found to be better dispersed than CFRP-CW and are capable of improving the interconnectivity of the void structure.

The variation in permeability was found to be similar to that of porosity and in order to confirm that permeability increased with increasing porosity, the porosity-permeability for control mix as well as GFRP and CFRP-CW reinforced mix were compared and depicted in Fig. 8. As observed, permeability increased with increasing porosity, while decreased with decreasing porosity.

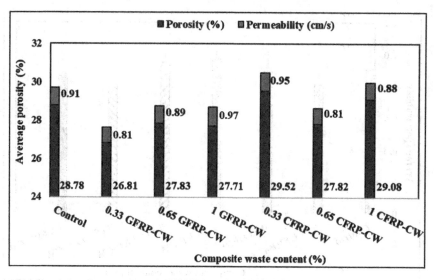

Fig. 8. CW content versus porosity-permeability

3.3 Compressive Strength

In general, the average compressive strength of GFRP-CW modified mix was higher than the control mix while the compressive strength of CFRP-CW modified mix was lower than the control mixes, as shown in Fig. 9.

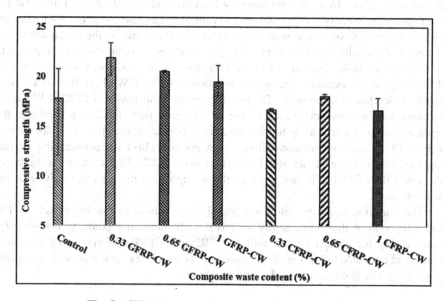

Fig. 9. CW content versus average compressive strength

The average compressive strength of GFRP-CW modified PC increased by around 10–23% while the average compressive strength of CFRP-CW modified mix decreased by around 5–6%. Further, the average compressive strength of PC decreased by around 5–6% with increasing GFRP-CW content. The compressive strength decreased with increasing GFRP-CW content because the porous nature of PC was more pronounced with increase in the GFRP-CW content. Likewise, the variation in the compressive strength due to increasing CFRP-CW also corresponded to the variation of porosity due to increasing CFRP-CW content. A similar finding was reported in the previous studies in which the compressive strength of PC decreased due to increase in the porosity (Ibrahim et al. 2019; Rangelov et al. 2016; Rodin et al. 2018).

4 Conclusions and Recommendations

The addition of CW without the extraction of fibers was found to be a suitable approach for the disposal of CFRP-CW and GFRP-CW in PC mixtures. The addition of CW resulted in an improvement in the hydrological properties of PC, without significantly compromising on the compressive strength. The following were the major findings of the study.

1. The optimum dosage for CFRP-CW was 0.65%, where the reduction in compressive strength was lower compared to the other dosages, its value being close to that of the control PC mix, and porosity and permeability being about 28% and 0.8 cm/s, respectively. These values are higher than those recommended by ACI and fall within the general range for PC mixtures reported in the literature.
2. For GFRP-CW, a dosage of 0.33% was found to be optimum as the highest compressive strength (>20 MPa) was achieved with porosity and permeability being well above the permissible limits prescribed by ACI and several other researchers in the past.
3. CW-PC materials have the potential to allow rapid drainage of surface runoff. Further, their porous nature is expected to impart high friction, which would reduce the risk of accidents due to hydroplaning. Thus, it can be concluded that recycling CW in PC is an efficient strategy to address the various environmental burdens associated with their disposal in landfills.

However, additional research must be undertaken to optimize the size, dosage, and proportions of CW in PC and other paving materials. Most importantly, extensive laboratory investigations are required to study the effect of aggregate gradations, w/c ratio, and CW content on the properties of PC. Additionally, the effects of incorporating CW in PC as a replacement of natural aggregates must also be explored. Incorporation methodologies that augment the interaction between CW and cement must be identified for development of high-performance PC mixtures. Such an approach is expected to minimize the energy consumption, greenhouse gas emissions, and costs involved in extracting the fibers from composite materials, while would facilitate conventional CW management practices, and ultimately promote the development of low-impact technologies for recycling CW in pavement infrastructure.

Acknowledgments. The authors gratefully acknowledge the National Aerospace Laboratories, Bangalore, India, for providing composite wastes. The authors are thankful to the Maxin Induatries, Coimbatore, India for shredding the composite wastes.

References

ACI. 522R-10: Report on pervious concrete. ACI Committee, 38 (2010)

Alves, S.M.C., da Silva, F.S., Donadon, M.V., Garcia, R.R., Corat, J.E.: Process and characterization of reclaimed carbon fiber composites by pyrolysis and oxidation, assisted by thermal plasma to avoid pollutants emissions. J. Compos. Mater. **52**(10), 1379–1398 (2017)

ASTM C39: Standard test method for compressive strength of cylindrical concrete specimens. ASTM International (2021)

ASTM C617: Standard practice for capping cylindrical concrete specimens. ASTM International (2015)

ASTM C1754: Standard test method for density and void content of hardened pervious concrete. ASTM International (2012)

Bank, L.C., et al.: Concepts for reusing composite materials from decommissioned wind turbine blades in affordable housing. Recycling **3**(1), 3–13 (2018)

Bledzki, A.K., Gassan, J.: Composites reinforced with cellulose based fibres. Prog. Polym. Sci. **24**(2), 221–274 (1999)

Gopalraj, S., Kärki, T.: A review on the recycling of waste carbon fibre/glass fibre-reinforced composites: fibre recovery, properties and life-cycle analysis. SN Appl. Sci. **2**(3), 433 (2020)

Haselbach, L., Boyer, M., Kevern, J.T., Schaefer, V.R.: Cyclic heat island impacts on traditional versus pervious concrete pavement systems. Transp. Res. Rec. J. Transp. Res. Board **2240**, 107–115 (2011)

Hesami, S., Ahmadi, S., Nematzadeh, M.: Effects of rice husk ash and fiber on mechanical properties of pervious concrete pavement. Constr. Build. Mater. **53**, 680–691 (2014)

Ibrahim, H.A., Mahdi, M.B., Abbas, B.J.: Performance evaluation of fiber and silica fume on pervious concrete pavements containing waste recycled concrete aggregate. Int. J. Adv. Technol. **10**(2), 1–9 (2019)

Kabir, M.M., Wang, H., Lau, K.T., Cardona, F.: Chemical treatments on plant-based natural fibre reinforced polymer composites: an overview. Compos. Part B: Eng. **43**(7), 2883–2892 (2012)

Kevern, J.T., Biddle, D., Cao, Q.: Effects of macrosynthetic fibers on pervious concrete properties. J. Mater. Civ. Eng. **27**(9), 1–6 (2015)

Kevern, J.T., Haselbach, L., Schaefer, V.R.: Hot weather comparative heat balances in pervious concrete and impervious concrete pavement systems. J. Heat Island Inst. Int. **7**(2), 231–237 (2012)

Kevern, J.T., Schaefer, V.R., Wang, K., Suleiman, M.T.: Pervious concrete mixture proportions for improved freeze-thaw durability. J. ASTM Int. **5**(2), 12 (2008)

Marolf, A., Neithalath, N., Sell, E., Wegner, K., Weiss, J., Olek, J.: Influence of aggregate size and gradation on the acoustic absorption of enhanced porosity concrete. ACI Mater. J. **101**(1), 82–91 (2004)

Monteiro, S.N., et al.: Fique Fabric: a promising reinforcement for polymer composites. Polymers **10**(3), 246 (2018)

Nassiri, S., Shareedah, O.A., Rodin, H.I., Englund, K.: Mechanical and durability characteristics of pervious concrete reinforced with mechanically recycled carbon fiber composite materials. Mater. Struct. **54**, 107–124 (2021)

Rangelov, M., Nassiri, S., Haselbach, L., Englund, K.: Using carbon fiber composites for reinforcing pervious concrete. Constr. Build. Mater. **126**, 875–885 (2016)

Rehder, B., Banh, K., Neithalath, N.: Fracture behavior of pervious concretes: the effects of pore structure and fibers. Eng. Fract. Mech. **118**, 1–16 (2014)

Rodin, H.I., Rangelov, M., Nassiri, S., Englund, K.: Enhancing mechanical properties of pervious concrete using carbon fiber composite reinforcement. J. Mater. Civ. Eng. **30**(3), 04018012 (2018)

Sawant, M., Dhande, D., Sabde, S., Bhalsing, R., Yadav, P., Ranjan, N.: Pervious concrete with glass fiber. Int. J. Res. Advent Technol. **5**(5), 26–29 (2017)

Shareedah, O.A., Nassiri, S., Chen, Z., Englund, K., Li, H., Fakron, O.: Field performance evaluation of pervious concrete pavement reinforced with novel discrete reinforcement. Case Stud. Constr. Mater. **10**, e00231 (2019)

Shende, A., et al.: Analysis of strength of pervious concrete by adding rice husk ash and glass fibre. Int. J. Res. Eng. Sci. Manag. **2**(3), 534–537 (2019)

Singh, A., Sampath, P.V., Biligiri, K.P.: A review of sustainable pervious concrete systems: emphasis on clogging, material characterization, and environmental aspects. Constr. Build. Mater. **261**, 120491 (2020)

Yang, Q., Hong, B., Lin, J., Wang, D., Zhong, J., Oeser, M.: Study on the reinforcement effect and the underlying mechanism of a bitumen reinforced with glass fibre chips. J. Clean. Prod. **251**, 119768 (2020)

Asphalt Modified with Recycled Waste Plastic in South Africa Encouraging Results of Trial Section Performance

Simon Tetley[1], Tony Lewis[1], Waynand Nortje[2], Deane Koekemoer[2], and Herman Visser[1(✉)]

[1] VNA Consulting, Pinetown, South Africa
{Simon.tetley,herman.visser}@vnac.co.za
[2] Shisalanga Construction, Durban, South Africa

Abstract. The use of waste plastic is seen as a means of enhancing the visco-elastic behaviour of the binder used in hot-mixed asphalt, as well as providing significant environment benefits in reducing the quantity of plastic pollution. This initiative has drawn wide attention as it not only utilises recycled waste plastic but has been shown to improve the properties of the asphalt. The trials were constructed in two sections, one using a stone skeletal SMA mix type and the other a high modulus EME mix. The performance of the two trial sections was assessed on two occasions, a year apart, and included detailed visual assessments as well as assessments of riding quality, rut depth, and texture. The structural performance was also assessed using deflection based mechanistic analyses. Both trials have performed well to date under the large percentage of heavy vehicles transporting material to and from a nearby asphalt plant, rock quarry and Ready mix facility. It is intended to continue monitoring the trial to assess both short and long-term performance of this innovative product. Using waste plastic in asphalt is seen as a cost-effective and environmentally friendly way of reducing the quantity of waste plastic that would otherwise require disposal at landfill sites; the intention of this paper is to share the technology used to produce the waste plastic modified asphalt as well as to present the performance that can be expected from the product.

Keywords: Asphalt · Waste plastic · Environmentally friendly · Full-scale trials · Performance

1 Introduction

It is evident from a literature study that over the past 10 to 15 years there is a growing worldwide interest in the use of various types of waste plastic in hot-mixed asphalt. While it appears that India is taking the lead with the number of kilometres of road paved with asphalt modified with waste plastic, several other countries have also embarked on this technology, the main purpose being to reduce the quantity of waste plastic that would otherwise fill landfills or pollute rivers and oceans.

A South African asphalt supplier carried out extensive research into assessing the properties of various types of waste plastics, to find the most suitable type for use as a hot-mix asphalt modifier. It was found that reclaimed high-density polyethylene (rHDPE), derived from used milk containers, caps and packaging, which is blended into penetration grade bitumen produced the best results in terms of quantity of rHDPE together with improvements to the asphalt performance characteristics.

This paper includes an overview of the properties of the rHDPE-modified binder as well as the properties of the respective asphalt mixes used in two trial sections. It covers the investigations of the trial sections carried out initially in May 2020, as well as the more extensive investigations undertaken a year later, in June 2021.

2 Properties of rHDPE Binder

As already discussed, the waste plastic used in the mix design for these trials was recycled from milk bottles, caps and packaging. This type of recycled plastic, High Density Polypropylene (HDPE), is supplied in pelletised form, equating to 118 two-litre milk bottles per ton of asphalt using a 6% dosage rate of rHDPE in the binder.

The double benefit from utilising a waste product while at the same time enhancing the properties of the asphalt mix, led to extensive testing culminating in a paper being presented at CAPSA'19 (Nortjé and Seforo 2019). This paper examines two ways of adding the rHDPE to the mix, firstly by adding it to the heated aggregate (known as the "dry" method) and secondly by the "wet" method, where the polymer is digested in the hot bitumen. Further research was carried out on the second technique and this was the method used in the preparation of the binder for the D755 trials. An important conclusion, after exhaustive testing of both binder and mix performance characteristics, was that 70/100 penetration grade bitumen modified with 6% rHDPE (by mass of binder) produces a binder with a near perfect visco-elastic behaviour. The Performance Grading (Bredenhann et al. 2019) of the rHDPE modified binder with other commonly used binders showed that the rHDPE modified binder's performance properties compare very well with those of bitumen modified with EVA and SBS, whilst fitting into a higher temperature range, with a Performance Grading of PG70V-22. Also notable was that the aging ratio, which is a good indicator of fatigue resistance of the HDPE modified binder, was found to be better than that of other standard modified binders, such as SBS modified binder, classified as A-P1 (TG1 Technical Guideline, November 2020).

Although the binder does not have any noticeable elastic recovery properties when tested using the ductility test, it does show very encouraging low temperature characteristics when tested using the Bending Beam Rheometer (BBR). The binder achieved a -22 °C low temperature performance grading value that would indicate high resistance to thermal cracking. It also achieves a high temperature performance grading value of 70 °C when tested with the Dynamic Shear Rheometer (DSR). The Performance Grading system takes traffic loading into account with the rHDPE modified binder falling into the "V" category (Sabita Manual 35/TRH8, June 2016) denoting its suitability for use in asphalt mixes under very heavy traffic loading conditions.

3 Hot-Mix Asphalt Design

The mix design work, including control designs, was carried out prior to the paving trials. In the case of the SMA type mix trial (hereafter referred to as Section 1) a control design was undertaken using SBS polymer modified binder (A-P1). The control mix for the EME type mix trial (hereafter referred as Section 2) was manufactured using a 10/20pen bitumen. Salient details of these mix designs are included in Table 1.

It is evident from Table 1 that the fatigue properties of EME type mix using 10/20pen bitumen and HDPE modified mix are similar. The SMA mix, with its higher binder content and film thickness compared to the EME mix achieves a far higher fatigue value. It was found that the HDPE modified binder tends to improve the compactibility of the asphalt mixes, resulting in lower void contents (VIM).

Table 1. SMA and EME type mix design – salient properties

Test/Property	SMA Control A-P1	rHDPE	EME 14 Control 10/20pen bitumen	rHDPE
Filler/Binder ratio	1.1	1.1	1.3	1.3
Optimum binder content (%)	6.3	6.4	5.3	5.4
Film thickness (µmm)	10.9	11.9	7.5	8.4
Richness modulus	–	4.1	3.5	3.4
Binder absorption (%)	0.3	0.2	0.4	0.3
Voids in mix (VIM) (%)	4.1	3.0	3.4	2.7
Voids in mixed aggregate (VMA) (%)	17.9	17.4	14.5	14.7
Voids filled with binder (VFB) (%)	75.9	81.6	78.8	79.9
ITS (kPa)	1164	1 406	–	–
Modified Lottman	0.86	0.947	0.902	0.925
Air permeability @ 7% voids	–	0.065	–	0.070
Schellenberg drainage	0.05	0.10	–	–
Stiffness modulus @ 20 °C (10 Hz)	15 564	16 223	21 500	18174
Fatigue 10 Hz, 260 µε @ 10 °C for EME and 200 µε for SMA	–	5 900 000	296	300
Hamburg Wheel Track (mm), 20 000 cycles at 50 °C	–	3.8	2.5	3.3

4 Trial Sections

The trials were undertaken on two sections of provincial road D755, the first, Section 1, which comprised a length of full width paving of around 240 m, is situated close to the asphalt plant. This was paved on 2 August 2019 and consists of Stone Mastic Asphalt (SMA) type mix that was paved to a thickness of approximately 40 mm, on top of the existing asphalt surface. This section is situated on a comparatively steep gradient of between 8% and 9%, furthermore it includes two speed bumps, which force vehicles to travel at low speeds. The second trial section, Section 2, was paved on 14 August 2019, on a moderate grade near the summit of D755. This section of D755 was purposely selected to test the rut-resistant properties of the rHDPE modified asphalt mix as the existing pavement was already heavily rutted (Fig. 1). Finally, the length of Section 1 was extended on 23 August 2019, using the same SMA mix as previously paved on this section.

Fig. 1. Trial Section 2 - severely rutted section of D755

Before paving commenced, areas on Section 2 exhibiting the most severe rutting were milled off to remove most of the rutting before the rHDPE modified asphalt was paved (Fig. 2). In this case the EME type asphalt mix was used. It was anticipated that some initial rutting would occur but that this would be a good test to find out whether the high-modulus asphalt could bridge the uneven surface and previously distressed pavement.

This trial section was also intended to check the low temperature crack resistance of the modified mix. To do this one of the lanes was paved a week before the other, purposely leaving the longitudinal joint uncut, as opposed to the normal good practice of trimming back the joint, before paving the adjacent lane.

Fig. 2. Trial Section 2 - paving the EME trial section after milling off severe rutting

5 Traffic Loading

As the trial sections are the sole access to a large asphalt plant they are trafficked by trucks hauling asphalt as well as delivering bitumen and aggregate. D755 is also on the NPC Sterkspruit Quarry's haul route, which adds to the extreme traffic loading. Besides producing crushed aggregate, a readymix concrete plant is also established in the quarry and D755 is therefore also trafficked by readymix trucks and large cement haulers.

The cumulative traffic loading for the period since the respective trial sections were paved, to date, has been calculated using weighbridge data from the asphalt plant and quarry as well as a classified 7-day traffic count. These traffic counts indicate that the trial sections had been subjected to approximately 200 000 Equivalent 80 kN Standard Axle Loads (E80s) from the time of their construction in August 2019 until June 2021.

6 Assessment of Performance

6.1 Scope of Assessments

The first assessment of the functional performance of the two trial sections was carried out on 7 May 2020 by means of a Network Survey Vehicle (NSV) which enables automated capture of rut depth, surface texture, and riding quality data. High-resolution photographs of the road are collected at predetermined intervals, in this instance every 20 m.

The second assessment of the two trial sections was undertaken on 5 June 2021, just over a year after the first assessment. The same parameters as the first trial were assessed however in addition deflection measurements were taken at 20 m intervals. An inspection pit was also excavated to a depth of 1 m in each of the trial sections. Besides establishing accurate layer thicknesses, disturbed samples were taken for laboratory testing and classification. It was then possible to utilise the information provided by the deflection measurements and inspection pits to carry out mechanistic modelling of the pavement structure.

The data from the NSV was analysed for both the trial sections using performance criteria from SAPEM, Chapter 6 (South African Pavement Engineering Manual 2014) for rut depth, texture (MPD) and riding quality (IRI).

6.2 Automated Data Capture: Rut Depth, Texture (MPD) and Roughness (IRI)

Analysis of Rut Depths

Rutting is an important parameter for gauging the condition of the pavement, and particularly the asphalt's ability to withstand the loads imposed by heavy, slow moving traffic. It is also a means of assessing the asphalt layer's ability to bridge weaker underlying layers. Rut depth measurements, undertaken by an NSV in May 2020 and again in June 2021, showed that only moderate rutting with average rut depths of 4.3 mm on Section 1 and 4.7 mm on Section 2 at June 2021.

Analysis of Mean Profile Depths

Surface texture tends to be initially related to the type of asphalt mix; in very broad terms stone skeletal type asphalt mixes, such as SMA, can be expected to exhibit higher Mean Profile Depths (MPD) than, for instance, continuously graded mixes. It is however possible for stone skeleton mixes to "close up" and lose texture under heavy traffic loadings. The texture depths are thus a useful measure of the asphalt mix's ability to withstand heavy traffic load.

The measurement of Mean Profile Depth was carried out using an NSV in May 2020 and again in June 2021. Whilst the texture of the SMA asphalt mix type on Section 1 showed little tendency to close up since construction, with average MPD values of 0.8 mm, some smoothing of the EME type mix on Section 2, with lower MPD values of 0.4 mm in the negative traffic lane, was evident. It should however be taken into account the EME type asphalt mixes are intended for base layers and not to act as a surfacing and slight smoothing under traffic could be expected.

Analysis of Roughness (IRI)

Riding quality or roughness, expressed in terms of International Roughness Index (IRI), is usually related to the quality of the paving process, particularly in the initial years following construction, such as in these trials. It would however be unrealistic to judge paving quality over such short trial sections and it is very unlikely that roughness could be atributed to longitudinal deformation of the asphalt mix. Instead, paving over rough surfaces most probably caused the rather high average IRI values 4.4 m/km., indicating poor riding quality.

6.3 Detailed Visual Assessment

Detailed visual assessments were undertaken using the criteria of both the TRH 12 (Flexible Pavement Rehabilitation Investigation and Design 1997) and TMH 9 (Pavement Management Systems: Standard Visual Assessment Manual for Flexible Pavements 1992). Very little deterioration of the trial sections was found in the visual assessment carried out in May 2020 however some moderate distress was evident in the visual assessment undertaken in June 2021.

Very little cracking has occurred on either of the trial sections, only one longitudinal crack, over a maximum length of 20 m, is evident in Section 2. The only other visual distress in either of the sections was minor rutting and aggregate loss.

6.4 Analysis of Deflection Measurements

As discussed previously in this paper, deflection measurements were undertaken using a Falling Weight Deflectometer (FWD) along each of the trial sections, at intervals of 20 m, staggered between the two traffic lanes. In Table 2 the deflection bowl of the two trial sections are benchmarked in accordance with the criteria suggested by (Horak 2008) into "sound", "warning" and "severe" categories for granular base type pavements. The 80th percentile deflection index values were used in the analysis.

The analysis uses the maximum deflection as well as three zones namely:
- Base Layer Index (BLI) = $D_0 - D_{300}$
- Middle Layer Index (MLI) = $D_{300} - D_{600}$
- Lower Layer Index (LLI) = $D_{600} - D_{900}$

Where: D_0 = Deflection under the load
D_{300} = Deflection at 300 mm sensor
D_{600} = Deflection at 600 mm sensor
D_{900} = Deflection at 900 mm sensor.

Table 2. Benchmarking of deflections indices for Sections 1 and 2 (Horak 2008)

Deflection Index	Section 1 – SMA type	Section 2 – EME type	Sound	Warning	Severe
Maximum (D_0)	462	440	<500	500 - 750	>750
Base layer (BLI)	199	136	<200	200 – 400	>400
Middle layer (MLI)	133	137	<100	100 – 200	>200
Lower layer (LLI)	137	77	<50	50 - 100	>100

It can be seen that, in accordance with the benchmarking system, whilst the upper layers of the pavement tend to be of adequate strength, the mid and lower pavement structure is relatively weak and is likely to contribute to most of the maximum deflection (D_0) at the surface. In essence the stiff upper pavement structure is able to distribute the loads imposed by the traffic without cracking or excessive rutting.

6.5 Mechanistic Analysis

An inspection pit was excavated to a depth of 1 m in each of the trial sections. Besides establishing accurate layer thicknesses disturbed samples were taken for laboratory testing and classification. It was then possible to utilise the information provided by the deflection measurements and inspection pits to carry out mechanistic modelling of the pavement structure. The quality of material found in the various layers is classified in accordance with TRH 14 (Guidelines for Road Construction Materials 1996). In the second horizon of Inspection Pit 2, the material appears to consist of low quality asphalt and has been regarded as aged, partly stripped asphalt in the mechanistic analysis. Table 3 summarises information from the inspection pit in Trial Section 1 while Table 4 summarises information from the inspection pit in Trial Section 2.

Deflection bowl measurements at the location of each test pits enabled the elastic modulus for each of the layers, as found in the respective inspection pits, to be back calculated. These stiffness values were used in conjunction with the pavement layer thicknesses and material properties to derive realistic pavement structures with assigned stiffness values. The structures were modeled using a mechanistic-empirical method to calculate the structural capacity of each pavement.

The estimated structural capacity, expressed in million E80 equivalent standard axle load repetitions (MESA), is identified as being the number of loadings until failure of the most critical layer in the pavement. The critical layer for Section 1 pavement is the 155 mm granular base with an axle capacity of 0.6 MESA, which, taking the estimated traffic loading into account would provide a structural design life of around 5 years. The critical layer for Section 2 is the 160 mm low quality asphalt layer, with structural an axle capacity of 2.96 MESA, which, given the anticipated future axle loading, equates to a structural design life of approximately 18 years.

The above notwithstanding, it should however be noted that the transfer functions used in these analyses do not take into consideration the superior fatigue properties of the rHDPE modified asphalt - as derived from the laboratory testing results - and it is distinctly possible that the future structural design capacity of these pavements could well be significantly in excess of the bearing capacity calculated for this investigation.

Table 3. Section 1: inspection pit 1

Layer thickness (mm)	Material Description	TRH14 Class
45	SMA type rHDPE modified asphalt	-
155	Pink quartzitic Sandstone – crushed rock	G4
150	Light brown weathered Granite – lightly cemented. Positive Phenolphthalein & HCl	EG5
50	Pale red weathered Granite	-
250	Dark yellow brown Silty Sand + quartzitic Sandstone gravel	G7
350	Dark brown Silty Sand	<G10

Table 4. Section 2: inspection pit 2

Layer thickness (mm)	Material Description	TRH14 Classification
70	EME type eHDPE modified asphalt	-
160	Aged, slightly stripped bitumen-bound weathered granite (low quality asphalt)	-
260	Light yellow brown weathered Granite	G7
440	Light yellow brown weathered Granite	G7
70	Pale red weathered Sandstone	-

7 Summary of Findings

7.1 General Description of the Trials

The main objective of this paper is to provide a detailed assessment of the performance of trial sections where the asphalt was modified with waste plastic. It covers research into the properties of recycled High Density Polypropylene plastic (rHDPE) modified with 70/100 penetration grade bitumen using 6% rHDPE content which results in performance grading of PG70V-22. It is important to note that the "wet" method, which entails digesting the plastic in hot bitumen, was used in the manufacture of the asphalt used in the trial sections. Comprehensive mix design work using rHDPE modified binder was carried out prior to the trials on both SMA and EME type asphalt mixes.

Two sections of road close to the supplier's asphalt plant, on a road known as D755, were selected for the trials, with a 40 mm layer of the rHDPE modified asphalt overlay being paving Section 1 while on Section 2 a depth of 70 mm of the severely rutting length was milled off before paving 70 mm of EME type rHDPE modified asphalt mix.

Based on weighbridge data from the asphalt plant and the quarry that utilize this road as a haul route, as well as a classified 7-day traffic count on D755, it is estimated that approximately 200 000 Equivalent 80 kN Standard Axle Loads (E80s) have been applied to these trial sections from date of construction in August 2019 to June 2021.

7.2 Findings: Section 1 SMA Type Asphalt Mix

The detailed visual inspection carried out on Section 1 shows that very little distress has occurred since construction, distress is limited to slight aggregate loss along the road's centreline as well as slight rutting in the wheel tracks.

The analyses of rut depth and Mean Profile Depth data captured by the NSV corroborate the visual inspection, 90th percentile rut depths are less than 10 mm and texture measurements show adequate MPD. It is also clear that little if any significant change in these properties has occurred between surveys carried out in May 2020 and 2021, respectively. The surveys of both occasions show that the riding quality to be is poor, however given the section's good performance in terms of rutting and texture the reason for the high IRI values is more likely to be caused by the poor riding quality of the original surface on which the 40 mm overlay was paved rather than to changes in the roughness of the trial section itself.

An analysis of FWD deflection bowl measurements on Section 1 indicates a relatively strong upper pavement structure compared to a weaker mid and lower structure. This finding confirms the inspection pit profile and results where poor-quality materials were found to be located from a depth of approximately 650 mm.

The deflection bowl measurement, together with the information on layer thicknesses and materials properties found in the inspection pit, was modeled to calculate the structural capacity the pavement. This was found to be approximately 5 years. It should however be noted that the transfer functions used in these analyses do not take into consideration the superior fatigue properties of the rHDPE modified SMA type asphalt, as found in the laboratory results, and it is probable that the structural design period of this pavement could be extended.

7.3 Findings: Section 2 EME Type Asphalt Mix

Section 2, which comprises a length of approximately 140 m, was purposely selected to evaluate the rut-resistant properties of the rHDPE modified asphalt mix as the existing pavement was already heavily rutted. Areas on Section 2 exhibiting the most severe rutting were milled off to remove most of the rutting before the 70 mm of the EME type rHDPE modified asphalt was paved. Although it was anticipated that some initial rutting would occur both the visual assessment and the automated rut depth measurements show only slight rutting after trafficking in the order of 200 000 E80s. At this stage this high modulus mix has been successful in bridging the uneven surface and previously rutted pavement.

This trial section was also intended to check the low temperature crack resistance of the modified mix. To do this one of the lanes was paved a week before the other, purposely leaving the longitudinal joint uncut, as opposed to the normal good practice of trimming back, before paving the second lane. It should be noted however that no evidence of cracking was found along the centreline longitudinal joint which would indicate good low temperature crack resistance.

The visual assessments and analyses of rut depth and Mean Profile Depth data captured by the NSV tend to agree with one another, 90th percentile rut depths are less than 10 mm. The analysis of MPD shows adequate texture depths except for the negative lane where smoothening is evident. No significant change in measurements between May 2019 and 2021 after additional traffic loading of approximately 100 E80s is evident between surveys carried out in May 2020 and June 2021, respectively. The roughness measurements indicate poor riding quality in both May 2020 and June 2021 surveys. This could be attributed to the uneven, partly milled surface onto which the 70 mm asphalt was paved rather than to changes in the roughness of the trial section itself.

It is likely that the severe rutting found on Section 2 can be attributed to the thick layer of low-quality asphalt, as found in Inspection Pit 2, and is shown to be the critical layer in the mechanistic analysis. The analysis nevertheless indicates an axle capacity of 2.96 MESA, which, taking into consideration the estimated traffic loading, would provide a structural design life of approximately 18 years.

8 Conclusions

The short-term performance of these full-scale trials in South Africa has illustrated that waste plastic can be effectively used in the manufacture of hot-mix asphalt. The "wet" process of digesting the recycled waste plastic in hot bitumen was successfully used in the preparation of the modified binder used in the manufacture of the asphalt in these trials. This technology offers exciting possibilities of utilizing waste plastic while at the same time enhancing the properties of asphalt.

Both trial sections, SMA and EME asphalt mix types described in this paper are performing well, with negligible defects after being subjected to approximately 200 000 80 kN equivalent standard loads.

The next step is to carry out a full scale trial on a urban expressway in the Durban Metro which carries an exceptionally heavy traffic loading. This expressway, known as the M7, which is trafficked by a large volume of heavy vehicles carrying goods from Durban's busy port to the hinterland, has been earmarked for this trial.

Preliminary investigation of this trial site, to assess the existing pavement structure and its condition, is already complete. It is intended to fully monitor all phases of the trial, using EME type mix as the base and SMA type mix as the surfacing, from the preparation of the modified binder, manufacturing and paving, as well as continual monitoring and testing of both their short (12 and 24 months) and long-term performance (+5 years), and it is intended to publish the results of the study at the appropriate time.

In the meanwhile, however, the good performance of initial trial sections of asphalt modified with waste high-density polyethylene plastic gives hope both in enhancing the properties of asphalt as well as in the environmental benefits that accrue from the effective utilisation of this waste plastic.

References

Bredenhann, S., Myburgh, P., Jenkins, K., O'Connell, J., Rowe, G., D'Angelo, J.: Implementation of a performance-grade bitumen specification in South Africa. J. S. Afr. Inst. Civ. Eng. **61**(3), 20–31 (2019). Published by South African Institute of Civil Engineering

Horak, E.: Benchmarking the structural condition of flexible pavements with deflection bowl parameters. J. S. Afr. Inst. Civ. Eng. **50**, 2–9 (2008). Published by South African Institute of Civil Engineering

Nortjé, W., Seforo, T.: Recycled high-density polyethylene in asphalt. In: Conference on Asphalt Pavements for Southern Africa, Sun City, South Africa (2019)

Sabita Manual 35/TRH8: Design & use of asphalt in road pavements, 2nd edn. Published by Southern African Bitumen Association (Sabita), South Africa, June 2016 (2016)

South African Pavement Engineering Manual, Chapter 6: Road prism and pavement investigation, 2nd edn. Published by South African Roads Agency SOC Ltd., South Africa (2014)

Technical Guideline TG1: The use of modified bituminous binders in road construction, 5th edn. Published by Southern African Bitumen Association (Sabita), South Africa, November 2020 (2020)

TMH 9: Pavement management systems: Standard visual assessment manual for flexible pavements. Published by Department of Transport, Pretoria, South Africa (1992)

TRH 12: Flexible pavement rehabilitation investigation and design. Published by Department of Transport for Committee of Land Transport Officials (COLTO), South Africa (1997)

TRH 14: Guidelines for road construction materials. Compiled by Committee of State Road Authorities, published by Department of Transport, South Africa (1996)

# Climate Resilient Road Design – 1

GIS Aided Vulnerability Assessment for Roads

Berna Çalışkan[1], Ali Osman Atahan[1(✉)], and Ali Sercan Kesten[2]

[1] Department of Civil Engineering, Transport Engineering, Istanbul Technical University, Istanbul, Turkey
atahana@itu.edu.tr
[2] Department of Civil Engineering, Istanbul Işık University, Istanbul, Turkey

Abstract. Road networks are vulnerable to natural disasters such as floods, earthquakes and forest fires which can adversely affect the travel on the network. However, not all road links equally affect the travel conditions in a given network; typically some links are more critical to the network functioning than the others.

The first stage of study involves the investigation of geological conditions. Image classification used for extracting information classes from 'Geological Map of Istanbul area' image file. The resulting raster layer used to create thematic map. A reclassification was performed for lithologic types. The second stage involves analyzing topological situation. A slope map prepared and classified according to percentage of slope values. The third phase is the analysis and interpretation of the accumulated data to establish suitable and applicable road vulnerability scores. The information in the source data for each vulnerability factor are classified into three different vulnerability scores: +2 (considerably increases vulnerability), +1 (increases vulnerability) and 0 (does not increase vulnerability) by using a vulnerability score table. The study area was categorized into three different traffic analysis zones as: (1) least favorable area; (2) favorable area; (3) most favorable area. Vulnerability values obtained to measure serviceability of critical links in dense urban road networks and applies them to the case of 'Beyoğlu' region.

Thematic layers were prepared using the Geographic Information System (GIS), and they were then combined to produce the serviceability of road links in the 'Beyoğlu' region. Consequently, A site specific vulnerability index is proposed, considering the serviceability of road links. A conceptual flowchart of the GIS processing steps taken to obtain the vulnerability index is illustrated.

Keywords: Road network vulnerability · Serviceability · Geological map · GIS Image classification · Road network vulnerability index

1 Introduction

Infrastructure is considered the backbone of our society and economy; it plays a crucial role in attaining the economic and social prosperity and sustainability of our world. Road networks are considered as one of the most important transport infrastructure systems, since they attain the economic and social prosperity of modern societies. In the last decades, vulnerability assessment studies for road networks and their assets gained great attention among the research community [1]. Vulnerability, exposure to risk or reliability are already well researched concepts in the field of transport.

The road transport system has much in common with the other technical infrastructures such as the electric power system, the water and sewage supply systems, the data and telecommunication systems, etc. [3]. Road network vulnerability can thus be seen as a special case of infrastructure vulnerability, and the interpretation of vulnerability should be much the same for all these systems.

2 Urban Road Network Vulnerability Assessment

The spatial characterization of the components (sub-systems) of the Infrastructure can be identified in three categories from a geometric point of view. In this study the second category is taking into consideration.

Point-like components (Critical facilities): single-site facilities whose importance for the functionality of the Infrastructure makes them critical, justifying a detailed description and analysis. Examples include hospitals, power-plants.

Line-like components (networks, lifelines): distributed systems comprising a number of vulnerable point-like sub-systems in their vertices, and strongly characterized by their flow-transmission function. Examples include Electric networks with vulnerable power plants, sub-stations, etc., or road networks with vulnerable bridges.

Area-like components: this is a special category specifically intended to model large populations of residential, office and commercial buildings, which cannot be treated individually. These buildings make up the largest proportion of the built environment and generally give the predominant contribution to the total direct loss due to physical damage [3].

The Road network is composed of a number of nodes and edges. It is a transportation network where edges can be directed (one-way) or undirected (two-way). All edges are in general vulnerable to seismic shaking or geotechnical hazards, with pavements that can rupture due to surface ground deformation. Some types of edges or road segments, like those identified below, have specific types of response to seismic action and associated vulnerability. The identified system components are:

Bridge, Tunnel, Embankment (road on), Trench (road in a), Unstable slope (road on, or running along).

Identification of critical road links is part of vulnerability analysis of transportation networks. Transportation system Includes roads, railways, waterways and airways that link a city or urban region to and from places beyond its boundaries and which are used for the transportation of people and goods.

Reasons for vulnerability:

- Increased temperatures pose a risk to the health of travellers and increase the demand of energy (e.g. increased use of air conditioning).
- Risk of accidents of travellers during extreme events is higher than under stable weather conditions. Accidents and extreme weather events can cause not only personal death/injury, but also interrupt the flow of transportation to and from urban areas.

- Increased pressure from extreme weather conditions may cause damage to transport infrastructures (e.g. melting asphalt, infrastructure heat buckling and destabilising, destruction by erosion, landslides, floods, or fallen trees after severe storm, etc.) and potentially alienate residents during and after extreme events.
- Electrical pumps in gasoline stations may fail if there is an interruption in the energy-supply system (due to extreme weather events), affecting the motorised vehicle operation.

3 Main Typologies for Vulnerability Assessment of Roadway Assets

This section includes the main vulnerability factors of roadway assets, i.e., the classification of soil and the functional hierarchical road categorization. The typologies of road pavements and geotechnical assets (embankments, cuts, trenches) are also described. Other structural assets, e.g., bridges, tunnels, and retaining walls are not included here and are described elsewhere [1].

Figure 1 shows our proposed general schematic for vulnerability analysis, derived from the specific method for road safety vulnerability studies suggested by Berdica, Bergh and Carlsson (2003).

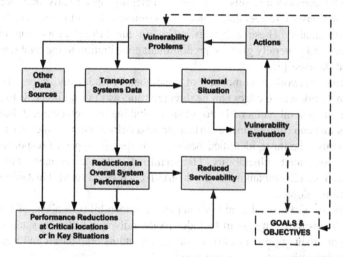

Fig. 1. Schematic for general vulnerability studies in transport networks [6]

3.1 Classification of Soil

Roadway assets are geotechnical structures, so the soil is considered a main factor for their vulnerability assessment. Various soil classifications are available to specify the type of the soil according to different variables and parameters. One of the most widely used classifications is the one given by Eurocode 8, which categorized the type of soil into four categories: soil A (rock), B (very dense sand), C (medium dense sand), D (loose-tomedium cohesion-less soil), and E (soil profile consisting of a surface alluvium layer) [1].

3.2 Functional Hierarchical Road Categorization

The classification of roads according to their functional hierarchy includes six types, which are local streets, minor collectors, major collectors, minor arterials, major arterials, and freeways. Different factors are used in this classification, such as accessibility, mobility, geometry, speed limits, and number of lanes [1].

3.3 Road Pavements

The major parameter of typology is the number of lanes [1].

3.4 Embankments, Slopes, and Trenches

The main parameters are the geometrical characteristics such as the height of embankments and slope angles. Some researchers used compaction quality of the embankments as an additional vulnerability factor [1].

3.5 Tunnels

The main factors that influence tunnels are the shape and dimensions of the structure, the main properties of the surrounding soil and/or rock (geological conditions), and the supporting system (e.g., steel, concrete, or masonry) [1].

3.6 Bridges and Retaining Walls

The damage of bridges is mainly related to the seismic performance of bridge components (i.e., deck, bearings, piers, shear keys, abutments, foundations, approach fills), which might be affected differently depending on the bridge type, the severity of the earthquake, and the local soil conditions. Moreover, the important typological characteristics for bridge abutments and retaining walls include the type of the abutment and its foundation, the soil conditions of foundation, and the fill material (backfill behind retaining walls/abutments) [1].

4 Weighting of Vulnerability Parameters

According to 'Seismic vulnerability classification of roads' study, Adafer and Bensaibi conducted the VI method by defining seven main items and nine main factors, which are divided into two main parameters, i.e., structural and hazard. Then, pairwise comparisons are performed and the validated weighting coefficients are determined for each parameter, item and factor as shown in Table 1.

Table 1. Weighting coefficients

Parameter	Weight (W_i)	Item	Weight (W_{ij})	Factor	Weight (W_{ijk})
Structural	0.25	Pavement	0.108	Number of lanes	0.667
				Pavement type	0.333
		Embankment	0.283	Height	0.648
				Compaction quality	0.122
				Slope	0.23
		Ground conditions	0.561	Ground type	0.2
				Landslides potential	0.8
		Maintenenace conditions	0.048	Pavement conditions	0.667
				Slope protection measures	0.333
Hazard	0.75	Seismic intensity	0.633	–	–
		Liquefaction potential	0.106	–	–
		Intersection with fault	0.261	–	–

Wij: weighting coefficient of the item. Wijk: weighting coefficient of the factor [2].

Adafer and Bensaibi proposed three different categories for the vulnerability of roads based on the range of the estimated VI as shown in Fig. 2: VR1 (safe: VI = 10–20), VR2 (moderately resistant: VI = 20–35), and VR3 (low resistant: VI = 35–50).

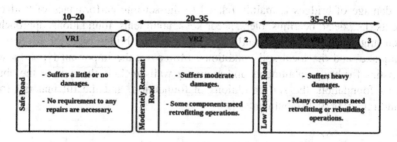

Fig. 2. Categories of vulnerability index developed from Adafer and Bensaibi [1]

The VI approach results in a rating of the assets' vulnerability based on different parameters and weighting factors. Therefore, it does not provide loss estimations, but is useful for prioritization purposes, in particular as part of periodical assessments of different components of the network. Hence, it is considered as an important tool toward efficient risk management and allocation of resources.

The main steps of the vulnerability index method is illustrated in Fig. 3.

Fig. 3. The main steps for vulnerability index method [1]

4.1 Geological and Topographical Conditions of the Study Area

The first stage of study involves the investigation of geological conditions. Image classification used for extracting information classes from 'Geological Map of Istanbul area' image file. The resulting raster layer used to create thematic map. Principal components evaluated in the Beyoğlu region are illustrated in Fig. 4.

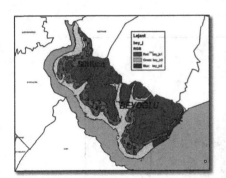
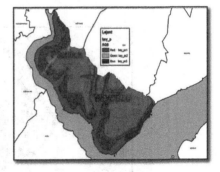

Fig. 4. Principal components evaluated in the Beyoğlu region

A reclassification was performed for lithologic types. And then geological map data converted to raster map layer which is shown in Fig. 5.

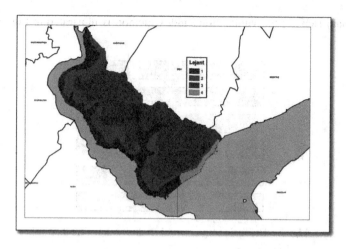

Fig. 5. Beyoğlu-geological map (raster format)

'Slope model' is generated from the digital elevation map (DEM). Slope values are categorized 0–5 (1), 6–10 (2), >10 (3) as a percentage. The slope of terrain is considered for railway route selection process as it directly influences the construction and operating costs. Digital topographical maps are used to create TIN, DEM and derivate layers. Slope map of Beyoğlu region is illustrated in Fig. 6.

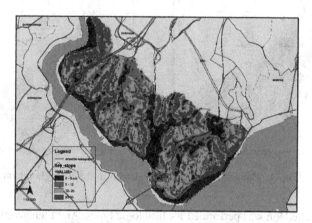

Fig. 6. Slope map of Beyoğlu region

In this paper, the network vulnerability is assessed by quantifying the importance of links within the network. The information in the source data for each vulnerability factor are classified into three different vulnerability scores: +2 (considerably increases vulnerability), +1 (increases vulnerability) and 0 (does not increase vulnerability) by using a vulnerability score table.

5 Vulnerability Scan for Meclis-i Mebusan Arterial Street

There are two distinct traditions with limited interaction. The first one could be characterised as topological vulnerability analysis of transport networks. A real transport network is then represented in the form of an abstract network (graph), i.e., an ordered pair G = (V, L) comprising a set V of nodes (or vertices) and a set L of links (or edges). The network could be undirected (no order is assumed between the nodes connected by a link) or directed (there is a start and end node of each link) and unweighted (all links have the same "length") or weighted (the links may have different "lengths"). The nodes and links in the abstract network may have different counterparts in the real network depending on the application in mind. The second tradition, which could be called system-based vulnerability analysis of transport networks, represents much more of the structure of the real transport system in the demand and supply models that are applied in the analysis. The transport network is still modelled as an abstract network (graph). Nodes and links then typically correspond to physical intersections and links in the real network. The network is usually weighted with link weights corresponding to actual lengths, travel times, costs or a combination of these in the form of generalised costs [7].

The study of topological properties of networks, including transport networks, is a growing researchfield. Digital Topographical maps and Geological map are taken from Istanbul Metropolitan Municipality [8, 9]. Elevation map of Beyoğlu region is illustrated in Fig. 7.

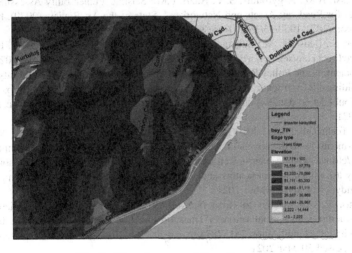

Fig. 7. Elevation map of Beyoğlu region

6 Conclusions

Thematic layers were prepared using the Geographic Information System (GIS), and they were then combined to produce the serviceability of road links in the 'Beyoğlu' region. Consequently, It proposes site specific vulnerability index considering the serviceability of road links and illustrates conceptual flowchart of the GIS processing steps taken to obtain the vulnerability index.

Vulnerability classification were done according to 5 classes. (1–2 Very High, 3–4 High, 5–6 Middle, 7–8 Low, 8–10 Very Low. Road network vulnerability is determined by GIS aided weighed overlay maps for geologic and topographical conditions of the study area. Spatial flood susceptibility evaluation is greatly influenced by elevation and slope.

When evaluating vulnerability, the vulnerable system itself and the expected adverse effects (threats) causing possible harm, including the possible type and extent of harm, need to be identified and made explicit. This should be informed by the objectives and purposes of the evaluation itself [4]. The applied frameworkdeveloped facilitates decision makers in understanding area specific vulnerabilities and adopting appropriatestrategies to improve vulnerability.

The temperature, the rainfall, the solar radiation, the permafrost distribution, the topography, the geotechnical types, the route design, the pavement structure combination, and the traffic distribution could also be studied for further studies. For a more precise assessment, surveys can be performed of each individual road in order to identify features that make the road liable to disruptions (underdimensioned drain pipes, wear and tear damage, danger of rockfall, etc.) [5].

References

1. El-Maissi, A.M., Argyroudis, S.A., Nazri, F.M.: Seismic Vulnerability Assessment Methodologies for Roadway Assets and Networks: A State-of-the-Art Review (2020). https://www.mdpi.com/2071-1050/13/1/61. Accessed 5 Apr 2021
2. Adafer, S., Bensaibi, M.: Seismic vulnerability classification of roads (2016). https://www.sciencedirect.com/science/article/pii/S1876610217356746. Accessed 15 Apr 2021
3. JRC Scientific and Policy Reports. Methodology for systemic seismic vulnerability assessment of buildings, infrastructures, networks and socio-economic impacts (2013). https://op.europa.eu/en/publication-detail/-/publication/869814b5-ec01-42ab-996b-922d7504a48c. Accessed 15 Jan 2021
4. ETC Technical Paper 2010/12 (2010) Urban Regions: Vulnerabilities, Vulnerability Assessments by Indicators and Adaptation Options for Climate Change Impacts (2010). https://climate-adapt.eea.europa.eu/metadata/publications/urban-regions-vulnerabilities-vulnerability-assessments-by-indicators-and-adaptation-options-for-climate-change-impacts-a-scoping-study. Accessed 15 May 2021
5. Jenelius, E.: Approaches to Road Network Vulnerability Analysis, Licentiate Thesis in Infrastructure with specialisation in Transport and Location Analysis (2007). https://www.researchgate.net/publication/251378740Approaches_to_Road_Network_Vulnerability_Analysis. Accessed 20 Mar 2021
6. Taylor, M., D'Este, G.M.: Concepts of network vulnerability and applications to the identification of critical elements of transport infrastructure (2003). https://www.australasiantransportresearchforum.org.au/sites/default/files/2003_Taylor_DEste.pd. Accessed 2 Mar 2021
7. Mattsson, L., Jenelius, E.: Vulnerability and resilience of transport systems – a discussion of recent research (2015). https://www.sciencedirect.com/science/article/abs/pii/S0965856415001603. Accessed 20 May 2021
8. Istanbul Metropolitan Municipality's Directorate of Geographic Information Systems, Digital Topographical maps for Istanbul, (2020)
9. Istanbul Metropolitan Municipality's Directorate of Directorate of Eartquake and Geotechnical Investigation, Istanbul Geological Map (1/100 000) (jpeg) (2020)

Investigation of Historical and Future Air Temperature Changes in the UAE

Reem N. Hassan, Waleed A. Zeiada[✉], Muamer Abuzwidah, Sham M. Mirou, and Ayat G. Ashour

University of Sharjah, Sharjah, UAE
wzeiada@sharjah.ac.ae

Abstract. Climate change refers to significant and long-term changes in the global climate. This climate change could alter different climate factors in future that affect the resilience of transportation systems such as temperature, precipitation, wind speed, solar radiation, and humidity. For example, global warming is expected to affect the design of the asphalt pavements and asphalt binder selection for roads construction in the UAE. This study aims at investigating potential historical trends of different climate factors using historical climate data collected from different weather stations from 1990 to 2020 to identify the factors that has more potential to change greatly in the future. This study considers historical and future air temperature records from automatic weather stations and airport weather stations. Initial analysis indicated that the historical air temperature trend seems to increase remarkably compared to other climate factors. The second part of this study focuses at investigating the future air temperature changes in the UAE. The prediction of future air temperature depends on 3 climate models used from the CMIP5 platform which is the fifth phase of The Intergovernmental Panel on Climate Change (IPCC). For all models used, 2 greenhouse gas concentration scenarios have been considered (RCP 4.5 and RCP 8.5). This study proposes required mitigation policies to minimize the global worming impact on the resilience of asphalt pavements in the UAE. And provide a climate profile for the country, which will be used in further research to predict pavement temperature, performance grade and pavement design, based on the MEPDG.

Keywords: Climate change · Climate models · Air temperature · Asphalt pavement

1 Introduction

The UAE's most widely used infrastructure for transporting people and goods comprises mainly of asphalt roadways. Climate changes may affect pavement layers quality so that they are vulnerable of accelerating traffic damage, therefore reducing their servicing life and has a significant negative influence on the nation's economy. An increase in pavement temperatures as a result of the rise in air temperature and rises in precipitation and flood conditions due to the vulnerability of the asphalt binder to such factors is significantly affecting the strength of asphalt concrete (AC) surface layers. It is therefore important to explore the effects of climate change on pavement design and performance (Chai et al. 2012). Between 1880 and 2012, the global surface temperature increased by 0.85 °C, and the rate of change in the last few decades was faster (Pachauri et al. 2014). Meanwhile, global sea levels and extreme weather events such as flooding, hurricanes and heatwaves have increased in many areas. At least in the near future, climate change is expected to continue. Whereas modern pavement planning practices implied that in the future the climate remains the same as before, climate scientists indicate that such predictions are now doubtful and use the General Circulation Models (GCMs) to suggest a rise in annual average temperatures, shifts in the average rainfall and changes to extremes in both (Knutti et al. 2013).

Climate changes have a substantial effect on the nature and the performance of the pavement. It has an impact on the stability, durability, and load capability of the pavement. Climate change are important variables in pavement performance along with considerations such as loads related to traffic, layer materials and maintenance and repair schemes. Temperature, precipitation, humidity and freeze-thaw cycles are the major climate variables covered by the Literature (Qiao et al. 2020). These variables usually lead to distress development, which contribute to degradation of the pavement and eventual collapse. Precipitation affects the water infiltrating the paving surface as well as the moisture level pavement section. Moisture content primarily affects the unbound and subgrade materials. The moisture modulus of unbound and subgrade materials will greatly reduce by increasing the degree of saturation. Moisture may also affect the subgrade rutting resistance. Also, the shear strength of the Unbound and subgrade material at high saturation may greatly reduce, and this can induce permanent deformation. Moreover, unbound and subgrade materials are more vulnerable to moisture degradation due to higher number of fine materials (Dempsey 1979). Temperature and moisture in cold areas induce the freeze-thaw phenomenon. Freeze-thaw cycles are concerned with growing ice beneath the pavement surface, leading to the development of a void and tensile strain on the paving surface. In previous studies, a drop in the resilient modulus from 48% to 63% was observed during freeze-thaw cycle. Which lead to extreme pavement distress during this period (Robyn V. McGregor.n.d.).

In addition to that, the performance of asphalt pavement depends heavily on its humidity and temperature, which influences distortion properties of unbound granular materials (Kodippily et al. 2020). The heat in the pavement originates mostly from air temperature and sunlight. Wind can also influence the convection of the surface and so modify the pavement's temperature. Groundwater levels and precipitation might be linked to moisture in the pavement (Chen et al. 2019). Temperature is one of the most important environmental variables, as an increase in average annual temperature and seasonal temperature change has led to increased asphalt rutting and thermal cracking (Hemed et al. 2020 and Stoner et al. 2019). As the temperature rises, the stiffness of asphalt materials can be reduced, which may change pavement stress-strain responses and may have long-term implications which may speed up the load-related degradation. Thermal cracking and surface distortion, including rutting, and shoving, is the main consequence of the influence of temperature on pavements. Moreover, Higher temperatures will result in faster ageing of asphalt mixture (Gilbert 1990). (Gudipudi et al 2017) evaluated the influence on Arizona, Maine, Montana and Virginia of climatic temperature rises. They have found that daily temperature rises should occur at mid-century (2040–2060) and that asphalt fatigue has grown 2% to 9% and rutting 9% to 40%.(Qiao et al. 2013) used the sensitivity of pavement performance to climate variables like temperature (average annual temperature and seasonal variability), precipitation, wind speed, sunshine percent, and groundwater level in specific locations to compare and measure paving service life experiences prior and post-climatic change in Virginia, USA. The study concluded that, Temperature is the most significant climatic element in the Mechanistic-Empirical Pavement Design Guide (MEPDG). All pavement distresses such as longitudinal cracking, fatigue cracking, and AC rutting are influenced by both an increase in the annual average temperature and seasonal variations. Minor (5%) rise in temperature, the pavement service life was found to be dramatically reduced (greater than 20%). (Chehab et al. 2019) investigated the adjustments that needs to be applied on the performance grade of pavement under high temperatures. This research aimed to analyse the compressive stress and strains in asphalt layers of the airfield pavements widely used for airfield pavements by means of mechanical inspection. The study has shown that the selection of binder PG grades for airfield pavements, in particular the deeper layers of asphalt (intermediate and base layers), have continuously high temperatures. For this purpose, PG bumping over the surface layer is advised in airfield pavements on intermediate and basic layers, in areas where the temperature is very high, and rutting is a serious distress. An engineering analysis was carried out by (Fletcher et al. 2016) to investigate changes in 20-year return values of extreme maximum pavement temperature in Canada. It was found that temperature is expected to rise by 1–3 C in the period of 2041–2070. As a result, profound shifts in the spatial distribution of PG are likely to arise from climate change with the severity of changes related directly to the severity of predicted temperatures. Many reports have described a premature decline in transport infrastructure as a potential issue because of climate change and temperature rises, but no previous studies have so far attempted to measure and plan the scale of the problem for the UAE.

1.1 Research Objective

The main objective of this research is to incorporates observations from actual climate events consisting of five climate factors (wind speed, humidity, sunshine, precipitation, and temperature) as well as global climate model outputs to analyse changes in the temperature covering a 30-year near future (2021–2050) and 30-year far future (2051–2080) periods. The aim of the project is to develop a methodology of quantitative assessment of extreme climate change. A set of three global climate models used from the CMIP5 platform which is the fifth phase of The Intergovernmental Panel on Climate Change (IPCC) is utilized in various climate-enforcement scenarios for anticipating future extremes. This study gives a national overview of the effects of climate change on flexible pavement and the focus on the region.

2 Study Area and Data

2.1 Study Area

The United Arab Emirates is an elective monarchy established by a seven-Emirates federation comprising Abu Dhabi, Ajman, Dubai, Fujairah, Ras Al Khaimah, Sharjah, and Umm Al Quwain. The current population of the UAE is 10,003,271, and a projected population of 11,054,579 in 2030, 12,207,333 in 2040 and 13,163,548 in 2050. The economy of the United Arab Emirates is the most diverse among Gulf Cooperation Council members. This research covered the entire area of the UAE, located at 23.4 and longitude 53.8. The location of the 20 stations that will be used in this research is shown in Fig. 1. They are well spread and will cover the whole area of the UAE.

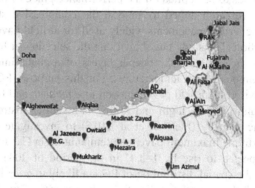

Fig. 1. Weather stations locations

2.2 Collection of Data

Five main climate parameters include air temperature, precipitation, wind speed, sunshine, relative humidity. These data for 14 weather stations (Alghweifat, Alquaa, Jabal Jais, Mezyed, Um azimul, Mukhariz, Razeen, Algazeera BG, Al Malaiha, Alqlaa, AlFaqaa, Madinat Zayed, Owtaid and Mezaira) and 6 airports (Sharjah, Abu Dhabi, Dubai, Ras AlKhaimah, Alain, and AlFujairah airport) in the UAE were collected by the Emirates National Center of Metrology. The duration of climatic data measurement is 15 min. The 20 weather stations and airports data involve 20 years (2001–2020). Air temperature data from global climate models were downloaded for UAE for historical (2001–2020) and future (2021–2080) periods considering different CO_2 emission scenarios.

3 Analysis of Historical Climate Factors in the UAE

Over the last two decades, climate change and adaptation evaluations have grown in number, consistency, and significance, as the effects of possible global warming have started to resonate with policy makers and the general population. Climate changes have a substantial effect on the nature and the performance of the asphalt pavement.

Different climate factors were studied at 14 ground weather stations from 2004–2018 and 6 airports from 2004–2018. The climate factors are taken from different 20 weather stations are air temperature, precipitation, wind speed, sunshine, and relative humidity.

3.1 Wind Speed

For the tested historical 15-year period, from 2004 till 2018, there was a minimal decrease in wind speed across UAE ground stations as shown in Fig. 2-a. The maximum annual wind speed was 12 m/h at alqlaa ground station in year 2011 while the minimum wind speed of 4 m/h was recorded in 2009 at Um Azimul. In addition, there was a minimal increase in wind speed value recorded by UAE airports. The minimum value of wind speed was at Ras AlKhaimah airport of 4 m/h during the year 2006. The maximum wind speed was recorded in year 2015 of 9.2 m/h at Alain airport as shown in Fig. 2-b.

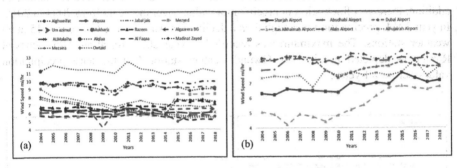

Fig. 2. Historical annual average wind speed: (a) automatic weather stations; and (b) airport stations

3.2 Relative Humidity

A slight decrease in humidity was observed between the years 2004 till 2018 at UAE ground stations and UAE airports. The maximum humidity of 80% was recorded in year 2007 in Alqlaa station as depicted in Fig. 3-a while the minimum relative humidity reached to 30% at Razeen in year 2009. The average relative humidity varied between 30% and 80%. The maximum humidity recorded by airports is 72% in year 2004 observed by Al Fujairah airport station while the minimum relative humidity reached to 38% at Alain airport in year 2012. the least recorded relative humidity was at Alain airport station since the existence of water bodies, mainly sea in other cities leads to higher relative humidity as shown in Fig. 3-b.

Investigation of Historical and Future Air Temperature Changes in the UAE 1153

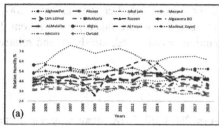
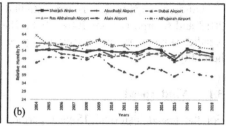

Fig. 3. Historical annual average relative humidity: (a) automatic weather stations; and (b) airport stations

3.3 Sunshine and Precipitation

Based on the trend observed in Fig. 4-a, there was negligible change throughout the 15 years (2004 to 2018) in the sunshine percentage. The average annual sunshine was 40% for all weather stations. The minimum observed was 22% at AlMezaira station. In addition, there was fluctuation in precipitation values over the years from 2004 to 2018. In year 2014, precipitation was high then a declination was observed in year 2015 at all weather stations. The maximum was recorded in year 2015, reaching 0.05 inches at Razeen as depicted in Fig. 4-b. Owtaid weather station had the least precipitation values across all years.

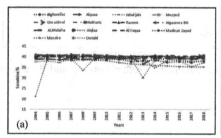
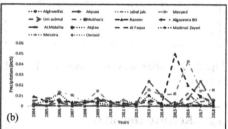

Fig. 4. (a) Historical annual average sunshine for automatic weather stations; (b) historical annual average precipitation for automatic weather stations

3.4 Air Temperature

For the past 17 years, from 2004 till 2018, there was an increase in temperature across UAE ground stations namely, Alghweifat, Alquaa, Jabal Jais, Mezyed, Um azimul, Mukhariz, Razeen, Algazeera BG, Al Malaiha, Alqlaa, AlFaqaa, Madinat Zayed, Owtaid and Mezaira as shown in Fig. 5-a. The maximum observed temperature of 31 °C was at Um azimul ground station in year 2017 and 2018 while the minimum temperature of 18.4 °C was in year 2006 at Jabal Jais. Figure 5-b depicts a significant increase in air temperature across UAE airports from 2004 to 2018, namely Sharjah, Abu Dhabi, Dubai, Ras AlKhaimah, Alain, and AlFujairah airport. The annual maximum air temperature ranged from 27.3–30.0°C. The maximum observed temperature of 30 °C was at Dubai Airport in year 2018 while the minimum temperature of 27.3 °C was in year 2008 at Sharjah Airport.

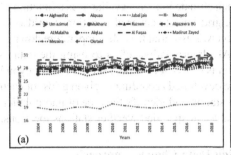 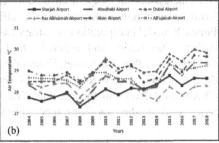

Fig. 5. Historical annual average air temperature: (a) automatic weather stations; and (b) airport stations

As many previous studies have shown, temperature is the most important factor. These studies have tested the increase in average annual temperature through a sensitivity analysis. The results shows that the most influential climatic factor in the pavement design is temperature. And, based on the analysis carried out in this section, the temperature factor has changed significantly during historical period compared to other investigated factors. This study focused on the change of air temperature during the years.

3.5 Future Climate Models Chosen for this Study

There were many climate models available on the European; Earth System Grid Federation (ESGF) Peer-to-Peer (P2P) company system, a partnership in the development, deployment, and maintenance software infrastructure for managing, disseminating, and analyzing model outputs and observational data website (esgf-node.llnl.gov). Only 3 models that provide temperature data every 3 h of observation which be more appropriate for the PG grade selection of asphalt binder and Mechanistic-Empirical (M-E) pavement design purposes. These 3 models are listed below. The first model was MRI-CGCM3 from the Meteorological Research Institute, Meteorological Agency, Japan. The second model was GFLD- ESM2G that facilitated by the Atmospheric and Oceanic Sciences Program, which was a collaborative program with Princeton University. The third model was GISS-E2R that was submitted by the Goddard Institute for Space Studies (GISS). Those 3 models were used to provide climate change prediction results including change in average surface air temperature and the sea level on a global base. In this analysis, different Co2 emission scenarios were considered (RCP 4.5 and RCP 8.5).

3.6 Interpolation to Compute Exact Air Temperature at Weather Stations

Using interpolation technique, location of each weather station and airport was defined and the exact air temperature values was calculated accordingly from the surrounded coordinates. The air temperature data for each climate model is available as a function of latitude and longitude according to specific grid lines to cover all locations worldwide. For each model interpolation or extrapolation was essential to scale temperature down to each weather station and airport in the UAE. Interpolation refers to using the data to predict data within the dataset. Extrapolation is the use of the data set to predict beyond the data set. For each model, grid maps were developed according to given grids for each climate model to know the location of each station and identify the type of interpolation required to get the exact air temperature values at specified weather stations locations.

3.7 Correction of Future Records Using Delta Change Approach

The delta change method is one of methods to verify and correct predicted future climate changes to make the output of climate change models ready to use for different applications. The method is based on the use of correction factors calculated as the average future records obtained from a specific model divided by corresponding historical recorded predicted also from the same model. For this purpose, historical temperature records every 3 h predicted by each model for the past 30 years were downloaded to be compared with the predicted future temperature records. The calculated delta factors are then used to predict the future climate records by multiplying the delta factor by the corresponding actual historical records collected from each weather station. In this study, the delta correction factors were calculated based on the average monthly temperatures calculated based on 30 years records (1991 to 2020). Delta method used to correct all data from different RCPs (4.5 and 8.5). The monthly delta factor was calculated for each month using Eq. 1.

$$\Delta \text{ month } i = \frac{\text{Future average monthly temperature for month } i \text{ (from model used)}}{\text{Historical average monthly temperature for month } i \text{ (from model used)}}$$

(1)

According to interpolation calculation, the air temperature data was calculated for 14 weather stations and 6 airports distributed across the UAE. For each station, 3 models were used to predict future air temperature. All air temperature values obtained will be corrected using the delta change method.

4 Interpolation Using ARC GIS PRO

Different interpolation methods were tested to visualize the effect of climate change on air temperature using the data from 20 weather stations covering the historical period from 2001–2020. These interpolation methods are discussed in some details in the as stated below.

4.1 Kriging Method

Kriging is a method of geostatistic interpolation that incorporates the distance and degree to which the values in unknown areas are measured. The weighted linear combination of the known sample values around that point to be estimated is a kriged estimate. The distance and directions of the sample points are assumed by Kriging to show a spatial correlation which could be utilized to explain surface variations. To determine the output value for each location, the Kriging Tool fits a mathematical function to a certain number of points or all points within a predefined radius. Kriging is a multi-stage technique, which comprises statistical exploration of the data, variogram modelling, surface building and (optional) variance surface exploration. Kriging method; however, doesn't pass any of the points values and results in greater or lower interpolated values than true values (Cao et al. 2009). Figure 6-a shows the historical (2001–2020) air temperature mapping using kriging method.

4.2 Natural Neighbor Method

Natural Neighbor Interpolation determines the nearest group of input data samples to a query location and uses weights to interpolate a value based on proportional regions. Also called Sibson or interpolation for "area-stealing" (Sibson et al. 1981). Figure 6-b shows the historical (2001–2020) air temperature mapping using natural neighbor method.

4.3 Trend Method

Trend is a statistical approach which identifies the surface to match the sample points with a least regression fit. It fits the whole surface using one polynomial equation. As a result, surface variation in response to input values is minimized. The area is designed such that the discrepancies between the actual values and the estimated values (i.e. the variance) are as low as possible at each input point (Yang et al. 2004). Figure 6-c shows historical (2001–2020) the air temperature mapping using trend method.

4.4 Spline Method

Spline calculates mathematical values, minimizing the total curvature of the surface resulting in a smooth area passing through the inputs. Concepts show that the curvature of the entire surface is minimized by bending a rubber plate through known spots. It adapts a mathematical function to a certain number of nearby input points as the sample points are passing through. However, Spline interpolation does not perform effectively when the sample points are close together and have large changes in their value. This is because Spline employs slope calculations to determine the form of the flexible rubber sheet (change distance) (Cao et al. 2009). Figure 6-d shows the historical (2001–2020) air temperature mapping using spline method.

4.5 Inverse Distance Weighted (IDW) Method

The interpolator for IDW implies that the local effect of every input point reduces with distance. It weighs the closer points than the furthest points to the processing cell. To compute the output value of each place, a certain number of points or all points within the defined radius can be utilized. Using this approach, the variable is mapped with the influence decreasing with the distance from the sampled place. The IDW technique is simply an average moving interpolator used in extremely variable inputs (Witteveen and Bijl 2009). The main advantages of using IDW is that it can estimate severe field changes, such as: Fault Lines, Cliffs. And dense even spaces are interpolated efficiently (flat areas with cliffs) (Maleika 2020).

The IDW is the adapted method for interpolating in this research because it is extremely versatile and adaptable to various sample data characteristics. It maintains the same minimum and maximum values of actual data. Also, the sample points are weighted during interpolation in the IDW interpolation technique in order to diminish the effect of one point compared to another by distance from the unknown. IDW also It enables very quick computations, The estimate integrates different distances, and the exponent weighing distance can accurately manage the impact of distances. Figure 6-e shows the historical (2001–2020) air temperature mapping using the IDW method.

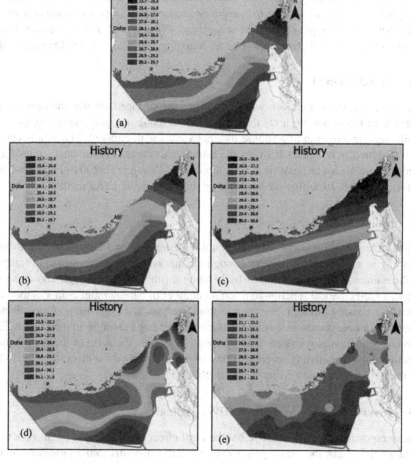

Fig. 6. (a) Kriging method; (b) Natural neighbor method; (c) Trend method (d) Spline method; (e) IDW method

The UAE is considered a part of the Middle East and it is in the Arabian Peninsula. It has a varied topography in its eastern portions but much of the rest of the country consists of flat lands, sand dunes, and large desert areas. The climate of the UAE is desert with the humidity of the Arabian Gulf, although it is cooler in the eastern areas that located at higher elevations. The country is hot in the summer (August). Therefore, it is necessary to take temperature values in consideration when it comes to PG selection and pavement design. Effect of climate change on air temperature in the past 20 years (2001–2020) was mapped using observed data from 14 weather stations and 6 airports covering the area of UAE. Figure 6-e shows the distribution of air temperature over the historical period.

The historical map shows that the maximum annual average observed temperature of 28.7–30 °C was at southern area while the minimum annual average temperature of 19–21 °C at Jabal Jais area, regardless of the interpolation method.

5 Analysis of Future Air Temperature in the UAE

Future projections are carried out based on two emission scenarios from three individual climate models are integrated. The medium emission scenario RCP 4.5 was used as the medium scenario of climate change. Whereas the emission scenario RCP 8.5 was used as the higher limit of climate change and representing the worst-case climate change scenarios. Figures 7-a and 7-b show the air temperature variation for near and far future periods using predicted data from model 1. The minimum annual average temperature using the medium emission scenario; RCP 4.5, was found to be 19.6 °C and 19.9 °C for near and far future, respectively. The stations with minimum air temperature are the northern stations such as Al Faqa, Al Malaiha, Fujairah as well as stations located in coastal areas such as Abu Dhabi airport. The least recorded temperature is the temperature of the mountainous area; Jabal Jais located in the northern area of the country. These values, in comparison with the historical minimum annual air temperature show an average increase of 1 °C which makes around 4% increase. whereas the maximum annual average temperature was 31.1 °C and 31.8 °C for near and far future respectively as shown in Figures 7-a and 7-b. These values, in comparison with the historical maximum annual average air temperature show an average increase of 1°C which makes around 4% increase. The stations with maximum air temperature are the southern stations, desert areas, such as (Mukhairaz, Um Azimul, Mezaira, Alquaa, Owtaid, Al Gazeera B.G).

In case of the high emission scenario RCP 8.5; the minimum annual average temperature was found to be 19.6 °C and 20.7 °C for near and far future respectively as shown in Fig. 7-c and 7-d. These values, in comparison with the historical minimum annual air temperature show around 6% increase. The stations with minimum air temperature are Al Faqa, Al Malaiha, Fujairah and AbuDhabi airport. The least recorded temperature is the temperature of the mountainous area; Jabal Jais located in the northern area of the country. whereas the maximum annual average temperature was 31. 3 °C and 33.4 °C for near and far future respectively as shown in Fig. 7-c and 7-d. These values, in comparison with the historical maximum annual average air temperature show an average increase of 7%. It is interesting to highlight that, areas showing a temperature rise increases in this scenario and the trend observed in the near and far future period is significantly different from the historical.

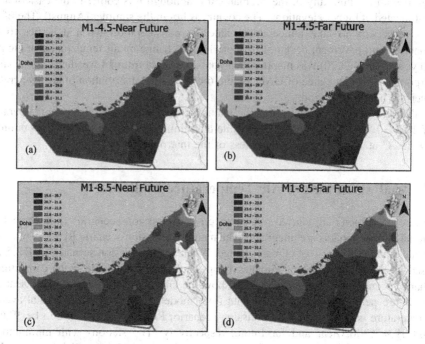

Fig. 7. (a) Model 1 RCP 4.5, near future air temperature; and (b) Model 1 RCP 4.5, far future air temperature (c) Model 1 RCP 8.5, near and far future air; and (d) Model 1 RCP 8.5, far future air temperature

Same analyses were carried out using data from Model 2 to predict air temperature in the near and far future. Figure 8 shows the air temperature variation for near and far future periods using predicted data from model 2. The minimum annual average temperature using the medium emission scenario; RCP 4.5, was found to be 19.4 °C and 19.9 °C for near and far future respectively as shown in Fig. 8-a and 8-b. The stations with minimum air temperature are the northern stations such as Al Faqa, Al Malaiha, Fujairah as well as stations located in coastal areas such as Abu Dhabi airport. The least recorded temperature is the temperature of the mountainous area; Jabal Jais located in the northern area of the country. These values, in comparison with the historical minimum annual air temperature show an average increase 1 °C which makes around 2% increase. whereas the maximum annual average temperature was 31.4 °C and 31.9 °C for near and far future respectively as shown in Fig. 8-a and 8-b. These values, in comparison with the historical maximum annual average air temperature show an average increase of 5%. The stations with maximum air temperature are the southern stations, desert areas, such as (Mukhairaz, Um Azimul, Mezaira, Alquaa, Owtaid, Al Gazeera B.G).

In case of the high emission scenario RCP 8.5; the minimum annual average temperature was found to be 19.9 °C and 20.9 °C for near and far future respectively as shown in Fig. 8-c and 8-d. These values, in comparison with the historical minimum annual air temperature show an average increase of 7%. The stations with minimum air temperature are Al Faqa, Al Malaiha, Fujairah and AbuDhabi airport. The least recorded temperature is the temperature of the mountainous area; Jabal Jais located in the northern area of the country. whereas the maximum annual average temperature was 31.9 °C and 33.6 °C for near and far future respectively as shown in Fig. 8-c and 8-d. These values, in comparison with the historical maximum annual average air temperature show an average increase of 8%. It is interesting to highlight that, areas showing a temperature rise increases in this scenario and the trend observed in the near and far future period is significantly different from the historical.

Air temperature mapping was also carried out based on values predicted from the third model. Model 3 showed similarity in the trend compared with models 1 and 2. Figure 9 shows the air temperature variation for near and far future periods using predicted data from model 3. The minimum annual average temperature using the medium emission scenario; RCP 4.5, was found to be 20.3 °C and 20.8 °C for near and far future respectively as shown in Fig. 9-a and 9-b. These values, in comparison with the historical minimum annual air temperature show an average increase of 7%. The stations with minimum air temperature are the northern stations such as Al Faqa, Al Malaiha, Fujairah as well as stations located in coastal areas such as Abu Dhabi airport. The least recorded temperature is the temperature of the mountainous area; Jabal Jais located in the northern area of the country. whereas the maximum annual average temperature was 32.2 °C and 32.9 °C for near and far future respectively as shown in Fig. 9-a and 9-b. These values, in comparison with the historical maximum annual average air temperature show an average increase of 7%. The stations with maximum air temperature are the southern stations, desert areas, such as (Mukhairaz, Um Azimul, Mezaira, Alquaa, Owtaid, Al Gazeera B.G).

In case of the high emission scenario RCP 8.5; the minimum annual average temperature was found to be 20.4 °C and 21.3 °C for near and far future respectively as shown in Figs. 9-c and 9-d. These values, in comparison with the historical minimum annual air temperature show an average increase of 9%. The stations with minimum air temperature are Al Faqa, Al Malaiha, Fujairah and AbuDhabi airport. The least recorded temperature is the temperature of the mountainous area; Jabal Jais located in the northern area of the country. whereas the maximum annual average temperature was 32.5 °C and 33.9 °C for near and far future respectively as shown in Figs. 9-c and 9-d These values, in comparison with the historical maximum annual average air temperature show an average increase of 9%. It is interesting to highlight that, areas showing a temperature rise increases in this scenario and the trend observed in the near and far future period is significantly different from the historical.

Air temperature mapping was also carried out based on values predicted from the third model. Model 3 showed similarity in the trend compared with models 1 and 2. Figure 9 shows the air temperature variation for near and far future periods using predicted data from model 3. The minimum annual average temperature using the medium emission scenario; RCP 4.5, was found to be 20.3 °C and 20.8 °C for near and far future respectively as shown in Figs. 9-a and 9-b. These values, in comparison with

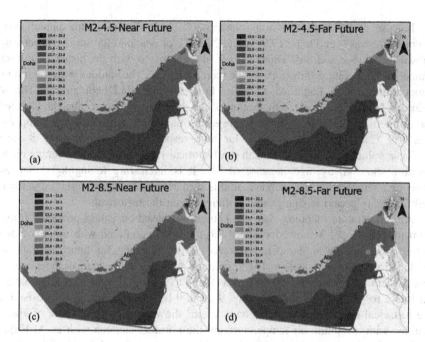

Fig. 8. Model 2 RCP 4.5, near future air temperature; and (b) Model 2 RCP 4.5, far future air temperature (c) Model 2 RCP 8.5, near and far future air; and (d) Model 2 RCP 8.5, far future air temperature

the historical minimum annual air temperature show an average increase of 7%. The stations with minimum air temperature are the northern stations such as Al Faqa, Al Malaiha, Fujairah as well as stations located in coastal areas such as Abu Dhabi airport. The least recorded temperature is the temperature of the mountainous area; Jabal Jais located in the northern area of the country. whereas the maximum annual average temperature was 32.2 °C and 32.9 °C for near and far future respectively as shown in Figs. 9-a and 9-b. These values, in comparison with the historical maximum annual average air temperature show an average increase of 7%. The stations with maximum air temperature are the southern stations, desert areas, such as (Mukhairaz, Um Azimul, Mezaira, Alquaa, Owtaid, Al Gazeera B.G).

In case of the high emission scenario RCP 8.5; the minimum annual average temperature was found to be 20.4 °C and 21.3 °C for near and far future respectively as shown in Figs. 9-c and 9-d. These values, in comparison with the historical minimum annual air temperature show an average increase of 9%. The stations with minimum air temperature are Al Faqa, Al Malaiha, Fujairah and AbuDhabi airport. The least recorded temperature is the temperature of the mountainous area; Jabal Jais located in the northern area of the country. whereas the maximum annual average temperature was 32.5 °C and 33.9 °C for near and far future respectively as shown in Figs. 9-c and 9-d These values, in comparison with the historical maximum annual average air temperature show an average increase of 9%. It is interesting to highlight that, areas showing a temperature rise increases in this scenario and the trend observed in the near and far future period is significantly different from the historical.

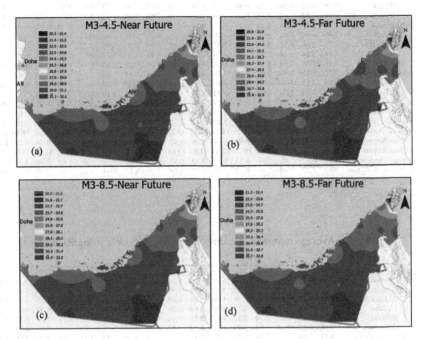

Fig. 9. Model 3 RCP 4.5, near future air temperature; and (b) Model 3 RCP 4.5, far future air temperature (c) Model 3 RCP 8.5, near and far future air; and (d) Model 3 RCP 8.5, far future air temperature

6 Minimum and Maximum Air Temperature Trends from All Models

Minimum annual average air temperature was calculated according to data recorded by 14 weather station distributed in all entire area of UAE for both historical and future periods. In the historical period, which includes 20 years from 2001 to 2020, using the average value every 5 years, all values of minimum annual average air temperature ranged from 3 °C to 9 °C. Jabal Jais area, which is located in the far north area, recorded the least value of the minimum air temperature of around 0 °C. The future Minimum annual average air temperatures were predicted using 3 different models each model contains 2 different seniors RCP 4.5 and RCP 8.5. The future data was calculated for both periods near future (2021–2050) and far future (2050–2080) to illustrate the minimum temperature change through the future. As per the results obtained, the minimum average air temperature will have a slight increase during near and far future, by an average of 0.5 °C using RCP 4.5 and only 1 °C using RCP 8.5 as shown in Tables 1 and 2.

Investigation of Historical and Future Air Temperature Changes in the UAE 1163

Table 1. Average minimum air temperature of all models using RCP 4.5

Stations Name/Year	2005	2010	2015	2020	2025	2030	2035	2040	2045	2050	2055	2060	2065	2070	2075	2080
Al ghewifat	3.76	4.34	6.18	5.96	4.60	6.51	6.30	4.64	6.56	6.36	4.77	6.77	6.55	4.78	6.81	6.61
Um Azimul	8.26	7.74	7.92	9.26	8.21	8.42	9.78	8.22	8.46	9.83	8.51	8.72	10.19	8.52	8.72	10.20
Al Fqaa	7.60	6.22	6.56	7.48	6.50	6.87	7.84	6.50	6.87	7.98	6.73	7.09	8.15	6.77	7.12	8.29
Mezaira	8.96	7.82	8.12	8.16	6.95	7.89	7.86	7.06	7.98	7.96	7.22	8.22	8.18	7.23	8.26	8.24
Maliaha	7.14	6.02	6.48	7.74	6.31	6.78	8.06	6.29	6.72	8.16	6.48	6.96	8.37	6.51	6.96	8.47
Mukhariz	5.36	6.00	5.84	5.94	6.26	6.20	6.26	6.31	6.27	6.38	6.50	6.43	6.55	6.53	6.47	6.65
Al Gazera B.G	3.28	3.64	4.74	5.06	3.86	5.03	5.36	3.85	5.06	5.45	3.98	5.22	5.58	3.95	5.24	5.69
Alquaa	5.94	5.14	5.84	7.56	5.44	6.24	7.96	5.42	6.34	8.03	5.72	6.55	8.39	5.64	6.58	8.39
Mezyed	7.16	6.08	7.54	8.70	6.43	8.10	9.10	6.38	8.08	9.25	6.63	8.41	9.54	6.64	8.41	9.66
Madinat Zayed	7.72	6.82	6.80	8.04	7.26	7.19	8.46	7.27	7.18	8.52	7.55	7.45	8.81	7.56	7.42	8.90
Jabel Jais	0.46	0.18	-0.06	-0.14	0.19	-0.06	-0.15	0.18	-0.06	-0.19	0.19	-0.07	-0.17	0.19	-0.07	-0.19
Razeen	6.38	4.76	5.06	6.90	5.03	5.34	7.28	4.99	5.37	7.34	5.19	5.53	7.56	5.16	5.57	7.58
Qlaa	8.30	7.90	8.80	9.28	8.32	9.28	9.69	8.32	9.33	9.83	8.61	9.61	10.14	8.58	9.63	10.23
Owtaid	5.88	6.06	6.42	6.80	6.54	6.76	7.16	6.58	6.78	7.27	6.78	7.02	7.45	6.82	7.06	7.58
Abu Dhabi Airport	9.32	8.78	10.24	11.46	6.29	7.49	7.12	6.72	7.74	8.67	6.53	7.78	7.37	6.96	8.05	9.08
Fujairah Airport	13.76	12.74	13.52	13.98	13.20	14.19	14.34	13.22	13.99	14.44	13.65	14.63	14.73	13.66	14.50	14.96
Al ain Airport	8.38	8.08	8.54	9.50	9.40	9.54	8.69	8.40	8.89	9.81	9.77	9.91	9.05	8.74	9.25	10.24
Dubai Airport	11.36	11.44	13.02	13.38	10.05	12.03	11.81	11.86	13.53	13.80	10.35	12.43	12.17	12.29	14.08	14.34
Sharjah Airport	8.30	6.52	7.92	8.64	6.34	8.25	8.59	6.77	8.58	8.88	6.51	8.49	8.87	7.00	8.88	9.22
Ras AlKhaimah Airport	7.92	7.22	7.48	7.90	6.62	7.96	8.27	7.50	7.86	8.17	6.94	8.33	8.66	7.90	8.27	8.61

Table 2. Average minimum air temperature of all models using RCP 8.5

Stations Name/Year	2005	2010	2015	2020	2025	2030	2035	2040	2045	2050	2055	2060	2065	2070	2075	2080
Al ghewifat	3.76	4.34	6.18	5.96	4.54	6.41	6.20	4.71	6.68	6.47	4.96	7.04	6.82	5.10	7.33	7.08
Um Azimul	8.26	7.74	7.92	9.26	8.21	8.42	9.81	8.37	8.61	10.05	8.85	9.06	10.59	9.31	9.46	11.08
Al Fqaa	7.60	6.22	6.56	7.48	6.55	6.90	8.04	6.49	6.86	7.93	6.93	7.30	8.45	7.25	7.60	8.81
Mezaira	8.96	7.82	8.12	8.16	8.11	8.60	8.66	8.29	8.80	8.87	8.74	9.27	9.32	9.00	9.60	9.64
Maliaha	7.14	6.02	6.48	7.74	6.32	6.77	8.14	6.43	6.87	8.33	6.73	7.22	8.70	7.00	7.51	9.08
Mukhariz	5.36	6.00	5.84	5.94	6.28	6.23	6.34	6.43	6.40	6.54	6.78	6.73	6.83	7.05	6.96	7.13
Al Gazera B.G	3.28	3.64	4.74	5.06	3.84	5.04	5.41	3.91	5.16	5.58	4.13	5.43	5.84	4.28	5.66	6.09
Alquaa	5.94	5.14	5.84	7.56	5.43	6.29	8.02	5.51	6.47	8.19	5.84	6.79	8.64	6.12	7.07	9.01
Mezyed	7.16	6.08	7.54	8.70	6.41	8.08	9.18	6.51	8.29	9.47	6.90	8.74	9.94	7.22	9.12	10.42
Madinat Zayed	7.72	6.82	6.80	8.04	7.27	7.18	8.51	7.41	7.30	8.73	7.83	7.71	9.18	8.22	8.06	9.58
Jabel Jais	0.46	0.18	-0.06	-0.14	0.19	-0.06	-0.17	0.19	-0.07	-0.19	0.20	-0.07	-0.18	0.21	-0.07	-0.18
Razeen	6.38	4.76	5.06	6.90	5.01	5.36	7.30	5.09	5.46	7.49	5.38	5.76	7.85	5.62	6.00	8.13
Qlaa	8.30	7.90	8.80	9.28	8.33	9.30	9.75	8.46	9.49	10.06	8.91	9.97	10.55	9.31	10.38	11.02
Owtaid	5.88	6.06	6.42	6.80	6.56	6.78	7.23	6.70	6.93	7.44	7.06	7.29	7.78	7.34	7.64	8.09
Abu Dhabi Airport	9.32	8.78	10.24	11.46	6.37	7.61	7.26	6.90	7.97	8.89	6.87	8.19	7.77	7.45	8.62	9.60
Fujairah Airport	13.76	12.74	13.52	13.98	13.36	14.31	14.40	13.46	14.25	14.73	14.39	15.43	15.51	14.52	15.40	15.91
Al ain Airport	8.38	8.08	8.54	9.50	9.48	9.58	8.80	8.55	9.04	10.00	10.30	10.45	9.53	9.33	9.86	10.86
Dubai Airport	11.36	11.44	13.02	13.38	10.10	12.11	11.88	12.08	13.78	14.08	10.86	13.07	12.76	13.05	14.90	15.20
Sharjah Airport	8.30	6.52	7.92	8.64	6.37	8.36	8.64	6.88	8.68	9.04	6.85	8.90	9.31	7.43	9.33	9.75
Ras AlKhaimah Airport	7.92	7.22	7.48	7.90	6.65	8.02	8.28	7.61	7.98	8.31	7.15	8.58	8.92	8.22	8.60	8.96

Figure 10 shows the annual maximum air temperature, an average trend for maximum air temperature through historical and future periods for 14 weather station records used in this study. It is interesting to observe that the maximum annual air temperature for all weather stations showed an increment trend through near and far future periods using both RCPS 4.5 and 8.5. For the 14-weather station considered, 12 showed an increment trend and all results were very close. The rate of increase for all weather stations ranged from 5 °C and 7 °C using RCP 4.5 and RCP 8.5, respectively. Alqlaa weather station, located in the coastal area showed an increasing trend but with values lower than other weather station. In addition, Jabal Jais area showed the least values of temperature, this is linked to the fact that the area is mountainous and has high altitudes. However, the maximum air temperature will increase in the near and far future. Based on the analysis and results of this study, the climate models predict a rise in the average future temperature of UAE especially maximum temperatures where significant increase was obtained. The maximum air temperature is expected to be greater on land than over coastal area which means that climate in the UAE will get warmer due to the abundance of desert areas. This warmer climate will cause a change in all climate factors leading to higher rate of evaporation causing the water cycle to speed up. This change will consequently cause more precipitation as well as changes in relative humidity.

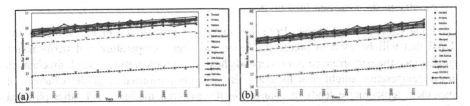

Fig. 10. (a) Average maximum air temperature of all models using RCP 4.5 (b) average maximum air temperature of all models using RCP 8.5

7 Conclusion

This research investigated the maximum and minimum climate zone for UAE. It is very important to study the effect of climate change on air temperature since in the design of the pavements, the temperature is a critical environmental component. Due to climate change, the binders to be chosen might vary, since the higher temperatures are more demanding, Climate change can cause significant changes in the spatial distribution of Performance Grade (PG) specification for selecting propose asphalt binders, as the magnitude of changes is directly related to the severity of the projected temperature. Temperature impacts the resiliency of the surface course in the case of asphalt roads. Asphalt layers lose their rigidity in excessively hot circumstances. Asphalt layers become weak at low temperatures and cracks are produced. conclusion from projections carried out using three climate models indicated that, generally, the hottest areas are in the south which is desert followed by the coastal area which overlooks the Arabian Gulf. Jabal Jais area at the north had the least maximum temperature of UAE and is in the far north-east. Also, the change of Maximum and Minimum Air temperature shows a general trend of rising from period to another. This result shows the significant warming by all models over UAE. Therefore, this research will serve as a guide for future road network planning and design in the UAE since temperature in the future shows significant variations from the historical temperature. This change will acquire adjustments to be done to accommodate the temperature rise. Main conclusions from this research state that:

- Climate change impacts different factors such as windspeed, humidity, precipitation, sunshine, and air temperature. It was found that the most affected factor in all periods is air temperature therefor, it was chosen to carry out further analysis.
- Different interpolation methods were tested to visualize the effect of climate change on air temperature and the IDW method showed best results and was chosen for further analyses.
- In the near and far future periods, it was found that areas of minimum temperature will continue to cool as well as areas of maximum temperature will show further rise in temperature.
- RCP 8.5; the worst-case climate scenario indicated that the temperature would continue to rise, and temperature increase will be observed by almost all areas of the country in the near and far future.

Finally, conclusions drawn from this study will provide a climate profile for the UAE that will be used in predicting future pavement temperature and performance grade since the performance grade depends heavily on temperature, and slight changes in air temperature might lead to major shifts in the performance grade. Changes of pavement temperature leads to changes in the material modules and other physical characteristics, which ultimately determine whether a pavement can operate under certain loading conditions. Climate change has the potential to enhance the incidence, length, and severity of degradation. Therefore, further research is currently being performed to address the impact of the predicted future climate on the pavement design based on the Mechanistic Empirical Pavement Design Guide (MEPDG).

Acknowledgements. Authors wish to acknowledge the ministry of energy and infrastructure for funding this research (Project no.130169) and the national center of meteorology research team for their valuable support concerning observed data supply. Special thanks are extended to Daniel Llort, Rafeeah Alali, Tala AbuShuqair and Mahra Naya Algafli. For their notable contribution in this project.

References

Baladi, G.Y.: Highway Pavement (NHI Course No. 13114. FHWA, McLean VA), May 1990

Buttlar, W.G., Roque, R.: Evaluation of empirical and theoretical models to determine asphalt mixture stiffnesses at low temperatures. In: Proceedings of the Technical Sessions, vol. 65, pp. 99–141. Asphalt Paving Technology: Association of Asphalt Paving Technologists (1996)

Cao, W., Hu, J., Yu, X.: A study on temperature interpolation methods based on GIS. In: 2009 17th International Conference on Geoinformatics, pp. 1–5. IEEE (2009)

Chai, G., van Staden, R., Guan, H., Loo, Y.: Impact of climate related changes in temperature on concrete pavement: a finite element study. In: 25th ARRB Conference Perth, Australia, pp. 1–12 (2012)

Chen, J., Wang, H., Xie, P.: Pavement temperature prediction: theoretical models and critical affecting factors. Appl. Therm. Eng. **158**, 113755 (2019)

Dempsey, B.J.: Moisture movement and moisture equilibria in pavement systems. Final Report (1979)

Gudipudi, P.P., Underwood, B.S., Zalghout, A.: Impact of climate change on pavement structural performance in the United States. Transp. Res. Part D: Transp. Environ. **57**, 172–184 (2017)

Knott, J.F., Sias, J.E., Dave, E.V., Jacobs, J.M.: Seasonal and long-term changes to pavement life caused by rising temperatures from climate change. Transp. Res. Rec. **2673**(6), 267–278 (2019)

Knutti, R., Masson, D.D., Gettelman, A.: Climate model genealogy: generation CMIP5 and how we got there, pp. 1194–1199 (2013)

Kodippily, S., Yeaman, J., Henning, T., Tighe, S.: Effects of extreme climatic conditions on pavement response. Road Mater. Pavement Des. **21**(5), 1413–1425 (2020)

Maleika, W.: Inverse distance weighting method optimization in the process of digital terrain model creation based on data collected from a multibeam echosounder. Appl. Geomat. **12**(4), 397–407 (2020). https://doi.org/10.1007/s12518-020-00307-6

Meyer, M., Flood, M., Dorney, C., Leonard, K., Hyman, R., Smith, J.: Synthesis of information on projections of climate change in regional climates and recommendation of analysis regions. National Cooperative Highway Research Project, NCHRP, pp. 20–83 (2013)

Monismith, C.L., Secor, G.A., Secor, K.E.: Temperature induced stresses and deformations in asphalt concrete. In: Association of Asphalt Paving Technologists Proceedings, vol. 34 (1965)

Qiao, Y., Dawson, A.R., Parry, T., Flintsch, G., Wang, W.: Flexible pavements and climate change: a comprehensive review and implications. Sustainability **12**(3), 1057 (2020)

Qiao, Y., Flintsch, G.W., Dawson, A.R., Parry, T.: Examining effects of climatic factors on flexible pavement performance and service life. Transp. Res. Rec. **2349**(1), 100–107 (2013)

Sibson, R., Bowyer, A., Osmond, C.: Studies in the robustness of multidimensional scaling: Euclidean models and simulation studies. J. Stat. Comput. Simul. **13**(3–4), 273–296 (1981)

Climate Teleconnections Contribution to Seasonal Precipitation Forecasts Using Hybrid Intelligent Model

Rim Ouachani[1,2(✉)], Zoubeida Bargaoui[2], and Taha Ouarda[3]

[1] Higher Institute of Transport and Logistics of Sousse, Sousse University, Sousse, Tunisia
rim.ouachani@istls.u-sousse.tn
[2] National Engineering School of Tunis, El Manar University, Tunis, Tunisia
[3] National Institute of Scientific Research -ETE, Québec, QC, Canada

Abstract. Long-term precipitation forecasts can provide valuable information to help mitigate some of the outcome of floods and enhance water manage. This study aims to extract significant information from oceanic-atmospheric oscillations that could enhance seasonal precipitation forecasting. A hybrid AI-type data-driven artificial neural network model called MWD-NARX based on a non-linear autoregressive network with exogenous inputs coupled to multiresolution wavelet decomposition (MWD) is then developed in this work. First, MWD is used to decompose climatic indices and precipitation data. Then the NARX ensemble model allowed to identify the statistical links between the decomposed indices and the decomposed precipitation according to temporal scales and to predict each precipitation decomposition. Ensemble precipitation forecasts are carried out over horizons ranging from 1 to 6 months. For operational forecasting, the forecasts obtained from the decompositions are summed to represent the true precipitation forecast value. The seasonal forecasts of average precipitation by sub-basins (SBV) of the Medjerda are carried out. Large scale teleconnections ENSO, PDO, NAO, AO and as well as Mediterranean Oscillation were used as inputs to the model. The forecasting model coupled to data pre-processing method made it possible to produce very satisfactory forecasts of non-stationary data by extracting modes of variability. The results indicate that exogenous inputs like climatic indices clearly improves the accuracy of forecasts on 82% of SBVs and increases the forecast lead-time up to 6 months. This research is the first of its kind using a hybrid prediction approach by means of indices related to ocean-atmospheric oscillations in North Africa.

Keywords: Seasonal forecasting · Precipitation · Climate teleconnection indices · Multiresolution wavelet decomposition · NARX · Neural networks

1 Introduction

A Precipitation forecasting is useful for water resources and hydraulic structures management. Extreme events have devastating consequences that disfigure the nature and cause thousands of casualties. For good management and forecasting of floods, it is essential to go through a long-term precipitation forecasting. Seasonal forecasting is useful for the agricultural sector which is among the key sectors of the Tunisian economy. A reliable forecast of the precipitation remains a challenge given that the precipitation is difficult to understand and to model because of the complexity of the phenomena and the atmospheric processes that generates (French et al. 1992). Several methods are used for forecasting rainfall from empirical models to the statistical models through the artificial intelligence models. Artificial neural networks are widely used in hydrological forecasting as they are very well-suited tools for the recognition of nonlinear relationships and their predictions. But their performance remains linked to several parameters from the network architecture with the number of hidden layers, the number of neurons or activation functions. Among the problems with neural networks is overfitting and poor generalization. The bootstrap method to produce ensemble forecasts by adding a noise to the outputs in the training period is among the methods which reduce the overfitting (Fitzgerald et al. 2013). However, recent work on the use of artificial intelligence in forecasting has emphasized the need for signal pre-processing to explore the explanatory variable as much as possible on scales of variability and to allow a selection of variables with predictive potential. Hence the need for hybrid models and spectral analysis methods. Improving the performance can be obtained with a preprocessing of the input and output data. Among the signal pre-processing methods we cite the Multiresolution Wavelet Decomposition (MWD) (Adamowski and Prokoph 2013). MWD is one the methods in the time-frequency domain that can process non-linear and non–stationary data. An application of this approach is presented in Muszkats et al. (2021), in which MWD was applied to many data analysis. Wang et al. (2009) have developed a hybrid model that combines wavelet analysis and artificial neural network (ANN), called "Wavelet Network", in order to take into account the non-linear character and the multi-temporal scale of the hydrological series. This model allows for short-term as well as long-term forecasting. Their model has proven that it can increase forecast accuracy and extend the forecast lead time. Kisi et al. (2014) reported the use of ANNs in forecasting hydrological variables as well as combined methods with ANNs. To cope with the long forecasting leadtime, several methods have been proposed such as the non-linear autoregressive network with exogenous inputs (NARX) (Leontaritis and Billings 1985a, b; Lin et al. 1996). The NARX network was used in forecast mode by, Ardalani-Farsa and Zolfaghari (2010); Chtourou et al. (2008); Narendra (1990). This network has made it possible to provide encouraging results. For example, Cadenas et al. (2016) used NARX model for wind speed fsssorecasting. The use of NARX networks isn't well developed in hydrological forecasting (Jiang et al. 2021, Di Nunno et al. 2021, Wang and Chen 2021). We cite, for example, Chang et al. (2014) who compare three models of ANNs: a static model, the backpropagation, and two dynamics (Elman and NARX) for forecasting stream levels. They conclude that the NARX network performs the best for real-time forecasting. A NARX network is based on recurrent neural networks and aims to provide better detection of long-term dependencies in forecasting. In addition, the NARX network is considered in this work in order to take into account the climate oscillation indices as inputs to the forecasting model. Through the use of ANNs, Abbot and Marohasy (2015) and Fallah-Ghalhari (2012) produce rainfall forecasts through a selection of predictive climatic indices.

In spite of the generalization ability of ANNs and due to the nonlinear and non-stationary nature of the rainfall time series, it is necessary the search for analysis alternatives that improve the accuracy of predictions.

Ouachani et al. (2013) studied the effect of climate variability on precipitation in the Medjerda basin and found that indices related to El Nino–Southern Oscillation (ENSO) as well as Mediterranean Oscillation have potential power in precipitation and streamflow forecasting. Large scale teleconnections ENSO, PDO, NAO, AO and as well as Mediterranean Oscillation were used as inputs to the model.

Could exogenous inputs such as climate indices, defined as difference between sea surface temperature or sea level pressure between two different localizations in the sea, add some additional information to an AI based MWD model is the question that we will try to respond.

This paper proposes a hybrid AI-type data-driven artificial neural network model called MWD-NARX based on a non-linear autoregressive network with exogenous inputs coupled to multiresolution wavelet decompositionfor long term rainfall forecasting. The exogenous inputs are climate indices. The preprocessing data method is applied to the inputs as well as the outputs before their forecast by the neural network model. The experiments are performed on observed monthly rainfall signal in the Medjerda basin in Tunisia. Ensemble precipitation forecasts are carried out over horizons ranging from 1 to 6 months. For operational forecasting, the forecasts obtained from the decompositions are summed to represent the true precipitation forecast value. The remaining part of the paper is organized as follows. The proposed methodology is detailed in Sect. 2, where a brief description of multiresolution wavelet decomposition (MWD), backpropagation scheme and NARX is developped as well as the hydro-climatic data. The obtained results are described in Sect. 3. Finally, the paper is concluded in Sect. 4.

2 Methodology

We opted for hybrid models of artificial intelligences with dynamic and recurrent artificial neural networks which appear to be promising for long-term predictions and can predict complex relationships between the explanatory variables and the target hydrological variable. Even if they suffer from a fairly limited power to physically explain hydrological processes for hydrological forecasting (Awchi 2014).

2.1 The Wavelet Multiresolution Decomposition

For analyzing non-stationary hydrologic series, Adamowski and Prokoph (2013) found MWD to be a potentially very useful method. In this study, the Mallat decomposition algorithm (Mallat 1989) is used. According to Mallat's theory, the original discrete-time series is decomposed into a series of linearly independent approximation signals and details.

As described by Kucuk and Ağirali-super (2006), the process includes a number of successive filtering steps as shown in Fig. 1 (b). Figure 1 (a) displays a full three-level MWD regime. The original signal $x(t)$ is first decomposed into an approximation and a detail. The decomposition process is thus reiterated, with a successive approximation then being decomposed in turn so that the original signal at the finest resolution is transformed into a coarse number of resolution components. The decomposition of $x(t)$ into A_1 and D_1 is a decomposition on the first scale. As shown in Fig. 1 (b), the approximation A_{i+1} is carried out by letting A_i pass through the low-pass filter H' and the subsampling by 2 (denoted by $\downarrow$) considering that the version Detailed D_{i+1} is obtained by letting A_i pass through the high-pass filter G' and the downsampling by two 2 $\uparrow$. The details then constitute the low scale, the high frequency components considering the approximations to be the large scale, low frequency components. Finally, the original signal $x(t)$ is decomposed into several detailed components and an approximation component that can reflect a trend in the raw series. The procedure provides raw data that can be broken down into $m+1$ components if m is set in the discrete wavelet transform. It is possible to reconstruct the original signal with the sum of the coefficients of the approximations and the details.

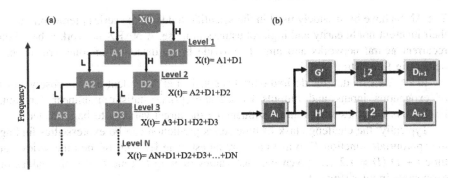

Fig. 1. Successive filtering using 2N to N scales successive details and a low frequency approximation

MWD allows us to decompose a signal x (t) according to Eq. 1:

$$x(t) = \sum_{j=1}^{J} D_j + A_J \qquad (1)$$

Where D_j constitutes the jth level of detail of a time series of length N, j = 1,..., J. A_J, a time series of length N, is the level approximation J. MWD requires setting the number of decomposition levels (scales) appropriate for the data. Often, the number of decomposition levels is chosen according to the length of the signal and given by the expression (2): (Tiwari and Chatterjee 2010)

$$J = \text{int}[\log(N)] \qquad (2)$$

where J is the level of decomposition and N is the number of samples. If we increase the level of decomposition, we tend to explore the signal at very low scales and conversely if the number of decompositions is very low, we will have a smoothing of the series but in any case we will never lose information with the decomposition by MWD as long as the reconstructed signal is the sum of the decompositions. We propose to decrease the number of levels so as to have less complicated series to predict. According to this methodology, in our study we have used 5 decomposition levels for monthly data (2^1 month mode, 2^2 month mode, ..., 2^5 month mode), because the optimum level of decomposition for monthly 2 and 3.

For the MWD, cross correlations with time lag are calculated between the decompositions of climate indices (exogenous inputs) and monthly precipitation (outputs) on the same scale. The indices generally having correlations largely exceeding the 95% significance level and generally exceeding 0.3 (Coulibaly and Burn 2004) for the small scales are retained as predictors.

2.2 The Non-linear Autoregressive Network with Exogenous Inputs (NARX) Artificial Intelligence Model

The ANNs have been widely used in the scientific field of time series prediction due to their inherent nonlinearity and high robustness in noise. A NARX network is based on recurrent neural networks and aims to provide better detection of long-term dependencies in forecasting.

In a NARX model, the desired output is considered dependent on past observations of exogenous inputs and outputs by an appropriate statistic nonlinear function. Exogenous variables are variables influencing the target variable to be predicted.

Typically, the challenge task of time series prediction can be expressed as finding the appropriate function F so as to acquire an estimate $\hat{y}(t+D)$ of the time series y at time $t + D$ ($D = 1,2 ...$) given the past values of y up to time t, plus the values of exogenous input x (Fig. 2).

A NARX model is formulated in discrete time as a recursive input-output Eq. (3) such that:

$$\hat{y}(t+D) = F(y(t), ..., y(t-d_y), x(t), ..., x(t-d_x)) \qquad (3)$$

where $y(t)$ and $x(t)$ represent the values of y and x in time t respectively. The variables d_y and d_x are the lag time parameters of model and in case of $D = 1$ we have the one step ahead prediction of time series y.

In this paper, we apply the backpropagation NN, which is one of the most popular NNs. The backpropagation learning algorithm includes four main steps as: Feed forward computation, Back propagation to the output layer, backpropagation to the hidden layer and weight updates. This forecasting engine has good abilities for dealing with nonlinear systems, such as forecasting problem of precipitation. In this problem, the next several hours' prediction is the main function of the forecast engine. The structure of a three layered back propagation NN is presented in Fig. 2. For each hidden neuron j, the input xj and output yj are defined as:

$$\hat{x}_i = \sum_{i=1}^{n} w_{ij} y_i \quad (4)$$

$$\hat{y}_j = F(\hat{x}_j + h_j) \quad (5)$$

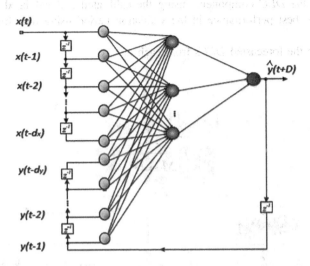

Fig. 2. NARX type neural network modified after (Mandal and Prabaharan 2006)

where w_{ij} is the weight between the ith neuron in the input layer and j^{th} neuron in the hidden layer; $F(.)$ is the activation function of the hidden neurons; $\hat{x}_i$ and $\hat{y}_j$ are the output of input neuron i and hidden neuron j; h_j is the bias of hidden neuron j. Shu and Ouarda (2007) propose a number of nodes in the hidden layer less than half the number of entries. In this work, the learning rule used to adjust the NAR weights is based on the Levenberg-Marquardt method, one of the BP algorithms. It is being more powerful and faster than the conventional gradient descent techniques.

2.3 Proposed MWD-NARX Forecasting Model

The overall procedure for MWD-NARX is given in Fig. 3 and described below:

(1) Data pre-processing is performed on the input and output data. We start by standardizing the hydrological series and climatic indices using Eq. 6.

$$x\_norm = (x - \bar{x})/(\sigma(x)) \quad (6)$$

(1) Decompose the standardized time series (x_norm_t) into a finite number of wavelet decompositions (*DECs*).
(2) Inputs are thus delayed in time according to the forecast lead-time. Figure 4 shows sequences of time-lagged inputs to predict a desired output (monthly

precipitation) at a given forecast lead-time. The time offset takes into account the system memory.

(3) Calibrate the NARX model using each DEC_i of the selected indices and precipitation as inputs to predict DEC_i of the precipitation. Validate the model using data from the validation period. Where the training data set represents 60% of the data series while 20% is for the validation of the model and the last 20% is for forecasting.

(4) Predict the *DEC* components using the calibrated and validated NARX model with the best performance in the validation period using data from forecasting period.

(5) Sum up the forecasted *DECs* from each model.

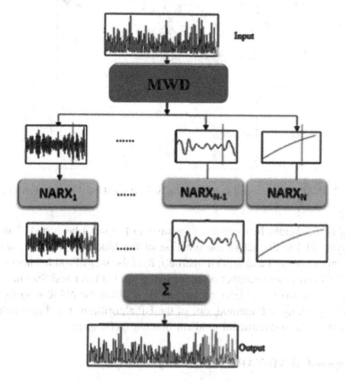

Fig. 3. Proposed methodology of the hybrid intelligent model MWD-NARX

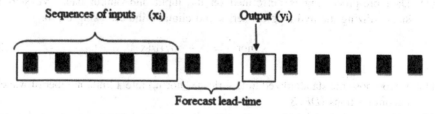

Fig. 4. Sequences of time-lagged inputs (x_i) to predict a desired output (y_i), here monthly precipitation, at a given forecast lead-time.

The NARX model is run 100 times using the bootstrap procedure that adds noise with known standard deviation equal to 0.2 to the output. The mean of the obtained ensembles represents the forecasting value. For all the 100 experiments, the performance measure of the mean is computed and thus a reliable estimate of the performance is obtained. Three layers are considered for the NARX model: one input layer, one hidden layer and one output layer. Several network architectures are tested by varying the number of nodes in the hidden layer from 5 to 20.

We compare the performance of the models in terms of performance criteria, for different numbers of neurons in the hidden layer, on validation set. The architecture with the best performance in the validation period is retained to make forecasts.

Among the model efficiency criteria in the literature, the coefficient of correlation (R^2) (Eq. 7) and the Nash–Sutcliffe coefficient (NASH) (Eq. 8) of efficiency are the most commonly employed performance evaluation criteria and are also found to be good evaluation criteria by experts (Crochemore et al. 2015).

The coefficient of correlation (R^2) is given by:

$$R^2 = \frac{\sum_{i=1}^{n}(y_i - \bar{y})(y_{pi} - y_p)}{\sqrt{\sum_{i=1}^{n}(y_i - \bar{y})^2 \sum_{i=1}^{n}(y_{pi} - y_p)^2}} \quad (7)$$

where n refers to the total number of observations; $\hat{y}_i$, y_{pi}, $\bar{y}$, $\bar{y}_p$ represent the predicted monthly precipitation, observed precipitation, mean observed precipitation data and the mean predicted precipitation value, respectively.

The Nash–Sutcliffe coefficient (NASH) is given by:

$$NASH = 1 - \frac{\sum_{i=1}^{n}(y_i - y_{pi})^2}{\sum_{i=1}^{n}(y_i - \bar{y})^2} \quad (8)$$

2.4 Hydro-Climatic Data

The analysis of the monthly precipitation is carried out using data from the Medjerda river basin, a trans-boundary river, located in northern Tunisia and which accounts for the Mediterranean water budget in the Blue Plan (Margat 2004). Rainfall data is provided by the National Water Resources Division of Tunisia (Ouachani 2016).

The Medjerda river basin has been discretized into a 5 km grid bounded by longitudes 8.18479594° and 10.21696646° and Latitudes 35.44324153° and 37.11214572°. Based on the entire observation network of Basin since 1899 and using the functional stations in 2010 (Fig. 5a), the average of the 3 closest observed neighbors is used to calculate the rainfall at the grid points. Then we estimated the average rainfall per sector (SBV 1 to 11) using for each the rainfall at the nodes inside the sector boundary (Fig. 5b).

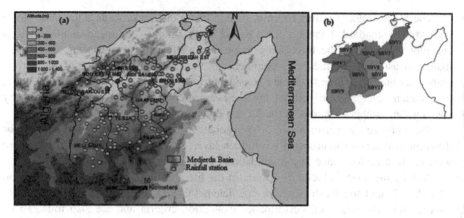

Fig. 5. Map of rainfall stations in the Medjerda river basin (a) and the 11 sectors from 1 to 10 for rainfall aggregation (b).

Eleven climate indices; the Arctic Oscillation, the Multivariate El Nino Southern Oscillation Index (MEI), Southern Oscillation Index (SOI), ENSO SSTs (NINO1+2 Niño3 Niño3.4 Niño3.4 North Atlantic Oscillation (NAO) and Mediterranean Oscillation Index (MOAC), Western Mediterranean Oscillation (WeMOI) and Pacific Decadal Oscillation (PDO) are used in this work as suggested by Ouachani et al. (2013). The common period 1950 to 2011 between time series is chosen. Before analysis, the precipitation and climate indices time series are standardized, by subtracting the mean and dividing by the standard deviation, to ensure a comparable scale.

3 Results

The seasonal forecasts of aggregated precipitation by sub-basins (SBV) of the Medjerda are carried out. Ensemble precipitation forecasts are carried out over 1 to 6 months lead time. For operational forecasting, the forecasts obtained from the decompositions are summed to represent the true precipitation forecast value. From the calibration period, the NARX ensemble neural network model is used to identify statistical relationships between the decomposed indices and precipitation decomposed on the same time scale by MWD.

Figure 6 illustrates the spatial variability of the Forecast Score R^2 of forecasts at the scale of SBVs by the MWD-NARX model using as inputs of the model only the climate indices with 1- and 3-months delay time. At 3 months lead time (Fig. 6b) the forecast score R^2 vary from 0.41 to 0.63 while for a shorter lead time of 1 month (Fig. 6-a) we obtain much more variable correlation coefficients from 0.18 on the side of the Ghardimaou basin (Algerian frontier) to 0.69 in the center of the Medjerda basin. This spatial variability is observed all the time. These results are similar to the study by Kashid and Maity (2012) on the monthly precipitation forecasting over homogeneous regions of the Indian Monsoon based on ENSO and the Indian Ocean Dipole (IOD). They obtain correlation coefficients varying according to the regions from 0.03 to 0.77. Across the Globe, Zeng et al. (2011) manage to make precipitation forecasts over Canada from a set of predictors formed by quasi-global SST anomalies and geopotential anomalies at 500 hPa over the northern hemisphere as well as six climatic indices (NINO3 .4, NAO, PNA, PDO, the Scandinavian mode and the East Atlantic mode. We can also observe that on the side of the Mediterranean Sea (east side of the basin), the far west and the Wadi Tessa basin, we have the lowest performance. We can conclude that the effect of climate oscillations on precipitation begins from 3 months of its observation.

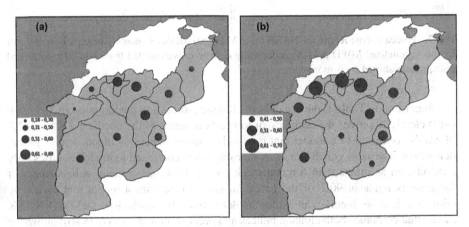

Fig. 6. Forecast score R^2 of monthly precipitation at the scale of SBVs by the MWD-NARX model at 1 month (a) and 3 months (b) lead time.

Figure 7 reports the forecast Scores R^2 (a) and NASH (b) maps for a model that combines both climate indices shifted from 1 to 20 months and precipitation shifted by 6 months. The aim of this experiment is to increase the prediction time frame by using climatic indices as a complement to the ground precipitation data 6 months back. It is clear that the forecast score R^2 is significantly improved in all the sub-basins. And the spatial variability of this score notes using only clime indices as predictors is non more observed when the precipitation of 6 months later are added to the predictors database. While low values of the NASH forecast score are obtained near the Mediterranean Sea and the Algerian frontier. In conclusion, the use of climate indices effectively improves the accuracy of forecasts in terms of the NASH and R^2 score forecasts.

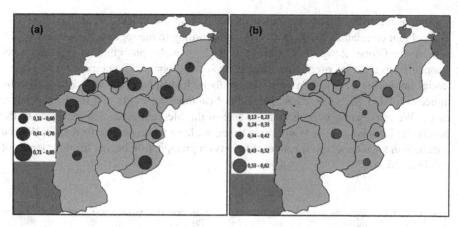

Fig. 7. Forecast score R^2 (a) and NASH (b) at Medjerda SBVs for monthly precipitation forecast by the regionalized MWD-NARX model using climate indices shifted from 1 to 20 months and precipitation shifted by 6 months

According to Table 1, the use of both climatic indices and precipitation (IC(+1) & P(+6)) clearly improves the accuracy of forecasts in terms of the coefficient R^2 on 82% of SBVs compared to a model that uses only climate indices as inputs with 1 month delay time. The same conclusion is made using when compared to model that uses only precipitation as inputs with 6 months delay time. However, a better enhancement of forecasts is made in 90% of the basins in comparison with a model that uses only climate indices as inputs with 3 months delay time. In conclusion, theMWD-NARX model that combines both climatic indices and precipitation (IC(+1) & P(+6)) allows an increase in the forecast lead-time.

Table 1. Statistics on R^2 forecast score in function of model predictors

R^2	IC(+1)	IC(+3)	P(+6)	IC(+1) & P(+6)
% SBVs enhanced	82	91	82	
Max (R^2)	0,69	0,63	0,68	0,79
Min (R^2)	0,18	0,41	0,17	0,51
Mean (R^2)	0,52	0,53	0,50	0,64

The regressions of the forecasts compared to the observations (Fig. 8) using the adopted model that combines both climate indices and precipitation as inputs already show that the point cloud is very little dispersed compared to the trend and is lower than the first bisector. As the points show, for example at SBV4 (Fig. 8d) good results are obtained as correlation coefficient is close to 0.79. Overall there is a good dynamic between the observations and the forecasts. The chronologies of the observations and those of the forecasts illustrate a good agreement (Ex. SBV 3 (Fig. 8c) & SBV4 (Fig. 8d)) and a very well reproduced dynamic. We also observe a very narrow spaghetti of the 100 simulations around the mean considered as final forecast. Regarding the flood peaks, we notice a very slight underestimation of the peaks and overestimation of low precipitation. In conclusion, forecasts by the MWD-ANN model are much less biased at a one-month lead-time.

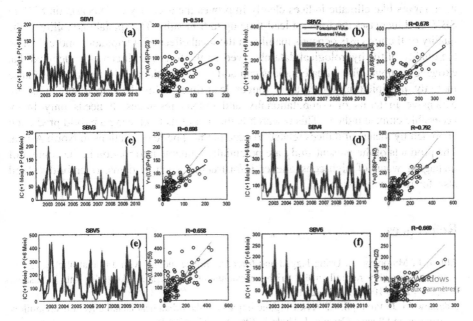

Fig. 8. Chronologies of monthly precipitation observed and forecasted at 6 SBVs (SBV1 to SBV6), respectively, from (a) to (f) using 1-month delayed climate indices and 6-month delayed precipitation using the regionalized MWD-NARX model. On the right the point cloud and the regression curve.

The poor reconstruction of the peaks is potentially due to the 6-month lag for the rainfall data, making the information on the rainy season lost because in general this season is 5 to six months long. The use of shorter lead-time seems to be necessary to obtain better forecasts of the precipitation peak values. Other studies have also noted the difficulties presented by forecasting models at the monthly scale for reconstructing peaks at 6 months lead times and more.

4 Conclusions

A hybrid intelligent forecasting model is proposed, which is based on Multiresolution Wavelet Decomposition (MWD) and Non-linear autoregressive network with exogenous inputs (NARX) scheme of ANN. The ensemble NARX based model allowed to identify the statistical links between the decomposed indices and the decomposed precipitation according to temporal scales and to predict each precipitation decomposition. The forecasting model coupled to data preprocessing method made it possible to produce very satisfactory forecasts of non-stationary data by extracting modes of variability. The MWD-NARX scheme improves the forecasting results and offers a simple approach for the stable prediction of non-stationary data. MWD allows extracting components to help reducing predictive uncertainty as well as improving forecasts of a Feed-Foreward neural network model. The results indicate that exogenous inputs like climatic indices clearly improves the accuracy of forecasts on 82% of SBVs and increases the forecast lead-time up to 6 months. The spatial variability of the quality of the forecasts depends mainly on the local effect of precipitation more than on the quality of the hydrological data observed on the forecasts. It can be concluded that exogenous inputs like climate indices can add some additional information to enhance monthly precipitation forecasts at longer lead-times. This developed model is not expensive in terms of computationally and human resources. It needs only freely accessible climate indices. This research is the first of its kind using a hybrid prediction approach by means of indices related to ocean-atmospheric oscillations in North Africa. The products of the present study are potentially valuable information in the context of managing water of high-consequence infrastructure such as dams, river bridges and also for road conception.

References

Abbot, J., Marohasy, J.: Using lagged and forecast climate indices with artificial intelligence to predict monthly rainfall in the Brisbane catchment, Queensland, Australia. Int. J. Sustain. Dev. Plann. **10**(1), 29–41 (2015)

Adamowski, J., Prokoph, A.: Assessing the impacts of the urban heat island effect on streamflow patterns in Ottawa, Canada. J. Hydrol. **496**, 225–237 (2013)

Ardalani-Farsa, M., Zolfaghari, S.: Chaotic time series prediction with residual analysis method using hybrid Elman–NARX neural networks. Neurocomputing **73**(13), 2540–2553 (2010)

Awchi, T.A.: River discharges forecasting in northern Iraq using different ANN techniques. Water Resour. Manage. **28**(3), 801–814 (2014). https://doi.org/10.1007/s11269-014-0516-3

Cadenas, E., Rivera, W., Campos-Amezcua, R., Cadenas, R.: Wind speed forecasting using the NARX model, case: La Mata, Oaxaca, Mexico. Neural Comput. Appl. **27**(8), 2417–2428 (2016). https://doi.org/10.1007/s00521-015-2012-y

Chang, F.J., Chen, P.A., Lu, Y.R., Huang, E., Chang, K.Y.: Real-time multi-step-ahead water level forecasting by recurrent neural networks for urban flood control. J. Hydrol. **517**, 836–846 (2014)

Chtourou, S., Chtourou, M., Hammami, O.: A hybrid approach for training recurrent neural networks: application to multi-step-ahead prediction of noisy and large data sets. Neural Comput. Appl. **17**(3), 245–254 (2008). https://doi.org/10.1007/s00521-007-0116-8

Crochemore, L., et al.: Comparing expert judgement and numerical criteria for hydrograph evaluation. Hydrol. Sci. J. **60**(3), 402–423 (2015)

Coulibaly, P., Burn, H.D.: Wavelet analysis of variability in annual Canadian streamflows. Water Resour. Res. **40**, W03105 (2004)

Di Nunno, F., Granata, F., Gargano, R., de Marinis, G.: Forecasting of extreme storm tide events using NARX neural network-based models. Atmosphere **12**(4), 512 (2021)

Fallah-Ghalhary, G.A.: Rainfall prediction using teleconnection patterns through the application of artificial neural networks. Mod Climatol **1**, 362–386 (2012)

French, M.N., Krajewski, W.F., Cuykendall, R.R.: Rainfall forecasting in space and time using neural network. J. Hydrol. **137**, 1–31 (1992)

Fitzgerald, J., Azad, R.M.A., Ryan, C.: A bootstrapping approach to reduce over-fitting in genetic programming. In: Proceedings of 15th Annual Conference on Genetic and Evolutionary Computation Conference (GECCO) (2013)

Jiang, F., et al.: Flood forecasting using an improved NARX network based on wavelet analysis coupled with uncertainty analysis by Monte Carlo simulations: a case study of Taihu Basin. China. J. Water Clim. Change **12**, 2674–2696 (2021)

Kashid, S.S., Maity, R.: Prediction of monthly rainfall on homogeneous monsoon regions of India based on large scale circulation patterns using Genetic Programming. J. Hydrol. **454**, 26–41 (2012)

Kisi, O., Latifoğlu, L., Latifoğlu, F.: Investigation of empirical mode decomposition in forecasting of hydrological time series. Water Resour. Manage. **28**, 4045–4057 (2014). https://doi.org/10.1007/s11269-014-0726-8

Kucuk, M., Ağirali-super, N.: Wavelet regression technique for streamflow prediction. J. Appl. Stat. **33**(9), 943–960 (2006)

Leontaritis, I.J., Billings, S.A.: Input output parametric models for non linear systems – Part I: deterministic non linear systems. Int. J. Control **41**(2), 303–328 (1985a)

Leontaritis, I.J., Billings, S.A.: Input output parametric models for non linear systems – Part II: stochastic non linear systems. Inter. J. Control **41**(2), 329–344 (1985b)

Lin, T., Horne, B.G., Tino, P., Giles, C.L.: Learning long-term dependencies in NARX recurrent neural networks. IEEE Trans. Neural Networks **7**(6), 1329–1338 (1996)

Mallat, S.G.: A theory for multiresolution signal decomposition: the wavelet representation. IEEE Trans. Pattern Anal. Mach. Intell. **11**(7), 674–693 (1989)

Mandal, S., Prabaharan, N.: Ocean wave forecasting using recurrent neural networks. Ocean Eng. **33**, 1401–1410 (2006)

Margat, J.: In the long term, will there be water shortage in Mediterranean Europe? In: Marquina, A. (ed.) Environmental Challenges in the Mediterranean 2000–2050. NATO Science Series, vol. 37, pp. 233–244. Springer, Dordrecht (2004). https://doi.org/10.1007/978-94-007-0973-7_14

Muszkats, J.P., Seminara, S., Troparevsky, M.: Applications of Wavelet Multiresolution Analysis. ICIAM 2019 SEMA SIMAI Springer Series. Springer, Cham (2021). https://doi.org/10.1007/978-3-030-61713-4

Narendra, K.S.: Adaptive control using neural networks. Neural networks for control, 3 (1990)

Ouachani, R.: Développement d'un modèle de prévision à long terme des apports en eaux de cours d'eau. Ph. D. Thesis, University of Tunis El Manar (2016)

Ouachani, R., Bargaoui, Z., Ouarda, T.: Power of teleconnection patterns on precipitation and streamflow variability of upper Medjerda Basin. Int. J. Climatol. **33**(1), 58–76 (2013)

Shu, Ch., Ouarda, T.B.M.J.: Flood frequency analysis at ungauged sites using artificial neural networks in canonical correlation analysis physiographic space. Water Resour. Res. **43**(7) (2007)

Tiwari, M.K., Chatterjee, C.: Development of an accurate and reliable hourly flood forecasting model using wavelet–bootstrap–ANN (WBANN) hybrid approach. J. Hydrol. **394**(3), 458–470 (2010)

Wang, J., Chen, Y.: Using NARX neural network to forecast droughts and floods over Yangtze River Basin. Nat. Hazards 1–22 (2021). https://doi.org/10.1007/s11069-021-04944-x

Wang, W., Jin, J., Li, Y.: Prediction of inflow at three gorges dam in Yangtze River with wavelet network model. Water Resour. Manage. **23**(13), 2791–2803 (2009). https://doi.org/10.1007/s11269-009-9409-2

Zeng, Z., Hsieh, W., Shabbar, A., Burrows, W.: Seasonal prediction of winter extreme precipitation over Canada by support vector regression. Hydrol. Earth Syst. Sci. **15**(1), 65–74 (2011)

Development of Pavement Temperature Prediction Models for Tropical Regions Incorporation into Flexible Pavement Design Framework

Chaitanya Gubbala[1], Krishna Prapoorna Biligiri[1(✉)], and Amarendra Kumar Sandra[2]

[1] Indian Institute of Technology Tirupati, Tirupati, India
{ce21d001, bkp}@iittp.ac.in
[2] Rajiv Gandhi University of Knowledge Technologies, Nuzividu, India
aksandra@rguktrkv.ac.in

Abstract. Asphalt pavements are affected by several factors such as traffic loading, materials properties, and environmental conditions. Of these, the environmental conditions have significantly influenced the long-term pavement performance since the variability in climatic factors affects the choice of materials used in pavement construction and rehabilitation. Therefore, collecting and using climatic data of a location becomes critical in selecting the right type of construction materials and the rational design of the pavement layers. Currently, the flexible pavement design guidelines in India do not incorporate real-time climatic data extracted from weather stations; instead, a standard climatic condition is assumed across the country that has diverse climates. To address this, this study was formulated with an object to develop a set of prediction models for estimating pavement temperatures over the entire duration of one full day, representative of any seasonal variation occurring annually in the tropical climate. The scope of the work encompassed: (a) development of mean monthly maximum (M_{max}) and minimum (M_{min}) temperature models based on the historical weather parameters and geographical coordinates collected from eight designated weather stations by the India Meteorological Department for ten years across the State of Andhra Pradesh, India covering 620 data points, (b) formulation of two different pavement temperature models corresponding to the half-day seasonal variation, and (c) establishment of generalized annual pavement temperature models for tropical regions for the developed M_{max} and M_{min} models. All the predicted models had high accuracy characterized by excellent statistical goodness of fit measures ($R^2_{adj} > 90\%$; $S_e/S_y < 0.350$).

Keywords: Pavement temperature prediction models · Tropical climate · Flexible pavement design · Environmental conditions · Long-term pavement performance

1 Introduction

The fundamental parameters required for the design of flexible asphalt pavements include traffic, subgrade properties, and modulus/stiffness of all pavement layers in the system. The design and construction of a high-quality pavement system would be successful if there is least variation in the materials properties of the various layers throughout the design service life. In reality, due to substantial variation in the environmental factors over the service life, the properties of materials used in construction will get changed compared to the design value leading to several types of distresses. Consequentially, the mechanical properties of asphalt concrete layer would get affected as a function of pavement surface temperature and other weather parameters. Therefore, an accurate prediction of these variations is necessary to arrive at a rational pavement design process that considers traffic loadings along with thermophysical properties.

Globally, several researchers have developed models for predicting the pavement temperatures based on climatic parameters such as air temperature, solar radiation, wind velocity, relative humidity, and thermal properties of materials (Thompson et al. 1987, Bosscher et al. 1998, Hassan et al. 2005, Gui et al. 2007, Loganathan and Souliman 2017, Li et al. 2018, Qin et al. 2019). During the development of the Strategic Highway Research Program (SHRP) in the USA, researchers proposed a parabolic equation to find the maximum pavement temperature based on the maximum air temperature and solar radiation and reported that pavement temperature could be estimated to an accuracy of up to 3 °C (Solaimanian and Kennedy 1993). However, this model accounted for maximum temperature and did not consider the temperatures occurring during winter. Later, Bosscher et al (1998) developed a pavement temperature prediction model as a function of time, depth from the pavement surface, and weather conditions. Emphasizing the importance of temperature on the behavior of asphalt concrete layers, Hassan et al. (2005) developed low and high pavement temperature prediction models for the selection of PG asphalt binder by considering air temperature, solar radiation, and duration of solar radiation as predictor variables. Diefenderfer et al. (2006) developed models for predicting daily maximum and minimum pavement temperatures highlighting the usefulness of temperature profile across the depth of asphalt concrete layers in selecting appropriate materials.

The impacts of climate change on pavements and pavement-related infrastructure were studied by Bizjak et al. (2014), who concluded that the effect of climate change on pavements depended upon local environmental conditions. Islam et al. (2015) developed a model to predict the pavement temperature at any depth using various parameters such as solar radiation, pavement surface temperature, and depth at which temperature is to be predicted. The limitation of this model was that the pavement surface temperature was predicted using average air temperature only. However, it is noteworthy that the surface pavement temperature may get affected by factors such as relative humidity, precipitation, and wind speed. Underscoring the importance of using low-temperature sensitive materials in urban areas to reduce Urban Heat Island (UHI), Chandrappa and Biligiri (2016) developed pavement surface temperature models based on meteorological factors to estimate heat energy flux of various types of pavement materials. In another attempt, Sreedhar and Biligiri (2016) developed pavement temperature prediction models based on thermophysical properties (thermal conductivity, specific heat capacity, and density) of different asphalt concrete mixes prepared in the laboratory and recorded climatological parameters (wind speed, air temperature, and relative humidity), while correlating the models with actual temperatures collected from the field pavement systems. Other researchers stressed the use of longer time series of weather and climate data (temperature, wind speed, relative humidity, precipitation, percentage of sunshine, and moisture index) that will help develop a framework and incorporate climatic changes into the mechanistic-empirical based pavement design (Mills et al. 2009; Hasan and Tarefder 2018; Stoner et al. 2019; Knott et al. 2019). Further, Assogba et al. (2020) recently indicated that the temperature profile along the depth of the pavement is nonlinear and directly affects the performance of pavement systems.

In gist, researchers have used different techniques and methodologies such as numerical analyses, analytical approaches, and statistical techniques to develop pavement temperature prediction models. Most of them also stressed the need for incorporating long-term real-time climatic parameters to formulate mechanistic pavement design process considering several thermophysical properties such as ambient temperature, solar radiation, location, wind speed, reflectance of pavement surface, and precipitation. However, the developed models were location-specific and are dependent on the local environmental conditions as well as materials used for construction. Furthermore, the performance of asphalt pavement was significantly influenced by the variation of pavement temperature during its service life in addition to traffic loadings. Currently, the flexible pavement design process in India does not incorporate real-time climatic data extracted from weather stations; instead, a standard climatic condition is assumed across the country that has diverse climates. Thus, there is a need to develop a comprehensive climatic database consisting of various parameters to formulate a robust pavement design framework for the conditions in India and similar climatic regions across the world.

In this direction, the major objective of this research was to develop a set of prediction models for estimating pavement temperatures over the entire duration of one full day, representative of any seasonal variation occurring annually in tropical climates. The scope of the work encompassed (Fig. 1): (a) development of mean monthly maximum and minimum temperature models based on the historical weather parameters and geographical coordinates collected from eight designated weather stations by the India Meteorological Department (IMD) for a period of 10 years across the State of Andhra Pradesh, India covering 620 data points, (b) formulation of two different pavement temperature models corresponding to the half-day seasonal variation using real-time pavement temperature, air temperature, solar radiations, and humidity, and (c) establishment of generalized annual pavement temperature models for tropical regions.

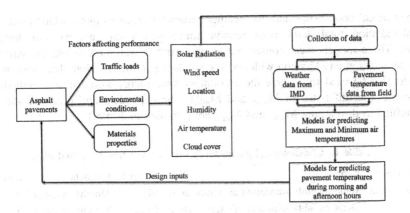

Fig. 1. Research outline.

2 Data Collection

Climatological and geographic coordinate data procured from IMD consisted of precipitation, sunshine, wind speed, surface air temperature, dry bulb and wet bulb temperatures, relative humidity, and latitude of the location of eight weather stations (locations) covering eight districts in the State of Andhra Pradesh, India. At each location, the data was obtained for a period of 10 years from January 2010 to December 2019, totaling 620 data points. All the data points were authenticated and arranged according to the geographical coordinate system. The collected datasets comprised the following parameters expressed as monthly average:

- Maximum air temperature (°C)
- Minimum air temperature (°C)
- Wind speed (km/h)
- Dry-bulb temperature at synoptic hours 3 & 12 (°C)
- Wet-bulb temperature at synoptic hours 3 & 12 (°C)
- Relative humidity (%)
- Total cloud cover (in Oktas)
- Daily total radiation (in MJ/m^2)

3 Data Analysis and Model Development

Two separate models were established to predict the pavement temperatures during morning (PTm) from 06.00 to 12.00 h and afternoon (PTa) from 13.00 to 19.00 h. The reasons for developing two different models were based on the observed increase in the solar radiation from zero to a peak value in the morning hours and a gradual reduction to zero in the afternoon hours. Prior to establishing models for predicting pavement temperature, models were developed to predict the mean monthly maximum air temperature (M_{max}) and mean monthly minimum air temperature (M_{min}) based on geographical coordinates and the weather parameters such as mean wind speed, daily total radiation, total cloud cover, relative humidity, dry bulb, and wet bulb temperatures. These models were considered as precursors towards predicting the pavement temperatures at any location with specific weather parameters. Note that these models could also be used to analyze the amount of heat energy stored in a pavement (for example, the difference of M_{max} and M_{min}). The dependent and predictor variables identified in developing M_{max} and M_{min} are provided in Table 1.

Table 1. Climatological parameters used for predicting M_{max} and M_{min}

Elements	Variable	Symbol	Frequency	Units
Dependent	Mean monthly maximum air temperature	M_{max}	Monthly average	°C
	Mean monthly minimum air temperature	M_{min}	Monthly average	°C
Predictor	Dry-bulb temperature	DBT	Monthly average	°C
	Wet-bulb temperature	WBT	Monthly average	°C
	Relative humidity	RH	Monthly average	%
	Total cloud cover	TOC	Monthly average	Oktas
	Daily total radiation	DTRAD	Monthly average	MJ/m^2
	Mean wind speed	MWS	Monthly average	km/h
	Latitude	Lat	N/A	degrees

The dependent variables M_{max} and M_{min} are expressed in degrees Celsius, and they were estimated using the following predictor variables which were highly correlated with response variables.

- MWS – mean wind speed expressed in km/h.
- DBT – dry bulb temperature expressed in degree Celsius.
- WBT – wet bulb temperature expressed in degree Celsius.
- RH – relative humidity expressed in percentage.
- TOC – total cloud cover expressed in Oktas (one Okta is equal to $1/8^{th}$ of sky covered with clouds).
- DTRAD – sum of daily total radiation from 05:00 to 20.00 h expressed in MJ/m^2.
- Lat – latitude of the location expressed in degrees.

3.1 Correlation Matrices for M_{max} and M_{min}

First, a correlation matrix was developed between dependent and all predictor variables to check the degree of correlations between them. Essentially, M_{max} was correlated to a few predictor variables while M_{min} was correlated to all predictor variables, as presented in Table 2. As observed, M_{max} positively correlated with DBT, TOC, and DTRAD while negatively correlated with Lat of the location. The negative correlation with latitude can be explained as follows. As the latitude increases, the intensity of solar radiation decreases due to the increasing inclination of solar rays. Similarly, M_{min} positively correlated with all predictor variables except MWS. The negative correlation with MWS was attributed to the decrease in the air temperature with increasing wind speed, mainly due to the convection (movement of traffic as well) process. In addition, the influence of predictor variables was further studied using one-tailed P-value significance test at 95% confidence interval. The P-values for each of the dependent and predictor variables were determined and presented in Table 2.

Table 2. Correlation matrix and one-tailed P- significance value (in parentheses) associated with M_{max} and M_{min} models.

Correlation (One-tailed P-value significance)	M_{max}	Lat	DBT	TOC	DTRAD		
M_{max}	1.000(N/A)	-	-	-	-		
Lat	-0.098(0.178)	1.000(N/A)	-	-	-		
DBT	0.960(0.000)	-0.172(0.051)	1.000(N/A)	-	-		
TOC	0.390(0.000)	-0.106(0.158)	0.531(0.000)	1.000(N/A)	-		
DTRAD	0.528(0.000)	-0.324(0.000)	0.448(0.000)	-0.165(0.059)	1.000(N/A)		
	M_{min}	Lat	MWS	WBT	RH	TOC	DTRAD
M_{min}	1.000(N/A)	-	-	-	-	-	-
Lat	0.055(0.303)	1.000(N/A)	-	-	-	-	-
MWS	-0.072(0.249)	-0.754(0.000)	1.000(N/A)	-	-	-	-
WBT	0.946(0.000)	-0.141(0.091)	-0.002(0.494)	1.000(N/A)	-	-	-
RH	0.357(0.000)	-0.005(0.481)	-0.254(0.008)	0.463(0.000)	1.000(N/A)	-	-
TOC	0.758(0.000)	-0.106(0.158)	-0.067(0.265)	0.731(0.000)	0.590(0.000)	1.000(N/A)	-
DTRAD	0.205(0.026)	-0.324(0.001)	0.237(0.012)	0.303(0.002)	-0.217(0.019)	-0.165(0.059)	1.000(N/A)

Based on the P-values, M_{max} was highly correlated with all variables, except Lat. Also, M_{min} was correlated to all predictor variables, except latitude and wind speed. Although latitude was not well correlated, it was included in the models to predict M_{max} and M_{min} as the weather parameters are location-dependent and would vary with a change in the location. Similarly, to address the effect of heat transfer during the convection process, MWS was included in the M_{min} prediction model, which was found to be linearly proportional to the coefficient of heat convection.

3.2 Correlation Matrices for PTm and PTa

Correlation matrices were also developed for the dependent variables PTm and PTa and the following predictor variables:

- AT – the air temperature in degree Celsius
- RH – the relative humidity in %
- Lat – latitude of the location expressed in degrees
- M_{rd} – the solar radiation in the morning hours from 6.00 to 12.00 h expressed in MJ/m^2
- A_{rd} – the solar radiation in the afternoon hours from 13.00 to 19.00 h expressed in MJ/m^2

Based on the correlation matrices, PTm had high correlations (>60%) with AT, RH, and reasonable correlations with Lat, and M_{rd}. Similarly, PTa had high correlations with AT, RH, while medium correlations (~20_40%) with Lat, and low correlation (<10%) with A_{rd}. Low correlation with A_{rd} was due to decreasing solar radiation in the afternoon hours. One-tailed P- significance values with 95% confidence interval also supported the above observation.

3.3 M_{max} and M_{min} Models

Multivariate regression analysis (MVRA) was employed to develop Mmax and Mmin predictive models. Further, curve fit estimation was used to check the variation of each predictor variable with each dependent variable (Mmax and Mmin). Multiple nonlinear regression models were developed considering Mmax as dependent variable and Lat, DBT, TOC, and DTRAD as predictor variables and illustrated as Eq. (1), and Mmin as dependent variable and Lat, WBT, TOC, RH, and DTRAD as predictor variables, as presented in Eq. (2).

$$M_{max} = 1.126 * ln(Lat) + 0.357 * DTRAD + 0.693 * DBT - 0.001 * TOC^3 + 3.370 \tag{1}$$

Where:

M_{max} = mean monthly maximum air temperature, in °C; Lat = latitude of the location, in degrees.

DTRAD = sum of daily total radiation from 05:00 to 20.00 h, in MJ/m²; DBT = dry bulb temperature, in °C; TOC = total cloud cover, in Oktas.

$$M_{min} = -0.007 * Lat + 0.505 * log_{10}MWS + 1.405 * WBT - 0.120 * RH + 0.273 * TOC \\ -0.186 * DTRAD - 1.367 \tag{2}$$

Where:

Lat = latitude of the location in degrees; MWS = mean wind speed in km/h; WBT = wet bulb temperature in °C; RH = relative humidity in %; TOC = total cloud cover in Oktas; DTRAD = sum of daily total radiation from 05:00 to 20.00 h in MJ/m².

The statistical package for the social sciences (SPSS) software was utilized for MVRA. The coefficients of predictive models M_{max} and M_{min} parameters, their standard error estimates, and upper and lower bound limits of the 95% confidence intervals are presented in Tables 3 and 4. In general, the most influential parameters for M_{max} were DTRAD and DBT as the correlations between them were high (>60%). Also, M_{min} was influenced by four parameters: WBT, RH, TOC and DTRAD with high correlations (>60%).

Table 3. Coefficients and standard error estimates of M_{max} model parameters.

Parameter	Estimate	Std. error	95% Confidence interval	
			Lower bound	Upper bound
Lat	1.126	0.431	0.280	1.971
DTRAD	0.357	0.027	0.304	0.411
DBT	0.693	0.023	0.648	0.739
TOC	−0.001	0.000	−0.002	0.000
Intercept	3.370	1.386	0.647	6.092

Table 4. Coefficients and standard error estimates of M_{min} model parameters.

Parameter	Estimate	Std. error	95% Confidence interval	
			Lower bound	Upper bound
Lat	−0.007	0.031	−0.069	0.054
MWS	0.505	0.166	0.179	0.832
WBT	1.405	0.034	1.338	1.472
RH	−0.120	0.006	−0.133	−0.107
TOC	0.273	0.027	0.219	0.326
DTRAD	−0.186	0.032	−0.249	−0.123
Intercept	−1.367	0.660	−2.664	−0.070

Using the developed models, M_{max} and M_{min} were predicted by simple substitution of weather parameters. Relationships were developed between the predicted M_{max} and M_{min} values as ordinates and the observed M_{max} and M_{min} values as abscissas and shown in Figs. 2 and 3. A 45° line was drawn to observe the distribution of the data points on either side of the ideal line. As noticed, the majority of the points were either close to or on the 45° equality line. In addition, statistically, M_{max} and M_{min} models had high accuracy for 620 data points, supported by $R^2_{adj.} \geq 90\%$ and $S_e/S_y \leq 0.350$.

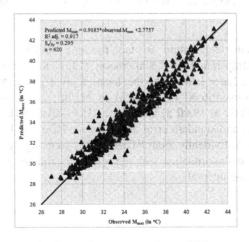

Fig. 2. Predicted versus observed M_{max}

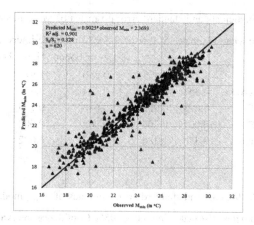

Fig. 3. Predicted versus observed M_{min}

3.4 Pavement Temperature Prediction Models

Next, the hourly pavement temperature data measured in the field along with AT, RH, WS, and hourly solar radiation was collected and assembled to develop the next set of equations: pavement temperature prediction models. The data was split into two categories to observe the effect of heating (morning hours from 06:00 to 12:00) and cooling (afternoon hours from 13:00 to 19:00) on the pavement temperature. Two pavement temperature prediction models PTm and PTa were developed to study the pavement temperature profile in the morning and afternoon hours, respectively. Based on the correlation matrix, AT, Lat, and M_{rd} were considered as predictor variables for PTm model and AT, Lat, A_{rd} and RH were considered as predictor variables for the PTa model. Nonlinear regression models were developed and are shown in Eqs. (3) and (4).

$$PTm = 1.368 * AT - 0.212 * lat + 0.044 * (ln(M_{rd}))^2 \qquad (3)$$

$$PTa = 1.349 * AT - 0.009 * lat^2 - 0.068 * A_{rd}^2 - 0.007 * RH \qquad (4)$$

Where:

PTm = pavement temperature in the morning (06:00 to 12:00 h) in °C; PTa = pavement temperature in the afternoon (13:00 to 19:00 h) in °C; AT = air temperature in °C; Lat = latitude of the location in degrees; M_{rd} = solar radiation during morning hours (06:00 to 12:00 h) in MJ/m^2; A_{rd} = solar radiation during afternoon hours (13:00 to 19:00 h) in MJ/m^2; RH = Relative Humidity in %.

The coefficients and standard error estimates for PTm and PTa model parameters are shown in Table 5. Evidently, both PTm and PTa were highly influenced by AT, as this parameter had high correlation with the response variable: pavement temperature, regardless of the time interval.

Table 5. Coefficients and standard error estimates of PTm and PTa model parameters

Category	Parameter	Estimate	Std. error	95% Confidence interval Lower bound	95% Confidence interval Upper bound
PTm model parameters	AT	1.368	0.011	1.346	1.391
	M_{rd}	0.044	0.014	0.015	0.073
	Lat	−0.212	0.020	−0.254	−0.171
PTa model parameters	AT	1.349	0.012	1.325	1.372
	Lat	−0.009	0.002	−0.013	−0.005
	A_{rd}	−0.068	0.032	−0.132	−0.004
	RH	−0.007	0.004	−0.015	0.000

Using the two developed models, PTm and PTa values were predicted, and plots were drawn between the observed and predicted values (Figs. 4 and 5). Since most of the points were in close proximity of the equality line, the developed models had high degree of accuracy with excellent statistical goodness of fit measures: $R^2_{adj.} \geq 99.7\%$ and $S_e/S_y \leq 0.051$, while also displaying high precision and low bias.

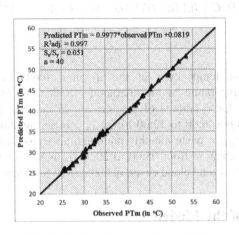

Fig. 4. Predicted versus observed PTm

As a final step, M_{max} and M_{min} models developed from the historical climatic data were substituted into the pavement prediction models and the recommended equations for predicting pavement temperatures in the morning and afternoon time intervals were arrived at, as in Eqs. (5) and (6).

$$PTm = 1.974 * M_{max} - 2.22 * ln(lat) - 0.212 * Lat - 0.71 * DTRAD + 0.044 \\ * (ln(M_{rd}))^2 + 0.002 * TOC^3 - 6.652 \quad (5)$$

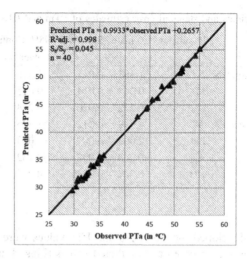

Fig. 5. Predicted versus observed PTa

$$PTa = 0.96 * M_{min} + 0.007 * lat - 0.009 * Lat^2 - 0.484 * log_{10}MWS + 0.108 * RH - 0.262 * TOC + 0.178 * DTRAD - 0.068 * A_{rd}^2 + 1.312$$

(6)

Where:

PTm = pavement temperature in the morning (06:00 to 12:00 h) in °C; PTa = pavement temperature in the afternoon (13:00 to 19:00 h) in °C; AT = air temperature in °C; Lat = latitude of the location in degrees; M_{rd} = solar radiation during morning hours (06:00 to 12:00 h); M_{max} = mean monthly maximum air temperature, in °C;); M_{min} = mean monthly minimum air temperature, in °C; DTRAD = sum of daily total radiation from 05:00 to 20.00 h, in MJ/m²; DBT = dry bulb temperature, in °C; TOC = total cloud cover, in Oktas; MWS = mean wind speed in km/h;

4 Validation of the Models

Model validation was carried out to check the efficiency of the prediction using the climatic parameters and pavement temperature data other than the stations that were used for the model development. The validation results of M_{max} and M_{min} models are shown in Fig. 6. Likewise, Fig. 7 presents validation for PTm and PTa models. As observed, the predictions were very accurate, represented by high R^2.

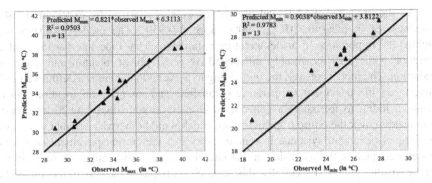

Fig. 6. Validation of M_{max} and M_{min} models

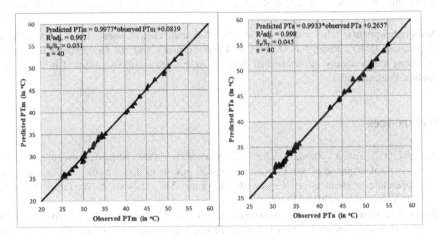

Fig. 7. Validation of PTa and PTm models

5 Conclusions

The major objective of this research was to develop a set of prediction models for estimating pavement temperatures over the entire duration of one full day, representative of any seasonal variation occurring annually in tropical climates. The scope of the effort encompassed the collection and assemblage of ten-year weather parameters and geographical coordinates from eight locations in the State of Andhra Pradesh, India, along with the measured thermophysical properties of a few locations covering a total of 620 data points, which assisted in the establishment of generalized annual pavement temperature models for tropical regions. The locations selected for the collection of weather parameters represented the variations in tropical climate such as near sea shore, moving away from the sea effect, and also the dry portion (continental interior). The Andhra Pradesh state representative of tropical climates by covering all these variations within its land mass.

All the predicted models had high accuracy characterized by excellent statistical goodness of fit measures ($R^2_{adj} > 90\%$; $S_e/S_y < 0.350$). The model estimates highlighted the importance of air temperature regardless of the time interval in a full-day duration while also remarkably influenced by geographical location and daily solar radiation. The models proved the underlying hypothesis that there is a definitive need to investigate the pavement temperature parameters and integrate all locations (in this context, India) to formulate a generalized climatic model (involving thermophysical parameters) over the entire tropical region, which would otherwise amount to irrational designs of pavement systems. It is noteworthy that the current pavement design procedures adapted in India consider a standard climatic condition, albeit there is diversity in the weather pattern. Essentially, this research study paved the way to recommend the development of a generic model that includes data from several weather stations, which will entail the creation of an integrated climatic database comprising varying environmental parameters, which will serve as key inputs in the pavement design process.

Acknowledgments. The authors thank the National Data Centre, India Meteorological Department (IMD), Pune; the Ministry of Earth Sciences, and the Government of India for providing the necessary climatological data for eight weather stations situated in the State of Andhra Pradesh, India.

References

Assogba, O.C., Tan, Y., Zhou, X., Zhang, C., Anato, J.N.: (2020). Numerical investigation of the mechanical response of semi-rigid base asphalt pavement under traffic load and nonlinear temperature gradient effect. Constr. Build. Mater. **235**, 117406 (2020). https://doi.org/10.1016/j.conbuildmat.2019.117406

Bizjak, F.K, Dawson, A., Hoff, I., Makkonen, L., Yihaisi, J.S., Carrera, A.: The impact of climate change on the European road network. In: Proceedings of the Institution of Civil Engineers – Transport, vol. 167, no. 5, pp. 281–295 (2014). https://doi.org/10.1680/tran.11.00072

Bosscher, P., Bahia, H., Thomas, S., Russell, J.: Relationship between pavement temperature and weather data: wisconsin field study to verify super pave algorithm. Transp. Res. Rec.: J. Transp. Res. Board **1609**, 1–11 (1998). https://journals.sagepub.com/doi/abs/10.3141/1609-01

Chandrappa, A.K., Biligiri, K.P.: Development of pavement-surface temperature predictive models: parametric approach. J. Mater. Civ. Eng. **27**(12), 0401514 (2016). https://doi.org/10.1061/(ASCE)MT.1943-5533.0001415

Diefenderfer, B.K., Al-Qadi, I.L., Diefenderfer, S.D.: Model to predict pavement temperature profile: development and validation. J. Transp. Eng. **132**(2), 162–167 (2006). https://doi.org/10.1061/(asce)0733-947x(2006)132:2(162)

Gui, J., Phelan, P.E., Kaloush, K.E., Golden, J.S.: Impact of pavement thermophysical properties on surface temperatures. J. Mater. Civ. Eng. **19**(8), 683–690 (2007). https://doi.org/10.1061/(ASCE)0899-1561(2007)19:8(683)

Hasan, M.A., Tarefder, R.A.: Development of temperature zone map for mechanistic empirical (ME) pavement design. Int. J. Pavement Res. Technol. **11** (2018). https://doi.org/10.1016/j.ijprt.2017.09.012

Hassan, H.F., Al-Nuaimi, A.S., Taha, R., Jafar, T.M.A.: Development of asphalt pavement temperature models for Oman. J. Eng. Res. **2**(1), 32–42 (2005). https://doi.org/10.24200/tjer.vol2iss1pp32-42

Islam, M.R., Ahsan, S., Tarefder, R.A.: Modeling temperature profile of hot-mix asphalt in flexible pavement. Int. J. Pavement Res. Technol. **8**(1), 47–52 (2015). https://doi.org/10.6135/ijprt.org.tw/2015.8(1).47

Knott, J.F., Sias, J.E., Dave, E.V., Jacobs, J.M.: Seasonal and long-term changes to pavement life caused by rising temperatures from climate change. Transp. Res. Rec. **2673**(6), 267–278 (2019). https://doi.org/10.1177/0361198119844249

Li, Y., Liu, L., Sun, L.: Temperature predictions for asphalt pavement with thick asphalt layer. Constr. Build. Mater. **160**(30), 802–809 (2018). https://doi.org/10.1016/j.conbuildmat.2017.12.145

Loganathan, K., Souliman, M.I.: Prediction of average annual surface temperature for both flexible and rigid pavements. J. Mater. Eng. Struct. **4**, 259–267 (2017). https://www.asjp.cerist.dz/en/article/32258

Mills, B.N., Tighe, S.K., Andrey, J., Smith, J.T., Huen J.: Climate change implications for flexible pavement design and performance in southern Canada. J. Transp. Eng. **135**(10) (2009). https://doi.org/10.1061/(ASCE)0733-947X(2009)135:10(773)

Qin, Y., Hiller, J.E., Meng, D.: Linearity between thermophysical properties and surface temperatures. J. Mater. Civ. Eng. **31**(11), 04019262 (2019). https://doi.org/10.1061/(ASCE)MT.1943-5533.0002890

Solaimanian, M., Kennedy, T.W.: Predicting maximum pavement surface temperature using maximum air temperature and hourly solar radiation. Transp. Res. Rec. **1417**, 1–11 (1993). http://onlinepubs.trb.org/Onlinepubs/trr/1993/1417/1417-001.pdf

Sreedhar, S., Biligiri, K.P.: Development of pavement temperature predictive models using thermophysical properties to assess urban climates in the built environment. Sustain. Cities Soc. **22**(3), 78–85 (2016). https://doi.org/10.1016/j.scs.2016.01.012

Stoner, A.M.K., Daniel, J.S., Jacobs, J.M., Hayhoe, K., Scott-Fleming, I.: Quantifying the impact of climate change on flexible pavement performance and lifetime in the United States. Transp. Res. Rec. **2673**(1), 110–122 (2019). https://doi.org/10.1177/0361198118821877

Thompson, M.R., Dempsey, B.J., Hill, H., Vogel, J.: Characterizing temperature effects for pavement analysis and design. Transp. Res. Rec. **1121**, 14–22 (1987). http://onlinepubs.trb.org/Onlinepubs/trr/1987/1121/1121-003.pdf

Impacts of Transport Investments

Experiences of High Capacity Transport in Finland

Vesa Männistö[✉]

Finnish Transport Infrastructure Agency, Helsinki, Finland
vesa.mannisto@vayla.fi

Abstract. The competitiveness of Finland's industry with respect to logistics has been improved by increasing the maximum dimensions and masses of heavy goods vehicles. The maximum height of vehicles was increased from 4.2 m to 4.4 m and the maximum mass from 60 to 76 tons in 2013. The maximum length of vehicle combinations was increased up to 34.5 m in 2019. These new dimensions can be utilized on all public and private roads unless separately signposted.

In order to prepare for this change, Finnish Transport and Communications Agency approved about 100 permits for testing of heavier and longer vehicles in 2013–2019. Fuel savings were up to 30% and cost of personnel can be up to 10–15% lower. No negative issues in traffic safety has been detected and bigger vehicle combinations are operating smoothly among other traffic.

The experiences show that 1.6–3.2 billion euros can be saved in logistics costs over the next twenty years. The reform will help to reduce the carbon dioxide emissions by two percent a year. However, there is a need to strengthen roads and bridges, especially when pavements are thin (less than 100 mm). Tight crossings and turning lanes need improvements to enable long vehicles to negotiate turnings safely. So far needs to improve the infrastructure have been modest.

Possibilities to increase maximum weights of vehicle combinations are under consideration, but this will need more investigations to ensure preservation of road condition and to develop a reliable access control system of vehicles.

Keywords: High capacity transport · High capacity vehicle · Fuel savings · Green house gas reduction

1 Introduction

The Finnish Transport Infrastructure Agency (FTIA) is responsible for the state-owned transport networks, i.e. roads, railways and waterways in Finland. The transport system is maintained and developed in co-operation with other actors, e.g. transportation industry and industries having a major role in land and sea transport demand.

The Finnish road network comprises of highways, municipal street networks and private roads. Together with the regional organization, the FTIA is responsible for the maintenance and development of the state-owned networks. There are 78,000 km of highways, 6,000 km's of railways and 16'000 km's of sea and coastal fairways maintained by the state in Finland. Funding for these comes fully from the state budget. The length of city street network is 26'000 km's and private roads 330'000 km.

Traffic system for goods transportation in Finland is based on seamless co-operation of different traffic modes. A major share of import and export transportation travels by sea. Goods from/to seaports utilize both railways and roads. Long distance raw material and export goods transportation are based on railways, supported by more agile road transportation. Short distance and changing transportation needs are mainly served by road transportation.

The economic welfare in Finland is based on export of goods and services. The most transportation intensive export industries are chemical, forest as well as steel and metal industries. Compared to competing industries in the middle of the Finland's main export market area (The European Union, 60% share of export volume in Finland), transportation distances in and from Finland to the market are by far higher. Therefore, there is a strong political will and economic incentives to increase the competitiveness of industries by different means.

2 High Capacity Transport

Road freight traffic is forecast to grow substantially in most countries. The anticipated increase in infrastructure capacity will not, on its own, be sufficient to accommodate increasing traffic volumes. Using bigger trucks, or high capacity vehicles (HCVs), is one of the most practical options for accommodating some of this growth.

The definition of high capacity transport (HCT) or high capacity vehicles (HCV) refer to vehicles, which exceed the measures and weights given by the local legislation (ITF 2019). Several countries, e.g. Australia, Finland, Sweden, and South Africa, have shown their interest in developing local regulations to accommodate the use of HCT vehicles in order to improve their logistic systems (Haavisto 2017). There are several reasons why HCT is seen as a transportation system, which should be further developed: (1) a need to decrease CO_2 emissions caused by heavy traffic, (2) improvement of land transportation logistic system; (3) improvement of traffic safety; and (4) to boost technical development of trucks and trailers.

HCVs are different from standard vehicles in that they are longer, heavier or both. Low density freight requires more cargo space resulting in extended length vehicles. Extending vehicle length beyond designated length limits will mean that such vehicles are not suitable for urban areas because they would have difficulty negotiation intersections and tight radius roadway curves. On the other hand, longer vehicles can operate safely on roadways free from geometric constraints like limited access roads or divided highway networks. Performance based standards (PBS) are very useful for determining vehicle turning characteristics and their suitability for specific roads. High density cargo requires a vehicle capable of increased weight capacity. This is achieved by adding axles to the vehicle ensuring that axle loads are not more than those of standard vehicles. Axle weight and configuration is controlled by road authority regulation to protect roads and bridges and HCVs can be configured to comply with these regulations. In addition, PBS can be used to ensure that the heavier vehicles have acceptable vehicle dynamic qualities with good vehicle stability and control characteristics. However, some bridges may not be suitable for HCVs and may require reinforcement (ITF 2019).

International Transport Forum has concluded, that it is beneficial to use the potential of High Capacity Vehicles to increase transport efficiency, reduce traffic volumes, lower emissions and achieve better safety outcomes. During the development phase, well-planned trials should be used to introduce High Capacity Vehicles on a road network. Moreover, high capacity vehicles should be configured for the specific area in which they will operate (ITF 2019).

3 Maximum Weight and Height of Vehicle Combinations in Finland

Since joining the European Union, Finland has applied the EMS (European Modular System) rules, where the combination maximum length is 25.25 m. In addition, the maximum gross weight of a vehicle combination has been 60 tons up to year 2013. The Ministry of Transport and Communications of Finland has aimed at to be a vanguard in HCT development and has therefore been active in planning to allow increased truck maximum weights and lengths. The most interest has risen in forest industry and daily goods market. The development process has been divided into two separate paths: (1) to increase maximum mass of combinations and (2) to increase maximum length of vehicle combinations.

The maximum authorized height for vehicles was increased from 4.2 m to 4.4 m and the gross weight from 60 tons to 76 tons. The regulation became effective on October 1, 2013, and it is applied on all public and private roads, if not separately restricted by signposting.

The competitiveness of especially export industries in Finland with respect to logistics was improved by increasing the maximum authorized dimensions and masses for heavy goods vehicles and combinations of vehicles. During the five-year transition period, the existing fleet could be used to transport loads that exceed the maximum mass and dimensions set for each vehicle, provided that also the safety requirements corresponding to the increased maximum authorized mass and dimensions are met by the vehicle. This enabled transportation companies to utilize the whole life cycle of their vehicles.

The goal of the reform was to improve Finland's competitiveness and reduce the gap between transport costs in Finland and Central Europe, which is the main export market of Finland. Due to the long distances involved, transportation costs in Finland are higher compared to most of other countries. Approximately 1.6–3.2 billion euros could be saved in logistics costs over the next twenty years by increasing the sizes of vehicle combinations. It was also estimated that the reform would help to reduce the carbon dioxide emissions from traffic by two per cent a year (Lapp and Iikkanen 2017, Männistö 2019). Forest industry has been the beneficiary of higher gross weights.

Introduction of the new dimensions was utilized extremely quickly by transportation companies. The trade of trucks and trailers had almost stopped while the companies were waiting for the details and timetable of the new regulation. Practically all the vehicle combinations purchased after the new decree became effective have been according to the new regulations. Moreover, the clients of transportation companies were also very agile in accommodating their business processes to utilize bigger vehicle dimensions.

4 Effect of New Dimensions to Road Networks

New regulation had a big impact on the FTIA, municipalities, regional road authorities, and trade and industry. Improvements to roads and bridges belonging to the road network maintained by the state had to begin in 2014, just after the regulation became effective. The municipalities were responsible for setting their own timeframes, and were free to plan the improvements to streets and bridges at their own pace.

As soon as the new dimensions were introduced the number of weight restricted bridges on public roads more than tripled from 150 bridges to around 500 bridges, and the increase in vehicle maximum height revealed some hundreds of bridges and road furniture with too low clearance. This lead to a huge effort of signposting of all the restricted bridges and portals in a very short timeframe. The situation was more problematic in cities and municipalities where the knowledge of bridge bearing capacities and clearances were modest. This lead in some cases to overreaction, and some cities even planned to restrict their road network from HCT only to be on the safe side.

The Ministry of Traffic and Communications (MINTC) allocated 55 million euros to improve the public road network to meet the new demands of bigger combinations. This funding was not adequate to fulfill all the needs and needs had to be prioritized. This prioritization problem was solved by very close co-operation with industries. Different questionnaires (e-mail, internet) were arranged to collect data on future heavy transportation needs both at central and regional level, and extensive discussions on future transportation needs were held with larger companies and confederations of industries. As a result of this fact finding mission, some 100 bridges were strengthened and 10–15 low clearance bridges were fixed. These operations have proved to be adequate for the industries, and improving of less important bridges could be avoided. Now, eight years after the maximum weight and height were raised, only a handful of problematic bridges need strengthening, and the needs are well managed in continuous co-operation with clients.

Higher masses have also a considerable effect on road structures. Field tests have shown that road structures are susceptible to higher loads, especially when the pavement thickness is less than 100 mm (Varin and Saarenketo 2014). Moreover, if the use of super single tires increases and the tire pressures are high, road life span will be shortened even down to one third of the planned life span (Varin and Saarenketo 2014). Therefore it is important that roads are strengthened e.g. by increasing pavement thickness before higher masses are introduced in a large scale. The use on weight restrictions and enforced use of twin tires or wide base super single tires are also important in protection of road structures. The use of the central tire inflation system (CTI) is also highly recommended (Korpilahti 2011).

Compensation of higher loads with more axles might, however, take partly care of accelerated road deterioration. Results in Finland show, however, that increased number of axles in the same vehicle causes the pore water pressures in road structures to rise, especially during the freeze-thaw period. Additionally, weak subgrades are not elastic enough to recover from more consecutive axles. Finally, increased number of axles in the same vehicle results in more tires in the same wheel path, which leads to higher rutting speed (Varin and Saarenketo 2014).

5 Piloting of Longer and Hevier Vehicle Combinations in 2013–2019

Although higher vehicle combination masses already improved the competiveness of Finnish industries, there was still a high demand for even bigger vehicle combinations for certain transportation needs. Forest industry would benefit of bigger truck dimensions in order to transport timber from forest to factories and sawmills, and further timber products to seaports (Venäläinen 2018). Marine transport companies need longer vehicle combinations to transport big sea containers (45 feet) to/from and between seaports. Wholesale and supermarket chains could utilize longer and higher combinations to transport goods between terminals and outlets.

From the technical point of view, Finnish trailer manufacturers, in co-operation with European truck manufacturers and goods transporting companies had shown interest in developing new technologies, and a large-scale pilot program would help them in large-scale testing. Moreover, HCT was developing rapidly in neighboring countries and competitiveness of industries as well as cross border traffic harmonization required active participation to international research and development of HCT.

To fulfill the needs, the Finnish Transport and Communications Agency (Traficom) started approving permits for testing of high capacity transport vehicles in 2013. These vehicles are of unique vehicle designs and either longer than 25.25 m or heavier than 76 tons, or both. About 100 pilot projects were executed between 2013 and 2019. Every pilot projects had a limited number of vehicles, usually only one. The use of each pilot vehicle was restricted to its own dedicated route, which was approved by the road owners (FTIA, municipalities, and private road owners). In total, the current routes for all the pilots comprised about 4000 km of public roads. The most common combination was DUO2, consisting of a tractor and two semitrailers (DUO2 2019). Another example of a typical combination is ETT (a tractor, a semitrailer, a dolly and another semitrailer). Combinations used in the pilots are illustrated in Figs. 1 and 2.

Fig. 1. Kalevi Huhtala's 100 ton, 34 m combination for wood chip transportation. Source: www.metsateho.fi/hct

Fig. 2. Kari Malmstedt's 92 ton, 32 m combination for timber transportation. Source: www.metsateho.fi/hct

Approval of a pilot required a pre-defined research plan from the piloting organization. There had to be a clear objective why a certain combination would be piloted. Improved economic competitiveness in short term was not a sufficient objective as such; research could be partly directed at economic benefits but also at technical matters, such as behavior of the combination in different traffic conditions, optimal size of combination, multi-purpose use of truck, optimal location of axles and bogies, etc. The results of pilots were reported to Traficom at least semi-annually and the main research results had to be open for publication.

Piloting rose very high interest among truck and trailer manufacturers as well as among transportation companies and companies with transportation needs. The most interested industry branches were forest, daily goods market, and sea container transportation. A typical set-up for research was co-operation between a transport company, a goods owner and academia.

The pilot projects were running for up to five years, and more than 18 million vehicle kilometers were driven during the pilots. According to the results gained, the following conclusions can be drawn:

- Transportation of two empty 45 feet sea containers instead of one at a time decreased fuel consumption for up to 30%. In transportation of wood chips, raw timber and cement, fuel savings were up to eight percent, if 84–90 ton vehicles are compared to a 76 ton combination (Lahti and Tanttu 2017)
- Import product transportation: restricted routes are less practical with import transportation, because import containers are usually distributed to many locations. Transportation of export containers is more effective, because the routes are mostly fixed, from terminals or factories to sea ports (Venäläinen and Poikela 2020)
- Personnel costs in transportation of parceled goods can be 10–15% lower when using HCT vehicle combinations. However, this depends strongly on how the total concept of the transportation of goods is planned. Typical bottlenecks are driving, loading and unloading in terminals and at clients' premises (Lahti and Tanttu 2017, Lahti 2018) as well as having a proper payload to both directions

- Timber transportation. Bigger truck combinations have worked well even in harsh winter conditions. Bigger combinations are especially effective in transportation between terminals and factories, and between factories and seaports. However, all bigger combinations cannot be applied from all loading locations along the smallest roads, thus high flexibility in truck and trailer arsenal is needed. Finnish forest industry also acknowledges that increase of both maximum length and height is both feasible and necessary for the competitiveness of the forest industry (Venäläinen and Poikela 2020)
- Modal split between highways and railways: no major changes have occurred nor are expected to occur in the future. It is estimated that maximum five percent of railways raw material transportation would be moved to HCT transportation. In practice, highway and railway systems complement each other and, for example, forest industry tries always to utilize railway systems if it is feasible, but because of the limited number of train operating companies, the competitiveness of railway transportation is not very high (Lapp and Iikkanen 2017)
- Road safety. There were concerns that longer and heavier vehicle combinations would be a road safety risk (Blanquart et al. 2016), but these pilots do not support these concerns. After 18 million vehicle kilometers in the pilots there is no evidence of realized increased safety risk (Lahti and Tanttu 2017). Big combinations are usually very stable at normal highway speeds (80 kph for heavy traffic). The critical issue is adequate winter maintenance (Lahti 2018). Other reckless drivers (overtakers) have caused most of the traffic safety problems with HCT vehicles
- Infrastructure managers have an important role in HCT traffic management. For example, digital information of hazard winter condition, traffic jams, and road works need to be available on line. Moreover, practical arrangements at road works to accommodate needs of HCT vehicles need professional action from parties responsible for work sites. Moreover, road accidents leading to diversions need proper and professional management also from the police and emergency services.

6 Maximum Length of Vehicles and Vehicle Combinations

The results from the pilots indicated, that increasing of maximum length would be more feasible in a short term than further increase of maximum masses of vehicle combinations. The maximum length of vehicle combinations was increased up to 34.5 m in January, 2019. Details of the new regulation can be found in (MINTC 2018). The requirements in the law are a mix of technical requirements and performance based measures, such as ability to negotiate certain types of crossings and roundabouts.

Increase of vehicle length applies, per se, to all the public, city and private roads, if not specifically restricted by the respective infrastructure managers. The transportation companies are responsible to verify possible problematic routes and crossroads in advance. There are also plans to develop a web application, which transport companies can use to inform the FTIA and other transport companies of problematic roads, crossroads, winter conditions, terminal yards and rest areas.

Expected annual savings of these longer vehicle combinations are still to be verified, but annual savings are estimated to be about 20 million euros in personnel costs and more than 20 million liters less diesel fuel consumed. Longer combinations do not have any negative impact on road infrastructure deterioration and is not therefore leading to higher maintenance costs. The only notable problems for the public road network in vehicle combination length increment are provision of adequate level of winter maintenance, behavior of combinations in crossroads, roundabouts and railway level crossings, and a small increase in road deterioration due to more axles following the same wheel path.

Most of the crossroads have been designed for much shorter combinations and longer vehicles could have problems to negotiate all of those. Other problems arising from longer vehicle combinations take place outside the public road network. Streets and private roads, as well as terminal and factory yards are often less suitable for longer vehicle combinations.

Utilization of new lengths has penetrated the market very quickly. As soon as transport companies have to invest in the new gear, they carefully consider how to utilize new measures. After the first two years of application of longer vehicles combinations, the following sectors have widely utilized new measures: 1) general volume based goods traffic, e.g. mail, 2) trunk traffic from central warehouses to hypermarkets, 3) long sea container transport, 4) forest industry side products (saw dust and wood chips) and 5) forest industry end products (plank and paperboard).

7 Future Development Plans

The next step in this development chain would naturally be the further increase of maximum masses of vehicles and vehicle combinations. Especially forest, building and mining industries are active in lobbying for higher masses. National plans for fossil-free transport also highlight the use of HCT (Andersson et al. 2020). Increase of maximum masses has, however, proven to be a more complicated issue. Contrary to the increase in length of combinations, there are several issues to be solved before higher masses can be introduced.

Effect of higher masses to road pavements and bridges is yet unclear. If the axle loads are not increased, there is no new major effect on pavements (Vuorimies et al. 2019). However, higher total masses will lead to higher number of weight restricted bridges. FTIA has estimated that number of weight restricted bridges will be instantly at least doubled in Finland, if maximum mass of 84 tons is introduced. There are also hundreds of bridges, where bearing capacity is not yet problematic, but their life span will be considerably shortened.

Use of higher masses cannot be allowed on all roads and streets, due to a high number of weakly or even non-constructed roads, e.g. most of the local roads and especially the long gravel road network. These roads often constitute the "last mile" of the transportation, and thus decrease possibilities to utilize higher loads. Solving of this will need special arrangements, e.g. mid-terminals along highways.

The needs of different industries sectors have to be identified before any route is opened for higher masses. It is anticipated that the "high mass" network will originally be rather short (some 5–10% of the highway network), and it will be gradually lengthened as soon as road and bridge conditions are investigated, and the real transportation needs are evaluated.

Signposting and access control. Due to a very long road network, which will not be allowed for higher masses, there is a need of signposting of that network. Due to an anticipated very short notice time before the law becomes effective, there is no time to execute physical signposting in the field. Therefore utilization of digital signposting is considered to solve this problem. This will require transport companies to download and use digital high mass network delivered by the FTIA to self-control their movements on road networks. Moreover, it is very probable that the access control of vehicles would be monitored using some kind of intelligent access program (IAP), such as the system utilized in Australia (Blanquart et al. 2019).

8 Conclusions

The competitiveness of Finland's industry with respect to logistics has been improved by increasing the maximum dimensions and masses of heavy goods vehicles. The maximum height of vehicles was increased from 4.2 m to 4.4 m and the maximum mass from 60 tons to 76 tons in 2013. The maximum length of vehicle combinations was increased up to 34.5 m in 2019. These new dimensions can be utilized on all public and private roads unless separately signposted.

In order to prepare for this change, The Finnish Transport and Communications Agency (Traficom) approved about 100 permits for testing of heavier and longer vehicles between 2013–2019. Pilots were very useful and effective in testing of different vehicle combinations in real life. The outcomes of pilots were duly reported to Traficom and the results were used in further development of the new decree.

The experiences show that approximately 1.6–3.2 billion euros can be saved in logistics costs over the next twenty years. Fuel savings were up to 30% and cost of personnel can be up to 10–15% lower. No changes in traffic safety has been detected so far and bigger vehicle combinations are operating smoothly among other traffic. The reform will help to reduce the carbon dioxide emissions by two percent a year. However, there is a need to strengthen roads and bridges, especially when pavements are thin (less than 100 mm). Tight crossings and turning lanes need improvements to enable long vehicles to negotiate turnings safely. So far needs to improve the infrastructure have been modest.

Possibilities to increase truck maximum weights are under consideration, but weight increase will need more investigations to ensure preservation of road condition and to develop a reliable access control system of vehicles.

Acknowledgements. The help of Mr Otto Lahti of the Finnish Transport and Communications Agency in providing necessary background information is greatly appreciated.

References

Andersson, A., Jääskeläinen, S., Saarinen, N., Mänttäri, J., Hokkanen, E.: Road Map for Fossil-free Transport - Working Group Final Report. Publications of the Ministry of Transport and Communications 2020, p. 17 (2020). http://urn.fi/URN:ISBN:978-952-243-598-9

Blanquart, C., Clausen, U., Jacob, B.: Towards Innovative Weight and Logistics, vol. 2, p. 2016. Wiley, New York (2016)

DUO2: We decrease the fuel consumption with up to 20 percent per transported unit load (2019). https://duo2.nu. Accessed 14 May 2021

Finnish Traffic Safety Agency: Kuljetusyrityksille myönnetyt luvat (2018). https://www.youtube.com/watch?v=7mlHnA5Ab2c. Accessed 21 Apr 2020

Haavisto, S.: HCT-ajoneuvoyhdistelmien vaatimukset tieympäristön liikennetekniseen mitoitukseen (in Finnish). Master's Thesis, University of Oulu, Faculty of Technology, Oulu (2017)

ITF: High Capacity Transport: Towards Efficient, Safe and Sustainable Road Freight. International Transport Forum Policy Papers, No. 69, OECD Publishing, Paris (2019)

Korpilahti, A.: Puukuljetusten kaluston kehittäminen, investoinnit ja kustannustehokkuus (in Finnish). Bioenergian Metsä –seminaari, Rovaniemi (2011)

Lahti, O., Tanttu, A.: Report on Wintertime High Capacity Transport (HCT) 2015–2016 (2018). https://www.trafi.fi/tieliikenne/luvat_ja_hyvaksynnat/hct-rekat/julkaisuja_ja_tutkimustuloksia. Accessed 11 May 2021

Lahti, O.: Experiences of HCT Pilot Tests After 13 Million Vehicle Kilometers – Lessons Learnt (2018). https://www.trafi.fi/tieliikenne/luvat_ja_hyvaksynnat/hct-rekat/julkaisuja_ja_tutkimustuloksia (in Finnish). Accessed 30 July 2020

Lapp, T., Iiikkanen, P.: Transport System Impact of HCT Vehicles (in Finnish). Research Report of the Finnish Transport Agency 57/2017. Finnish Transport Agency (2017)

Ministry of Transport and Communications (MINTC): Luonnos valtioneuvoston asetukseksi ajoneuvojen käytöstä tiellä annetun valtioneuvoston asetuksen muuttamisesta LVM009:00/2018 (in Finnish) (2018). Accessed 28 Apr 2021

Männistö, V.: Recent High Capacity Transport Development in Finland (2019). https://trid.trb.org/view/1572394

Rissanen, J.: The implementation of vehicle combinations longer than 25.25 meters in Finland. Master of Science Thesis, Tampere University Master's Degree Programme in Civil Engineering, Tampere (2020)

Trafikverket: Årsrapport High Capacity Transport. Ett FoI program inom Closer vid Lindholmen Science Park (in Swedish) (2016). https://closer.lindholmen.se/en/news/high-capacity-transport-annual-report-2016. Accessed 14 Apr 2021

Varin, P., Saarenketo, T.: The Effect of New Tyre Type, Tyre Pressure and Axle Configurations of Heavy Trucks on Asphalt Pavement Lifetime. In: 6th Eurasphalt & Eurobitume Congress, Prague (2014)

Venäläinen, P.: Study on Impacts of Increased Lengths on Timber and Chip Transportation in Finland (in Finnish). Metsätehon raportteja 264. Helsinki (2018)

Venäläinen, P., Poikela, A.: Puutavara-ja hakeautojen massojen noston vaikutukset –aiheen 2. väliraportti. Metsätehon raportti 258. Saatavissa (2020). http://www.metsateho.fi/wp-content/uploads/Raportti-258-Puutavara-ja-hakeajoneuvojen-massojen.pdf

Vuorimies, N., Kurki, A., Kolisoja, P., Varin, P., Saarenketo T, Pekkala, V., Haataja, M.: Road Structure Strain on Test Loads of More than 76 Tons HCT combinations in 2015–2018. Finnish Transport Infrastructure Agency. Helsinki 2019. Research reports of the FTIA 21/2019 (2019)

Impacts of Transportation Infrastructure Investments and Options for Sustainable Funding

Daniel J. Findley[1](✉), Steven A. Bert[1], Weston Head[1], Nicolas Norboge[2], and Kelly Fuller[3]

[1] Institute for Transportation Research and Education, NC State University, Raleigh, NC, USA
daniel_findley@ncsu.edu
[2] State of North Carolina Office of State Budget and Management, Raleigh, NC, USA
[3] NC Chamber Foundation, Raleigh, NC, USA

Abstract. The long-term viability of the motor fuels tax as a primary funding mechanism for transportation has been questioned for its effectiveness into the future. Vehicle fuel economy improvements, coinciding with substantive real purchasing power losses of the federal motor fuels tax in the United States (the federal tax rate has not been adjusted for inflation since 1993), have created uncertainty about how the motor fuels tax could sustainably finance the transportation system needs. Transportation agencies face historical revenue shortfalls, and important parts of the infrastructure require maintenance to sustain or improve their conditions.

The State of North Carolina currently invests approximately $5 billion annually in its transportation system. This investment enables the state to achieve an overall infrastructure rating of mediocre. In this condition, the state is facing serious challenges affecting driver safety and economic productivity. Our economic analysis demonstrated that highway construction projects lead to immediate positive economic impacts in as few as two years after project completion. One key finding illustrated that the number of businesses within one mile of NC highway projects increased by 73% a rate that is 48% higher than the growth of business establishments within one mile of unimproved NC highways. Other studies have shown that the time to make capital investments, such as infrastructure, often has the most value in times of recession with increased spending and job creation.

Keywords: Funding · Investment · Economic impact · Transportation · Infrastructure · Revenue

1 Introduction

The long-term viability of the motor fuels tax as a primary funding mechanism for transportation has been questioned for its effectiveness into the future. Vehicle fuel economy improvements, coinciding with substantive real purchasing power losses of the federal motor fuels tax in the United States (the federal tax rate has not been adjusted for inflation since 1993), have created uncertainty about how the motor fuels tax could sustainably finance the transportation system needs. Transportation agencies face historical revenue shortfalls, and important parts of the infrastructure require maintenance to sustain or improve their conditions.

In North Carolina, the General Assembly revised the motor fuels tax formula to better align with the state's transportation system needs (implemented January 1, 2017). Though effective as a stop-gap measure, vehicle fuel economy improvements and the increasing use of alternative fuels continue to challenge the efficacy of the motor fuels tax, even with its revised form.

During NCDOT's financial recovery, the onset of COVID-19 has created a historical "first-of-its-kind" impact on the state's transportation system that has exacerbated funding concerns. North Carolina's traffic volumes fell dramatically throughout the COVID-19 period, with a decrease of approximately 38% during the periods of greatest caution (Shuman 2020). COVID-19 travel reductions have led to a sharp decline in revenue, such that NCDOT fell below its statutorily mandated cash floor of $293 million (NCDOT 2020). According to state law, once the department dips below the cash floor, it can no longer enter into new transportation project contracts (NCDOT 2020).

These fiscal issues directly affect drivers. With project delays and the absence of sustained funding, North Carolina's infrastructure and mobility conditions are likely to worsen. As a result of COVID-19, America's transportation agencies are facing a projected $35–$50 billion shortfall due to the drop in motor fuels tax revenue (AASHTO 2020; Duncan 2020). Traffic has rebounded on the nation's roads; however, vehicle miles travelled stayed well below normal levels for several months (Duncan 2020). The prospect of continued funding disruptions through the pandemic, vehicle fuel efficiency improvements, and federal tax policy challenges, raises questions about the future viability of the motor fuels tax as a primary funding mechanism.

The convergence of systemic transportation financing issues and external shocks have placed states in a challenging fiscal environment. This paper presents a set of recommendations for potential revenue generation options and provides implementation strategies to achieve a more sustainable portfolio of transportation funding mechanisms. Additionally, this paper evaluates land use changes and economic development outcomes associated with highway investments occurring from 2001 to 2016 in North Carolina. This paper seeks to answer two questions: 1) do investments in roadway infrastructure produce worthwhile economic development outcomes and 2) if so, how can the state sustainably fund the necessary improvements and maintenance into the future.

2 Economic Importance of Infrastructure Investment

When roads are built, improved, or widened, regional markets expand to new groups of people, potentially spurring additional development, which results in further economic impacts on the surrounding communities. New roads create valuable real estate for commercial businesses, provide the necessary infrastructure and connection for industry, and connect commuters with access to further employment centers, prompting residential development. All of these types of development create jobs, economic contributions, and tax revenues for the region where the road is located. Historical data demonstrates that every $1 billion of transportation investment in North Carolina generates 14,300 jobs, $10.3 billion in wages, and $10.8 billion in gross state product (ITRE 2015).

For this research, our team developed a methodology to quantify the development surrounding road projects that occurred within the state of North Carolina over the time period of 2001 to 2016 and compared that to the developments surrounding roads that had not been built or improved over that time period. The research team utilized the road classification system maintained by the North Carolina Department of Transportation (NCDOT), to isolate the impacts of different types of roads on development. The analysis included three roadway types – interstates, US routes, and NC highways.

2.1 Methodology

To quantify the development surrounding road projects in North Carolina, the research team utilized land use data from the National Land Cover Database (NLCD), ESRI's Business Analyst as well as geospatial analysis capabilities provided by ESRI's ArcGIS. ESRI's Business Analyst is a dataset maintained by ESRI using InfoUSA data that provides the point location, as well as, business specific information for a significant portion of businesses in the United States. The dataset provides the industry classification, the number of employees, an estimated annual revenue for the location, and the street address for each business point location. These two datasets, combined with the geospatial analysis techniques made possible by ESRI's ArcGIS, were the tools used to quantify the development that occurred around road projects within North Carolina between 2001 and 2016.

For our analysis, development was classified as two distinct measures, the percentage of growth in the number of businesses and jobs within a one-mile buffer of the road project between 2001 and 2016, gathered from the ESRI Business Analyst data, and the square footage of land that developed within a one-mile buffer of a road project between 2001 and 2016. Both measures were included since the number of businesses and jobs provides a direct economic impact created by the road project, and the change in land use development provides an estimate of the total amount of development that occurred, including residential developments, in addition to providing an estimate of the size of the developments that occurred. Using both of these measures, it is possible to gain an understanding of the type and scale of development that occurred around road projects in the state.

2.2 Identifying Road Projects

The research team identified roads that had either been built, widened, or had some form of major road improvement over the time period of 2001 to 2016. For example, the 2001 to 2016 land cover changes for the area surrounding the I-87 and I-440 interchange east of Raleigh are shown in Fig. 2. Areas shaded in red indicate development, with the darker the shade indicating a higher intensity of development. In 2001, before I-87 was connected to I-440, the area where I-87 would be located is still green, indicating a mostly forested, undeveloped area. By 2016, this area is now a dark shade of red, and looking at the full picture, it can be seen that the dark red is showing the path of I-87. To identify roads that had been built, widened, or had some form of major road improvement over the period of 2001 to 2016, such as the I-87 and I-440 interchange, the research team identified points in the land cover dataset that were undeveloped in 2001 and then developed in 2016 (changing from some other color to the color red). Since land development is not isolated to only road projects, the research team used a buffer analysis, in which the buffer width was the width of a typical road to isolate only the points associated with a road project that was longer than one-fourth of a mile.

To determine the percentage increase in the number of businesses and jobs between 2001 and 2016 associated with road projects in the state, the research team created one-mile buffers around each road project. A one-mile buffer was chosen in an attempt to fully capture any development that may be resulting from a road project, but also to avoid picking up non-related development locations such as city centers. For comparison purposes, a one-mile buffer was also created for all other road segments in the state that were not built or improved between 2001 and 2016 to show the differences between development resulting from a road project and simply development resulting from being located near a road. The research team also quantified the square footage of land that was transformed from some form of undeveloped land cover to developed land associated with road projects within the timeframe of 2001 to 2016 with a one-mile buffer of the roadways.

2.3 Findings

Across roadway types, the road projects led to a significant increase in the surrounding development. Table 1 shows information related to areas within one mile of a road project and Table 2 shows the same information for areas that are not within one mile of a road project. NC highway road projects had the greatest impact on development, including the following changes from 2001 to 2016 (Fig. 1):

- The number of businesses within one mile of an NC highway project increased by 73%, 48% more than unimproved NC highways.
- The number of jobs within one mile of an NC highway project increased by 35%, 16% more than unimproved NC highways.
- NC highway projects also led to an increase of 7.8 million square feet of development per mile of road, more than 17 times the amount of development associated with unimproved NC highway road segments.

The magnitude of the impact of NC highways likely has to do with the amount of street front business property associated with NC highways. Interstates had the least nominal amount of square footage of development within one mile compared to the other road classifications, which is likely due to the limited access nature of most interstates. Despite being the least nominally developed, interstate segments with road projects still developed more than three times the amount of square footage than non-road project interstate segments, showing that the construction or improvement of an interstate segment has significant impacts on surrounding development. As shown with

Impacts of Transportation Infrastructure Investments and Options 1211

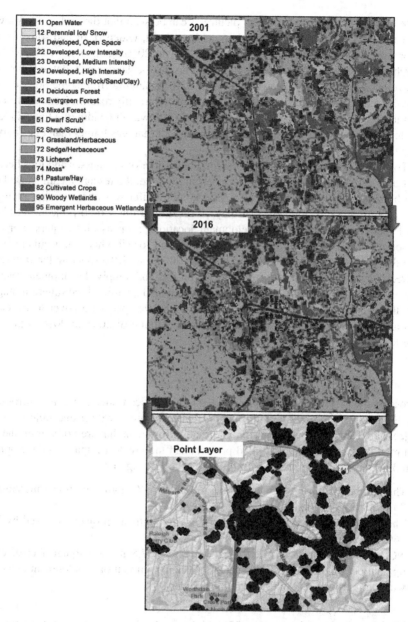

Fig. 1. Land cover classification: 2001 to 2016 for example project (I-87 and I-440 interchanges)

this analysis, transportation infrastructure projects typically increase the value of nearby land and grant additional benefits to firms within distance. There are examples of methods, known as value capture, that can be used to generate funding from the increase in value of the property. Value capture is the use of mechanisms to gain financing for infrastructure projects from land owners/users who stand to benefit from the projects, primarily land developers. There are eight common value capture techniques, but four are more prevalent: tax increment financing, special assessments, development impact fees, and joint development. Currently 48 states allow for the use of value capture for financing public infrastructure projects.

Road improvements help catalyze economic activity by providing greater access between and within communities. New construction, road widening, and other improvements can also spur contingent development. Road projects in North Carolina have led to significant amounts of both new business establishments and square footage of development in their surrounding areas compared to similar locations without nearby road projects.

Table 1. Change in firms, jobs, and land development within one mile of road projects (2001–2016)

Impact factor	Interstates	US highways	NC highways
Increase in the number of businesses within one mile of the road	3,560 to 6,050 (+69%)	17,250 to 24,250 (+41%)	15,950 to 27,520 (+73%)
Increase in the number of jobs within one mile of the road	47,700 to 84,100 (+76%)	195,000 to 251,400 (+29%)	231,000 to 313,000 (+35%)
Square footage of development within one mile of the road (per mile of road)	2,335,000	4,649,000	7,888,000

Table 2. Change in firms, jobs, and land development within one mile of roads without projects (2001–2016)

Impact factor	Interstates	US highways	NC highways
Increase in the number of businesses within one mile of the road	79,700 to 113,700 (+42%)	185,000 to 236,400 (+28%)	165,100 to 205,900 (+25%)
Increase in the number of jobs within one mile of the road	1,096,000 to 1,418,000 (+29%)	2,119,000 to 2,555,000 (+21%)	1,786,000 to 2,122,000 (+19%)
Square footage of development within one mile of the road (per mile of road)	788,000	496,000	441,000

3 Revenue Generation Options

Businesses depend on state-of-the-art transportation infrastructure to efficiently transport necessary components and final goods to their destinations. As shown in the previous section, there is a strong connection between economic factors and highway investment. North Carolina currently relies heavily on user fees to fund the construction, maintenance, and operation of the state's transportation system. User fees, in which fees are derived from the use of the transportation network to then fund it, serve as a primary method for transportation funding. In fiscal year 2019, NCDOT collected a total of $5.0 billion in revenue–most of which was derived from user-based sources such as the state gas tax and vehicle registration fees. Specifically, the state collected $2 billion in state motor fuel tax, $800 million in highway use tax, $600 million in vehicle registrations, $100 million in license fees, $200 million in vehicle title fees. The state also received $1.2 billion from federal funding sources (which are also derived from user-based methods such as the 18.4-cents per gallon federal gas tax). Finally, in 2019, the state also received $72.3 million in federal grants (NCDOT 2019a). North Carolina uses these sources of funding to support a wide variety of uses – most of which are dedicated toward the construction and maintenance of the state's highway network. According to the uses of 2018–2019 NCDOT appropriations, more than half (50.5% or $2.5 billion) goes toward construction activities, whereas $1.3 billion supports roadway maintenance. Other appropriations of transportation funding include transfers to other state agencies ($207.3 million), debt service ($194.6 million), administration, and other uses ($364.5 million). As is the case with many other states, a relatively small amount of funding, approximately 7.2%, supports non-highway modes (e.g., aviation, rail, public transit, ferry service, and bike and pedestrian services) (NCDOT 2019b).

The following list identifies 16 possible options for generating revenue for transportation. Those revenue options were evaluated based on six criteria (Table 3), including: yield adequacy, stability, implementation and administration, equity, economic efficiency, and public acceptance. Each option was given a score of 1 to 5 in these six categories, an overall score, and ranked on the basis of their overall scores (Fig. 2). Among the mechanisms evaluated, four options demonstrate the most promise for new or revised use in North Carolina: the road user charge, the state motor fuels tax, the statewide sales tax, and the highway use tax. The options considered for revenue generation included the following:

- Vehicle Miles Traveled Fee
- Fine-based Fees
- Heavy Vehicle Fees
- Severance Fees
- Vehicle Title, Registration, Vanity Plate Fees
- Highway Use Tax
- State Motor Fuels Tax
- Flat-rate Tolling
- High-Occupancy Toll Lanes
- Cordon Pricing (Priced Zones)
- Statewide Sales Tax

Table 3. Criteria for revenue options

Criteria	Rationale
Yield adequacy	Ultimately, the amount of revenue a funding option is able to draw will be the major determining factor in the scope of any project. There must be enough money available to support the project in its entirety, over what may potentially be a very long period of time
Stability	Hand-in-hand with 'Yield Adequacy', the reliability of the funding option to provide the revenue expected to move the project forward as planned is of utmost importance. Not only must the option provide revenue consistently, but must also be *expected* to do so to allow for good planning and room for flexibility. A high degree of reliability translates to good credit, which can be leveraged for financing
Implementation and administration	The funding option should be easy to set up and maintain. Reducing extraneous costs in time, money, and effort associated with initiating and supporting an option allows the maximum amount of potential revenue to go towards its intended purpose. By minimizing deadweight losses, these costs arising from structural inefficiencies, revenue sources can be streamlined and applied to their greatest effect
Equity	The most significant non-pecuniary criteria: equity in the distribution of the cost burden is seen as universally important not only in garnering 'Public Acceptance' but also in fulfilling obligations to populations which require the benefits of infrastructure projects to improve their economic station, in which presently they are unable to afford such improvements alone. This criterion subdivides into distribution of costs based on the user's ability to sustain the cost and on users paying for their respective derived benefits
Economic efficiency	This criterion ensures the funding option creates clear economic signals that the revenue being generated is truly going towards its intended purpose and that there is a measurable relationship between the costs incurred and the benefits derived. Further, it seeks to minimize the adverse impacts the cost burden might have on the populations which the project is intended to benefit by ensuring the funding option does not distort local markets
Public acceptance and feasibility	This criterion is used to assess how a proposed revenue option aligns with the public and political perceptions of its implementation. If a revenue mechanism is largely contentious throughout the sociopolitical spectrum and is universally unpopular, it would not be feasible to expect implementation. However, if the option is apolitical and not subject to social apprehension, it would be feasible as an implementation option

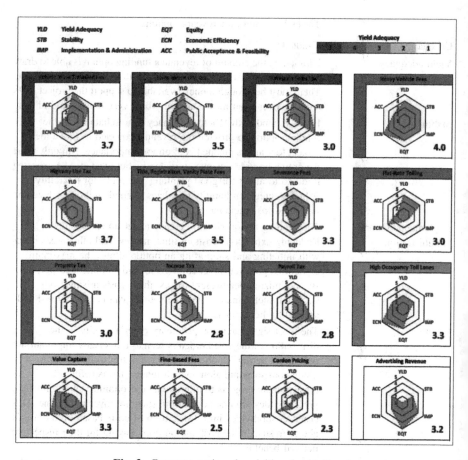

Fig. 2. Revenue options by yield and overall rank

- Income Tax
- Property Tax
- Payroll Tax
- Advertising Revenue
- Value Capture

3.1 Implementation Recommendations

Transportation infrastructure is deeply connected with economic development. From providing access to jobs, education, and healthcare, to moving the goods and services relied on by residents and businesses, a high-functioning transportation system is needed to create opportunities that grow the economy. Systemic transportation financing issues and external shocks from severe weather events and COVID-19 have placed transportation agencies in an exceedingly difficult position to finance its transportation system needs.

Looking into the future, per capita motor fuels tax receipts are expected to decline in real terms as a result of vehicle fuel economy improvements and flat-lining federal appropriations. Approximately 64.3% of North Carolina's budget is financed through motor fuels tax receipts with the largest funding share coming from the state motor fuels tax (40.4%) and the second largest coming from the federal motor fuels tax (23.9%) (NCDOT 2020). In 2015, the Diversifying Revenues to Improve Commerce and Economic Prosperity report called into question the long-term viability of the motor fuels tax for funding North Carolina's transportation system needs (ITRE 2015). Five years later, the motor fuels tax continues to operate as our primary funding mechanism, despite its loss in efficacy.

Across the nation, other innovative states are diversifying their transportation revenue streams to achieve healthy and sustainable portfolios. Oregon and Utah have established permanent road user charge programs, which charge road users who have registered in the program by the mile instead of by the gallon. Eight states have completed RUC pilot programs and three are developing RUC programs. Seven states have implemented statewide sales tax measures that ensure a portion of sales tax revenues is dedicated for transportation funding projects. Virginia and Kansas received upwards of $530 million in sales tax revenue for transportation projects in FY2019 (VDOT 2020; KDOT 2020). Forty-six states tax the sale of motor vehicles. North Carolina issues a 3% rate through its Highway Use Tax. Of all the states that assess a tax, North Carolina has the lowest effective rate. Four states (Delaware, New Hampshire, Montana, Oregon) do not have a sales tax on vehicles (World Population Review 2021).

At a rate of 36.2 cents per gallon, North Carolina has the eighth highest motor fuels tax rate in the nation, which may make adjusting the gas tax challenging. However, this rate is not substantially greater than the national average (although it is higher than North Carolina's neighboring states). Conversely, road user charges and statewide sales taxes measures have gained traction in the last decade to become established revenue options in the states that have implemented them.

North Carolina's Highway Use Tax is responsible for approximately 16% of North Carolina's transportation budget; however, a notable portion of its yield is not allocated to transportation uses. In 2019, $74.4 million in tax receipts from short-term leases and rentals were diverted to the North Carolina General Fund (NCDOR 2020). Generating additional revenue is a necessity for modernizing North Carolina's transportation system and supporting the state's economy. As North Carolina charges the lowest effective vehicle sales tax rate of the 46 states that assess a fee, adjusting the HUT and directing all of its revenue to transportation purposes could be a good starting place.

The North Carolina Department of Transportation conducted a strategic transportation plan that identifies transportation needs and projected revenues available to meet those needs through the year 2050 (NCDOT 2020). Transportation revenues and needs through 2050 indicate that North Carolina's transportation funding gap (the difference between projected revenue income and projected infrastructure costs) could be as high as $45 billion from 2020 to 2050. That report's findings, in conjunction with the NCDOT 2040 Plan and the American Society of Civil Engineer's infrastructure report card update, were used to estimate the level of investment required to obtain excellent, good, mediocre, poor, and very poor infrastructure rankings. Table 4 demonstrates the infrastructure condition that is forecasted to exist based on its associated level of annual investment in 2019 dollars. North Carolina currently invests approximately $5 billion annually in its transportation system, which enables the state to achieve an overall infrastructure rating of mediocre. The following subsections provide recommendations that can be undertaken collectively, or individually, to transition away from the motor fuels tax and establish a sustainable portfolio for transportation investment in North Carolina.

Table 4. System condition per level of investment

Infrastructure condition	Annual system needs ($ Billions)
Very Poor (F)	<$3.8
Poor (D)	$3.8–$4.8
Mediocre (C)	$4.8–6.3
Good (B)	$6.3–8.2
Excellent (A)	>$8.2

3.1.1 Recommendation 1 - Road User Charge Pilot Program

In 2019, North Carolinians drove 121.1 billion miles, consumed 5.5 billion gallons of gasoline, and generated.

$2.0 billion in state gas tax receipts (NCDOR 2020; NCDOT 2020). To offset the revenue generated by the state motor fuels tax, a 1.8 cents per-mile-fee would need to be assessed to each driver. This 1.8 cent fee does not yet account for administrative expenses, which are more involved for a Road User Charge program. Assuming an administrative cost of 10%, the minimum per-mile-fee to offset North Carolina's existing motor fuels tax revenues would need to be 1.98 cents-per-mile ($290 annually for the average driver). Unaccommodated by other measures, this would require a 3.0–5.0 cents-per-mile charge ($417–$695 annually for the average driver). To understand how a road user charge program might be implemented in North Carolina, best practices from RUC pioneer states were reviewed (Oregon, Utah, California, Washington, Colorado, Minnesota, and Delaware). It became apparent that for an RUC program to be successful, an RUC pilot to overcome administrative, technological, and privacy concerns was essential. Interstate travel and pricing policies were also found to be important considerations to address.

North Carolina can leverage the best practices from other states to implement a road user charge pilot program.

Its road user charge pilot can be used to distribute costs more equitably among road users. Currently, older and less fuel-efficient vehicles pay a larger share of motor fuels taxes. These vehicles are often driven by those who cannot afford to pay for electric vehicles and fuel-efficient alternatives. On average, drivers pay $23.00 per month on motor fuels taxes for low-efficiency vehicles. Meanwhile, fuel-efficient vehicles achieving 45 miles-per-gallon cost drivers $6.57 per month in motor fuels taxes and electric vehicle drivers pay $0.00 (I-95 2019). North Carolina should begin working to establish an RUC Fee Task Force. It can develop a grass roots base similar to the contexts of Oregon, Washington State, and California. An RUC task force can help build buy-in from the public and work to develop the administrative requirements for a successful program. The OReGO program utilizes the support of a volunteer coordinator and vendors to develop marketing strategies that reach volunteers from varying backgrounds. ODOT chose to target residents, businesses, and governments who operated vehicles at different miles per gallon categories. The California government took a thoughtful approach to determine if a road user charge would be acceptable to its residents (CSTA 2017). From the onset of the planning stage, the state legislature included provisions to ensure personal information was protected. Tracking mileage was done by using vendors from the private sector to stimulate innovation and diversity for applications (CSTA 2017).

3.1.2 Recommendation 2 - Adjust the Statewide Sales Tax

From 2013 to 2020, a total of 26 bills related to the use of sales taxes for transportation system funding have been enacted (NCSL 2020). At least 19 states have implemented statewide or local tax options to support transportation funding, including North Carolina (ITRE 2020). Counties that pass a referendum for Article 43, levy a half-cent sales and use tax, which is dedicated to financing public transportation. The quarter-cent Article 46 tax is general purpose and can be used for transportation. These local tax options are great mechanisms for funding local initiatives that are generally transit focused. However, multi-county projects that support statewide economic development require additional resources. It is recommended that North Carolina increase its statewide sales tax by 0.5 to 1.0% and dedicate these revenues for transportation purposes. This would align North Carolina with other states who are leaders in statewide sales tax implementation. It is anticipated that a 0.5% sales tax increase would generate an additional $782.7 million in transportation revenue and a 1.0% increase would raise $1.5 billion (ITRE Analysis sourced from NCDOR Annual State Sales and Use Tax Statistics 2019). Virginia, Kansas, Texas, and Idaho are national leaders in their use of sales tax revenue for transportation projects. Virginia directs 1.75% of its sales tax receipts to transportation, which equated to approximately $833.5 million in transportation revenue in fiscal year 2019 (VDOT 2019). Kansas directed $533 million in sales tax receipts to its state highway fund in fiscal year 2019 (KDOT 2019). Meanwhile, Texas sets aside sales tax revenue to fund highways. Specifically, $2.5 billion of the state sales tax revenue will be reserved for transportation, so long as overall sales tax receipts are at least $28 billion (Transportation for America 2020). Idaho imposes a 1% state sales tax to support transportation revenue (Goble 2020). This rate may increase to 2%, if the Idaho General Assembly can gain the support of the governor (Idaho Legislature 2020).

3.1.3 Recommendation 3 - Improve the Highway Use Tax

In October 2019, an issue brief was released on behalf of the NC First Commission entitled, "The North Carolina Highway Use Tax." It documented the history of North Carolina's HUT and discussed expanding its revenue capacity. The brief demonstrated that vehicle buyers in North Carolina pay significantly less tax on vehicle sales than in all neighboring states. For example, the typical North Carolinian car buyer pays $372 in highway use tax, while car buyers in neighboring states are assessed $500 to $868 (NC First Commission 2019). Upon examination of other states, North Carolina assesses the second-lowest effective tax rate on vehicles (World Population Review 2021). It is recommended that North Carolina raise its Highway Use Tax from 3 to 4%. This would generate approximately $275 million in additional transportation revenue (NC First Commission 2019). There are also tax revenue redirects that could be implemented within the framework of an HUT adjustment, which would benefit North Carolina's transportation system. For example, lawmakers may consider eliminating or raising the $2,000 cap on recreational vehicles and commercial vehicles. Additionally, if all the proceeds from short-term vehicle leases were directed to the Highway Fund, this would increase revenues by approximately $74 million annually (NCDOR 2020). Currently, approximately $10 million is dedicated for transportation uses.

3.1.4 Recommendation 4 - Phase-Out of the Gas Tax

To coincide with a permanent RUC program, increasing the state's gas tax rate, either on a per-gallon basis or changing how it is indexed, could be one option for policymakers to consider as a temporary measure to generate additional transportation revenue. However, at a rate of 36.2 cents per gallon, North Carolina has the eighth highest motor fuels tax rate in the nation, which may make raising the gas tax a significant challenge (Federation of Tax Administrators 2020). North Carolina's motor fuels tax rate is well-positioned compared to other states, largely due to an adjustment in the motor fuels tax rate formula. It is recommended that North Carolina supplements its motor fuels tax revenue with a statewide sales tax increase of 0.5 to 1.0%. This would align North Carolina with states that have implemented sales taxes measures for transportation funding, including Virginia, Kansas, Nebraska, Connecticut, Texas, Idaho, and Arkansas. It is also recommended that North Carolina increase its Highway Use Tax from 3 to 4% to better align with what other states charge for vehicle sales taxes.

4 Conclusions

Road projects led to a significant increase in the surrounding development from 2001 to 2016 compared to roadways without improvements. These changes included increases in jobs, businesses, and buildings. This study showed that road improvements help catalyze economic activity by providing greater access between and within communities. New construction, road widening, and other improvements can also spur contingent development. Road projects in North Carolina have led to significant amounts of both new business establishments and the construction of buildings in their surrounding areas compared to similar locations without nearby road projects. Therefore, road and transportation development can be considered as a useful economic development tool.

However, per capita motor fuels tax receipts are expected to decline in real terms as a result of vehicle fuel economy improvements and flat-lining federal appropriations.

Table 5. Schedule of rates and revenues that can be implemented to replace the motor fuels tax

Revenue policy	Level D[a] Rate	Level D[a] ($ millions)	Level C[b] Rate	Level C[b] ($ millions)	Existing[c] Rate	Existing[c] ($ millions)	LOS B[d] Rate	LOS B[d] ($ millions)	LOS A[e] Rate	LOS A[e] ($ millions)
Replace Motor Fuels Tax	—	—	—	—	—	—	—	—	—	—
Implement Road User Charge (cents per mile)	0.5	$441	1.1	$1,050	1.3	$1,292	2.1	$2,159	3.7	$3,667
Allocate General Sales Tax Revenue for Transportation	—	—	0.25%	$391	0.25%	$391	0.50%	$783	0.75%	$1,174
Adjust the Highway Use Tax	4.00%	$1,071	4.00%	$1,071	4.00%	$1,071	4.00%	$1,071	4.00%	$1,071
Direct All Short-term Vehicle Lease and Rental Revenue to Highway Fund	8.00%	$84	8.00%	$84	8.00%	$84	8.00%	$84	8.00%	$84
All Other NCDOT Revenue to Accumulate with No Changes[f]	—	$2,204	—	$2,204	—	$2,204	—	$2,204	—	$2,204

[a]Grade Level D assumes replacing the motor fuels tax with a 0.5 cents-per-mile RUC, increasing the HUT from 3–4%, and redirecting of all short-term vehicle lease and rental revenue to the Highway Fund.
[b]Grade Level C assumes replacing the motor fuels tax with a 1.1 cents-per-mile RUC, directing 0.25% of North Carolina's revenue from total taxable sales to the Highway Fund, an increase of the HUT from 3–4%, and redirecting all short-term vehicle lease and rental revenue to the Highway Fund.
[c]Achieving existing conditions assumes replacing the motor fuels tax with a 1.3 cents-per-mile RUC, directing 0.25% of North Carolina's revenue from total taxable sales to the Highway Fund, increasing the HUT from 3–4%, and redirecting of all short-term vehicle lease and rental revenue to the Highway Fund.
[d]Grade Level B assumes replacing the motor fuels tax with a 2.1 cents-per-mile RUC, directing 0.5% of North Carolina's revenue from total taxable sales to the Highway Fund, increasing the HUT from 3–4%, and redirecting of all short-term vehicle lease and rental revenue to the Highway Fund.
[e]Grade Level A assumes replacing the motor fuels tax with a 3.7 cents-per-mile RUC, directing 0.75% of North Carolina's revenue from total taxable sales to the Highway Fund, increasing the HUT from 3–4%, and redirecting of all short-term vehicle lease and rental revenue to the Highway Fund.
[f]Based on NCDOT fiscal budget for FY2019-2020.

Approximately 64.3% of North Carolina's budget is financed through motor fuels tax receipts with the largest funding share coming from the state motor fuels tax (40.4%) and the second largest coming from the federal motor fuels tax (23.9%) (NCDOT 2020). In 2015, the Diversifying Revenues to Improve Commerce and Economic Prosperity report called into question the long-term viability of the motor fuels tax for funding North Carolina's transportation system needs (ITRE 2015). Five years later, the motor fuels tax continues to operate as our primary funding mechanism, despite its loss in efficacy. To sustain the state's transportation system, revenue generation options must be considered. Those revenue options were evaluated based on six criteria, including: yield adequacy, stability, implementation and administration, equity, economic efficiency, and public acceptance. From this evaluation, sales tax and highway use tax recommendations can alleviate the short- to medium-term financial needs of North Carolina's transportation system. Meanwhile, an RUC pilot program is essential for laying the groundwork for the long-term stability of North Carolina's transportation system. A menu of revenue mechanisms and the associated revenue generating potential that can be used to modernize North Carolina's transportation infrastructure are shown in Table 5. The table also shows the levels of investments that could be used to change the state's level of service (from a low of Level D to a high of Level A).

Acknowledgements. This study was funded and supported by the North Carolina Chamber Foundation.

References

AASHTO: AASHTO Letter to Congress on COVID-19 (2020). https://www.enotrans.org/wp-content/uploads/2020/04/2020-04-06-AASHTO-Letter-to-Congress-on-COVID-19-Phase-4-FINAL.pdf

CSTA: California Road Charge Pilot Program. California State Transportation Agency (2017). https://dot.ca.gov/-/media/dot-media/programs/road-charge/documents/rcpp-final-report-a11y.pdf

Duncan, I.: Reeling from the loss of gas tax revenue during pandemic, states are deferring billions of dollars of transportation projects. Washington Post, July 2020 (2020). https://www.washingtonpost.com/local/trafficandcommuting/reeling-from-the-loss-of-gas-tax-revenue-during-pandemic-states-are-deferring-billions-of-dollars-of-transportation-projects/2020/07/09/b01c87da-b705-11ea-a510-55bf26485c93_story.html

Federation of Tax Administrators: Tax Rates/Surveys – Tax Rates (2020). https://www.taxadmin.org/tax-rates#mf

Goble, K.: Idaho bills would boost transportation revenue. Land Line (2020). https://landline.media/idaho-bills-would-boost-transportation-revenue/

Idaho Legislature: House Bill 325 (2020). https://legislature.idaho.gov/sessioninfo/2020/legislation/H0325/

ITRE: Meta-analysis of Statewide and Local Tax Options (2020)

ITRE: Diversifying Revenues to Improve Commerce and Economic Prosperity (2015). https://ncchamber.com/wp-content/uploads/NCCFEconomicStudy_ITRE_20150120.pdf

I-95 Corridor Coalition MBUF: Equity and Fairness Considerations in a Mileage Based User Fee System (2019). https://static1.squarespace.com/static/5a600479ccc5c5e5c8598516/t/5ee8d7aa5a77d7165cc68a08/1592317872316/I95_CC_MBUF_Task_3.2_Equity_and_Fairness_Considerations_Tech_Memo_Final%5B1%5D.pdf

KDOT: KDOT 101. Kansas Department of Transportation, January 2019 (2020). http://kslegislature.org/li/b2019_20/committees/ctte_h_trnsprt_1/documents/testimony/20190131_01.pdf. Accessed 20 Mar 2020

National Conference of State Legislatures: Recent Legislative Actions Likely to Change Gas Tax (2019). https://www.ncsl.org/research/transportation/2013-and-2014-legislative-actions-likely-to-change-gas-taxes.aspx. Accessed 23 Aug 2019

NCDOR: Monthly Sales and Use Tax Statistics (2020). https://www.ncdor.gov/news/reports-and-statistics/monthly-sales-and-use-tax-statistics

NCDOT: Finance & Budget–State Funding Sources 2019. North Carolina Department of Transportation (2019a). https://www.ncdot.gov/about-us/how-we-operate/finance-budget/Documents/ncdot- revenue-sources.pdf. Accessed 9 Apr 2020

NCDOT: Finance & Budget–NCDOT Funding Distribution 2018–2019. North Carolina Department of Transportation (2019b). https://www.ncdot.gov/about-us/how-we-operate/finance-budget/Documents/ncdot-funding-distribution.pdf. Accessed 9 Apr 2020

NCDOT: Taxes (2020). https://www.ncdot.gov/dmv/title-registration/taxes/Pages/default.aspx. Accessed 6 Apr 2020

NCDOT: NC MOVES 2050 Plan (2020). https://www.ncdot.gov/initiatives-policies/Transportation/nc-2050-plan/Documents/draft-technical-summary-report.pdf

NC First Commission: The North Carolina Highway Use Tax. NCDOT (2019). https://www.ncdot.gov/about-us/how-we-operate/finance-budget/nc-first/Documents/nc-first-brief-edition-3.pdf

Shuman, R.: INRIX U.S. National Traffic Volume Synopsis: Issue #6 (18–24 April 2020). Inrix (2020). https://inrix.com/blog/2020/04/covid19-us-traffic-volume-synopsis-6/

Transportation for America: The states that successfully raised revenue since 2012. Transportation for America (2020). http://t4america.org/maps-tools/state-transportation-funding/. Accessed 30 Mar 2020

VDOT: Virginia Department of Transportation: Fiscal Year 2020: VDOT Annual Budget, June 2019 (2019). https://www.virginiadot.org/about/resources/budget/Final_VDOT_Budget,_6-18-2019.pdf

World Population Review: Car Sales Tax by State 2021 (2021). https://worldpopulationreview.com/state-rankings/car-sales-tax-by-state

Climate Resilient Road Design – 2

Climate Resilient Urban Mobility by Non-motorized Transport

Kigozi Joseph(✉)

Business Development Division, PROME Consultants Limited, Kampala, Uganda
kgzjoseph@gmail.com

Abstract. African cities have begun to suffer climate change effects. In most African cities, populations are increasing rapidly and the reliance on Non-Motorized transport (NMT) is high, but dedicated NMT infrastructure remains underdeveloped. In all cities and towns across Uganda, the use of private vehicles has risen steadily over the years and has congested these cities, poisoned the air and killed NMT users at exceptionally high rates. This paper seeks to answer the question whether NMT projects are economically viable and how cities can maximize benefits of NMT for Climate conscious economic growth. This paper presents an Economic analysis of the NMT pilot project in Kampala using the Non-Motorized Transport Project Assessment Tool (NMT-PAT) to quantitatively and qualitatively analyze the expected impacts (benefits and costs) with focus on Environmental and Health Benefits. The results of the analysis indicate that considering a design life of 15 years, Kampala city will experience reductions in emissions to the tune of 675,000 tons for Carbon dioxide, 13.81 tons of Particulate matter and 2536 tons of Nitrogen dioxide. The health benefits in terms of reduction in accidents valued at Uganda shillings 4,163,611,405,517.35 (USD 1,134,499,020) will also be realized. A general improvement in journey quality, security and livability will also be achieved as well as a reduction in the noise levels by about 3.75 dB. To encapsulate by implementing the proposed NMT infrastructure, a Net present Value of 14 trillion shillings (USD 3 Billion) shall be realized thus demonstrating that NMT investment is viable.

Keywords: Non-motorized transport · Environmental benefits · Economic benefits

1 Introduction

In most African cities, population is increasing rapidly and the reliance on non-Motorized transport (NMT) is high, but dedicated NMT infrastructure remains underdeveloped. Globally, road crashes have been taking a higher toll in the past years, with over 1.35 million people killed on the world's roads annually and another 20–50 million seriously injured (World Health Orgsnisation 2021), making road safety a public health priority.

At a national level, the use of private vehicles has risen steadily over the years and has congested most towns and cities, poisoned the air and killed NMT users at exceptionally high rates. The private sector which is the major provider of transport services is extremely fragmented and undercapitalized. This is coupled with a regulatory system that in inadequate, ineffective and marred with political interference thus causing the mess we see on the streets of most African cites.

The table below extracted from the Uganda Police Annual crime report of 2019 shows the fatalities per road user category for the years 2018 and 2019 (Table 1).

Table 1. Annual fatalities (Uganda Police 2019)

Road User Category	2019	2018
Driver	194	202
Motorcyclist	1064	878
Pedal Cyclist	**136**	**160**
Passenger on motorcycle	422	380
Passenger in Light omnibus	82	93
Passenger in Medium omnibus	08	30
Passenger in Heavy Omnibus	27	48
Passengers in other vehicles	462	474
Pedestrians	**1485**	**1424**

It is apparent from the data presented in the table above that pedestrian constitute the largest percentage of fatalities and thus action needs to be taken to make the roads safer for pedestrians. The key issues that have brought about the high rate of pedestrian's fatalities is the deadly combination of high vehicle speeds, human behaviour, and lack of forgiving road infrastructure in the sense of complexity and ambiguity. Well-designed NMT infrastructure has the potential to reduce the probability that pedestrian and cyclist crashes occur and to reduce the severity of injury when a crash actually occurs. Implementation of NMT infrastructure is therefore one of the key actions that can make roads safer for pedestrians and cyclists in-line with the United Nations Decade of Road Safety Action 2021–2030, which aims to reduce road fatalities and injuries by 50% by 2030.

Across Africa, a significant percentage of Urban dwellers are of a low-income status and as such heavily depend on Non-Motorized Transport and Para-transit modes for mobility. The average transport expenditures of the urban poor accounts for 10–20% of their household incomes (World Bank 2005). Uganda's real gross domestic product (GDP) grew at 2.9% in 2020, less than half of the 6.8% recorded in 2019 due to the macroeconomic effects of COVID-19 (World Bank 2021). It is therefore evident that due to the contraction of the economy, the already small incomes of the urban poor have become smaller thus continuance of the high incidence of poverty among this section of the population. Non-Motorized transport creates the opportunity for the users to reduce their expenditure to up to 5% of their Household incomes thus reducing the expenditure burden and freeing up funds for investment in business. It is the view of this paper that NMT infrastructure improvements can increase urban productivity and thus reduce the levels of poverty.

African cities have waited too long to deal with the existential threat of Climate change and thus have begun to suffer its effects. With a global car fleet predicted to triple by 2050 (over 80% of that in the developing world) we have to find a way to reconcile the need for increased mobility with an ambitious reduction in emissions along with improved air quality (UN Environment 2019). Climate change should therefore be central when it comes to transport planning. Investment in NMT infrastructure is presenting Kampala and the Uganda at large with an opportunity to achieve the collective climate related goals and the Nationally Determined Contributions (NDCs).

City authorities and Decision Makers currently do not consider NMT infrastructure investments as a priority because there is limited knowledge on the return on investment that would be achieved from these investments. Transport professionals therefore have a big role to play in order to change this mindset. Financial institutions like commercial banks have been observed not to be very forth coming when it comes to NMT related investments. Long term commitment from the donor agencies will therefore be instrumental in enabling cites sustain efforts in the planning and implementation of NMT infrastructure.

2 Proposed NMT Design for Kampala Central Business District

The Non-motorized Transport (NMT) project in Kampala aims at transforming selected road infrastructure corridors to ensure all inclusive, safe, secure, enjoyable movement of pedestrians and cyclists to and within the Central Business District (CBD). The project scope covers a 15-km NMT corridor for Pedestrians and Cyclists along the Kampala–Namanve railway reserve connecting to 4 km of Pedestrianized streets within the Kampala Central Business District (See Table 2 Below).

3 Evaluation Methodology

Over view of Non-Motorized Transport Project Assessment Tool (NMT-PAT)

Whereas various research conducted on NMT in various parts Europe like the Netherlands and some parts of Africa indicates that NMT investments have an economic benefit, in Uganda there is not an existing, published and comprehensive method for the economic appraisal of NMT infrastructure. This has made it difficult for decision makers to undertake technical appraisals that entail combination of economic, engineering and socio-environmental aspects that is required to assess whether the an NMT project should be carried forward for more detailed investigations, planning and Implementation.

The Non-Motorized Transport Project Assessment Tool (NMT-PAT), developed by a joint venture between United Nations Environment Programme (UNEP) and the Centre for Transport Studies was adopted to analyze the Economic costs and benefits of

Table 2. Proposed NMT design

| Existing MGR Railway Corridor with pedestrians dangerously sharing the space with the trains moving along the Railway | Proposed design of NMT infrastructure along the MGR Railway corridor with dedicated pedestrian and bicycle ways physically separated from the railway line. |

| Current Typical Street within the CBD | Typical CBD Street After Re-design |

the proposed NMT project in Kampala. NMT-PAT is a Microsoft-excel based tool that is designed to evaluate the costs and benefits associated with walking and cycling. NMT PAT consists of worksheets each dedicated to a specific evaluation criterion providing input and output functionality for the user (UNEP & Centre for Transport Studies, n.d.). With NMT PAT we have been able to generate meaningful insight regarding the economic viability of NMT in Kampala whist taking into consideration Environment, Health, Economic and Social aspects. The key inputs to the NMT-PAT model are described in here below.

Key Model Input 1: Regional Economic Data

The NMT-PAT economic evaluation model requires regional data as an input. Regional economic data includes; modal split, metropolitan population, economic growth rate, average inflation rate, GDP per capita and the discount rate. The regional economic data applicable for Kampala also shown in the Table 3 below was obtained from the following sources;

i. The Statistical Abstract 2019 and 2021 by the Uganda Bureau of Statistics;
ii. Transport Note No. TRN-6 by the World Bank; and
iii. Smart Moving Kampala by Iganga Foundation.

Table 3. Regional economic data (Source: KCCA traffic data)

Indicator		Value
Modal split	Walking	0.0%
	Cycling	0.0%
	Bus	0.5%
	BRT	0.0%
	Minibus taxi	21.0%
	LRT	0.0%
	Cars	37.0%
	Motorcycles	41.0%
	Taxi	0.0%
	Other	0.5%
Metropolitan population	2021	1,507,080
	2036	4,000,000
Economic growth rate		6.50%
Average inflation rate		2.60%
Currency		Ugandan shillings
GDP/capita		Sh 3,248,600.00
Discount rate		12%

Key Model Input 2: Estimated Project Investment Costs

The estimated project implementation costs were determined through a unit rate analysis of the major cost elements of the project i.e. Equipment costs, Labour costs, Materials costs and Overheads. The anticipated project investment costs for the NMT project considering 2021 as the financial year for design and 2022 as the financial year for construction were input in to the model as shown in Table 4 below.

Table 4. Estimated project investment costs (Source: NMT design reports)

Project cost item	Financial year	Value	Net present value
Construction costs	2022	Sh 132,706,107,983.00	Sh 129,343,185,168.62
Design costs	2021	Sh 12,970,610,798.30	Sh 12,970,610,798.30

4 Evaluation Results

<u>Environmental Benefits</u>

The mode shift from Motorized transport to Non-motorized transport contributes to the reduction in air pollution trough reduced greenhouse gas emissions, noise, road deaths and injuries resulting from motor accidents, congestion and social isolation. The Nitrogen dioxide contained in vehicle emissions increases the symptoms of the people suffering from respiratory and heart disease.

The Logic of estimating environmental benefits is that provision of NMT infrastructure creates a mode shift to zero emission NMT modes thus improving air quality and improving the flow of motorized traffic. The improvement in air quality results in a reduction in pollution induced respiratory incidents such as asthma and Chronic Obstructive Pulmonary Disease (COPD).

Three major indicators were considered using the NMT-PAT modal; Emissions, Energy use and Noise pollution. Suffice to mention that this analysis was originally adopted from the Transportation Emissions Evaluation Model for Projects (TEEMP) and incorporated into NMT-PAT. The amount of petrol and diesel used and their respective emissions were calculated by considering the total number of trips per day, dominant fuel type, average occupancy per mode and average trip length. Three types of emissions were calculated i.e., Carbon dioxide (CO_2), Nitrogen Oxides (NOX) and Particulate matter (PM). The summary of results of the analysis are as shown in Table 5 below for the with-project and business as usual scenarios.

Table 5. Estimated reduction in emissions (Source: NMT-PAT model)

Emissions (tons)	Business as usual	With project	Savings
	Tons	Tons	Tons
CO_2	2,854,000.72	2,178,542.67	675,458.05
PM	1,567.67	1,553.86	13.81
NOx	12,551.91	10,015.74	2,536.17

<u>Health Benefits</u>

Cycling and walking has potential to improve health for both individuals and society at large resulting from the increased level of physical activity and reduction in traffic accidents.

The Health Economic Assessment Tool (HEAT) developed by the World Health Organization is designed to enable users to conduct economic assessments of the health impacts of walking and cycling. HEAT estimates the value of reduced motility that results from specified amounts of walking and cycling (World Health Organization 2014). HEAT can be used for assessments such as;

(i) Assessment of current or past levels of cycling and walking
(ii) Assessment of changes overtime i.e. before and after comparisons
(iii) Evaluation of new or existing projects

The NMT-PAT Model consists of a simplified version of the HEAT model. For the NMT project under assessment two indicators were considered i.e., Physical activity and Traffic accidents. The HEAT assessment calculates the monetary value associated with change with the Relative Risk of Mortality (RRM) of a new user due to the change in the level of physical activity putting into consideration the Value of a Statistical life. The results obtained from the HEAT assessment are shown in the Table 6 below.

Table 6. HEAT assessment results (Source: NMT-PAT model)

Indicator	Difference in PKT (Km)	No. of new users	PKT per new user per week (Km)	Physical activity per week (minutes)	Relative risk of mortality	Net present value of physical activity
Walking	93,995.67	25,000.00	26.32	328.98	0.7846	Sh 3,811,673,251,545.58
Cycling	299,077.13	25,000.00	83.74	358.89	0.6052	Sh 6,985,757,522,832.62
Total						Sh 10,797,430,774,378.20

The Health benefits related to the reduction of traffic accidents were calculated based on the prediction of the reduction of NMT fatalities and injuries between the business and usual and with project scenarios. The results of this assessment are shown the Table 7 below.

Table 7. Economic value of accident reduction (Source: NMT-PAT model)

Indicator	Total health value of accidents		Net present value of reduced accidents
	Business as usual - 2036	With project - 2036	
NMT fatalities	25,255,277,219,440.50	2,625,041,933,057.80	Sh 2,307,496,765,709.29
NMT injuries	20,315,766,980,869.30	10,156,201,443,160.20	Sh 1,856,114,639,808.06
Total			Sh 4,163,611,405,517.35

5 Other Benefits

Savings on User Costs

User participation is an important ingredient when it comes to the Success of projects such as NMT. The perceptions and views of the users are important in making an inventory of problems faced and identifying the most appropriate solutions. In this regard, an assessment of the road user costs per kilometre for both Motorized and Non-

Motorized modes within the Central business district was conducted. Key informant interviews with pedestrians and motorists along the following streets were conducted;

i. Dastur Street;
ii. Market Square Road;
iii. Market Service Road; and
iv. Along the Railway Corridor to Namanve.

The interviewees were selected at random and the purpose of the interviews was to determine how the people have travelled to their destinations, frequency of visits, travel time, purpose of visits and the costs incurred in making the visit (spending patterns). The responses provided were analyzed, summarized and an average cost of travel was determined as shown in the Table 8 below;

Table 8. Average cost per km using motorized transport means. (Source: Stakeholder consultations)

Mode of transportation	Average estimated cost per Km (UGX)	Average estimated cost per Km (USD)
14-seater Taxi	400	0.108
Boda-Boda (motorcycle)	1500	0.407
Private car	1000	0.271
Private Taxi	2000	0.542

A road user travelling by a 14-Seater Taxi from Namanve to Kampala CBD a distance of 15 km for work spends an average of UGX 12000 or USD 3.3 per day to travel to and from work. If the road user shifts to a bicycle for the same trip a saving of USD 1200 per year is made.

Time Savings

In addition to the anticipated savings on transport costs, the travel time that will be saved due to the implementation of the NMT project was determined by estimating the share of trips that will be affected by the project and time saving per trip for each mode. An economic value was attached to the time saved to determine the Net present value of the total time saved (Table 9).

Increase In VAT Tax Revenue

It is envisaged that the implementation of the NMT project will bring about a decrease in the number of motorized trips and thus there shall be lost fuel tax revenue as road users will spend less on fuel. The money that will be saved by the users will be spent on other taxable consumables and thus there shall be an increase in Value Added Tax (VAT) revenue. This scenario has been analyzed in NMT-PAT and is illustrated in Table 10 below.

Table 9. Value of time saved (Source: NMT-PAT model)

Indicator	Share of trips directly affected by project	Estimated time saving due to project	Value of time	Value of time saving per day	Total amount of time saving	Net present value of time saving
	%	minutes	Sh/minute	Sh	Person years	Sh
Walking	2.0%	2.00	29.09	87,263.60	31.25	71,742,297.97
Cycling	2.0%	10.00	29.09	436,317.98	156.25	358,711,489.83
Bus	2.0%	2.00	24.75	75,583.38	31.82	62,139,607.23
BRT	5.0%	5.00	24.75	463,994.87	195.31	381,465,576.87
Minibus taxi	5.0%	−15.00	24.75	−3,842,530.09	−1,617.46	−3,159,071,469.68
LRT & rail	0.0%	0.00	24.75	–	–	–
Cars	5.0%	−10.00	29.09	−3,097,719.81	−1,109.33	−2,546,738,224.96
Motorcycles	1.0%	5.00	29.09	271,032.53	97.06	222,824,833.71
Taxi	1.0%	−10.00	20.40	−153,036.90	−78.13	−125,816,716.58
Other	0.0%	0.00	29.09	–	–	–
Total				−5,759,094.45	−2,293.22	−4,734,742,605.62

Table 10. Increase in VAT revenue. (Source: NMT-PAT model)

Indicator	Fuel levy	Petrol price per Litre	Diesel price per litre	Lost fuel tax revenue per year	Net present value of lost fuel tax revenue
	(%)	Sh	Sh	Sh	Sh
Fuel tax	28%	4,000.00	3,800.00	334,609,293,018	1,111,639,891,512
Indicator	Value Added Tax (VAT)	Money saved per person per year	Percentage of saved money spent on taxed consumables	Tax gained per year	Net present value of tax gained
	(%)	Sh	(%)	Sh	Sh
Value Added Tax	18%	1,800,000.0	20%	178,429,392,000.0	488,975,742,270.7

6 Conclusions

The Summary of the Economic evaluation is detailed in Appendix 1. All the indicators discussed above have been monetized and a Net Present Value (NPV) for each indicator determined. The Sum of the NPVs has been calculated as the Net Present Benefit (NPB). The Net present value of the entire project has been calculated by subtracting the Net present Cost from the Net Present Benefit (See Table 11 below).

Table 11. Project Net Present Value and Cost-Benefit Ratio (Source: NMT-PAT Model)

Net Present Benefit (NPB)	Sh. 14,333,643,288,048.30
Net Present Cost (NPC)	Sh. 142,313,795,966.92
Net Present Value (NPV)	Sh. 14,191,329,492,081.40
Cost-Benefit Ratio (CBR)*	100.72

It is evident from the **positive** Net Present Value (NPV) and the Benefit Cost Ratio that the NMT pilot project is Economically viable and attractive. The economic parameters obtained give us confidence that;

i. NMT is an Environmentally Sustainable transport mode for African Cities;
ii. NMT is an Economically feasible venture considering all the other potential benefits;
iii. The NMT project in Kampala will be able to cover its operating costs over its life time;
iv. The sensitivity analysis conducted shows that the project can adapt to economic fluctuations; and can be incorporated into the wider economic framework of the country;

v. The possibility of using the Spaces provided along the project corridor for alternative uses such as exhibitions, Sports events and Markets contributes greatly to the overall project economic attractiveness.

7 Challenges Facing NMT Project Implementation

Whereas Non-Motorized Transport is an attractive and viable investment for African cites, the following challenges are foreseen as African cities incorporate NMT in to their transport anatomy;

i. The unavailability of space in the urban centers renders prioritization of NMT very disruptive to motor vehicle flow.
ii. Absence of NMT Design standards custom made for the African Setting.
iii. Poor Driving Habits of Motorists that endanger Non-Motorized Transport Users.
iv. Encroachment of newly built pedestrian spaces by street vendors and thus reducing the capacity and aesthetic appearance of these facilities.
v. Poor land use planning without linkages and coordination to urban transport planning.

8 Recommendations

The Economic analysis results give us confidence that indeed NMT projects constitute part of the future of Urban Transport Systems which have greatly been tested by the 21st Century mobility challenges exacerbated by the COVID-19 pandemic. The introduction, promotion, acceptance and adoption of NMT modes of travel by an already car-centric generation living in African cities will require the following to be put in place.

i. Preparation of mobility plans for pedestrians, cyclists and other forms of non-Motorized transport.
ii. Modification of City Road infrastructure to make it more accommodative to Non-motorized Transport.
iii. Build Special Infrastructure for pedestrians and cyclists making special considerations for vulnerable groups like Women, children and people with disabilities.
iv. Introduce city wide traffic calming to reduce vehicle speeds across the city road network.
v. Enhance and promote private sector participation in interventions and business ventures that can increase bicycle ownership and use eg. Bicycle share services, publicity and, media campaigns.
vi. Training and Sensitization of NMT users and Motorists on how to safely share the road even in situations where exclusive NMT facilities are not available.

vii. Many city authorities are working with limited resources therefore NMT infrastructure improvement should put into consideration use of low-cost equipment and labor-intensive methods that involve as much as possible the local community.
viii. Continued research and dissemination of Knowledge in an organized, effective and economic way so that transport professionals across Africa and the world can access it and use it to develop NMT programs in their own areas.

Acknowledgements. Great appreciation goes to the Management and Staff of PROME Consultants Limited that gave me the opportunity be part of the Design team of the NMT Design Project for Kampala Central Business from which I picked the inspiration to undertake this research and prepare this paper.

References

Ministry of Finance Planning and Economic Development: Semi-Annual Budget Monitoring Report 2018/19, s.l.: s.n (2019)
UBOS: Uganda National Household Survey 2012/2013. UBOS, Kampala Uganda (2014)
UBOS: Statistical Abstract, s.l.: s.n (2019)
UN Environment: Share the Road Programme Annual Report 2018, s.l.: UN Environment (2019)
UNEP & Centre For Transport Studies: NMT Project Assesment Tool: User Guide, s.l.: s.n (n.d.)
World Bank: Assessment of the Non-Motorized Transport Program, s.l.: s.n (2002)
World Bank: Non-Motorized Transport in African Cities; Lessons From Experience in Kenya and Tanzania, s.l.: Sub-Saharan Africa Transport Policy Program (2005)
World Bank: The World Bank in Uganda (2021). worldbank.org/en/country/uganda/overview. Accessed 12 August 2021
World Health Organization: Health Economic assesment tools for walking and for cycling (2014). http://www.heatwalkingcycling.org//heatwalkingcycling.org/index.php?pg+cycling&act=more1. Accessed September 2019

Author Index

A

Abdelkader, Ahmed F., 395
Abulkhair, Myasar, 622
Abuzwidah, Muamer, 303, 897, 1148
Akhnoukh, Amin, 232, 450
Akhnoukh, Amin K., 997
Al Khateeb, Ghazi, 87
Albuquerque, Francisco Daniel B., 915
Alhelo, Lina, 933
Al-Khateeb, Ghazi, 622, 1083
Almbaidheen, Khalil, 622
Alvarez, Allex E., 72
Andriejauskas, Tadas, 379
Antunes, António, 379
Appelt, Veit, 365
Asadollahi Pajouh, Mojdeh, 868
Ashour, Ayat G., 1148
Aslyamov, Timur, 257
Atahan, Ali Osman, 667, 1139
Atahan, Ali, 933
Aziz, Shawon, 511

B

Bakioglu, Gozde, 667
Barabino, Benedetto, 525
Barateiro, José, 379
Bargaoui, Zoubeida, 1167
Baumgardner, Gregory M., 269
Bayya, Radhika, 647
Belhaj, Majda, 46
Bert, Steven A., 287, 1207

Bhusari, Shubham, 379
Bielenberg, Robert, 755, 772, 789, 805, 821
Biligiri, Krishna Prapoorna, 647, 1098, 1112, 1181
Birdsall, James, 425
Biswas, Sukalpa, 379
Bolshakov, Tikhon, 257
Bonera, Michela, 525
Bonilla, Minerva, 997
Boyapati, Rama Krishna, 269
Boychenko, Elena A., 602
Branzi, Valentina, 493
Burghardt, Tomasz E., 676
Burlacu, Alina Florentina, 117
Burlacu, Florentina Alina, 134
Burmistrov, Igor, 1071
Burmistrov, Igor N., 602

C

Caballero, Rafael, 171
Çalışkan, Berna, 1139
Chandio, Imtiaz Ahmed, 1026
Charak, Akhil, 1112
Chompooming, Kridayuth, 187
Clute, Khyle, 789
Coclanis, Peter, 287

D

Dagaeva, Maria, 257, 881
de Franco Peixoto, Creso, 837
Dep, Linus, 3
Dib, Joseph, 97
Doveh, Etti, 317

E
Ekhande, Tejan, 450
Ergin, Mahmut Esad, 950
Erian, Nihal, 357
Eustace, Deogratias, 345
Ezzat, Helal, 87, 1083

F
Faizrakhmanov, Emil, 257
Faller, Ron, 755, 772
Faller, Ronald, 805, 821, 850, 868
Fang, Chen, 850
Fares, Mohamad Yaman, 303
Faria, Susana, 474
Findley, Daniel, 997
Findley, Daniel J., 287, 1207
Françoso, Maria Teresa, 837
Freitas, Elisabete, 474
Friedman, Robert D., 395
Fuller, Kelly, 1207
Fullerton, Clare, 997

G
Gitelman, Victoria, 317
Gorshkov, Nikolay V., 602
Gubbala, Chaitanya, 1181
Gücüyener, Murat, 243

H
Halleman, Brendan, 357
Hassan, Reem N., 1148
He, Liang, 46
Head, Weston, 1207
Himeno, Kenji, 589
Hofko, Bernhard, 35
Hosoi, Yusuke, 575
Hosokai, Takeo, 975
Huq, Armana Sabiha, 542
Huytook, Attaphon, 407

I
Ilıcalı, Mustafa, 950
Ismail, Mohammed, 1083

J
Joseph, Kigozi, 1225
Joseph, Owino, 634

K
Kaloush, Kamil, 735
Kaloush, Kamil E., 611
Kamaraj, Amudha Varshini, 269
Kaptan, Aşkın Kaan, 243
Kawakami, Atsushi, 589
Kawamura, Hinari, 575
Kawauchi, Kensaku, 975
Kelvin, Msechu, 634
Kesten, Ali Sercan, 1139
Khalil, Mahmoud, 897
Khalil, Nariman J., 97
Khijniak, Dmitri, 691
Kildeeva, Sofya, 881
Kiselev, Nikolay V., 602
Kishi, Yuki, 458
Kligis, Maris, 416
Kobayashi, Hiromasa, 458
Koekemoer, Deane, 1125
Kokot, Darko, 379
Kolesnikov, Evgeny A., 602
Konno, Koji, 575
Kornprobst, Tim Lukas, 1053
Kraikuan, Juti, 202, 217

L
Lechtenberg, Karla, 772, 850, 868
Lei, Huanyu, 1007
Lertworawanich, Ponlathep, 407, 719
Lewis, Tony, 19, 1125
Lingenfelter, Jessica, 789
List, George, 287
Liu, Heng, 1007
Lukasik, Dan, 691

M
Magliola, Dana, 287
Mahajan, Gauri R., 647
Makhmutova, Alisa, 881
Mamulat, Stanislav, 1071
Mamulat, Stanislav L., 602
Mamulat, Yuriy, 1071
Männistö, Vesa, 963, 1197
Marecos, Vânia, 379
Maternini, Giulio, 525
Mbakisya, Onyango, 634
Medina, Jose, 735
Medina, Jose R., 611
Memon, Irfan Ahmed, 1026
Meocci, Monica, 493
Metlenkin, Dmitry, 1071
Minnikhanov, Rifkat, 257, 881
Mirou, Sham M., 1148
Mizushima, Yuichi, 975

Author Index

Montha, Arak, 407
Moolla, Tasneem, 19

N
Nagao, Yukio, 458
Naidoo, Kaslyn, 19
Nakamura, Taishi, 575
Nakashima, Hirotaka, 575
Napiah, Madzlan, 1026
Neves, João, 335
Norboge, Nicolas, 997, 1207
Nortje, Waynand, 1125
N-Sang, Seng Hkawn, 735

O
Obando, Carlos J., 611
Ochola, Evans Omondi, 151
Odoki, Jennaro Boniface, 151
Oliveira, Joel, 474
Ouachani, Rim, 1167
Ouarda, Taha, 1167
Ozen, Halit, 950

P
Paala, Miguel Enrico III Cabag, 134
Paliotto, Andrea, 493
Pandurangan, Venkataraman, 1112
Pashkevich, Anton, 676
Pérez-Grau, Francisco Javier, 171
Petrović, Jelena, 379
Petrus, Ángel, 171
Phutantikul, Peerapat, 187
Pillay, Justin, 19
Proust, John, 379
Punthutaecha, Koonnamas, 187

Q
Qawasmeh, Baraah, 345
Qureshi, Sabeen, 1026

R
Raffo, Veronica Ines, 134
Rakmark, Tatree, 407
Ram, Sewa, 511
Rasdorf, William, 997
Rasmussen, Jennifer, 868
Ritthiruth, Pawin, 217
Rocha, Ana Maria A. C., 474
Rosenbaugh, Scott, 805, 821

S
Saarenketo, Timo, 963
Salem Al-Samahi, Soughah, 986
Sampson, Udeh, 634
Sandra, Amarendra Kumar, 1181
Santos, Adriana, 474
Saroufim, Edwina, 97
Sawangsuriya, Auckpath, 441
Seracino, Rudolf, 232
Shabib, Ahmed, 897
She, Shiying, 1007
Shekhovtsova, Svetlana, 1071
Shirai, Yu, 589
Shotz, Howard, 425
Siddiqui, Abdul Basith, 1039
Silva, Telma, 335
Singh, Avishreshth, 1098, 1112
Soomro, Ubedullah, 1026
Steinbergs, Raitis, 416
Stevens, Ryan, 57, 559
Stolle, Cody, 755, 789
Sukkari, Alaa, 87
Surehali, Sahil, 1098

T
Talpur, Mir Aftab Hussain, 1026
Tan, Emily, 117
Terada, Masaru, 589
Tetley, Simon, 19, 1125
Thongprapha, Setthaphong, 187
Thriscutt, Mark J., 707
Thüer, Ulrich, 1053
Thungappa, Rajprabhu, 232
Tipagornwong, Chawalit, 187
Torres, Juan Miguel Velasquez, 134
Troxler Jr., William F., 3
Troxler, Robert E., 3
Trujillo, Miguel Ángel, 171
Tumakova, Yana, 1053

U
Usuda, Yukio, 458

V
Vacková, Pavla, 46
Valentin, Jan, 46
Van Geem, Carl, 379
Varela Soto, Fernando, 986
Vega, Lady D., 72
Viguria, Antidio, 171
Visser, Herman, 19, 1125

W
Wachiraprakarnpong, Apichai, 202, 217
Watanabe, Yumi, 975

Watcharakornyotin, Jirasak, 202
Wiboonsarun, Thanwa, 202
Wright, Alex, 379

X
Xiang, Yanling, 1007

Y
Yu, Wichai, 217

Z
Zeiada, Waleed A., 1083, 1148
Zheng, Meng, 1007
Ziada, Waleed, 87, 622